MW01098668

HANDBOOK OF FIELD SURVEYING

Staff of Research and Education Association

Research and Education Association
505 Eighth Avenue
New York, N. Y. 10018

HANDBOOK OF FIELD SURVEYING

Printed in the United States of America

Library of Congress Catalog Card Number 83-061836

International Standard Book Number 0-87891-530-3

PREFACE

The practice of surveying is generally carried out in current times with a high degree of accuracy, due to the high value of land that is often encountered. Boundaries of land used for homes, offices and factories must generally be determined with high precision to avoid conflicts with neighboring owners and the costly difficulties that may arise in making adjustments afterwards. This is particularly the case where a boundary line that needs to be altered affects pre-existing construction.

As a result, the practice of surveying must usually be carried out with expertise by the land surveyor, engineer, and scientist. This involves following prescribed procedural steps methodically, and checking measurements with much patience. Time-consuming calculations must often be performed repeatedly to achieve a feeling of assurance that the survey is recorded with satisfactory accuracy.

This handbook is intended to fulfill these objectives by providing instructional guidance in the principles and procedures of field surveying, and in the use, care, and adjustment of surveying instruments. Discussions are included on computations and adjustments, calibrations, and graphic portrayals of survey data.

Procedures are described in detail with comments on how to avoid errors. The correct procedures are emphasized, and at the same time, those procedures are illustrated which experience has shown provide unsatisfactory results.

In preparing this handbook, considerable use was made of information originated by the U.S. Dept. of the Army, Dept. of the Interior--Bureau of Land Management, National Oceanic and Atmospheric Administration, and the Naval Education and Training Command. Their contributions are gratefully acknowledged.

Dr. Max Fogiel
Technical Director

TABLE OF CONTENTS

1. INTRODUCTION
General, 1
Field work, 3
Office work, 6

2. UNIVERSAL EQUIPMENT, 10

3. LEVELS
Hand levels, 20
Engineer level, 21
Semiprecise level, 25
Precise level, 30
Level accessories, 34
Meteorological equipment, 37

4. DIRECTION AND ANGLE EQUIPMENT
Engineer transit, 48
One-minute theodolite, 60
One-second theodolite, 68
Two-tenth-second theodolite, 74
One-tenth-second theodolite, 79
Gyro-azimuth surveying instrument, 81
Compasses, 83
Direction and angle measuring accessories, 85
The solar transit, 89

5. DISTANCE MEASURING EQUIPMENT
Introduction, 96
Tapes, 96
Microwave instruments, 102
Lightwave instruments, 108

6. DISTANCE MEASURING
Taping, 113
Microwave, 121
Lightwave, 124
Indirect methods, 127

7. PLANETABLE TOPOGRAPHY
Equipment, 132
Field procedure, 138
Strip topography and profiles, 147

8. LEVELING
Introduction, 149
Differential leveling, 150
Trigonometric leveling, 161
Barometric leveling, 165

9. ANGLE AND DIRECTION OBSERVING
Description, 172
Field procedure, 177

10. TRAVERSE
Description and preperation, 182
Marking lines, 185
Distance and angle measuring, 186
Completion, 190

11. TRIANGULATION
Description, 194
Field procedures, 196

12. SPECIAL SURVEYS
Introduction, 200
Description, 200
Photogrammetry, 203

13. LOCATION OF DETAIL AND AREA DETERMINATION
Location of detail, 211
Area determination, 215

14. LAND SURVEYS
Introduction, 220
Surveys of private lands, 223
Surveys of public lands, 226
Construction of monuments, 237
Corner accessories, 238

APPENDIX

A. PRACTICAL PROBLEMS, 241

B. PRECISION: ERRORS, ACCURACY, AND PROBABILITY, 249

C. UNITS OF MEASURMENT, 256

D. TABLES, 260

E. SUMMARY OF COMMERCIAL EQUIPMENT AVAILABLE FOR SURVEYORS, 298

F. GLOSSARY, 328

INDEX, 338

CHAPTER 1

INTRODUCTION

Section I. GENERAL

1–1. Surveying

Surveying is the science of determining relative positions of points on, under, or near the earth's surface. These points may be cultural, hydrographic, or terrain features for mapping; or points needed to locate or lay out roads, airports, and structures of all kinds. The relative horizontal and vertical positions of these points are determined from distances and directions measured in the field and computations made on a reference spheroid, ellipsoid or plane relative to the chosen datum. The field measurements are made to a specified degree of accuracy which is usually dependent upon the use and purpose of the control points being established. Accuracy is the prime consideration in surveying. Instruments for each type of survey are employed, using prescribed techniques, methods, and procedures to achieve a designated accuracy. Less accuracy than specified results in a survey which will prove useless for its intended purpose; more accuracy, on the other hand, tends to waste time, effort, and money, and may not improve the final results. However, sometimes it is easier, with no appreciable increase in time or money, to use methods that achieve more accuracy than that required. In such cases the higher accuracy is desirable.

1–2. Surveying Classification

Surveying can be classed in a technical sense or functional sense as described below.

a. Technically, surveys are classed as—

(1) *Geodetic*—A survey in which the figure and size of the earth is considered. It is applicable for large areas and long lines and is used for the precise location of basic points suitable for controlling other surveys.

(2) *Plane*—A survey in which the surface of the earth is considered a plane. For small areas, precise results may be obtained with plane-surveying methods, but the accuracy and precision of such results will decrease as the area surveyed increases in size.

b. Functionally, surveys are classed as—

(1) *Construction surveys.* These surveys are conducted to obtain data essential to planning and cost estimating to locate or lay out engineering works or projects. These surveys normally cover areas considered small enough to use the plane surveying techniques.

(2) *Military surveys.*

(*a*) Field artillery fire-control surveys are made to determine relative positions of weapons and targets, and from these and other unassociated entries the correct firing data may be determined. Although relatively large areas and long distances are often involved, plane survey techniques are used. The required accuracy for these surveys does not warrant the use of geodetic survey procedures.

(*b*) Air defense employs highly sophisticated guided missiles that usually require highly accurate position and azimuth determinations of both launcher and targets. These data are usually determined by geodetic survey techniques.

(3) *Topographic surveys.* Topographic surveys are conducted to establish horizontal and/or vertical positions of points in the field, which are identifiable on aerial photographs for use by photocompilers in the compilation of maps. Since the control stations are usually distributed over comparatively large areas, their relative positions are determined by topographic survey procedures. Topographic surveying requires the establishment of horizontal and/or vertical control of third and lower order accuracies.

(4) *Basic control (geodetic) surveys.* A survey which provides positions, horizontal and/or

vertical, of points to which supplementary surveys are adjusted. The fundamental control survey of the United States provides geographic positions and plane coordinates of triangulation and traverse stations and the elevations of bench marks which are used as the basis for hydrographic surveys of the coastal waters, for the control of the topographic survey of the United States, and for the control of many state, city and private surveys.

(5) *Special surveys.* These surveys are conducted either with special equipment or for a special purpose.

(*a*) *Astronomic surveys.* The celestial determination of latitude and longitude. Separations are calculated by computing distances corresponding to measured angular displacements along the reference spheroid.

(*b*) *Satellite surveys.* These surveys are conducted by using an artificial earth satellite for long line surveys where the distance between stations is from 100 to 1500 miles. They are conducted for worldwide surveys for intercontinental, interdatum, and interisland geodetic ties. Astronomic, topographic, and basic control surveys are usually used in conjunction with satellite surveys. Special project instructions are written to detail methods, techniques, equipment and procedures to be used in these surveys.

(*c*) *Hydrographic surveys.* A survey made in relation to any considerable body of water, such as a bay, harbor, lake, or river for the purposes of determination of channel depths for navigation, location of rocks, sand bars, lights, and bouys; and in the case of rivers, made for flood control, power development, navigation, water supply, and water storage.

(*d*) *Field classification and inspection.* Field inspection and identification of features which a map compiler is unable to delineate; identification and delineation of political boundary lines, place names, road classifications, buildings hidden by trees, and so forth. Field classification may be included as part of the control survey effort and is normally completed prior to actual stereocompilation phase. Also, the process of comparing aerial photographs with conditions as they exist on the ground, and of obtaining information to supplement or clarify that which is not readily discernible on the photographs themselves.

(*e*) *Gravity surveys.* The collection of gravity survey data requires measurements of the intensity of the gravitation force at or near the earth's surface. While measurements on land are made at discrete points, those made in the air or over the ocean are recorded continuously during motion. The ultimate goal is to obtain a good density and distribution of gravity observations over the entire surface of earth. Gravity observations are divided, generally, into absolute and relative measurements. Absolute gravity determinations are those which measure the entire magnitude of gravity at a point while relative measurements detect the intensity variations of the gravity field from point to point.

(*f*) *Land surveys.* The process of determining boundaries and areas of tracts of land. The term cadastral survey is sometimes used to designate a land survey, but in this country its use should be restricted to the surveys of public lands of the United States (ch. 14).

(*g*) *Engineering surveys.* A survey executed for the purpose of obtaining information that is essential for planning an engineering project or development and estimating its cost. The information obtained may, in part, be recorded in the form of an engineering map.

1–3. Geodetic and Plane Surveys

The figure of the earth is not considered a sphere, but a *spheroid* (an ellipsoid of revolution), flattened at the poles and bulging at the equator. There are various determinations of the size and shape of the spheroid, which are normally differentiated by their degree of flattening. Different projections are based on these spheroids and each projection has unique characteristics and serve different purposes. The curvature of the earth's surface is very similar to that of a sphere. Because of this curvature, surveys are divided basically into geodetic and plane surveys. The basic difference between geodetic and plane surveys can best be expressed in terms of established points: in plane surveys all established points are referenced to a flat plane, while in geodetic surveys all established points are referenced to the curved surface of a spheroid. Both technical classifications involve the same types of measurements, made by the same basic methods, and there are few basic differences in the field operation.

a. Geodetic Surveys. In geodetic surveys, the stations are normally long distances apart and more precise instruments and procedures are required, than those of plane surveys. These distances or areas measured on the surface of the earth are not along straight lines or planes but on a curved surface, and allowances must be made for this in the computations. To accomplish this, the earth's major and minor diameters are computed accurately, and from these, a spheroid of reference. The position of each geodetic station is related to this spheroid. These positions are expressed as *latitudes* (angular distance north or south of the equator), and *longitudes* (angular distance east or west of the prime meridian), or as *northings* and *eastings* on a rectangular grid system which is correlated with latitude and longitude.

b. Plane Surveys. When the extent of the survey becomes small (less than 250 sq km in area) and when only limited accuracy is required, the effect of curvature can be ignored. These surveys are treated as if the measurements were made on a plane and are known as plane surveys. Highway and railroad surveys, which may extend for hundreds of kilometers (miles), are usually in a narrow strip and are considered as plane surveys. However, a limited computation for earth's curvature is necessary in this case. The computation methods used in plane surveys are described in appropriate chapters of this book.

1-4. Survey Networks

Horizontal and vertical survey control within a country is usually established by a network of control arcs, which are all referenced to a single datum and are therefore related in position and elevation to each other, regardless of their distance apart. In the United States these networks are referred to as basic, supplementary, and auxiliary networks. These are all referenced to the North American 1927 Datum for horizontal control points and the sea level datum of 1929 for vertical control points. Within the continental United States the definitions listed below are used. Each country uses similar networks tailored to meet that country's need and terrain.

a. Basic Network.

(1) The basic horizontal control is usually established by first order triangulation or precise electronic traverses in a network of arcs about 96 kilometers apart throughout a country.

(2) The basic vertical control is established by first order differential leveling in a network of arcs spaced from 96 to 160 kilometers apart throughout the country.

b. Supplementary Network.

(1) The supplementary horizontal control is usually established by modified second or second order survey procedures. The supplementary arcs are used to fill in the areas between the basic control arcs. The ultimate objective is to place stations of either the basic or supplementary network at intervals between 6 to 16 kilometers apart.

(2) The supplementary vertical control is established by second order differential leveling, within the basic control arcs to provide a planned line spacing of about 10 kilometers. Along these lines bench marks are placed about every 2 kilometers.

c. Auxiliary Network.

(1) These auxiliary (or additional) horizontal networks are usually established by second or third order survey procedures in order to provide horizontal control in the immediate areas. This horizontal control is used by the surveyors for construction and engineering surveys, and mapping projects.

(2) These vertical networks are established by third order differential leveling and are used to provide vertical control in the immediate areas. While the horizontal control networks are being established, usually vertical control is also carried throughout these networks by trigonometric leveling. The elevations established by this method is referred to as "trig levels" and is used for -struction and engineering surveys, and mapping projects.

Section II. FIELD WORK

1-5. Operations

Field work in surveying consists of making and recording measurements. The operations are in general, as follows:

a. Measuring distances and/or angles for—

(1) Establish points and lines of reference for locating detail such as boundary lines, roads, buildings, fences, rivers, bridges, and other existing features.

(2) Staking out or locating roads, buildings, landing strips, and other construction projects.

(3) Establish triangulation and traverse stations for basic, supplementary, and auxiliary control networks.

(4) Establish picture points for mapping projects.

(5) Establish lines parallel or at right angles to other lines, determine the area of tracts of land, measure inaccessible distances, extend straight lines beyond obstacles, and any other work which may require use of geometric or trigonometric principles.

b. Measuring differences in elevations and determining elevations—

(1) To establish elevation reference points, called *bench marks.*

(2) Of terrain along a selected line for plotting profiles and computing grade lines.

(3) To stake out grades, cuts, and fills for earthmoving and other construction projects.

(4) For trig elevations on triangulation and traverse stations for control nets and mapping projects.

(5) For any other work which may require use of geometric or trigonometric principles.

c. Record field notes to provide a permanent record of the field work in the form of—

(1) Planetable sheets.

(2) Land survey plans.

(3) Property plans.

(4) Field recording booklets.

(5) Single sheet recording forms.

(6) Recovery and establishment control cards.

(7) Control diagrams, showing relative location, methods, and type of control established and/or recovered.

1–6. Factors Affecting Field Work

The surveyor in the field must constantly be alert to the different conditions he encounters and the requirements of the survey. The weather, terrain, personnel, equipment, purpose, accuracy of the survey, systematic procedures, and the expected rate of progress are some of the factors which will affect the work.

a. Weather and Terrain. Physical factors such as weather and terrain will affect each field survey in varying degrees. Measurements using telescopes can be stopped by fog, mist, or smog. Swamps and flood plains under high water will impede taping surveys. Lengths measured over open water or fields of flat, unbroken terrain can create ambiguities in measurements when using microwave equipment. The longest length that can be determined by lightwave distances measurements is reduced in bright sunlight. Generally, reconnaissance will predetermine the conditions and alert the survey party to the best time and method to use and the rate of progress to be expected.

b. Personnel. The status of training of the personnel is another factor that affects field work. Experience in handling the instruments being used for a survey can shorten survey time without introducing excessive errors which would require resurvey. The personnel factor is variable and will have an affect on the rate of progress.

c. Equipment. In addition to the reasons cited in *a* and *b* above the equipment used in the survey will have an affect on the rate of progress. Because of the very nature of surveying, repair and/or replacement of broken instruments or parts is sometimes the biggest factor in the slowing down or stopping of a field survey. Therefore, reliability of equipment must be taken into consideration when setting completion dates.

d. Purpose. The purpose and type of the survey are primary factors in determining the accuracy requirements. First order triangulation, traverse, or leveling, which becomes the basic or "control" of future surveys, must be made to the required accuracy standards. At the other extreme, cuts and fills for a highway survey requires accuracies of much lower standards. In some surveys inaccessible distances must be computed. The distance is computed by trigonometry using angles and distances which can be measured. Therefore these measurements in many cases must be made to a high degree of precision in order to maintain the accuracy in the computed distance.

e. Accuracy. As stated in *d* above, the purpose of the survey will determine the accuracy requirements. The required accuracy, in turn, will influence the selection of instruments and procedures. For instance, comparatively rough procedures can be used in measuring for earthmoving, but grade and alinement of a highway must be much more precise and require more accurate measurements. Each increase in precision also increases the time required to make the measurement, since greater care and more observations must be made.

f. Systematic Procedure. Each survey measurement will be in error to the extent that no measurement is ever exact. The errors classified as systematic and accidental are discussed in appendix C. Besides errors, survey measurements are susceptible to mistakes or blunders. These arise from misunderstanding of the problem, poor judgment, confusion, or simply from an oversight. By working out a systematic procedure, the surveyor will often detect a mistake when some operation seems out of place. The system will be an advantage in setting up, in making observations, in recording field notes, and in making computations.

g. Speed. Survey speed is not the result of hurrying; it is the result of saving time through the following:

(1) The skill of the surveyor in handling his instruments.

(2) The intelligent planning and preparation of the work.

(3) The process of making only those measurements that are consistent with accuracy requirements.

1-7. Field Notes

The field notes of the surveyor must contain a complete record of all measurements made during the survey, when necessary sketches, diagrams, and narrations should be made to clarify the notes. The best field survey is of little value if the notes are not complete and clear. The field notes are the only record that is left after the survey party departs the field survey site.

a. Recording. All field notes must be lettered neat and legible. The lettering should be in free hand, Gothic style as illustrated in fig. 3-14. All notes should be in black or blue-black ink, suitable for photographic copying. However, in special cases a 3H or 4H pencil may be used. Numerals and decimal points should be legible and permit only one interpretation. Notes must be kept on the standard survey forms and not on scraps of paper for later transcription.

b. Field Notes. The survey notes are usually kept in a field notebook. These notebooks are of two types: the permanent bound book and the loose leaf. The following information must appear in each book no matter which type is used.

(1) Instruction for return of book, if lost.

(2) Index of field notes contained in the book.

(3) List of party personnel and their duties.

(4) List of instruments used, to include, serial numbers, calibration data, and dates used.

(5) A generalized sketch and description of project.

(6) The survey *measurement* notes. On each page containing the measurement notes, the heading must be filled out to include, station name, date, instrument man, recorder, instrument used and where pertinent the weather. The body will contain all pertinent measurement notes. Each page must have the instrument man's initials on the bottom to indicate that he has checked the page for errors or omissions.

c. Forms of Recording. Field note recording takes three general forms; tabulation, sketches, and descriptions.

(1) *Tabulation.* The numerical measurements are recorded in columns according to a prescribed plan depending upon the instrument used, order of accuracy of the survey, and type of measurement.

(2) *Sketches.* Sketches add much to clarify field notes and should be used liberally. They may be drawn to scale or approximate scale or exaggerated for clarity. A planetable sheet is an example of a sketch drawn to scale. A control card sketch should be drawn to approximate scale, and exaggerated when necessary to show important details needed for clarity. The measurements should be added directly on the sketch or keyed in some way to the tabular data. A very important requirement of a sketch is legibility.

(3) *Descriptions.* Tabulations with or without added sketches can also be supplemented with descriptions. The description may only be one or two words to clarify the recorded measurements. It may also be a lengthy narration if it is to be used at some future date, possibly years later, to locate a survey monument.

d. Abbreviations and Symbols. It is recommended that standard abbreviations, signs, and symbols may be used in field notes. If there is any doubt as to the meaning or interpretation of a symbol or abbreviation the words must be spell ed out.

e. Corrections. ERASURES ARE NOT PERMITTED IN FIELD NOTES. Individual numbers recorded incorrectly will be lined out by a single diagonal line and the correct values added above. The circumstance of the correction of all original figures should be explained in the remarks column, except for obvious mathematical errors. No position will be voided or rejected in the field notes, except in the case of bumping the instrument or observing the wrong target, and then a note must be made in the remarks column stating the reason for void. Pages that are voided or rejected must be referenced to a substitute page. *The procedure for corrections is mandatory* since the field notes are considered legal evidence.

f. Waterproofing. When working in parts of the world that are subject to high humidity and/or rain, it has been found that the waterproofing of field notes can be accomplished in the field. This waterproofing can be accomplished by spraying a thin coat of acrylic clear plastic on the field record. This spray may be applied before the recording and it will make the paper waterproof, and it may still be written on with ordinary writing instruments. The field notes can be sprayed again after the recording and the plastic then fixes the writing and prevents water damage to the records. One such spray is "Krylon, Workable Fixatif"; however, there are many other sprays on the market and any of them may be used.

1–8. Hand and Voice Signals

Members of a survey party will find themselves at some distance apart at times and have to communicate with each other. During taping operations, this distance would be a tape length where voice communications are adequate. When making angle measurements, this distance may be several kilometers and require radio communication. Between these two extremes, many operations are at distances far enough apart where voice signals could not be heard and hand signals must be used.

a. Hand Signals. The hand signals have been standardized and are recommended for surveyors. The most common hand signals used are illustrated in figure 1–1 and figure 1–2. There are many other signals that may be used, but they must be agreed upon and understood by all members of the party. Each signal is given while facing the person being signaled.

Figure 1-1. Hand signals for numbers.

b. Voice Communications. A radio is used for voice communications on long line surveys.

Section III. OFFICE WORK

1–9. Operations

Office work in surveying consists of converting the field measurements into a more usable form. The conversion or computation may be required immediately to continue the work or it may be held until a series of field measurements is completed before it can be compiled and adjusted. This is called office work even though some of the operations may be performed in the field during lapses between measurements to save time. Some office work requires special equipment such as calculators, conversion tables, or drafting equipment. This equipment may not be available in the field and the work must be done in the office.

a. Working Up Field Notes. During survey operations, many field measurements require some form of arithmetical computation. It may be a simple addition of several full lengths and a partial tape length to record a total distance between two points. It may be adding or subtracting differences in elevation to determine height of instrument or elevation during leveling, or it may be checking angles to see that the allowable error is not exceeded. Since the process differs for each type of surveying operation, the method will be discussed under recording for each operation.

b. Computing. Office computing converts distances, angles, and rod readings into a more usable form. The measurements may end up as a computed volume of dirt to be moved for a highway cut or fill, an area of land needed for a construction project, or an adjusted position of some point or mark from which other measurements can be made.

(1) *Distances.* The field measured distances may be measured on a slope or in short horizontal sections. For higher accuracy, corrections must be applied for temperature, tension, and sag of the

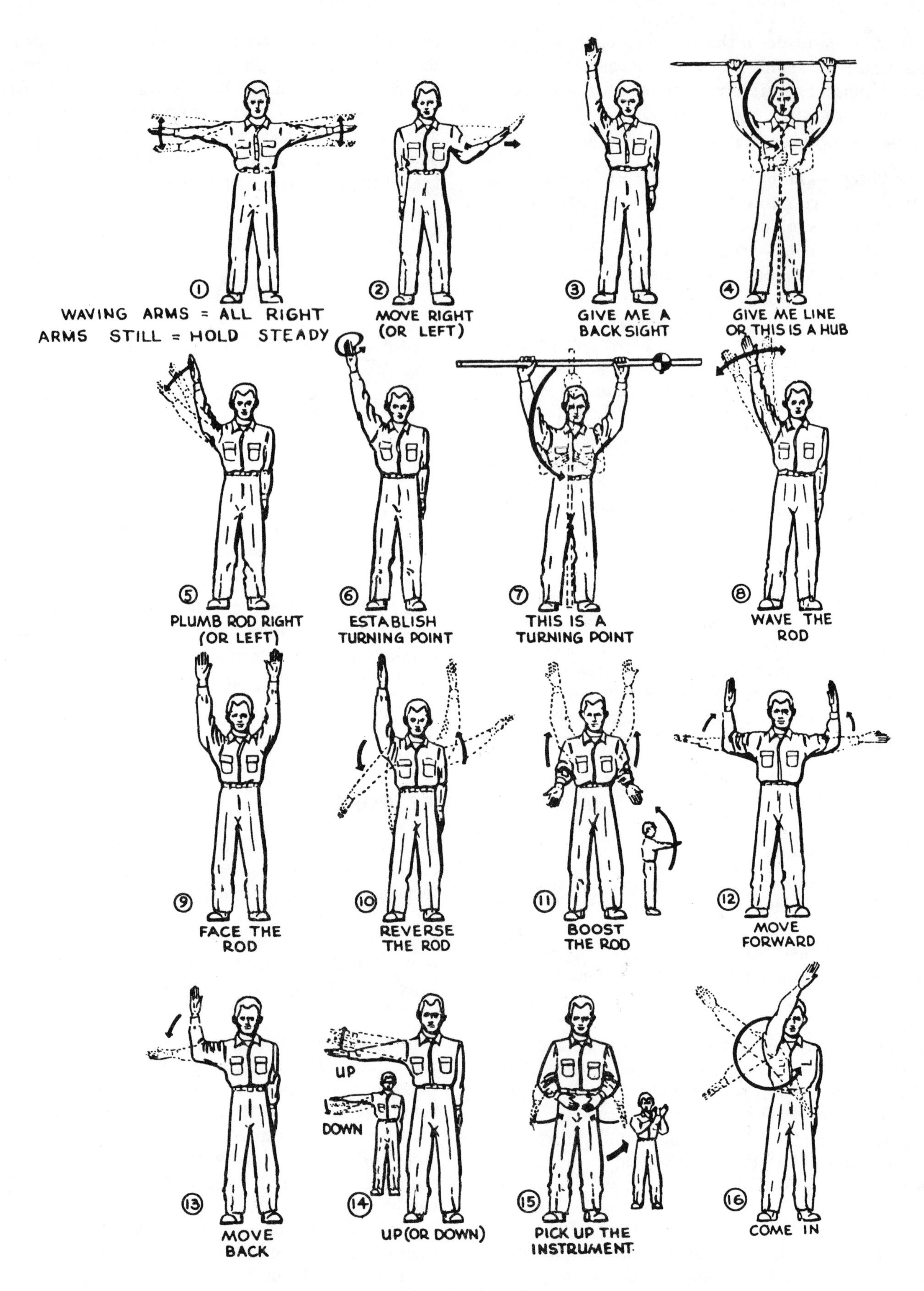

Figure 1-2. Common hand signals.

tape. The desired result is the horizontal distance between points. In electronic distance measuring, the distance is almost always on a slope and has to be reduced to the equivalent horizontal distance.

(2) *Bearings and azimuths.* In many operations, the measured angles are converted into directions of a line away from a north or south line. These directions are called bearings or azimuths of the line.

(3) *Relative positions.* The distance and direction of a line between two points determines the position of one point relative to the other point. If the direction is given as a bearing or azimuth, a trigonometric formula using sine or cosine of the angle, multiplied by the distance, will result in a coordinate difference between the two points.

(4) *Road and railroad work.* The computations determine the grade and position of a road or railroad, and the volume of earth to be moved to prepare the bed. Airport and airstrip surveys are quite similar to road surveys in alining and grading the runways.

(5) *Areas.* Area computations are required at times after the positions of points and distances between them have been determined. This is performed in the office either by drafting the survey to scale and measuring graphically or by computing the area by trigonometry or by arithmetic.

c. Adjusting. Some survey techniques are not complete until one more procedure in computing, known as adjusting, is performed.

(1) *Definition.* Adjusting is the determination and application of corrections to the data being adjusted thereby making the data consistent within themselves and providing the most probable value for the determination of points to some reference point. Small errors which are not apparent during individual measurements can accumulate to a sizeable amount. For example, assume that 100 measurements were made to the nearest unit for the accuracy required. This requires estimating the nearest one-half unit during measurement. At the end of the course, an error of +4 units results. Adjusting this means each measurement is reduced 0.04 unit. Since the measurements were only read to the nearest unit, this adjustment would not be measurable at any point and the adjusted result would be correct. Some of the more precise surveys require least square adjustments.

(2) *Traverse.* Traverse is the measurement of the lengths and directions of a series of lines between points on the earth, and is used to determine the positions of the points. When finally computed, the accumulated closing error shows up as a position displacement and is distributed among the traverse points.

(3) *Triangulation.* Adjusting triangulation consists of adjusting the three angles of each triangle to equal 180°, and then computing the lengths of the unknown sides starting with a measured or computed base line. These sides become the known distance values of adjoining triangles. Computing continues through a net until a new measured or previously determined base line is reached. The closure (difference between the measured base line distance and the distance computed through the triangles) must be within prescribed limits. This amount of closure is then adjusted back through the net affecting the positions of the stations.

(4) *Elevations.* Normally, elevations on a level line are computed in the field as the measurements are made. When the line is closed or tied to a known elevation, the difference in elevation between the measured and known elevations is adjusted over all the stations in the line.

1–10. Computations

a. Recording. Office computations reduce the field notes to tabular or graphic form for a permanent record or for continuation of field work. Standard forms, printed on letter size sheets are available for many survey computations. The computer is not limited to the use of these forms and can improvise new ones when needed. All computations pertaining to one survey are bound into a single file and referenced to the book and page number of the field book. Each page of the file also contains the names of the computer and the person who checked the computations. Usually, enough of the field notes are transcribed to the computation file so that further reference to field notebooks is unnecessary.

b. Checking. Most mathematical problems can be solved by more than one method. In checking a set of computations, a method which differs from the original computation is used, if possible. An inverse solution, starting with the computed value and solving for the field data, is one possibility. The slide rule, the planimeter, and the protractor are also used for approximate checking. A graphical solution can be used, when feasible, especially if it takes less time than an arithmetical or logarithmic solution. Each step that cannot be checked by any other means must be recomputed. When an error or mistake is found, the computation should be rechecked before the correction is accepted.

1–11. Significant Figures

The term significant figures refers to those digits in a number which have meaning, that is, whose values are definitely known to be correct.

a. Measured Quantity. In a measured quantity, the number of significant figures is determined by the accuracy of the measurement. For example, a roughly measured distance of 193 meters has three significant figures. More carefully measured, the same distance, 192.7 meters, has four significant figures. If measured still more accurately, 192.68 meters has five significant figures.

b. Surveying. In surveying, the significant figures should reflect the allowable error or tolerance in the measurements. For example, suppose a measurement of 941.26 units is made with a probable error of ±0.03 unit. The ±0.03 casts some doubt on the fifth digit which can vary from 3 to 9, but the fourth digit will still remain 2. We can say that 941.26 has five significant figures and from the allowable error we know the fifth digit is doubtful. However, if the probable error were ±0.07, the fourth digit would be affected. The number could vary from 941.33 to 941.19, and the fourth digit could be read 1, 2, or 3. The fifth digit in this measurement is meaningless. The number has only four significant figures and should be written as such.

c. Zeros. The significant figures in a number ending in one or more zeros are unknown unless more information is given. The zeros may have been added to show the location of the decimal point. *For example*: 73200 may have three, four, or five significant figures depending on whether the true value is accurate to 100, 10, or 1 units. If the number is written 73200.0. it indicates accuracy is carried to the tenth of a unit and is considered to have six significant figures.

d. Decimals. Using decimals, the significant figure is not always the number of digits. A zero may or may not be significant depending on its position with respect to the decimal and the digits. As mentioned in *c* above, zeros may have been added to show the position of the decimal point. Some examples are given below.

0.000047 — two significant figures
0.0100470 — six significant figures
2.0100470 — eight signicant figures

e. In Long Computations. The values are carried out to one more digit than required in the result. The number is rounded off to the required number of digits as a final step.

1–12. Rounding off Numbers

Rounding off is the process of dropping one or more digits and replacing them with zeros, if necessary, to explain the correct number of significant figures. Numbers are rounded off according to the following rules:

a. When the digit to be dropped is less than 5, the number is written without the digit or any others that follow it (.054 becomes .05).

b. When the digit is equal to 5, the nearest even number is substituted for the *preceding* digit (.055 becomes .06; .045 becomes .04).

c. When the digit to be dropped is greater than 5, the preceding digit is increased by one (.047 becomes .05).

d. Dropped digits to the left of the decimal points are replaced by zeros.

e. Dropped digits to the right of the decimal point are never replaced. *Examples*:
2738.649 to five significant figures equals 2738.6
792.850 to four significant figures equals 792.8
792.750 to four significant figures equals 792.8
675823. to four significant figures equals 675800
675863. to four significant figures equals 675900
4896.3 to four significant figures equals 4896
4896.7 to four significant figures equals 4897

CHAPTER 2

UNIVERSAL EQUIPMENT

2–1. Introduction

Surveying instruments are the devices with which measurements are made. Many of these devices have similar features. These include tripods to hold the instrument steady at a convenient height and over a mark while measuring; level vials and leveling heads to make some definite part of the instrument horizontal and/or vertical; telescopes to magnify the image being viewed at some distance; crosshairs to point on the image; and either plumb bobs or optical plummets to set the instrument exactly over a selected point on the ground. Other features which are unique to only one or two instruments will be discussed in following chapters that deal with the individual instrument.

NOTE

> Where the word "crosshair" or "crosswire" is used in reference to the reticle of a telescope, it must be understood that these may be actual spider hairs, or they may be lines engraved on glass. The make and type of instrument is the controlling factor on what is used. Therefore these words will be used interchangeably within this book when referring to the reticle of the different telescopes. The appropriate maintenance manual should be consulted for exact nomenclature (fig. 2–1).

2–2. Tripod

The tripod is the base or foundation which supports the survey instrument and keeps it stable during observations. It consists of a tripod head to which the instrument is attached, three wooden or metal legs which are hinged at the head, and metal pointed shoes on each leg to press or anchor into the ground to achieve a firm setup. The leg hinge is adjusted so that the leg will just begin to fall slowly when it is raised to an angle of about 45°. The tripod head may have screw threads on which the instrument is mounted directly; it may have a screw projecting upward through the plate; or it may have a hole or slot through which a special bolt is inserted to attach the instrument.

a. Types. Two types of tripods are furnished to surveyors—the fixed leg tripod and the extension leg tripod. A leg of the fixed leg tripod consists of two lengths of wood as a unit or one solid piece of wood, attached to a tripod head with a hinge and fitted with metal tipped shoes. At points along the length, the two piece legs are perpendicularly braced to give great stability. A leg of the extension leg tripod is made of two sections which slide up and down. The lower section slides into or out of the upper section, enabling the surveyor to adjust the height of tripod as desired. On rough terrain, the legs can be adjusted to different lengths to establish a horizontal tripod head or to set the instrument at the most comfortable working height. The fixed legs must be swung in or out in varying amounts to level the head. Therefore the instruments height is not as easily controlled with the fixed legs as with the extension legs, and the observer must learn the correct spread of the fixed legs to get the desired observing height. It is

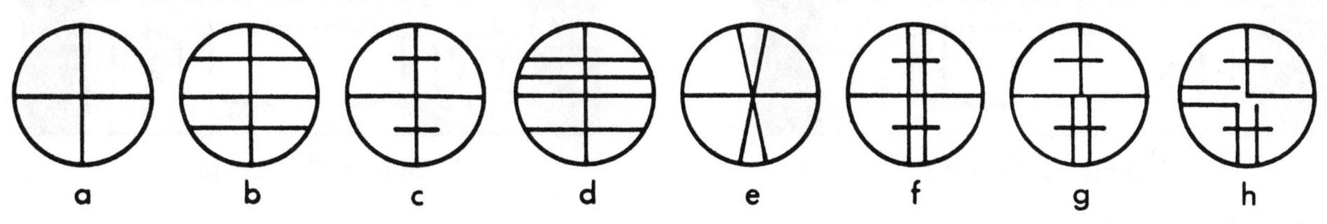

Figure 2-1. Arrangement of cross hairs.

recommended that the extension leg tripod be used when the surveying is to be done over rough terrain.

b. Mounting Instruments. When mounting the survey instrument on the tripod it must be gripped firmly to avoid dropping. The transit is held by the standard opposite the vertical circle while it is being mounted. The engineer level is held at the center of the telescope and usually theodolites and precise levels are gripped near the base of the instrument. The instruments must be screwed down to a firm bearing, but not so tightly that they will bind or that the screw threads will strip.

2–3. Setting the Tripod

The tripod legs must be properly placed to achieve a stable setup. The legs must also be anchored by firmly embedding, tying together, or plastering down.

a. General. On level terrain the legs should be made to form an angle of about 60° with the ground surface. Loosen the restraining strap from around the legs and secure it around one leg. An effective way to set the tripod down is to grip it with two of the legs close to the body while standing over the point where the setup is required. Using one hand, push the third leg out away from the body until it is about ½ to ¾ of a meter away from the point. Lower the tripod until this third leg is on the ground. Place one hand on each of the first two legs and spread them while taking a short backward step, using the third leg as a support point. When the two legs look about as far away from the mark as the third one, lower the two legs and press them into the ground, if possible. Any slight adjustment to level the head further is made by moving the third leg in or out before embedding. On extension leg tripods, this final adjustment can be made by sliding the proper leg up or down.

b. Smooth Surfaces. On smooth or slippery paved or rock surfaces, the tripod leg hinges should be tightened while setting up to prevent

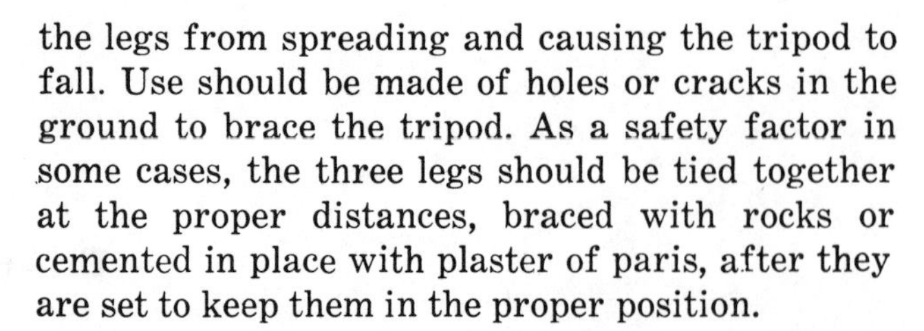

the legs from spreading and causing the tripod to fall. Use should be made of holes or cracks in the ground to brace the tripod. As a safety factor in some cases, the three legs should be tied together at the proper distances, braced with rocks or cemented in place with plaster of paris, after they are set to keep them in the proper position.

c. Sloping surfaces. When setting up on steeply sloping surfaces, the third leg is placed uphill and at a greater distance from the mark. The other *two legs are set downhill* as *a* above. Before releasing the downhill legs they should be checked for stability to see that the weight of the instrument on the tripod head will not overbalance and cause the tripod to slip or fall.

2–4. Care of Tripod

When setting the legs in the ground, care must be taken to apply pressure longitudinally. Pressure across the leg can crack the wooden pieces. The hinged joint should be adjusted (para 2–2) and not overtightened to cause strain on the joint, or to strip or lock the metal threads. The machined tripod head must be kept covered when not in use, and should not be scratched or burred by mishandling. The head cover should be attached to the tripod leg when not in use. Any damage to it can be transferred to the tripod head. Mud, clay, or sand adhering to the tripod must be removed and the tripod wiped with a damp cloth and dried. The metal parts should be coated with a light film of oil and wiped with a clean cloth. Foreign matter can get into the hinged joints or on the machined surfaces and cause wear. Stability is the tripod's greatest asset. Instability, wear, or damaged bearing surfaces on the tripod can evolve into unexplainable errors in the final survey results.

2–5. Level Vial

Many surveying instruments are equipped with one or more level vials to indicate the horizontal plane, and to assist in keeping parts of the instrument horizontal and others vertical while the instrument is in use. This is very important since it

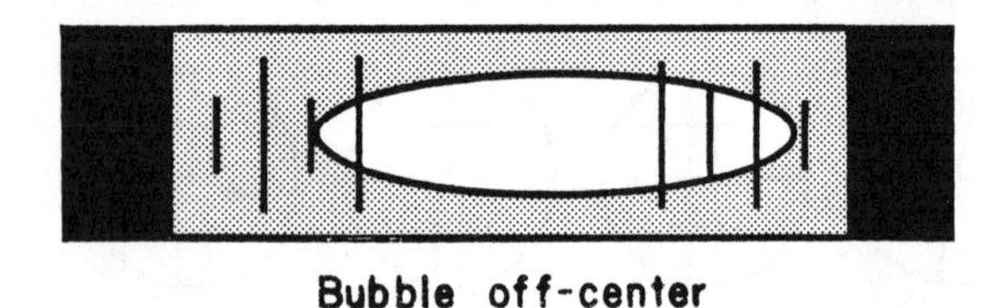

Bubble off-center

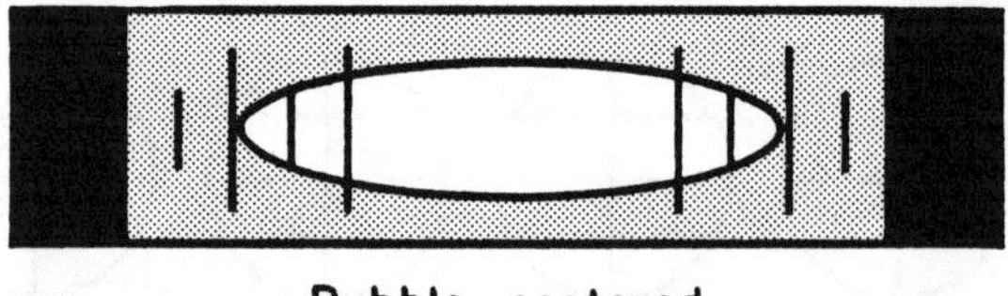

Bubble centered

Figure 2-2. Level vial.

defines the vertical axis and horizontal plane at the instrument from which measurements are made.

a. Description. A level vial is a glass tube with the inside surface along the long axis ground as an arc of a circle. The vial is filled almost completely with liquid and sealed at both ends. The liquid is normally either alcohol or ether which do not freeze at low temperatures. The level vial is sometimes referred to as a spirit level. When the vial is held horizontally, the remaining unfilled space takes the form of an air bubble and seeks the highest part of the arc. The vial is usually marked with graduations in both directions from the center or from one end to the other. The bubble is centered in the middle of the vial of a well adjusted and leveled instrument (fig. 2–2). The vial is usually mounted in a protective metal housing that is attached to the instrument with screws to permit its removal. The level vial is adjusted by capstan head screws that extend through the housing and work directly on the vial.

b. Sensitivity. The sensitivity of a bubble is a measure of the amount it moves for a given inclination, and depends on the construction of the vial. The longer the radius of the arc in the vial, the flatter the curve and the greater the sensitivity, since a smaller vertical movement of the end of the level vial will start the bubble moving. The markings on the vial are not standardized on all instruments and sensitivity is described by the radius of the bubble arc. This radius is about 6 meters for a transit plate level, about 8 meters for the alidade levels, about 21 meters for the engineer level, and about 200 meters for precise levels. The high sensitivity vials are used on the instruments for precise work. High sensitivity vials on less precise instruments wastes time in centering and the quality of the work is not necessarily improved. The exact amount of bubble movement measured in divisions of the vial to unit of arc is determined by calibration.

c. Calibration. The level vials on very precise instruments are extremely sensitive and bringing the bubble into exact center would be time consuming. The vial, therefore, is graduated so that the inclination may be read and a correction computed. This level correction is determined by calibration of the level vial in seconds of arc per graduation. There are many methods used in calibrating level vials.

d. Alinement. When a bubble vial must be replaced, the new vial can be replaced as follows:

(1) Select a fairly level surface to work on.

(2) Set the vial housing on the surface and mark its position.

(3) Insert the vial in the housing and let it come to rest.

(4) Note the amount that the bubble is off center (①, fig. 2–3).

(5) Lift housing and vial carefully, turn it end for end, and set it down reversed, but in the same place as marked in (2) above.

(6) Note the amount that the bubble is off center now (②, fig. 2–3).

(7) Shim the lower end of the vial up until the bubble is halfway between the first and second positions noted in (4) and (6) above (③, fig. 2–3).

(8) Repeat (4) and (6) above, as a check. If bubble returns to the same position both direct and reverse (not necessarily centered), the vial is parallel to the holder. The bubble may not be centered because the surface which was selected is out of level by the amount that the bubble is off center.

(9) Mount the bubble vial permanently in housing.

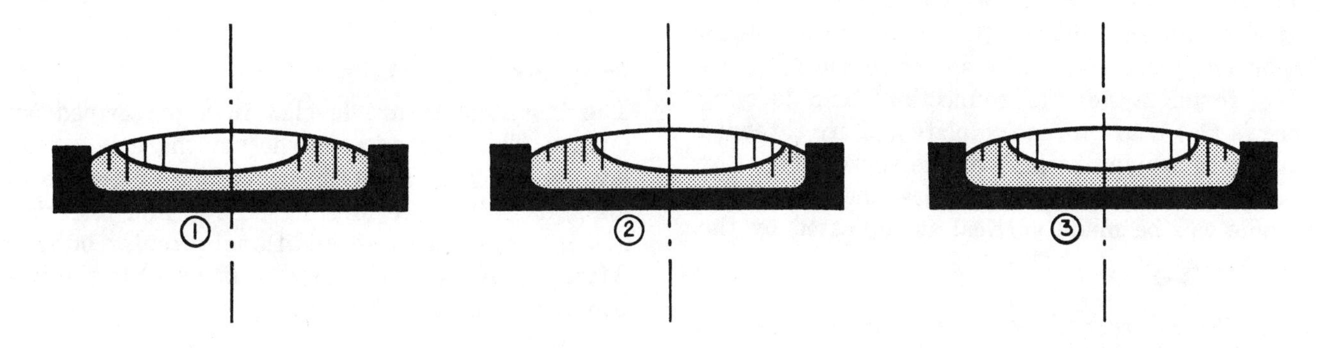

Figure 2-3. Replacing level vial.

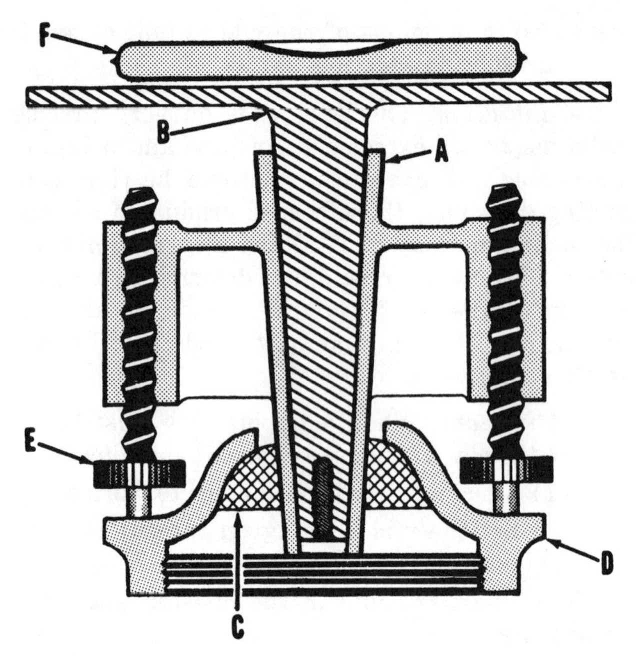

Figure 2-4. Cross section of four-screw leveling head.

2–6. Circular Level

Some instruments have a circular level vial in addition to the plate level. The sensitivity is usually lower than the long vial bubbles, but it is sufficient for quick or approximate leveling.

2–7. Leveling Heads

Many surveying instruments have an assembly called a leveling head which supports the instrument on the tripod and is used to level the instruments. Two types of heads are in use; the four-screw and three-screw type. This refers to the number of leveling screws used to level the instrument.

a. Four-Screw Type. The four screws are spaced equally 90° apart around the head. A cross section is shown in figure 2–4. The leveling head contains a socket (A) in which a spindle (B) rotates. A ball-and-socket joint, consisting of a hemispherical piece (C) and a socket in the footplate (D), forms a flexible connection. The leveling screws (E) bear on the footplate and are threaded into arms which are part of the socket. Thus, by manipulating the screws in pairs, the axis of the spindle can be made vertical as indicated by the level vial (F).

b. Three-Screw Type. The three screws are spaced equally 120° apart and the screws can be manipulated individually or in conjunction with

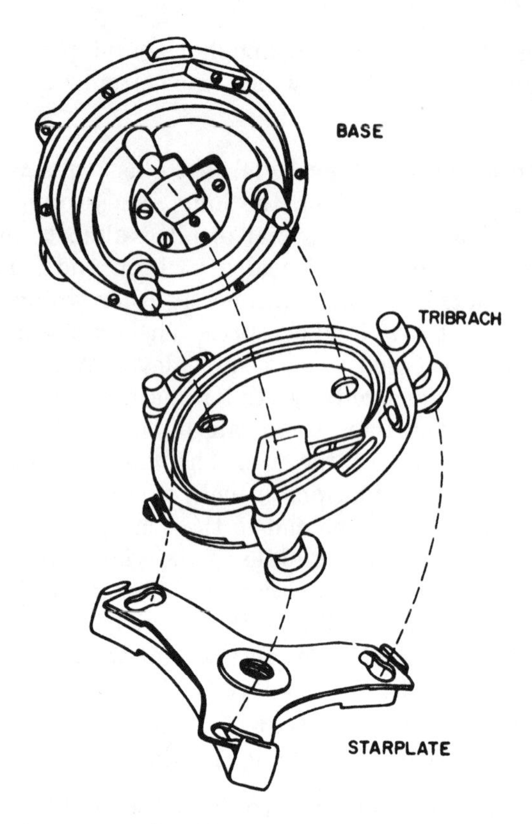

Figure 2-5. Three-screw leveling head.

one or both other screws to level the head. The lower end of the screw may bear on the footplate as in the four-screw type, be held against the footplate by a starplate, or have a tribrach arrangement (fig. 2–5). The tribrach is an assembly which consists of the leveling screws, the optical plumbing assembly and circular level, and is detachable from the instrument head. This latter combination permits interchangeability of instruments without moving the tripod. After observations are completed, the instrument is detached at its base and moved to the next location to another tribrach, while the tripod and tribrach remain at their set position. The next instrument or target is then snapped into place, checked for level, and is ready for its operation.

2–8. Leveling the Instrument

The instruments are leveled in a prescribed sequence. This depends on whether there is a four leveling-screw head with either one or two level vials, or only three leveling screws and one vial, and in many cases, an additional circular bubble. After the instrument is firmly attached to the tripod, it is leveled as follows:

a. Four Leveling-Screw and One Level.

(1) Turn the instrument about its vertical

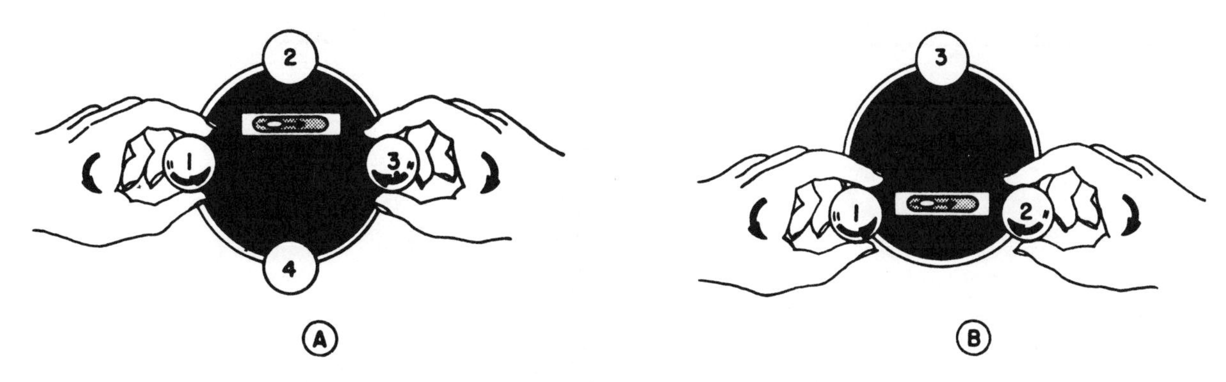

Figure 2-6. Manipulating the leveling screws.

axis until the level vial is approximately parallel to two opposite leveling screws, 1 and 3, Ⓐ, figure 2–6. With these screws, tilt the instrument until the level bubble is approximately centered in the vial. Note that the level bubble moves in the same direction as the left thumb. The two screws are turned in opposite directions at the same rate, keeping a slight pressure between the leveling screw feet and the footplate.

(2) Turn the instrument 90° until the level vial is parallel to the other pair of screws, 2 and 4, Ⓐ, figure 2–6. Once again, bring the level bubble to the approximate center.

(3) Turn the instrument back to its original position as in (1) above, and recenter the bubble.

(4) Repeat (2) and (3) above, until the bubble remains at center in both positions.

(5) Turn the instrument 180° around from its position at (1) above. If the bubble remains at center, the instrument is leveled along this axis. If not, bring the bubble halfway back to center using the leveling screws. This will bring the instrument to level, but the vial is out of adjustment.

(6) Repeat procedures in (5) above, using position in (2) above, as the starting point and again compensate for the difference in centering. Note that the bubble will be in the same off-center position for any pointing of the instrument since the vial is out of adjustment.

(7) When leveling is completed, all four leveling screws must bear firmly against the footplate without the screws binding.

(8) The level vial may be adjusted by using the procedure in (5) above, except that the bubble is brought back halfway to center with the vial adjusting screw(s) and then completely centered by the leveling screws. By returning to the position used in (1) above, the bubble should remain centered. If not, the process is repeated being a little more careful in bringing the bubble back halfway to center with the level vial adjusting screw(s). The bubble should be kept in good adjustment, but since it is extremely difficult to get a perfect adjustment, the procedure in (5) above, should always be followed as the last step.

b. Four Leveling-Screws and Two Levels. The two level vials are set at an angle of 90° apart in these instruments. The leveling is similar to *a* above, but the instrument does not have to be turned as much. The instrument is first turned so that the vials are parallel to pairs of opposing screws, 1 and 3, 2 and 4, Ⓐ, figure 2–6. Each bubble is centered using the appropriate pair of screws. The instrument is turned 180° and if either bubble is off center, it is brought halfway back with the appropriate pair of screws. Either bubble may be brought to better adjustment by using the vial adjusting screw(s) and bringing the bubble the remaining distance to center. After making any adjustment, the centering must be rechecked.

c. Three Leveling-Screws and One Level. These instruments are leveled in a manner similar to *a* above. The instrument is turned until the level vial is parallel to a line joining two of the leveling screws, 1 and 2, Ⓑ, figure 2–6. The bubble is centered by turning one or both of these screws. The instrument is turned 90° and the bubble is now centered using the third screw, 3, Ⓑ, figure 2–6. Centering of the bubble is checked by turning the instrument 180° from each of the previous positions and compensating as described in *a* above.

d. Circular Bubble. The circular level vial has a small circle marked on the glass. The bubble, which is round, is centered when it is concentric with this circle. The instrument is turned and the

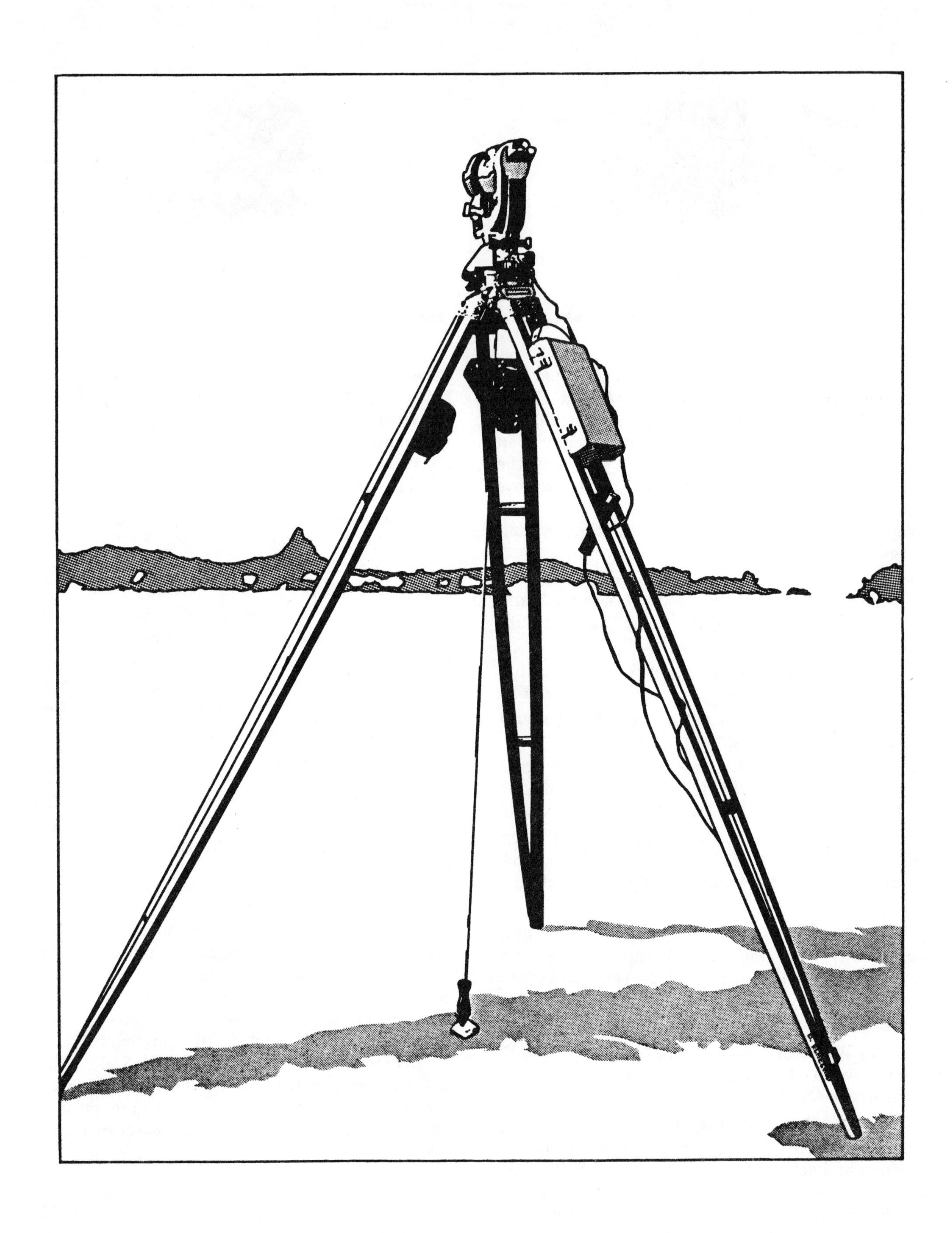

Figure 2-7. Instrument set over station.

screws are manipulated as in *c* above. The bubble is centered on an axis parallel to two screws and then at right angles using the third screw. Since the circular bubble is not very sensitive, leveling is completed using the plate level vial. The circular level is adjusted as follows:

(1) Bring the bubble of circular level to approximate center by using the leveling screws.

(2) Level the instrument using the plate level as described in *c* above. If circular bubble is not centered in the marked circle when instrument is level, the circular level is out of adjustment.

(3) Mark this unlevel position of the bubble on the vial window with a grease pencil or other suitable marking device for glass. Remove the instrument from the tripod. Place the instrument on a steady support so that the circular level housing projects over the edge of the support. This will provide access to the three circular level adjusting screws in the base of the level housing. Adjust the level of the instrument with suitable shim material until the circular level bubble is alined with the temporary reference mark placed on the vial window. Now adjust the three spring-loaded screws in the base of the housing until the bubble is centered in the regular marked circle. The three adjusting screws should not be tightened all the way. All of them should "float" the level on the adjusting springs.

(4) Remove the temporary reference mark previously placed on the vial window. Return instrument to tripod and bring to exact level with the plate level. Observe position of circular level bubble. If it is centered in the marked circle, the circular level is correctly adjusted. If bubble is not centered, repeat the procedure in (3) above until level bubble is correctly centered.

2–9. Plumb Bobs

The plumb bob usually has a body and a removable cap made of brass with a replaceable point made of hardened steel. It is a weight tapered so its point will hang directly below a supporting string. For normal tripod work, a movable slide is mounted on the string, and is used to adjust the height of the plumb bob above the point. When the string is hung under the center of the instrument, the point of a freely suspended plumb bob will be directly below this center (fig. 2–7). The vertical axis of the leveled instrument can be considered to be exactly over the point or conversely, the point on the ground to be exactly beneath the vertical axis of the instrument. In taping, the plumb bob may be used to "drop" a measurement

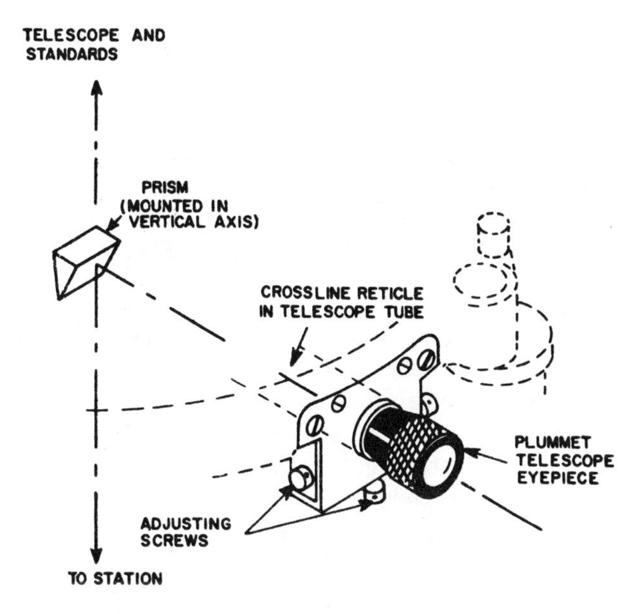

Figure 2-8. Optical plumbing assembly.

from the tape to the ground or to find the reading on the tape directly above a specific point on the ground. Plumb bobs come in varying sizes and weights, depending upon the need.

2–10. Optical Plumbing Assembly

The optical plumbing assembly or plummet (fig. 2–8) is a device built into the alidade or the tribrach of some of the instruments to center the instrument over a point. It consists of a small prismatic telescope with a crosswire or marked circle reticle adjusted to be in line with the vertical axis of the instrument. After the instrument is leveled, a sighting through the plummet will check the centering over a point very quickly. The plummet's advantage over the plumb bob are that it permits the observer to center over a point from the height of the instrument stand, and that it is not affected by wind. The plummet is especially useful for work on high stands or towers. A plumb bob requires someone at ground level to steady it and to inform the observer on the platform how to move his instrument and when he is exactly over the point. With the plummet, the centering and checking is done by the observer.

2–11. Centering the Instrument

Centering over a station and maintaining this position is a very important operation in surveying. The precision is dictated by the accuracy with which angles are to be measured. Horizontal angles, measured with an off-center instrument, will apply to the point which is directly under the instrument and not to the station mark itself, thereby causing the angles to be in error in reference to the station mark.

a. The tripod is set up and the instrument firmly attached. A plumb bob is suspended below the instrument and the string adjusted until the plumb bob is about 2 centimeters above the station mark.

b. If the instrument is not initially set up over the station, the tripod, instrument, and plumb bob are picked up keeping the tripod legs in the same relative position, and set down over the station mark. The instrument is leveled approximately and shifted on the footplate or tripod head until the plumb bob is over the mark.

c. The loosened screw is tightened and the leveling completed. The plumb bob is lowered to 0.5 centimeter above the mark and the centering checked. If necessary, centering is improved by loosening the screw again and shifting the instrument on the tripod head until the plumb bob is again exactly over the mark. The instrument is then checked for level, releveling if necessary. This procedure is followed until the instrument is both level and centered exactly over station mark.

d. If the instrument has an optical plumbing assembly, the plumb bob is removed and the plummet used for the final centering. The leveling is checked whenever the instrument is moved while centering. It is not only good practice, but necessary for precise surveys, to check the leveling and centering at intervals, before, during and after each series of direction (angle) measurements. There is always the possibility of tripod settling, or pressure created during instrument manipulation, or even an accidental bump causing the instrument to go out of level or off-center or both.

e. When final centering is done with a plumb bob, it is frequently necessary to shield the plumb bob and string from the wind to prevent them from swaying. An effective method is to stand on the side from which the wind blows and use the body and legs as a shield. In a mild breeze, placing one leg close to, but not touching, the string and resting the foot near the station mark will form a good shield. In high winds, a wind screen or observing tent should be erected around the instrument and station.

f. Not centering the instrument creates the largest errors in directions (angles) when the distance from the instrument to the sighted targets is small. For short sights, extra care must be exercised in centering.

2–12. Telescopes

The telescope is that part of the transit, theodolite, level, or alidade that acts as a sighting device and enlarges the image of any viewed object. It contains lenses to magnify the sighted image and a reticle to define the line of sight. The viewing system consists of an erecting or inverting eyepiece assembly, an adjustable reticle with a series of fine lines, and an objective lens assembly of either the internal or external focusing type. The telescope is a combination of optical and mechanical systems mounted in a tube using spacers, retainers, and screws. The combination is designed to maintain the relative spacing and alinement of the lenses and reticle to a high degree of precision (fig. 2–9).

a. Eyepiece. The eyepiece is essentially an adjustable microscope used to observe the lines on the reticle which appear superimposed on an enlarged view of the object or target.

(1) An inverting eyepiece lens turns all objects upside down and reverses them left to right. It is used on high precision instruments to take advantage of better optical qualities. The erecting eyepiece allows the observer to see the objects as they appear naturally.

(2) The eyepiece must be adjusted to permit the observer to see the crosshairs or crosslines in sharp focus. This adjustment depends on the observer's eye and is ordinarily made each time an observer uses the instrument. The telescope is pointed toward the sky and the eyepiece is rotated until the crosshairs appear very sharp and intensely black. Rotating the eyepiece moves it in a helical mount toward or away from the crosshairs for focusing. The final motion should be toward the crosshairs and should stop when they appear the sharpest.

b. Reticle. The reticle is a ring which holds the crosshairs that define the line of sight through the telescope. The ring is held in place by adjusting screws which permit the reticle to be exactly centered in the axis of the telescope. These can be four capstan-head or three fillister-head adjusting screws depending on the type of instrument. The crosshairs may be fine spider threads, fine platinum wire, glass fiber, or fine lines etched or applied to a glass plate. The two principal crosshairs are at right angles to each other and are set in the telescope so that one is vertical and the other horizontal. Some instruments have additional lines or ticks which are used for special purposes. Stadia hairs or lines are two horizontal lines spaced at a precise distance above and below the center hair. Some stadia reticles contain cross ticks to subdivide the distance between stadia hairs into tenths. Some theodolites use reticles with a "split" hair for one-half of the vertical crosshair. This permits centering distant objects between the hairs. If the single hair is used, the object should be bisected by the single hair (fig. 2–10).

c. *Objective Lens.* The objective lens is located at the forward end of the telescope and consists of two or more lens elements designed to give proper magnification and an undistorted field of view.

(1) In some telescopes, the objective lens assembly is moved as a unit to focus the object being observed. Other telescopes use the rear element of the objective plus an internal-field lens assembly which moves as a unit longitudinally in the tube for focusing. When properly focused, the objective lens forms an image of the object in the plane of the crosshairs. The image formed by the objective lens and the crosshairs are viewed as one image through the eyepiece.

(2) After the eyepiece has been focused, the telescope is pointed toward the object to be observed. Focusing the objective consists of turning either a knob or a knurled tube on the telescope depending upon the type of instrument. The knob operates a rack and pinion to move the objective lens assembly while the tube moves the assembly on a helical track to focus the lens. This brings the image into the plane of the crosshairs.

(3) To test for focusing, the observer must move his head from side to side or up and down while sighting through the eyepiece. Any apparent movement of the crosshairs in relation to the object image means that *parallax* is present. The image is not exactly in the plane of the crosshairs and the objective focusing must be refined. Since this condition can occur each time the objective lens is focused, a parallax check must be made whenever a new object is observed.

2–13. Care of the Telescope

Normally, the only parts of the telescope that are visible to the surveyor are a tube, the faces of the objective and eyepiece lenses, the focusing devices such as knobs or knurled rings, and the heads of the reticle adjusting screws. The telescope is never taken apart in the field.

a. Dust and dirt which accumulate on the exposed metal surfaces can be injurious if allowed to get into the operating parts. A clean cloth is used to remove the dust from the metal by gently wiping the surfaces. Avoid pushing it under the sleeves or onto the threads of the screws.

b. *Never* use a cloth on the glass surfaces. First, remove the dust carefully by using the camel's hair brush. Then, and only if it is still necessary, use the chamois or lens tissue and gently wipe the face of the lens. *Avoid rubbing* with the chamois or tissue, since this can remove the lens coating. Do not touch the lens surfaces, oil on the skin leaves a deposit which attracts dust if the surface is touched, first the brush and then the chamois or tissue are used to remove the deposit.

c. For subfreezing weather or arctic use, a winterizing kit is furnished. Rubber knob covers permit instrument handling without the skin adhering to the metal. Orange sticks gently passed over the lens surfaces will pick up any moisture which

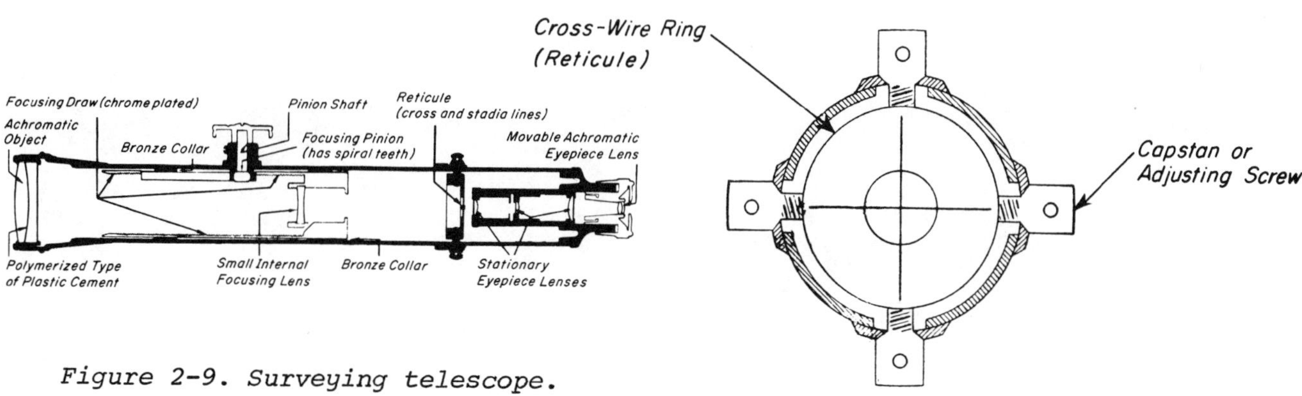

Figure 2-9. Surveying telescope. Keuffel & Esser Co.

Figure 2-10. Cross-Wire ring.

condenses on the surface. The orange stick will fit into areas that are difficult to reach where the metal and glass join.

d. Oiling is a maintenance procedure, and should not be attempted in the field, unless absolutely necessary. Use of oil may loosen a sticking or hard-to-turn knob or sleeve, but it does not remove the cause. It may drive the obstruction deeper into the assembly and cause extensive damage.

e. The appropriate technical manual that governs the operation and maintenance of the instrument must be consulted before any cleaning, oiling or adjustment is undertaken.

CHAPTER 3

LEVELS

Section I. HAND LEVELS

3–1. Locator Hand Level

The hand level, like all surveying levels, is an instrument which combines a level vial and a sighting device (fig. 3–1 and 2). The locator hand level (fig. 3–1) is so called because in use it is hand held in front of the eye. For greater stability, it may be rested against a tree, rod, or range pole. A horizontal line is provided in the sight tube as a reference line. The level vial is mounted atop a slot in the sight tube in which a reflector is set at a 45° angle. This permits the observer, in sighting through the tube, to see the landscape or object, the position of the level bubble in the vial, and the index line at the same time. The distance, over which a hand level is sighted, is comparatively short and no magnification is provided for the sighting.

3–2. Abney Topographic Hand Level

This more specialized type of hand level, called the Abney topographic hand level, adds a graduated arc so that the vertical angle and the percent of grade can be measured. This topographic level (fig. 3–2) has a reversible arc assembly mounted on one side. The arc is graduated in degrees on one side and percent of grade on the other. The level vial is attached at the axis of rotation of the index arm. The bubble is centered by moving the arc and not the sight tube as in the locator level. Thus, the difference between the line of sight and the level bubble axis can be read in degrees or percent of grade from the position of the index arm on the arc. The 45° reflector and the sighting principle with its view of the landscape, bubble, and index line are the same as in the locator hand level.

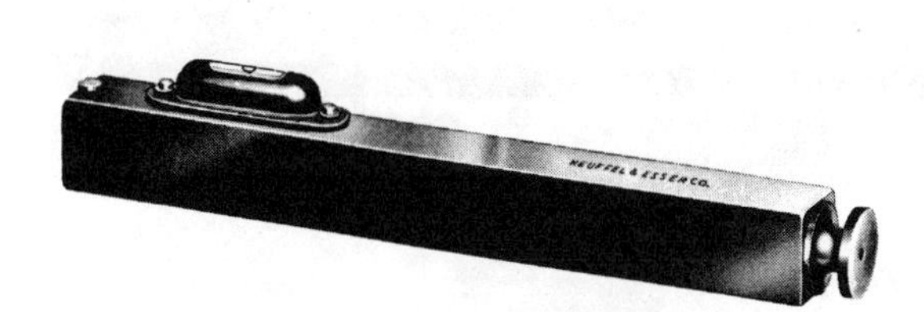

Figure 3-1. Hand level.

Figure 3-2. Abney hand level and clinometer.

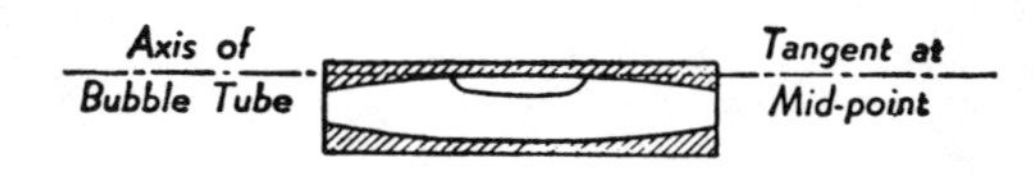

Figure 3-3. Bubble tube.

Section II. ENGINEER LEVEL

3-3. Description

The sighting device of an engineer level (fig. 3-4) is a 30 ±3, variable power telescope, maximum length 18 inches, with an erecting eyepiece. Some models use internal focusing, while others use external focusing objective lens assemblies. The reticle has two crosshairs at right angles to each other and some models have added stadia hairs. The telescope is rigidly attached to the level bar which supports an adjustable, highly sensitive level vial. This rigidly supported instrument is known as the "dumpy" type. The telescope and level bar assembly are mounted on a spindle which permits the unit to be revolved only in a horizontal plane. It cannot be elevated or depressed. A clamp and tangent screw act on the spindle for small motions to permit accurate centering. The spindle mounts in a four-screw leveling head which rests on a footplate. The footplate screws onto the threads on the tripod head. When the instrument is properly adjusted and leveled, the line of sight, defined by the horizontal crosshair will describe a horizontal plane (fig. 3-5).

3-4. Care

Care of the level, as with any other instrument, starts with an inspection of the instrument when it is issued.

a. Initial. Remove the level from the box and mount it on the tripod head. Examine it for missing parts or any sign of damage caused by bumps or jars.

(1) Check the bubble to see that it is not broken, then look through the telescope to see that the crosshairs are in place. These two hairs are used for the leveling operation and if either is missing or broken, the level cannot be used and must be replaced.

(2) Try all the clamps, motions, and screws, checking for freedom and smoothness of action. Determine the cause of any binding, friction, or malfunction, and whether it can be corrected at once. DO NOT TRY TO USE A FAULTY LEVEL.

(3) Wipe all exterior metal surfaces to remove dust and any excess grease or oil. Polished surfaces, such as the bearing surface or the footplate, should be cleaned carefully and thoroughly. Dirt or grit can scratch these surfaces and cause undue wear which will affect results.

(4) Check the lenses for chips and cracks. If dirty, clean lens using the camel's-hair brush first to remove the dust. Wipe gently with the chamois to remove any film. If this is insufficient, get an approved lens cleaning solution and clean the lens. Any accidental finger marks on the lens should be wiped or cleaned off immediately.

b. Care During Operations. Always carry the level stored in its carrying case between the office and starting point of a survey. The level should be

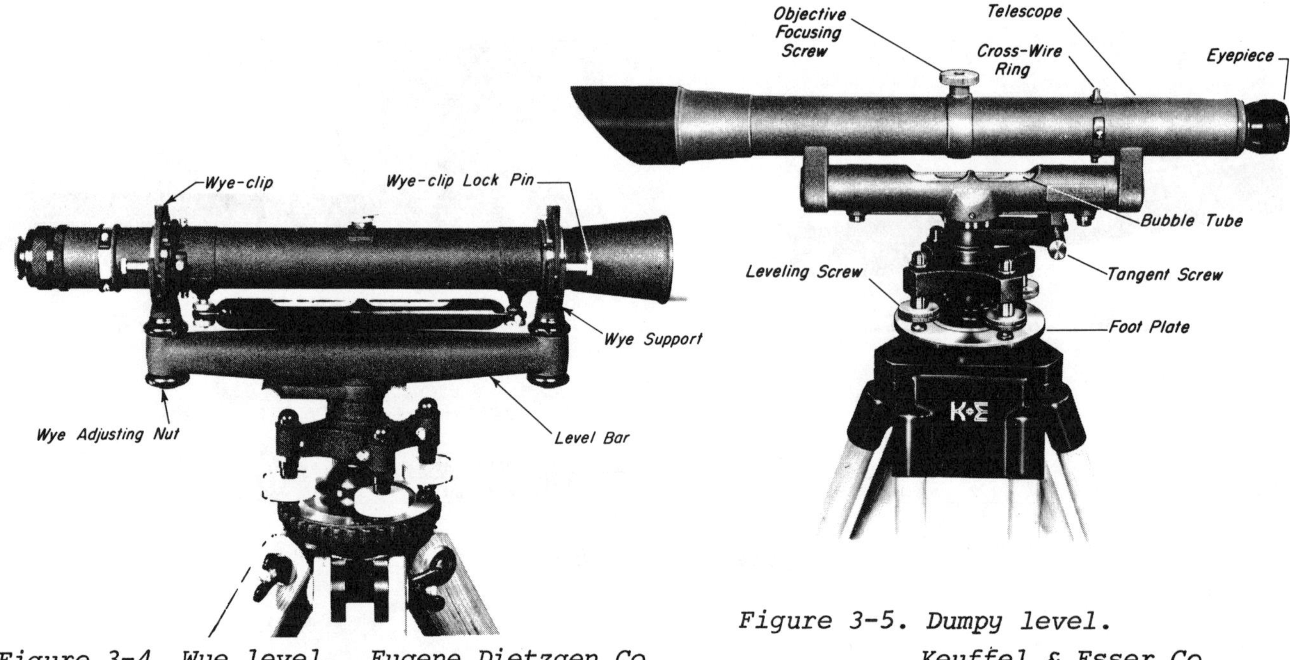

Figure 3-4. Wye level. Eugene Dietzgen Co.

Figure 3-5. Dumpy level. Keuffel & Esser Co.

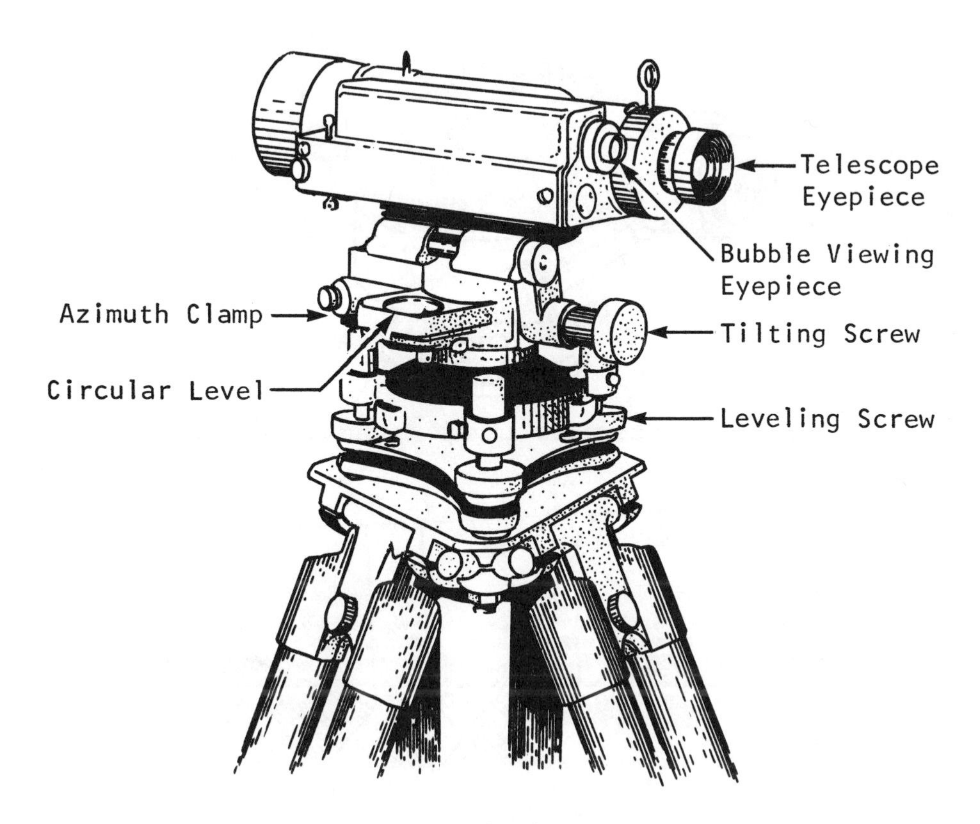

Figure 3-6. Tilting level.

Figure 3-7. Tilting level with micrometer screw.

Figure 3-8. Automatic level. Keuffel & Esser Co.

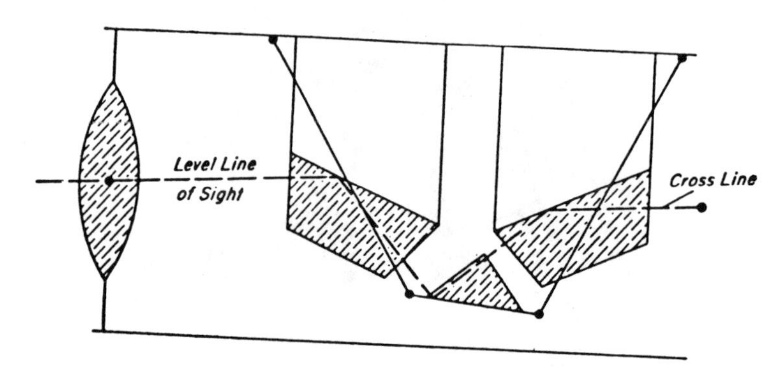

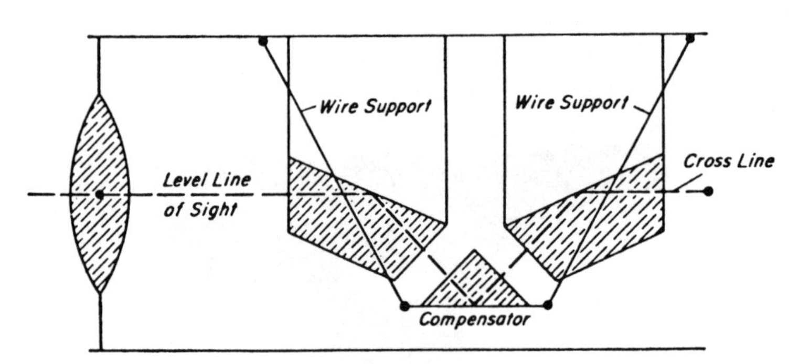

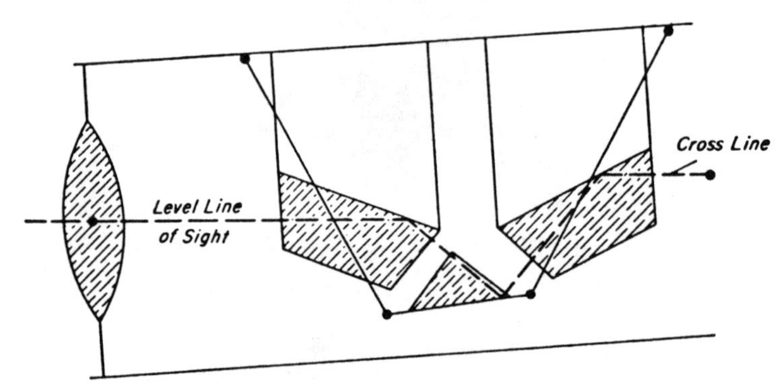

Figure 3-9. Compensator of self-leveling level.

hand carried in its case when transporting it over long distances or rough terrain during survey operations. Handle the case gently to avoid jolts or vibrations which can damage or misaline the delicate parts. When moving from point to point during actual surveys, it is permissible to carry the level on its tripod. Tighten the clamp screw to prevent motion, but do not overtighten. In dusty or salt water spray areas, remove the sunshade and cover the objective lens with the dustcap. Carry the level and tripod under the arm with the instrument cradled in front especially in uncleared or rough terrain. The level, like all survey instruments, is highly susceptible to expansion and contraction if exposed to in the sun. If the sun strikes only part of the instrument, unequal expansion will cause changes in the adjustments and introduce errors in the results. It is good practice to select shaded locations for instrument setups or to carry an umbrella to shade the instrument. There are three cardinal rules that all surveyors abide by:

(1) NEVER set up the tripod without finding good footings and setting the legs firmly to prevent collapse of the tripod.

(2) NEVER leave the setup instrument unattended.

(3) NEVER straddle the tripod legs.

3–5. Adjustment

A check of the instrument's adjustment should be made when it is first issued and before it is taken to the field. It is necessary to check the adjustment every day before starting work, at any time the instrument is bumped or jolted, and at the end of each day's work. The instrument should be set up and approximately leveled over both pairs of screws. Since the check will also include the optical assembly, the crosshairs and objective should be focused sharply, using a well-defined object about 50 meters away, and then the parallax removed (para 2–12*c*(3)). The check and adjustment are made in three steps and in the order listed.

a. Level Vial. Adjustment of the level vial (fig. 3–10) makes the axis of the level bubble perpendicular to the axis of rotation (vertical axis).

(1) Set the instrument over diametrically opposite leveling screws, and center the bubble carefully (①, fig. 3–10).

(2) Rotate the telescope 180° and note the movement of the bubble away from center, if instrument is unadjusted (②, fig. 3–10).

(3) Bring the bubble half the distance back to the center of the vial by turning the capstan screws at the end of the vial (③, fig. 3–10).

(4) Relevel with leveling screws (④, fig. 3–10) and rotate the instrument 180°. Repeat (3) above, if the bubble does not remain at the center of the vial.

(5) Check the final adjustment by noting that the bubble remains in the center of the vial during the entire revolution about the vertical axis.

b. Horizontal Crosshairs. The horizontal crosshairs are adjusted (fig. 3–11) to make the horizontal hair lie in a plane perpendicular to the vertical axis.

(1) Level the instrument carefully.

(2) Sight one end of the horizontal crosshair

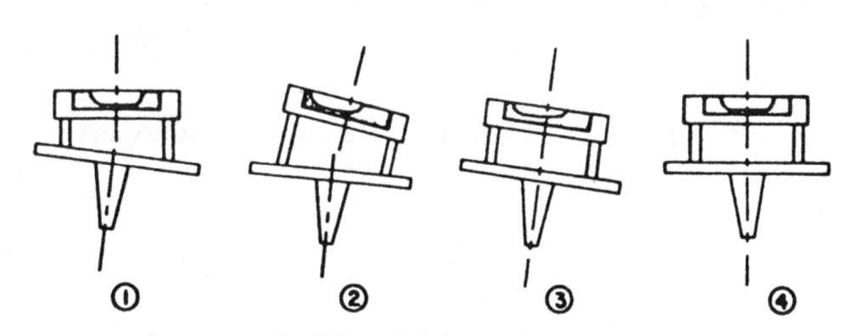

Figure 3-10. Adjustment of the level vial.

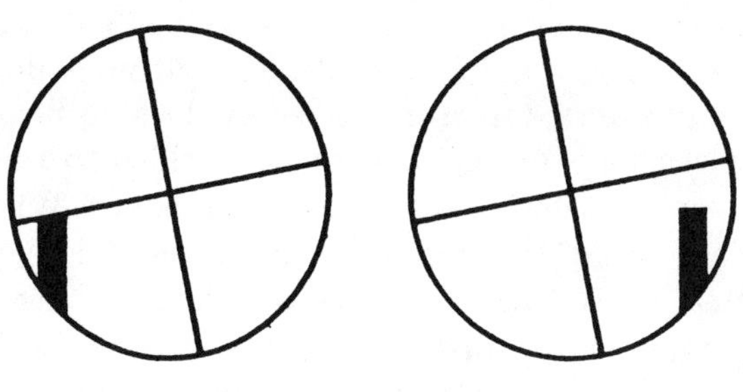

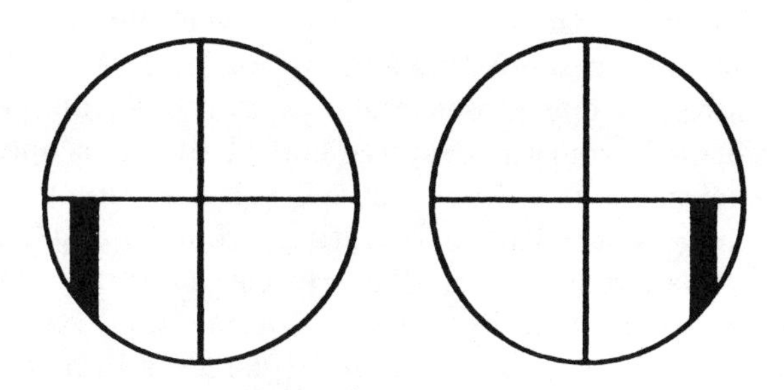

Figure 3-11. Adjustment of horizontal cross hairs.

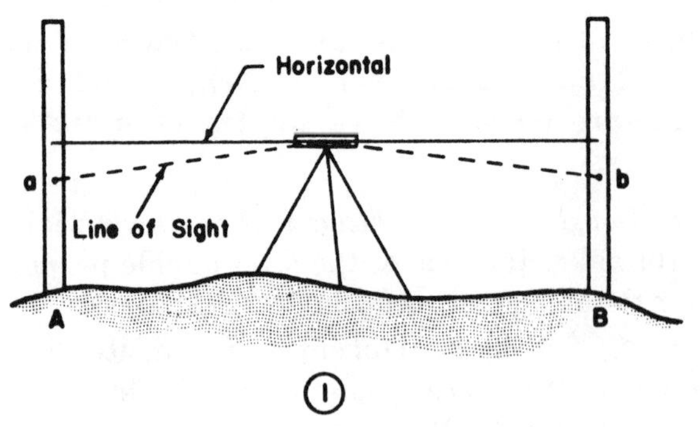

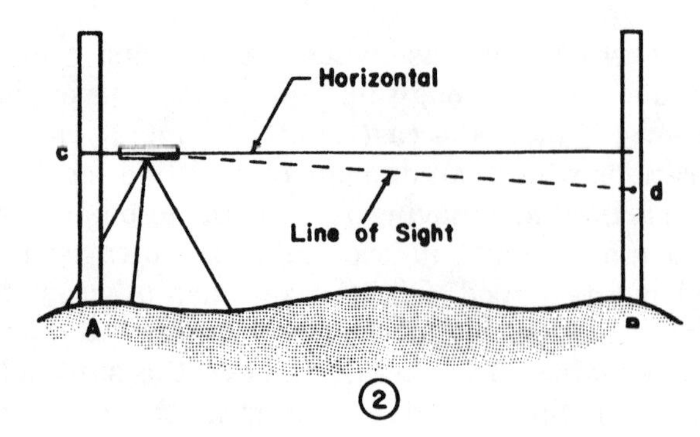

Figure 3-12. Two peg test.

on a well-defined point about 50 meters away. Turn the telescope slowly on its vertical axis, using the slow motion screw. If the crosshair is in adjustment, the hair will stay on the point through its entire length.

(3) If it does not, loosen the reticle adjusting screw and turn the reticle by lightly tapping two opposite screws.

(4) Sight on the point again and if the horizontal hair does not follow the point through its entire length, turn the ring again.

(5) Repeat this procedure until the condition is satisfied.

c. Line of Sight. The adjustment **(fig. 3–12)** makes the line of sight parallel to the axis of the level vial. This method is known as the "two peg" test.

(1) Set up the instrument (①, fig. **3–12)**; drive in a stake A about 50 meters away; drive in a stake B at the same distance in the opposite direction.

(2) Take a rod reading "a" on stake A and a rod reading "b" on stake B. With the instrument exactly halfway between the two stakes, (b-a) is the true difference in elevation between the stakes.

(3) Move the instrument close to stake A (②, fig. **3–12)** so that the eyepiece swings within 10mm of the rod.

(4) Take a rod reading "c" on stake A through the objective lens, and a rod reading "d" on the stake B in the normal manner. If the instrument is in adjustment (d-c) will equal (b-a).

(5) If the instrument is out of adjustment, calculate what the correct rod reading "e" should be, (e = b+c-a). Move the horizontal crosshair to the correct reading by loosening the correct vertical capstan screw and tightening the opposite screw.

(6) Check the horizontal crosshair adjustment (*b* above). The ring may have turned during this adjustment.

(7) Rerun the peg test to check the adjustment.

Section III. SEMIPRECISE LEVEL

3–6. Description

The level in figure 3–13 is designed for more precise work than the engineer level and is referred to as a semiprecise or prism level. All visual leveling operations and observations can be performed from the eyepiece end of the telescope. The telescope is a 30 power, 10-inch long, interior focusing type with an inverting eyepiece and an enclosed fixed reticle. The reticle contains the crosswires and stadia wires and is mounted internally and cannot be moved or adjusted as in many other instruments. Focusing the objective is done by an internal field lens through a rack and pinion controlled by a knob on the upper right-hand side of the telescope. The telescope and level can be tilted through a small angle in the horizontal plane about the vertical axis. This maintains a line of sight perpendicular to the vertical axis just before the rod reading is made. The tilting is done with a graduated micrometer screw located below the telescope eyepiece. A cam is provided to raise or lower the telescope off the tilting device. When the cam is in the raised position the telescope is lifted off the tilting device and holds it firmly when moving the instrument. The cam is in the lowered position only when observing with the instrument. The eyepiece located to the left of the telescope eyepiece is used for viewing the level

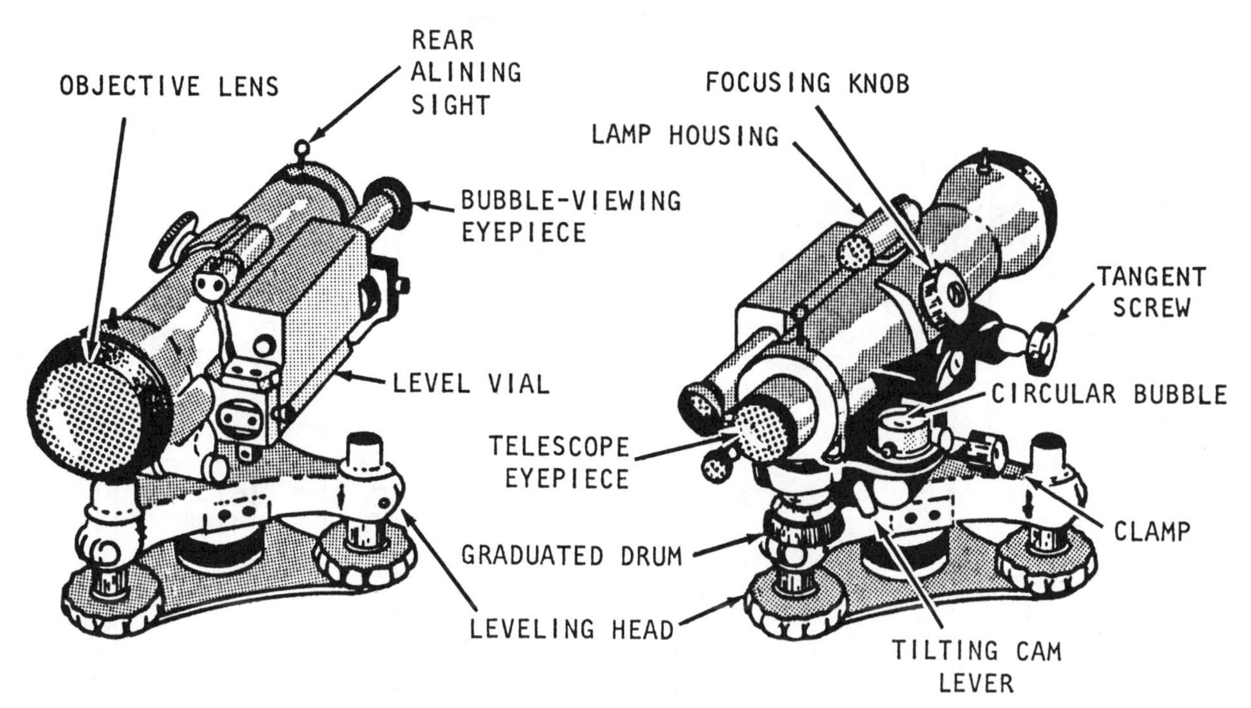

Figure 3-13. Semiprecise level.

vial through a prism system which brings both ends of the level bubble into view. The ends of this level bubble must be brought and maintained in coincidence during the observation. The level vial is not located directly under the telescope, but to the left and below, directly in line with the capstan screws under the level vial eyepiece. Its sensitivity is given as 30 second of arc per 2 millimeter spacing. A circular bubble which is viewed through a 45° mirror is provided for approximately leveling the instrument before using the level vial. For night work, battery-powered lights illuminate the level vial, the reticle, and the circular level. The clamping screw and horizontal-motion tangent screw are located on the right-hand side; the former near the spindle and the latter below the objective lens. The instrument has a three-screw leveling head. The tripod for this level is a fixed leg type for more stability during setup.

3–7. Care

a. Initial. The initial care of this level is similar to the engineer level (para 3–4) concerning the inspection, trying the motions, wiping the metal surfaces, and cleaning the lenses. However, the prisms, the level vial microscope, the tilting knob, and lighting equipment are additional components that must be examined.

b. During Operations. The added precision of this instrument dictates additional care in handling. Jolts, bumps, and vibrations can cause the instrument to go out of adjustment (para 3–8). When moving the level, the tilting cam level must always be in the raised position. The method of carrying and transporting is the same as for the engineer level. Protection from sunlight is usually required when using this instrument, and therefore a surveying umbrella should be considered necessary equipment. Cloudiness is not protection from the uneven expansion and contraction of the instrument. The cardinal rules of NEVER setting up an instrument without making sure of a firm footing and NEVER leaving the setup instrument unattended, and NEVER straddle the tripod legs apply.

3–8. Instrument Test and Adjustments

Semiprecise and precise level measurements require a constant check on the instruments to eliminate any possible errors caused by the instrument being out of adjustment. The circular level, the level vial, and parallax tests must be made before the level is used. In addition, the determination of a stadia factor and "C" factor must be accomplished.

a. Circular Level. The circular level is checked and adjusted as described in paragraph 2–8*d*.

b. Parallax. The parallax must be removed as described in paragraph 2–12*c*, before the tests and adjustments are started.

3 WIRE LEVELING

PROJECT	LOCATION	ORGANIZATION	SUN	WIND	WEATHER
30/US/12LL/S/169	Belvoir, Va.	Belvoir Gas	7	0	Clear, Calm, & Warm

OBSERVER	RECORDER	INSTRUMENT
J. Doe	E. Gehly	Wild N-3 # 1268

FROM	TO	DATE	TIME	LINE OR NET	PAGE NO.	NO. OF PGS.
		25 Apr '8_	0730	Stadia Constant	1	

STATION	BACKSIGHT FACE OF ROD	MEAN	BACK OF ROD	INTERVAL	SUM OF INTERVALS	FORESIGHT FACE OF ROD	MEAN	BACK OF ROD	INTERVAL	SUM OF INTERVALS	REMARKS
	1355										Rod# 778 (R)
25	1230 (R)			125'				25'	÷	0.100'	Rod# 780
	1105			125'	250'			250'			
	1425										
35	1250			175'				35'	÷	0.100'	
	1075			175'	350'			350'			
					600'						
	1486										
45	1260 (R)			226'				45'	÷	0.100'	
	1034			226'	452'			452'			
					1052'						
	1505										
55	1230			275'				55'	÷	0.100'	
	0955			275'	550'			550'			
					1602'						
	1595										
65	1270 (R)			325'				65'	÷	0.100'	
	0945			325'	650'			650'			
					2252'						
	1675										
75	1300			375'				75'	÷	0.100'	
	0926			374'	749'			749'			
					3001'						
Total = 300'										0.600' ÷ 6' =	
		300' ÷ 3001'	= 0.100'			Check				0.100	

FOR SUN: 0 = CLOUDY WITH SUNSHINE, SHOW DIRECTION OF PROGRESS IN RELATION TO THE SUN BY NUMBERS IN THE DIAGRAM.

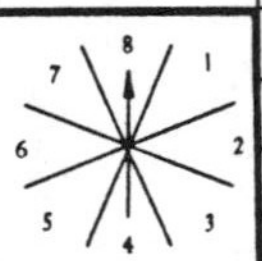

FOR WIND: 0 = CALM, 1 = MODERATE, 2 = STRONG

Figure 3-14. Determination of stadia constant.

c. Determination of Stadia Constant. The stadia constant factor of the instrument should be carefully determined. This factor is required in the computation of the length from the stadia intervals and in computing the allowable error for a level line. This determination must be done for all levels equipped with stadia wires. Should new cross wires be installed in the field, a redetermination of the constant must be made before using the level in a survey. The observations and computations for this determination must be recorded as a permanent record along with the time and date and kept with the project files. The determination is made by comparing the stadia intervals observed over a known distance course. The course should be laid out on a reasonably level track, roadway, or sidewalk, and nails or other marks placed in a straight line of measured distances, of 0, 25, 35, 45, 55, 65, and 75 meters. The plumbing point of the instrument and zero point are the same on this level; therefore the instrument is plumbed directly over the zero stake. However, this is not true for all instruments and the manufacturer's manual should be checked for the zero point correction. Read the rod at each of the six points and record the intervals as shown in figure 3-14. The level bubble need not be accurately centered, but should at least be free of the ends of the tube. The half-wire intervals should be computed as a check against erroneous readings. The sum of the total intervals for the six readings should also be computed. The stadia constant coefficient is the sum of these measured distances (300 meters) divided by the sum of the six total wire intervals. As a check against gross errors each separate observation should be computed. The average of the six separate computations serves as a numerical check. Any tendency for the six computed values to creep in one direction will be good evidence that some error in the measurement of the distance has been made from the center of instrument to the zero point of the instrument and then to the first of the six rod points.

d. Determination of "C" Factor. Each day, just before the leveling is begun or immediately after the beginning of the day's observations, and immediately following any instance when the instrument is subjected to unusual shock, the error of the instrument ("C" factor) must be determined. This determination may be done during the regular course of leveling or over a special course; in either case the recording of the observations must be done on a separate page of the recording notes

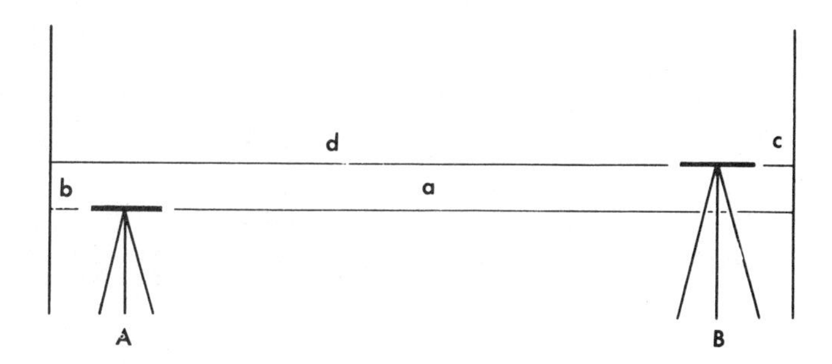

Figure 3-15. Procedure for "C" factor.

with all computations shown. The following procedure is used for the observation and computation.

(1) Set two rods (fig. 3–15) about 100 meters apart and the instrument about 10 meters forward of the rear rod (A); read the front rod (a), then the rear rod (b); move the instrument to a position (B) about 10 meters behind the front rod; read front rod (c) and then the rear rod (d). The two instrument stations must be between the two rod points. The readings must be made with the level bubble carefully centered (in coincidence) and then all three wires are read for each rod reading.

(2) The required "C" factor determination is basically the ratio of the required rod reading correction to the corresponding subtended interval, or

$$C = \frac{(\text{sum of near-rod readings}) - (\text{sum of distant-rod intervals})}{(\text{sum of distant-rod intervals}) - (\text{sum of near-rod intervals})}$$

The total correction for curvature and refraction (para 8–5) must be applied to *each* distant rod reading before using them in the above formula. It must be remembered that the sum of the rod intervals must be multiplied by the stadia constant in order to obtain the actual distance before determining the curvature and refraction correction. The recording and computation of this factor is shown in figure 3–16. This figure shows the before and after adjustment computations. The maximum permissable "C" factor varies with the stadia constant of the instrument. The instrument must be adjusted at once if the "C" factor is greater than 0.004 for instruments with a stadia constant of approximately 1/100, 0.007 for stadia constants of 1/200, and 0.010 for stadia constants of 1/333. It is desirable to have the determination of the "C" factor made under the anticipated con-

PROJECT 30/US/2LL/5/69		LOCATION Belvoir, Va.		ORGANIZATION Belvoir Gas		3 WIRE LEVELING		
OBSERVER J. Doe		RECORDER E. Gehly		INSTRUMENT Wild N-3 # 1268		SUN 7	WIND 0	WEATHER Clear, Calm & Warm
FROM	TO	DATE 25 APR '8-	TIME 0800	LINE OR NET "C" Factor Check			PAGE NO. 2	NO. OF PGS.

STATION	BACKSIGHT FACE OF ROD	MEAN	BACK OF ROD	INTERVAL	SUM OF INTERVALS	FORESIGHT FACE OF ROD	MEAN	BACK OF ROD	INTERVAL	SUM OF INTERVALS	REMARKS
A	1613					2802					Rod #778 (R)
	1565(R)	15650'		48'		2337	23370'		465'		Rod #780
	1517			48'		1872			465'		
	4695'				96'	7011'		Ref =	-0.6'	930'	
											Stadia
B	1775					1428					Constant =
	1717	17167'		58'		0974(R)	09743'		454'		0.100
	1658			59'	117'	0521			453'	907'	
	9845'	32817'			213'	9934'	33113'	Ref =	-0.6'	1837'	
							- 1.2' =	Tot. Ref		- 213'	
							33101'			1624'	
							32817'				
							- 28.4' ÷ 1624' = -0.017' = C				
		AFTER	ADJUSTMENT								
B	1774					1411					
	1715	17153'		59'		0959(R)	09583'		452'		
	1657			58'		0505			454'		
	5146'				117'	2875'		Ref =	-0.6'	906'	
A	1602					2759					
	1546(R)	15467'		56'		2303	23033'		456'		
	1492			54'	110'	1848			455'	911'	
	9786'	32620'			227'	9785'	32616'	Ref =	-0.6'	1817'	
		32604'					- 1.2' =	Tot. Ref		- 227'	
		+1.6' ÷ 1590' = +0.001' = C					32604'			1590'	

FOR SUN: 0 = CLOUDY WITH SUNSHINE, SHOW DIRECTION OF PROGRESS IN RELATION TO THE SUN BY NUMBERS IN THE DIAGRAM. (1 2 3 4 5 6 7 8)

FOR WIND 0 = CALM 1 = MODERATE 2 = STRONG

JED

Figure 3-16. Recording and computation of "C" factor.

ditions as to length of sight, character of ground, and elevation of line of sight above the ground. The date and time must be recorded for each "C" factor determination, since this information is essential in computing the level corrections when making office computations for the particular day's work.

e. Adjustment of Semiprecise Level. The level is adjusted if the "C" factor exceeds the allowable limits for the instrument stadia constant. The adjustment is made as follows: point to the distant rod, with bubble centered, and read the middle wire. Move telescope, by means of the micrometer screw, so as to change the reading at the middle wire by an amount equal to "C" times the total rod interval in millimeters. The middle wire should be moved upward on the rod if "C" is plus or downward on the rod if "C" is negative. Then with the telescope held in this position bring the bubble to center by raising or lowering the end of the level vial, using the capstan screws at the eyepiece end of the level vial. If the left bubble end image is too low, the lower capstan screw is loosened slightly and the upper one tightened by an equal amount. Both screws are turned in the same direction. The adjustment is continued gradually until the bubbles ends are in coincidence. After the adjustment is completed all screws should be moderately tight, otherwise the adjustment will not hold. The "C" factor is redetermined **(fig. 3–16)** to check that it now meets the allowable limits.

f. Two-Peg Test. When using the semiprecise level for lower order surveys, the level error may be determined and corrected by use of the two-peg test as described in paragraph 3–5*c*.

Section IV. PRECISE LEVEL

3–9. General

The precise level, also known as the geodetic level, is designed to achieve the highest precision in leveling measurements. There are many precise levels and any of them may be used as long as they are equivalent in accuracy to the level described. The level vial sensitivity, telescopic power, focusing distance, and size of the objective lens are considered in the design. These factors may vary individually, but when one factor is weakened, another factor must be strengthened to maintain the specified accuracy. The telescope power may be 32 to 42 power. The focus may be brought down to 2.0 meters. The level vial sensitivity is 1.7 to 1.9 seconds of arc per 2-millimeter graduation and the length of the bubble is controlled by an air chamber. The specification for a precise level is given as a capability to run 1 kilometer of level lines to an accuracy of ±3mm k, where k is the distance in kilometers. The level is then tested and rated on its ability to maintain this required precision. This type of level is normally used only for second or higher orders of accuracy in connection with topographic and geodetic surveying

3–10. Wild N3 Level

The numbers in parenthesis in this paragraph and succeeding paragraphs of this section are referenced to fig 3–17&18. The star shaped base plate (17) has the standard thread, enabling the level to be attached to any sufficiently stable tripod. The three footscrews (1) and circular bubble (19) are used for the pre-leveling of the instrument. The rotatable upper part of the instrument consists chiefly of the telescope and the tubular level. The telescope can be clamped in any desired direction by tightening the horizontal clamp (14). The horizontal slow motion tangent screw (13) is used, with the clamp tightened, to obtain the fine pointing. The rotatable eyepiece (7) of the telescope has a diaptric scale, enabling the observer to make an immediate setting corresponding to his own eyesight, when focusing the reticle. The focusing knob (8) is used for focusing the telescope. Between the telescope objective and the cover glass (10) is the parallel glass plate, which is activated by the micrometer drive knob (9). The reticle has so-called wedge shaped crosswires and stadia wires, with a stadia constant of usually 1/100. At distances of more than 10 meters this instrument has an additive constant of —20cm (para 3–12*c*). The bubble ends of the tubular level, magnified 2½×, are observed in the eyepiece (21). The tilting screw (4) is used to obtain the coincidence setting of the bubble ends, viewed in the eyepiece (21). If the bubble ends are completely out of the field of view an arrow is seen (fig. 3–19), which indicates plainly the direction in which the tilting screw must be turned in order to obtain the coincidence setting.

a. Focusing. Point the telescope towards a uniformly light colored surface, or a sheet of white paper, and turn the telescope eyepiece until the reticle crosswires appear sharp and black. The ring is then turned slowly counterclockwise until the image starts to go out of focus. A small clockwise rotation will refocus the wires correctly. This setting is constant for one observer and the personal dioptric scale number should be noted for future use.

b. Sighting. After loosening the horizontal clamp the telescope is pointed, over the peep sights, at the level rod on the measuring station. The clamp is tightened gently, and the vertical wire of the reticle cross is set along the middle of the rod, using the slow motion screw. The focusing knob (8) is turned until the image of the rod graduations are seen sharply. The observer should move his eye up and down and from side to side to insure that no parallax exists. If parallax is noted the instrument should be refocused (*a* above).

3–11. Care

The precise level must be accorded the care and attention that any delicate scientific instrument merits. Special attention must be given to prevent any sudden shocks, jolts and bumps which will require retesting of the instrument. The level must be kept in its case when not in use and when transporting it to and from working areas. The clamp screw should be set to prevent the instrument from swinging, but not overtightened. The lenses and working parts must be kept clean and free of dirt particles. To avoid the effects of sunlight, a surveyors umbrella should always be used. If there is any great difference between the working and storage temperature, the instrument should be allowed to adjust itself to the actual working conditions for about 15 minutes before observations are started. This is particularly important in low temperatures.

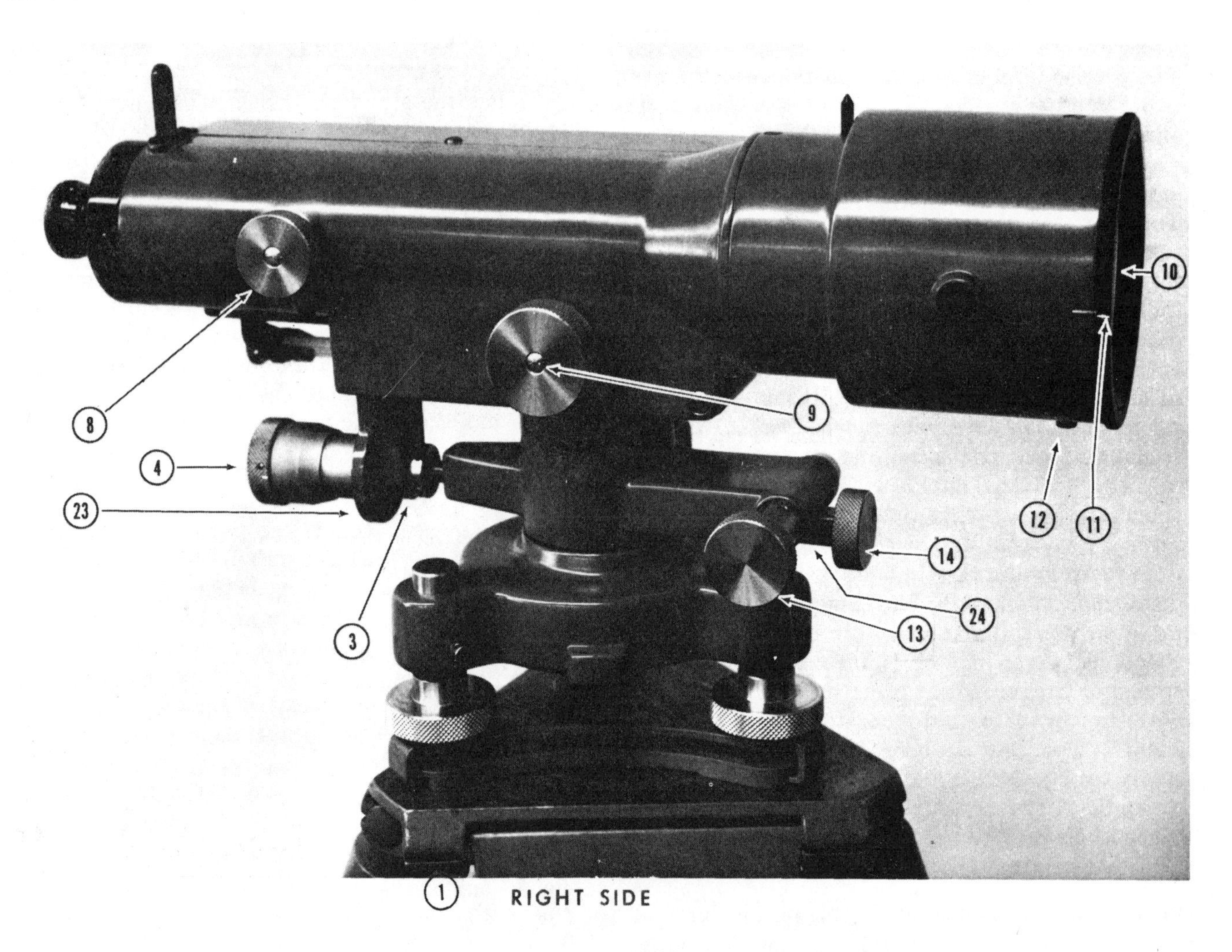

Figure 3-17. Precise level (Wild N3), right side.

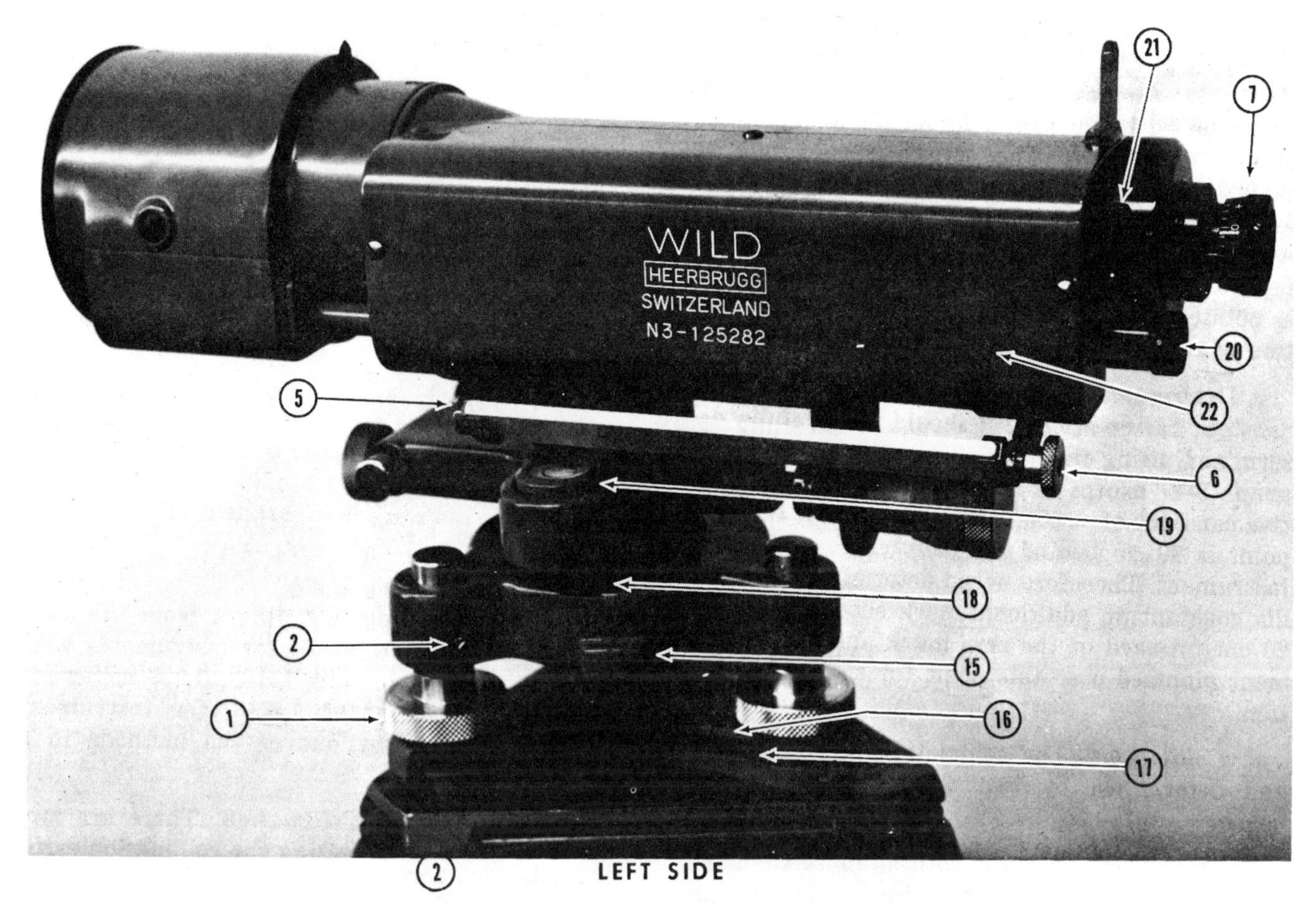

Figure 3-18. Precise level(Wild N3), left side.

3–12. Tests and Instruments

Precise level measurements require a constant check, and the following tests and adjustments should be accomplished at specified intervals for the N3.

a. Screws. The footscrews should have a smooth and adequate run. This can be checked only when the instrument is not screwed onto the tripod. The adjustment is made by using the adjusting pin to turn the adjustment screws (2) found on the side of each footscrew, until the required run is obtained. The run of the tilting screw (4) can be regulated by means of the adjustable nut (3). A similar adjustable nut (24) is used to regulate the slow motion screw.

b. Circular Bubble. The instrument is attached to the tripod and set up on firm ground. Using the footscrews, the circular bubble is centered exactly in the middle of the black setting circle. The telescope is rotated through 180°. If the bubble changes position by more than 1mm, it should be adjusted. When doing this the bubble's glass cover must not be touched. Half of the bubble displacement is taken out with the footscrews and the other half with the adjustment screws (18). As an adjustment screw is loosened the bubble runs toward it and, as it is tightened, the bubble runs away from it. The first screw to be turned, therefore, is the one that is the nearest to being in line

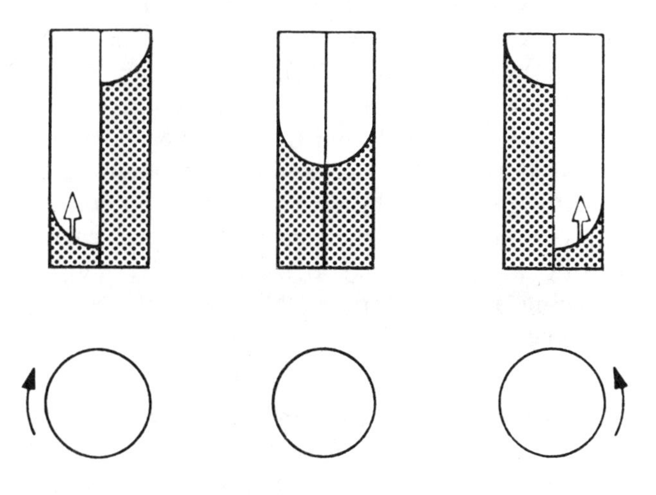

Figure 3-19. Leveling the tubular level vial.

with the middle of the bubble and the center of the black circle. This screw is to be turned until the bubble reaches the center of the circle or until it can be set in the center by means of one of the other two screws. The screws should not be turned more than is necessary to complete the adjustment. The bubble is correctly adjusted when it stays in the center of the black circle regardless of the direction in which the telescope is pointed. This should be checked upon completion of adjustment.

c. Determination of Stadia Constant. The stadia constant factor of the N3 should be carefully determined, using the procedure explained in paragraph 3–8*c* except as noted. The N3 has an additive constant of —20cm, that means that the zero point is 20 cm behind the plumbing point of the instrument. Therefore, when determining the stadia constant an additional mark should be placed 20 cm forward of the zero mark and the instrument plumbed over this point for the determination.

d. Collimation. The collimation may be tested and determined by two methods as described below.

(1) The collimation is nothing more than the level error, "C" factor, as explained in paragraph 3–8*c*. The method described in referenced paragraph is the preferred method in many surveys.

(2) Another method of determining this error is as follows: Lay off a course about 60 meters long on reasonably flat terrain (fig. 3–20). Divide the course into three equal sections (using a tape) of length d. An invar rod is set up at the intermediate points, B and C, respectively. The instrument is set up, in turn, over the terminal points A and D, respectively. With the tubular level in coincidence at A, readings a'_1 and a'_2 are taken on the rods, held at B and C. With the instrument leveled at D, readings a'_3 and a'_4 are taken. If the line of sight is absolutely horizontal (no collimation error), these readings will be the correct reading a_1, a_2, a_3, and a_4, respectively, and the following relationship is valid, as can be seen from the figure:

$$a_4 - a_1 = a_3 - a_2$$

However, if this relationship is not true, then the line of sight is inclined to the horizontal plane by the small angle, δ. If an imaginary line, parallel to $a'_1 - a'_2$, is projected through a'_3, it will cut the rod at B in the correct position a_4, thus giving the required value for a true, horizontal line

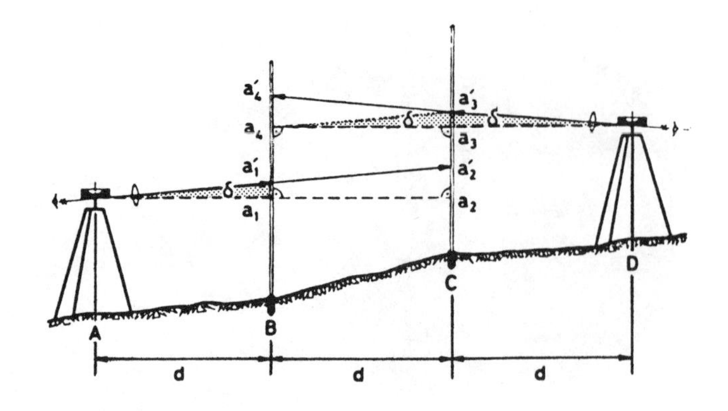

Figure 3-20. Collimation test.

of sight through D. This is seen plainly in figure 3-20. As the darkened triangles are all congruent and, as a_4, a'_1, a'_2, a'_3 is a parallelogram,

$$a_4 - a'_1 = a'_3 - a'_2$$

giving, $a_4 = a'_1 - a'_2 + a'_3$.

If the actual reading, a'_4, differs from the computed value by 0.004 meter for instruments with a stadia of approximately 1/100, 0.007 meter for 1/200, 0.010 meter for 1/333, the instrument must be adjusted by one of the methods in *e* below.

e. Adjustment of Collimation. There are two methods used for correcting the collimation error of the N3 level. The method used depends on the amount of collimation error.

(1) *Front cover glass.* If the a'_4 reading differs from the computed value by only a few millimeters (about 4), then the adjustment is made by turning the cover glass (10). Loosen the locking screw (12) about half a turn, then the micrometer reading is set to the computed value of a_4 and the cover glass mounting is turned until the required rod reading is set on the center crosswire. The tubular level must be in coincidence during this procedure and should be checked after making the correction. After checking the rod readings and level the locking screw is tightened.

(2) *Adjusting screws.* If the a'_4 reading is greater than the computed value a_4, more than about 4 millimeters, the adjustment is made using the adjusting screws. The micrometer reading is set to the computed value of a_4 and the tilting screw (4) is turned until the required rod graduation is on the center crosswire. The ends of the tubular level should then be restored to perfect coincidence by means of the vertical adjustment screws of the level vial holder. First, however, the outer horizontal screw must be loosened slightly, so that the bubble ends can follow the

movements of the vertical screws. If the left bubble end is too low, the lower adjusting screw is loosened slightly and upper one tightened by an equal amount, turning both screws to the left as seen from the objective end of the telescope. The adjustment is done a little at a time until the bubble ends are in coincidence. The outer horizontal screw is then retightened. All screws should be moderately tight, otherwise the adjustment may not hold. After the adjustment is completed, the collimation is checked to see if it now meets the accuracy requirement.

f. Cross Error. The cross error (wind parallelism) is defined as the difference between the direction of the tabular level and the direction of the line of sight of the telescope, i.e. the amount of horizontal nonparallelity between the two. It is checked by sighting a rod with the telescope positioned in such a way that it lies over one footscrew and at a right angle to the line connecting the other two footscrews. The circular bubble is centered using the footscrews, the tubular level bubble is brought into coincidence using the tilting screw and a rod reading is taken and recorded. The footscrew to the left of the eyepiece is now turned a half turn. The right footscrew is then turned until the recorded rod reading is again on the crosswires. If the tubular level bubble ends are no longer in coincidence, the level vial must be adjusted for cross error. The adjustment is done using the horizontal adjusting screws on the objective end of the level vial housing. These screws are turned equal amounts, in the same direction, until the bubble ends are again in coincidence. If this adjustment is made, the horizontal collimation (*d* above) must be checked.

Section V. LEVEL ACCESSORIES

3–13. Standard Leveling Rod

A leveling rod is in essence a tape supported vertically and is used to measure the vertical distance (difference in elevation) between a line of sight and a required point above or below it. This point may be a permanent elevation station, known as a *bench mark*, or it may be some natural or constructed surface.

a. The standard Philadelphia leveling rod (①, fig. 3–21) is a graduated wooden rod made of two sections. The foot rod can be extended from 7.1 feet to 13.1 feet. The graduations on the rod are feet, tenths of feet, and hundredths of feet. Instead of using a line or tick to mark each hundredth, the distance between alternate graduations are painted black on a white background. Thus, the value for each hundredth is the line between the colors; the top of the black, even values; the bottom of the black, odd values. The tenths and feet are numbered in black and red, respectively. The observer usually reads the rod directly while sighting through the telescope. This rod may be used with the level, transit, theodolite, and on occasion, with the hand level to measure the difference in elevation. The metric rod (②, fig. 3-21) can be extended from 2.0 meters to 3.7 meters. This rod is graduated in meters, decimeters, and centimeters.

b. The targets furnished with the foot rod and metric rod have a vernier which permits reading the scale to one-thousandth of a foot and one millimeter respectively. The description of the verniers and how to read them can be found in paragraph 4-3*d*.

c. Care of the leveling rod consists of keeping it clean, free of sand and dirt, unwarped, and readable. It must be carried over the shoulder or under the arm from point to point. Dragging it through the brush or along the ground will wear away or chip the paint. When not in use, the leveling rod is stored in its case to prevent warping. The cases are designed to support the rod either flat or on its side. The rod must not be leaned against a wall or remain on damp ground for any extended period since this can produce a curvature in the rod and result in unpredictable random and systematic errors in leveling.

3–14. Precise Level Rods

Precise level (fig. 3-22) rods are required when running second or higher order level lines. There are many precise rods and any of them may be used as long as they are equivalent in accuracy to the rod described. The rods must be of the one piece, invar strip type, with the least graduation on the invar strip of 1 centimeter. The invar strip is 25 millimeters wide and 1 millimeter thick and is mounted in a shallow groove in a single piece of well-seasoned wood. The groove is slightly wider than the strip and deep enough so that the strip is free to move. The rod has a metal footpiece to which the strip is securely fastened. In order to eliminate any error due to sagging of the invar

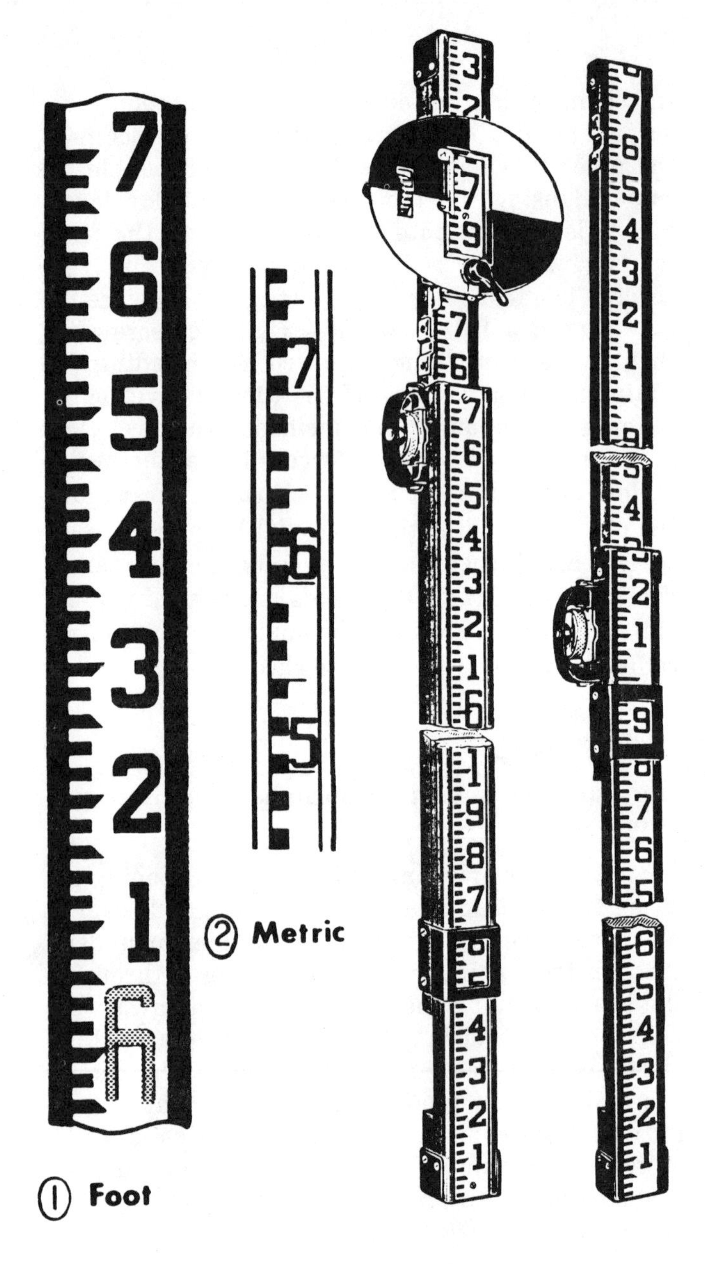

Figure 3-21. Philidelphia rod.

strip when the rod is held erect, it is placed under tension by a stiff spring, set into a recess in the top of the staff and bearing against a small brass angle plate attached to the top of the strip. This ension spring also allows the strip to expand or contract because of temperature changes. The front of the rod (①, fig. 3-22) is graduated in meters, decimeters, and centimeters on the invar strip. The back of the rod (②, fig. 3-22) must be graduated in feet and tenths of feet, or yards and tenths of yards. Mounted on the rear of the rod is a circular level vial and a thermometer (just back of the invar strip). The thermometer is needed to determine the temperature for correcting the strip length due to expansion or contraction. These rods must be standardized by the National Bureau of Standards (or its foreign counterpart) and their index and length corrections determined. The rods with similar characteristics are paired and marked. These pairings must be maintained throughout a line of higher order levels. The invar strips should be checked periodically against a standard to determine any changes which may affect their accuracy. The precise rod is a scientific instrument and must be treated as such, not only when in use but during storage and transporting. When in use the rods should be alternately carried by the handles and face up on the shoulder. If the rods are always carried with the face down or up, they will become slightly bowed and the reading will be too large. When not in use they must be stored in their shipping containers to avoid damage. The footpiece should be inspected frequently to make sure that it has not been bent or otherwise damaged. The precise rod

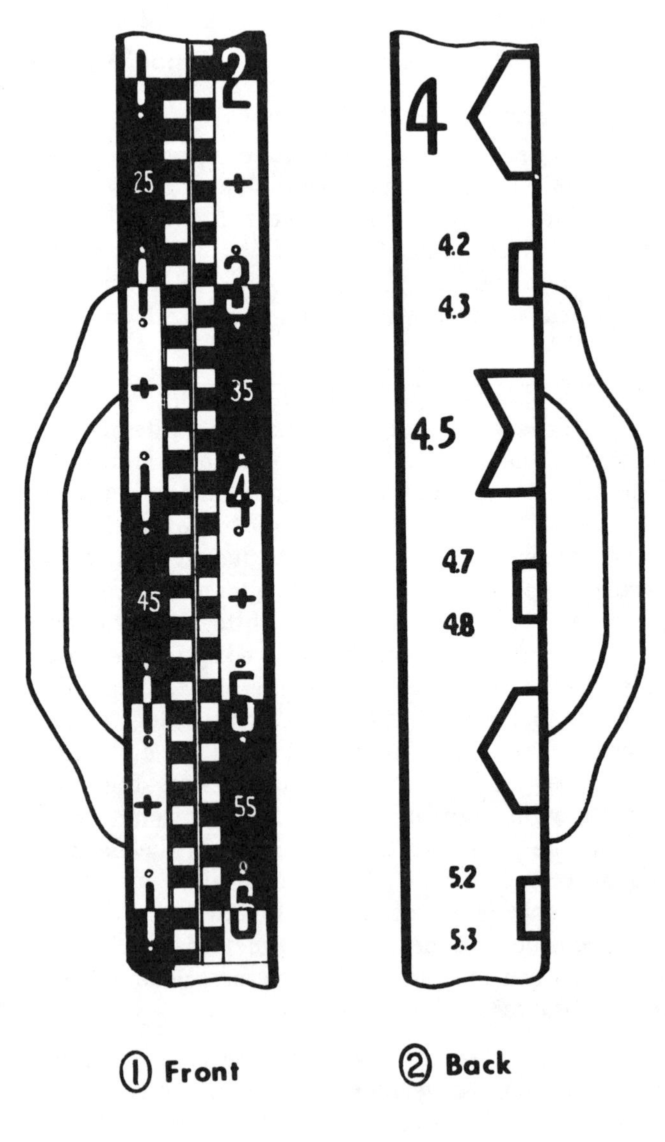

Figure 3-22. Precise rod.

should never be used for lower order leveling unless absolutely necessary.

3–15. Stadia Rods

The stadia rods (fig. 3–23) are wooden boards about 12 feet or 4 meters long, hinged at the center for quick folding and carrying. The rod has painted graduations in feet, tenths, and half-tenths (1/20) of a foot or meters, decimeters, and centimeters. In reading the foot rod, the half-tenths can be estimated to hundredths of a foot. The stadia rod is read directly without using a target.

3–16. Rod Targets

A leveling rod is usually read directly by the instrumentperson. Conditions which hinder direct readings, such as poor visibility, long sights, and partially obstructed sights, as through brush or leaves, sometimes make it necessary to use targets. The target is also used to mark a rod reading when setting numerous points to the same elevation from one instrument setup.

a. Targets for the Philadelphia rod are usually oval, with the long axis at right angles to the rod, and the quadrants of the target painted alternately red and white. The target is held in place on the rod by a thumb screw. A lever on the face of the target is used for fine adjustment of the target to the line of sight of the level. The targets have rectangular openings approximately the width of the rod and 0.15 foot high, through which the face of the rod may be seen. A linear vernier scale (para 4–3) is mounted on the edge of the opening with the zero on the horizontal line of the target for reading to thousandths of a foot.

b. Targets for the precise rod are used for exceptionally long sights across inaccessible areas, such as rivers, gorges, or ravines. The targets are about 0.3 meters long and 0.15 meters wide. A white strip about 2 centimeters wide is painted on a black background. A small rectangular hole with an index line centered in the white stripe is used for reading the rod. A thumb screw on the back holds the target in place.

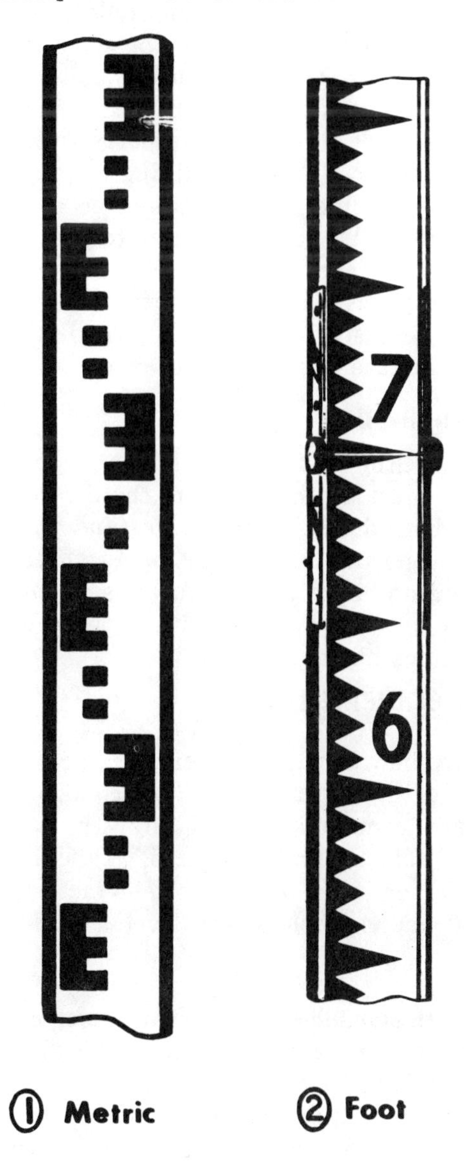

Figure 3-23. Stadia rod.

3–17. Rod Levels

For accuracy in leveling, it is extremely important that the leveling rod is held vertically when the reading is made. This makes the rod perpendicular to the horizontal line of sight and gives the correct reading. Any deviation from vertical makes the rod reading too large and introduces an error. Spirit levels (fig. 3–24), mounted on brass angles and held against a corner of the rod, are efficient in keeping the rod vertical. The angles may be hand held against the rod or attached to it temporarily. The levels may use one circular or two long vials at right angles. The circular level is preferred because only one bubble has to be watched by the rodperson.

3–18. Turning Pins, Pedestals, and Wooden Stakes

The point on which a leveling rod is held between a foresight and the next backsight while the instrument is being moved to the next setup is called a *turning point.* It must be sufficiently stable to maintain the accuracy of the level line. Where the proper natural features or manmade construction

Figure 3-24. Rod levels.

are not available, a turning pin, a turning plate or pedestal, or a wooden stake is used. These not only furnish the solid footing, but also identify the same position for both sightings. Normally, the pins or the plates are used for short periods and are taken up for future use as soon as the instrument readings are completed. Wooden stakes are used for longer periods except when wood is scarce or local regulations require their removal.

a. Turning Pins. The turning pin (fig. 3–25), is a tapered steel spike with a round top and a ring through the shaft for ease in pulling. It is driven into the ground with a mallet. After turning pins have served their purpose at one point, they are pulled and carried to the next turning point.

b. Turning Plates or Pedestals. These are triangular metal plates with turned-down corners or added spikes which form prongs, and have a projection or bump in the center to accept the rod. They are devised for use in loose, sandy, or unstable soils. The plate is set by placing it on the ground, points down, and stepping on it to press it to a firm bearing. After use, it is lifted, shaken free of dirt and mud, and carried forward to the next turning point.

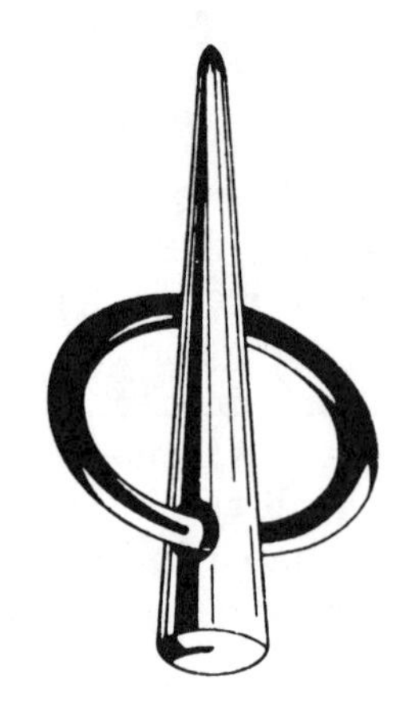

Figure 3-25. Turning pin.

c. Wooden Stakes. Wooden stakes are used when a point must be marked for some time, but not permanently, or when pins or plates are not available. The stakes may be of precut and shaped lumber, or made by the survey party from locally available timber. Stakes remaining for a considerable length of time in areas subjected to freezing and thawing must be long enough to prevent them from being worked out of the ground. Wooden stakes are placed in locations where they are not likely to be disturbed, or else, are driven flush with the surface of the ground.

3–19. Umbrella

For all leveling of third or higher order the instrument must be shaded from the direct rays of the sun, both during observations and when moving from one instrument station to another. The surveying umbrella is used for this purpose. The umbrella can also be used with good results as a windscreen when necessary.

Section VI. METEOROLOGICAL EQUIPMENT

3–20. General

Meteorological equipment is used for barometric leveling only for small scale maps (1:250,000) to determine differences in elevation. However, meteorological equipment is used extensively in topographic, and geodetic surveys in conjunction with the electronic distance measuring equipment (EDME), covered in chapter 6 of this book.

3–21. Altimeter

The surveying altimeter is an aneroid-type barometer which measures atmospheric pressure changes. For survey purposes, the scale is calibrated to show elevation differences, pressure differences or both, under standard conditions. An altimeter commonly used by the surveyor is the Wallace and Tiernan, type FS–199, designed for use up to 4,500 meters above and 300 meters below sea level (fig. 3–26).

a. Description. The aneroid barometer measures air pressure by means of a sealed metal cylinder or capsule from which the air has been removed. The cylinder is designed to expand or contract in one dimension as the air pressure increases or decreases. A series of levers and gears translates this longitudinal motion into a needle movement

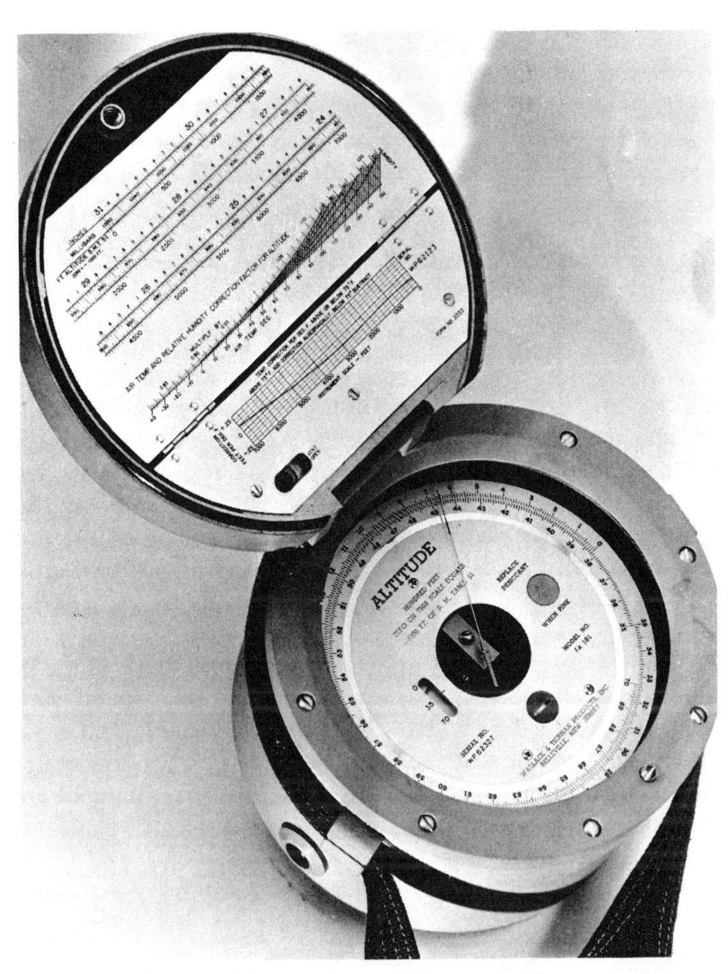

Figure 3-26. Surveying altimeter.

around a circular dial. The dial is subdivided to indicate the changes in pressure.

b. Principle. The basic principle of altimetry is that the pressure created by the weight of a column of air decreases as the observer rises in altitude. Since air is compressible, the ratio of pressure change to altitude rise is not a constant. In addition, temperature and relative humidity variations, and to a small extent, gravity, change the density of the air, and in turn, the weight of the air column and the barometric pressure. Because of these variables, the altimeter cannot be calibrated to read elevations directly. The dial is subdivided into elevation differences based on a selected standard pressure-altitude relationship. The readings are corrected for the variables. Air temperature and relative humidity are measured, and from prepared charts, corrections for these factors are read and applied to the readings. The corrected values represent the actual differences in elevation between the points. If the true elevation of one of the points is known, the corrected altimeter differences will give the elevations of all the points. One other factor is considered in altimetry. Barometric pressure is continually changing due to the movement of air masses over the world. This pressure change causes fluctuation in the altimeter readings even when set up on one point. The extent to which this fluctuation can be observed and evaluated determines the accuracy of the survey. Since direct sunlight can create uneven heating within the instrument, and therefore, inaccurate readings, the altimeter is always set in the shade or shaded during operation. During unstable weather conditions and periods of high gusty winds, pressure changes are very erratic, and an altimeter survey during these times will yield inaccurate results. The Beaufort Scale (table 3–1) was devised to describe wind velocities by a series of numbers from 0 to 12. Each number defines a velocity range and is described by the visual conditions which are created by these velocities. Altimeter surveys should be confined to days when the wind force does not exceed number 3, and should never be conducted when the force exceeds number 4.

c. Reading. The dial on the 4,500-meter altimeter is graduated on four scales with 2-meter intervals, and is numbered every 50 meters. Readings can be estimated to 0.2 meters. The readings are from 0 to 4,800 meters with 0 corresponding to —300 meters. During periods of low atmospheric pressure and when working near sea level, the scale gives positive readings and avoids plus and minus values. The final result is corrected by subtracting the 300 meters. A revolution counter on the dial face shows which scale is to be read. A mirror behind the dial face is used to eliminate parallax in the readings. When reading the altimeter, it should be set as level as possible and the pointer is viewed so that the direct image is superimposed on the reflected one. The dial may be set to agree with other instruments by turning the adjusting screw located under the glass window. This screw must *not* be turned beyond the limits marked in red on the dial. A thermometer in the face of the dial gives the instrument temperature when needed. A change in the instrument temperatures will change the reading of the altimeter.

d. Care. The surveying altimeter is a precision device with its dial individually calibrated for the particular instrument. The dial and internal mechanism are matched and are not interchangeable with similar parts of other instruments. Field care and maintenance is limited to the adjustment and procedures described.

Table 3–1. Beaufort Scale of Wind Forces

Beaufort No.	Wind description	Wind velocity (miles/hour)	Wind velocity Knots	Visual conditions during this range
0	Calm	Less than 1	Less than 1	Smoke rises vertically.
1	Light air	1–3	1–2½	Smoke drifts in general direction of wind. Ordinary wind vanes not affected.
2	Light breeze	4–7	2½–6	Wind is felt on face; leaves rustle. Ordinary wind vanes move.
3	Gentle breeze	8–12	6–10	Leaves and twigs in constant motion. Small flags extended.
4	Moderate breeze	13–18	10–16	Dust and loose paper raised. Small branches sway.
5	Fresh breeze	19–24	16–20	Small trees sway. Waves on inland waters show crest.
6	Strong breeze	25–31	20–27	Large branches in motion. Wires whistle.
7	Moderate gale	32–38	27–32	Whole trees sway. Walking against wind uncomfortable.
8	Fresh gale	39–46	32–40	Twigs break off trees. Walking against wind difficult.
9	Strong gale	47–54	40–47	Structures in areas begin to show slight damage. Shingles blow off and chimney tops begin to topple.
10	Whole gale	55–63	47–54	Trees uproot. Structures begin to show considerable damage.
11	Storm	64–75	54–65	Widespread damage.
12	Hurricane	Above 75	Above 65	

(1) *Handling.* The altimeter contains a mechanism which translates small expansion and contraction movements of a drum into a circular motion of a pointer on a dial. The gears, levers, and pivots that accomplish this are very delicate. Sunlight shining on the instrument can heat the mechanism unevenly and create false readings. Thus, the altimeter should be shaded as much as possible during transportation and operation. A seemingly light bump or jolt can bend or break some of the fine mechanism, or at least, throw the gear teeth out of mesh. The altimeter is mounted with rubber shocks to absorb some of the bumps and vibration. For short distance transportation in a vehicle, it should be hand held, or at least placed where vehicle motion will not be transmitted to the mechanism. The altimeter should be lifted and set down carefully to avoid jarring. It should be handled in the same manner as a fine watch.

(2) *Renewing the desiccant.* When the indicator turns pink, the silica gel must be removed and either dried or replaced with fresh desiccant. The desiccant tube cap is unscrewed and the tube removed. After renewing the desiccant, the tube and cap are replaced.

(3) *Replacement of lamps and batteries.* The batteries must be removed when the altimeter is not in use or is to be stored to prevent instrument damage from corrosion. The lamps and batteries can be removed or replaced by unscrewing the caps in the case.

e. Adjustment. Altimeter surveys are always performed with sets of altimeters. All of the instruments must be compared to a standard, usually a mercury column at a weather station. After comparison, the scales are reset or the difference in readings is recorded as a correction factor. It is not necessary to transport all the altimeters to a weather station. One instrument, arbitrarily selected as the master, is used for the comparison, and then returned to adjust the remaining altimeters. The date and place of the comparison should be recorded. The adjustment of the altimeters consists of comparing their readings to the master and either resetting the needle or recording the difference in readings as a correction factor.

f. Calibration. The adjustment or standardization (*e* above) checks the readings at only one point along the scale. Friction differences along the scale due to design, wear, or mechanical imperfections do not allow the needle to move equal amounts for equal changes in air pressure. The correction for this differential movement is determined by calibrating the altimeters before the field survey. A sample calibration recording is shown in figure 8–13.

(1) Calibration consists of transporting the instruments up a hill or mountain that is high enough to cover the expected range of altitudes in the survey. This registers the instruments' movement due to altitude (or air pressure) changes. At every 50 meters (or 100 feet) of altitude, the altimeters are set down and each one, in turn, is read and the readings recorded. The measurements are repeated at 50-meter (or 100-foot) intervals until the altitude range is covered. On the return trip, down the mountain, the same points are occupied and the altimeters read again. To establish a good mean, the entire process is repeated. Each altimeter will have four readings at each point for the final calibration.

(2) With a portable test kit, as many as six altimeters can be calibrated without moving the instruments. The altimeters are connected to the pump and the pump operator changes the pressure until the 50-meter (or 100-foot) altitude increment is reached on the master. At this time, a "tip" is called and the observers at each instrument record the reading on their respective altimeters. This process is continued over the expected altitude range twice. In either method, any instrument that is radically out of adjustment, is withdrawn. The readings from the remaining instruments are meaned for each stop. This becomes the calibrated mean and is used as the value from which all the altimeter correction factors are determined. Calibration must be repeated any time after—

1. Three months have elapsed since the last calibration.

2. The instruments have been shipped over a considerable distance.

3. Any instrument in the group has been dropped or jolted.

4. Any instrument is to be used with a group in which it has not been calibrated.

(3) Only by placing an altimeter in a pressure chamber and comparing it to a mercury column (changing the pressure systematically to make the comparison) can absolute values of pressure be determined for the altimeter readings. Absolute values in inches or millimeters of mercury are required for refractive index of air in electronic distance measurements. The method in (2) above could compare an altimeter to one that was compared to a mercury column. The method described in (1) above shows whether several instruments operate alike, but does not show that any are correct in absolute terms.

3–22. Psychrometer

The psychrometer is an instrument containing both wet and dry bulb thermometers. It is used to determine the temperature and saturation pressure, and for correcting the altimeter (barometer) readings. Both temperature readings are used as arguments in determining the total effective atmospheric pressure and the mean atmospheric pressure. The dry bulb is again used as an argument and combined with total effective pressure to determine the refractive index. The thermometers should be as well matched as possible and should be calibrated or have their index errors determined by immersion in a solution of pure melting ice. The wick on the wet bulb should be kept clean and replaced when it gets dirty. Only distilled water should be used on the wick, and water should not be allowed to get on the dry bulb. There are two types of psychrometers; the battery operated type for second or higher orders of accuracy; and the sling type for third and lower orders of accuracy.

a. Battery Operated. There are many models and types of battery operated psychrometers and almost any type or model may be used as long as it is equivalent to the one described here and shown in figure 3–27. It must be a self-contained, portable instrument which, when assembled with batteries installed, includes all the necessary components for measuring wet and dry bulb temperatures without the use of external power sources or additional equipment. The thermometer tubes should be illuminated. When power is introduced to the fan motor, through the switch, the fan draws air in the air intake, across the thermometer bulbs, through the air duct in the housing, and out the exhaust ports. This air flow must be of proper velocity to ventilate the wet and dry bulb thermometers for the psychrometric observations. The air intake and both exhaust ports must be entirely free of obstruction when in operation.

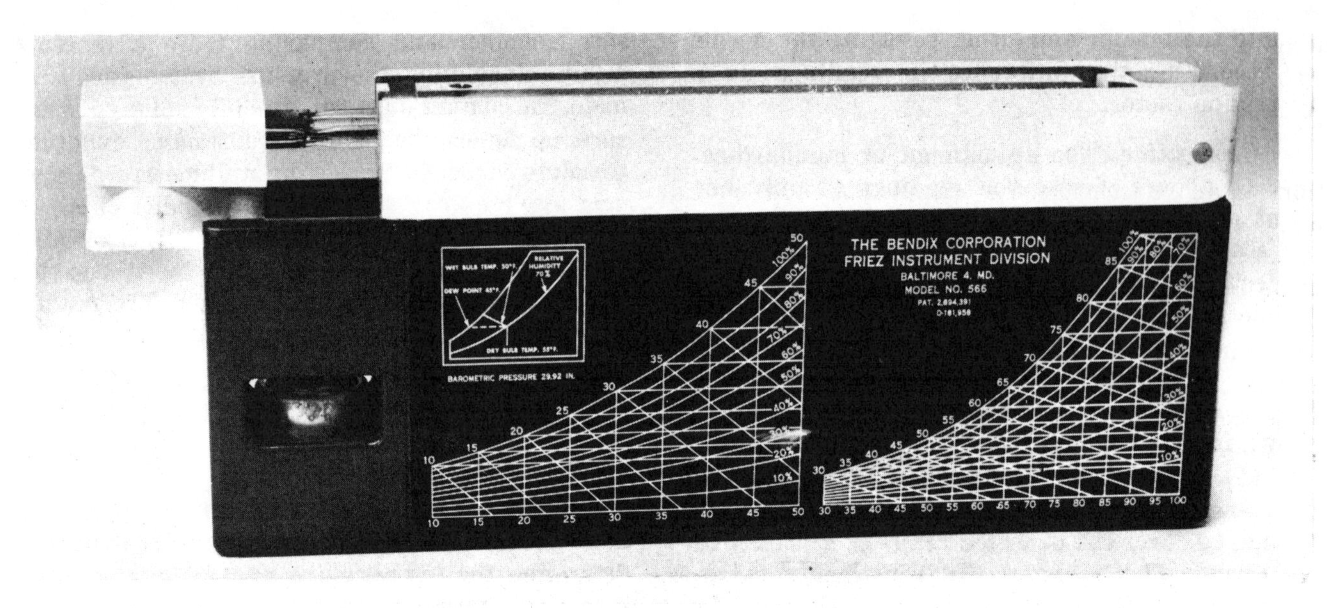

③ BACK VIEW

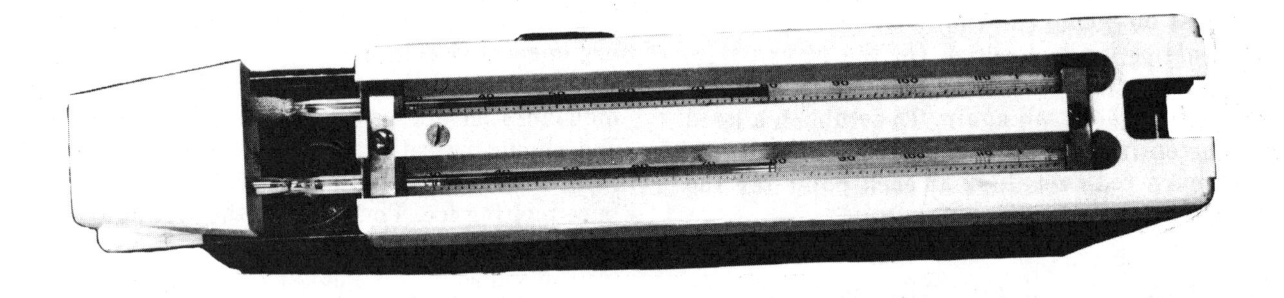

② TOP VIEW

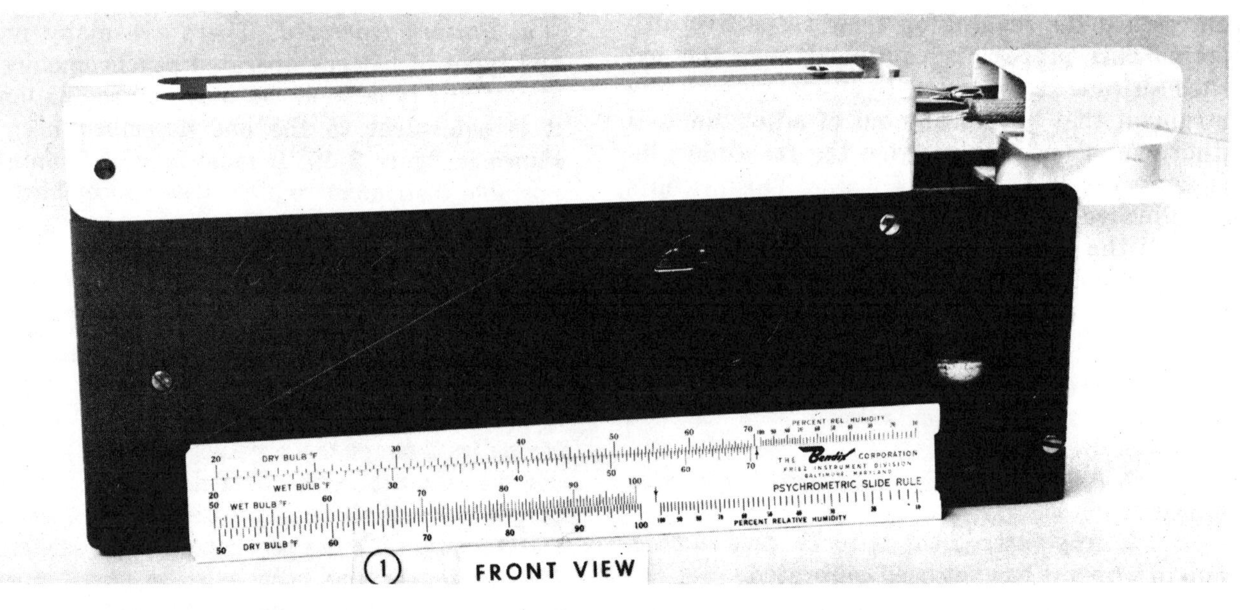

① FRONT VIEW

Figure 3-27. Battery operated psychrometer.

Since the instrument samples the air to which it is exposed, care must be taken to use the psychrometer far enough away from any source that may cause erroneous thermometer readings. It should be operated about five feet above ground level.

b. Sling. The sling psychrometer is an accessory of the 4500-meter altimeter. Two thermometers are mounted in a sling and whirled rapidly by hand in the air (about 120 revolutions per minute) to force air past both bulbs. The two thermometers must be read immediately after the whirling stops because the temperature begins to rise as soon as the whirling stops. The sling psychrometer should be used for only third or lower order surveys because it cannot be depended upon for stable readings.

3–23. Thermometer

A separate thermometer is not needed when using the altimeter for most EDME's except the model 2A geodimeter. For first order base lines the air temperature must be determined by a thermometer, other than the thermometer in the psychrometer. This thermometer must be calibrated or have the index error determined by immersion in a solution of pure melting ice and must be one that can be read or estimated to 0.1°F.

Figure 3-28. Reversible compensator level.

Figure 3-29. Compensator level in operation.

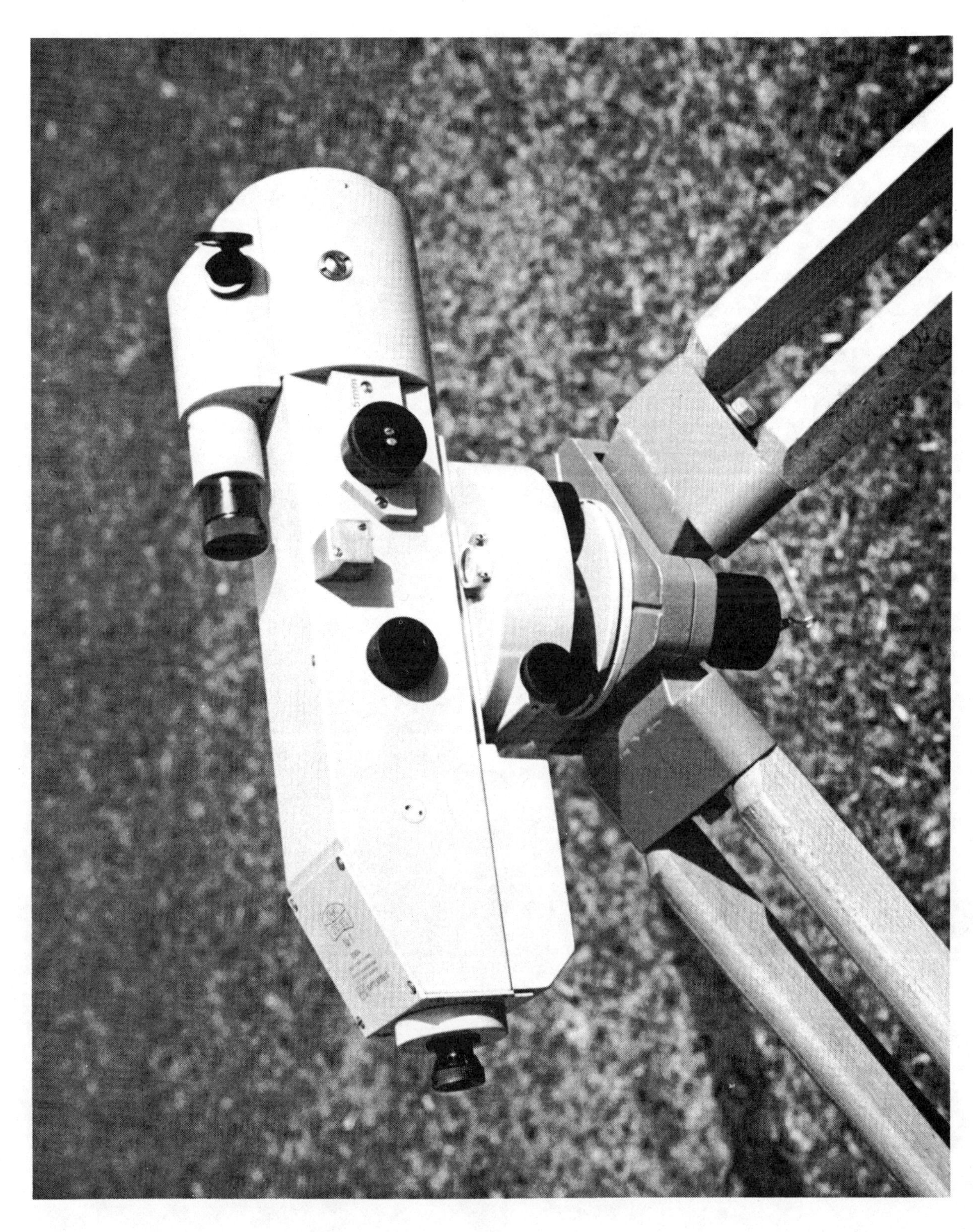

Figure 3-30. Compensator level.

Figure 3-31. Double-scaled rod with braces.

Figure 3-32. Double-scaled rod on turning pin.

Figure 3-33. Calculator, recorder, and programmable terminal.

CHAPTER 4

DIRECTION AND ANGLE EQUIPMENT

Section I. ENGINEER TRANSIT

4-1. Introduction

A primary survey field operation consists of many components, among which are included the measuring of horizontal and vertical angles or directions, and extending straight lines. The instruments which the surveyor uses to perform these functions have additional refinements and can be used for other operations. For instance, to measure the vertical angles, a level device is needed to establish a horizontal plane of reference. Once this plane is established and the vertical motion clamped, the transit is, in effect, a level and can be used for differential (direct) leveling (para 8–4). Angle measuring instruments fall into two basic categories—the transit and theodolite. In recent years, manufacturing improvements have permitted construction of direct reading theodolites which are replacing many of the vernier reading transits now being used. However, some transits are still purchased and are being used effectively in construction surveying. The engineer or surveyor's transit (fig. 4–1) was often called the universal surveying instrument because of its many uses. It may be used for observing horizontal angles and/or directions, vertical angles, and differences in elevation; for prolonging straight lines; and for measuring distances by stadia. Although the transits of the various manufacturers differ in appearance, they are alike in their essential parts and operation. Two types which are common are the 1-minute and the 20-second transits. This refers to the least reading that can be made on the horizontal scales. The vertical circle is identical in both, reading to 1 minute.

4–2. Description

The transit contains several hundred parts. For descriptive purposes, these may be grouped into the leveling head, the lower plate, and the upper plate or alidade assemblies (fig. 4–2).

a. Leveling Head. The leveling head assembly of the transit normally is the four-screw type, constructed so that the instrument can be shifted on the footplate for centering over a marked point on the ground.

b. Lower Plate. The lower plate assembly of the transit consists of a hollow spindle perpendicular to the center of a circular plate, and accurately fitted to the socket in the leveling head. The lower plate contains the graduated horizontal circle on which the values of horizontal angles are read with the aid of two verniers, A and B, set on opposite sides of the circle. A clamp controls the rotation of the lower plate, and provides a means for locking it in place. A slow motion tangent screw is used to slowly rotate the lower plate a small amount with respect to the leveling head. The rotation accomplished by the use of the low clamp and tangent screw is known as the *lower motion.*

c. Upper Plate. The upper plate, alidade, or vernier plate assembly consists of a spindle attached to a circular plate that carries the verniers, telescope standards, plate level vials, and a magnetic compass. The spindle is accurately fitted to coincide with the socket in the lower plate spindle. A clamp is tightened to hold the two plates together, or loosened to permit the upper plate to rotate relative to the lower plate. A tangent screw permits the upper plate to be slowly moved a small amount, which is known as the *upper motion.* The standards support two pivots with adjustable bearings that hold the horizontal axis and permit telescope movement in a vertical plane. The vertical circle moves with the telescope. A clamp and tangent screw are provided to control this vertical movement. The vernier for the vertical circle is attached to the left standard. The telescope is an erecting type and magnifies the image about 18 to 25 times. The reticle contains stadia wires in addition to the crosswires. A magnetic compass is mounted on

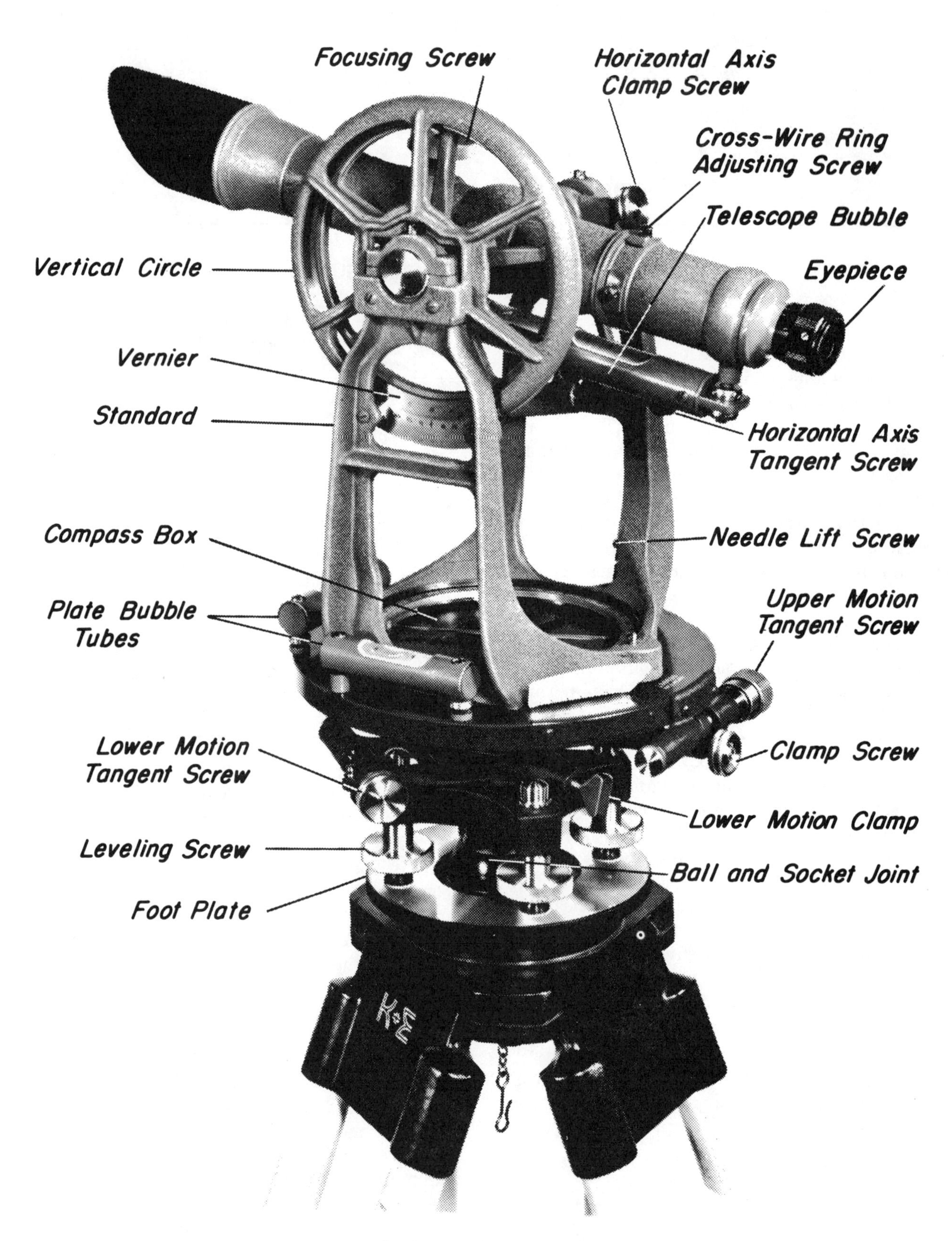

Figure 4-1. Transit.

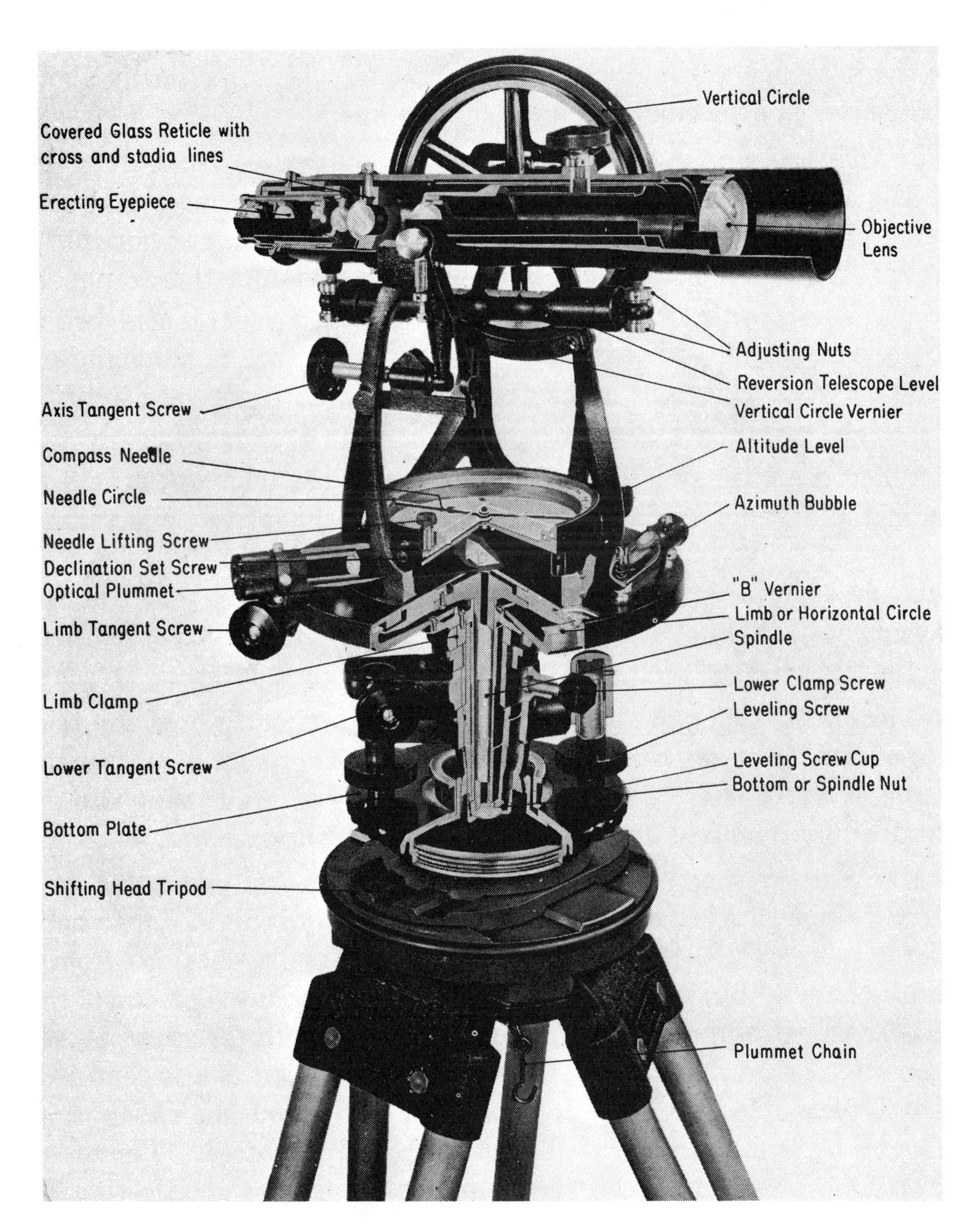

Figure 4-2. Transit, sectioned view.

the upper plate, between the two standards, and consists of a magnitized needle pivot on a jeweled bearing at the center of a graduated circle. A means is provided for lifting the needle off the pivot to protect the bearing when not in use.

d. Level Vials. Two plate level vials are placed at a right angle to each other. On many transits, one plate level vial is mounted on the left or vertical circle standard under the vernier. The other vial is then parallel to the axis of rotation for the vertical motion. The sensitivity of the plate level vial bubbles is about 70 seconds of arc for 2 millimeters of movement. Most engineer transits have a level vial mounted on the telescope to level it. The sensitivity of this bubble is about 30 seconds of arc per 2 millimeters of movement.

4–3. Circles and Verniers

The horizontal and vertical circles and their verniers are the parts of the engineer transit by

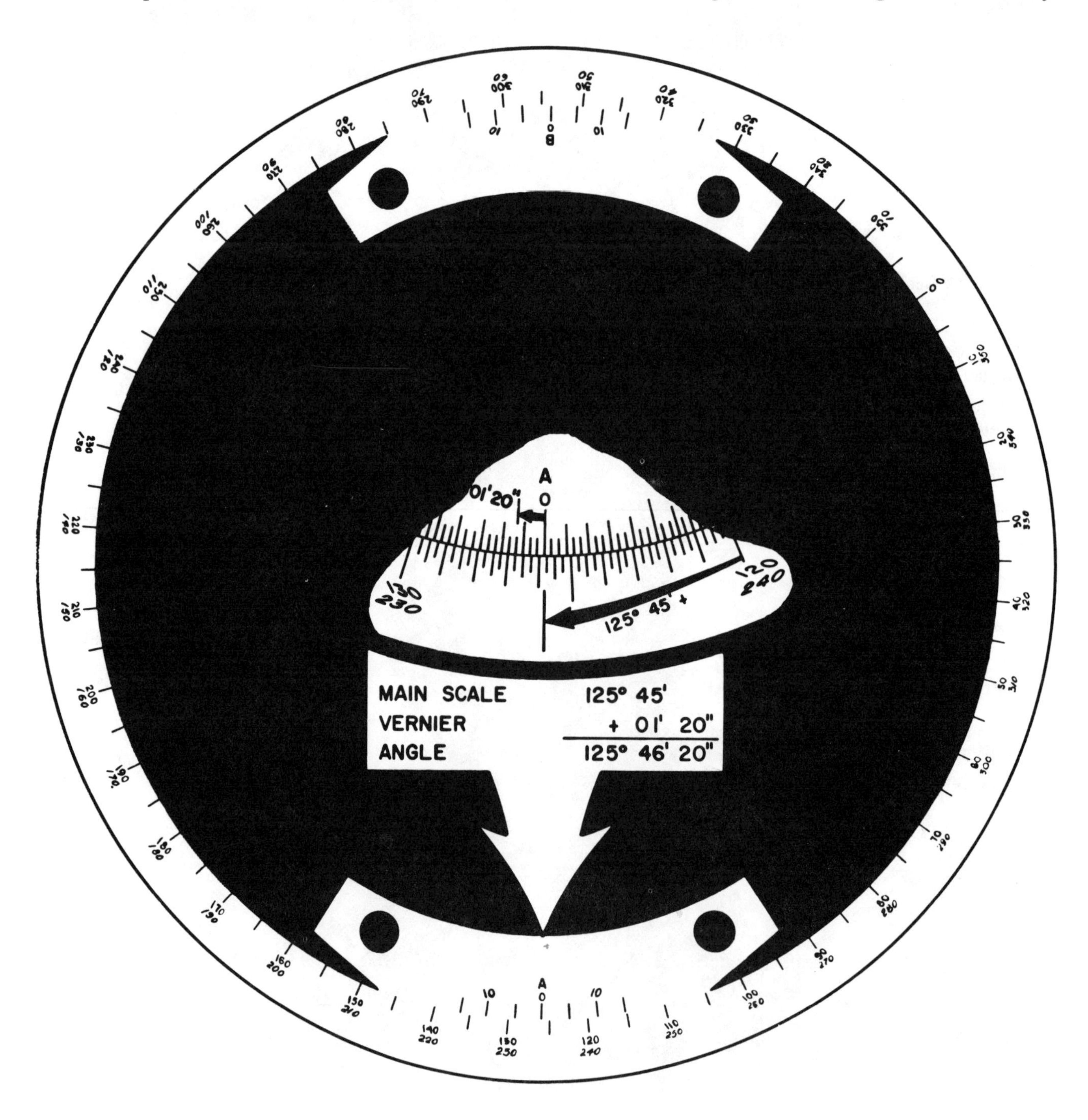

Figure 4-3. Horizontal scales, 20-second transit.

which the values of the horizontal and vertical angles are determined. A stadia arc is also included with the vertical circle on some transits.

a. Horizontal Circle and Vernier. Transits generally have the horizontal circle and verniers graduated to give least readings of either 1 minute or 20 seconds of arc. The horizontal circle is mounted on the lower plate. It is graduated to 15 minutes for the 20-second (fig. 4–3) and 30 minutes for the 1-minute (fig. 4–4) instrument. The plates are numbered from 0° to 360°, starting with a common point and running both ways around the circle. Two double verniers, known as A and B verniers are mounted on the upper plate with their indexes at circle readings 180° apart. A double vernier is one that can be read in both directions from the index line. The verniers reduce the circle graduations to the final reading of either 20 seconds or 1 minute.

b. Vertical Circle and Vernier. The vertical circle of the transit (fig. 4–5) is fixed to the horizontal axis so that it will rotate with the telescope. The vertical circle normally is graduated to 30 minutes, with 10° numbering. Each quadrant is numbered from 0° to 90°; the 0° graduations define a horizontal plane, and the 90° graduations

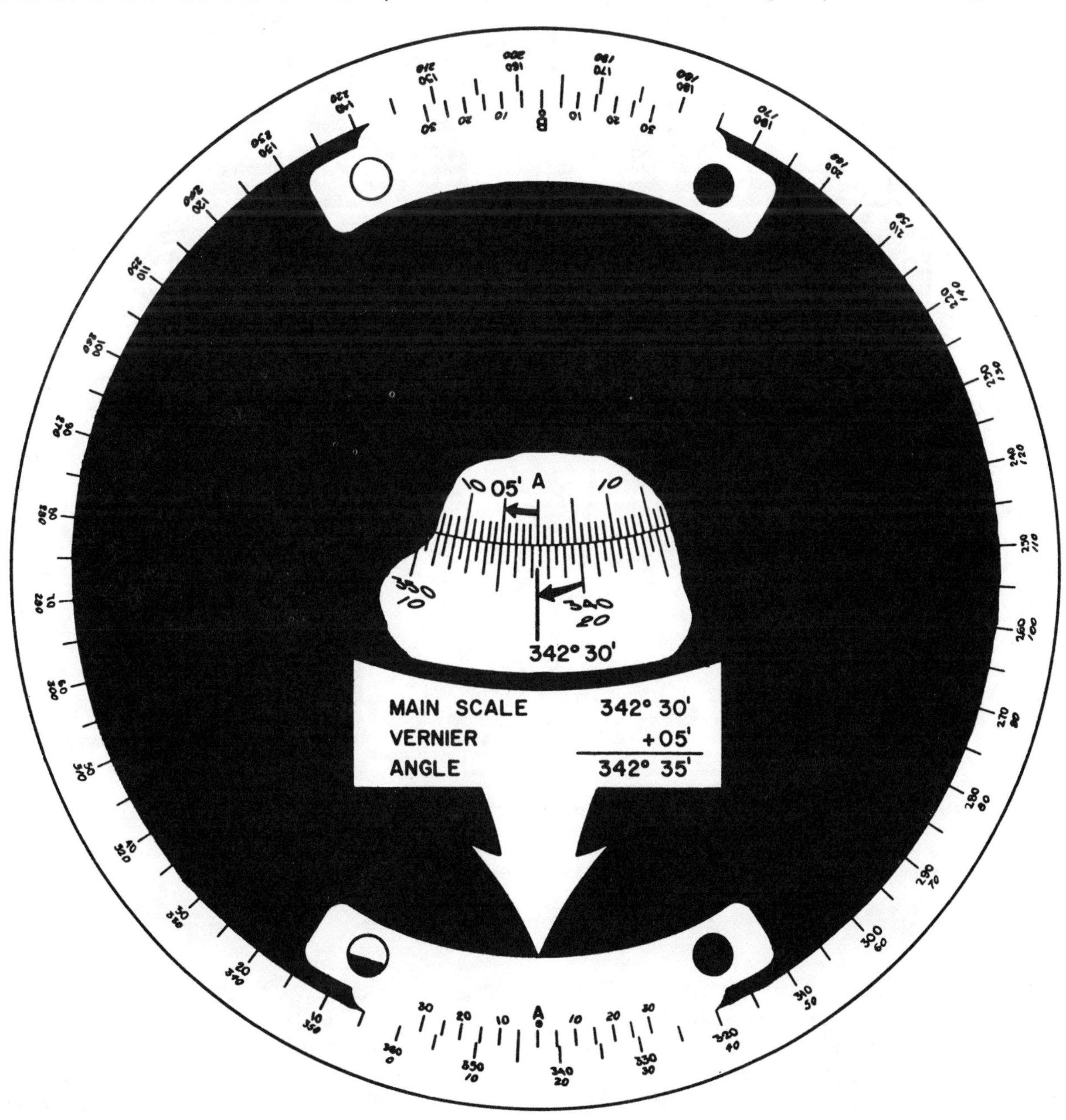

Figure 4-4. Horizontal scales, 1-minute transit.

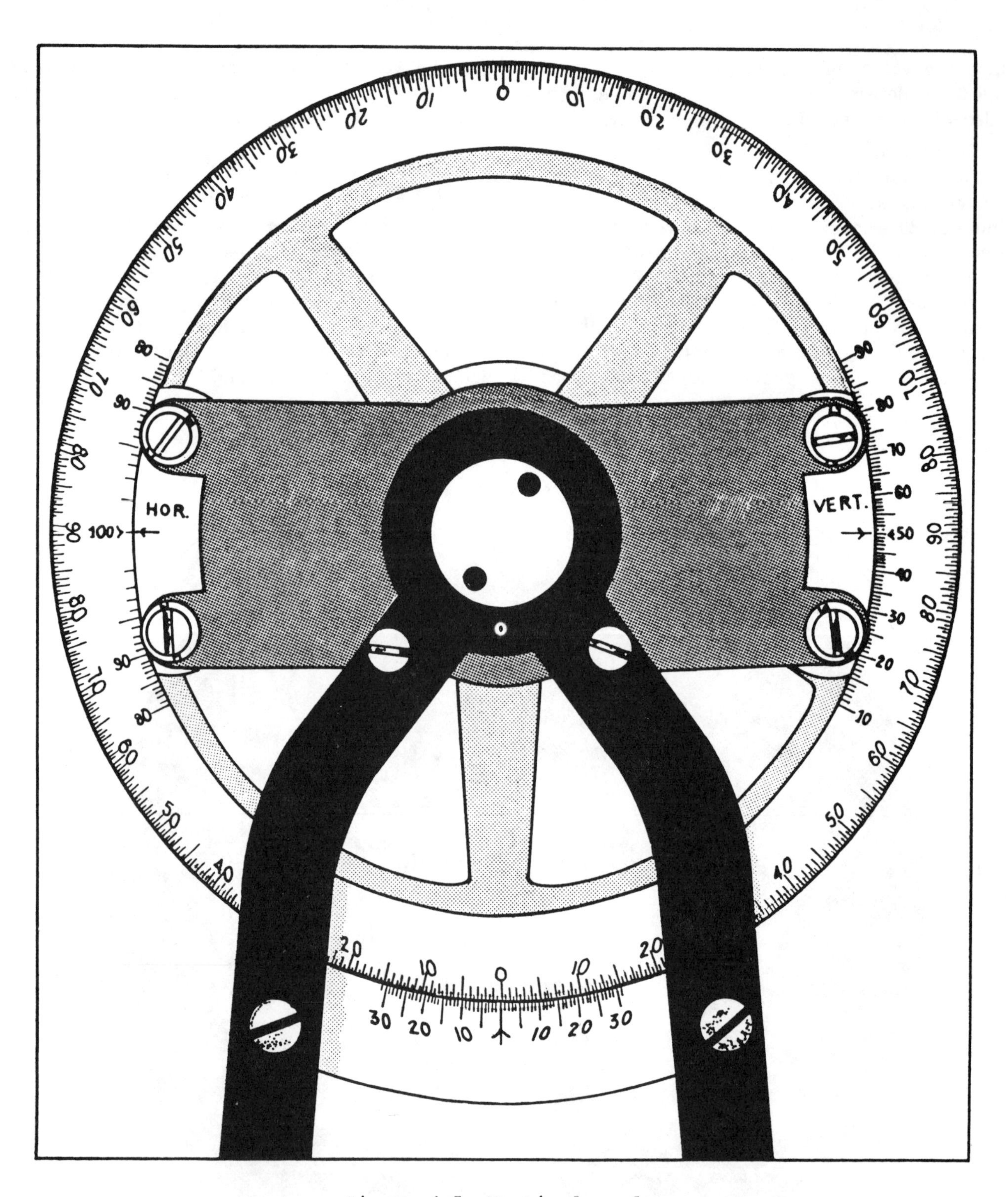

Figure 4-5. Vertical scales.

lie in the vertical plane of the instrument. The double vernier used with this circle is attached to the left standard of the transit, and its least reading is 1 minute. The left half of the double vernier is used for reading angles of depression, and the right half of this vernier is used for reading angles of elevation. Care must be taken to read the correct vernier for the type of angle being observed.

c. Stadia Arc Divisions. In addition to the vernier, the vertical circle may have a H and V, or HOR and VERT series of graduations, called a stadia arc (fig. 4–5). The H scale is adjusted to read 100 when the line of sight is level, and grad-

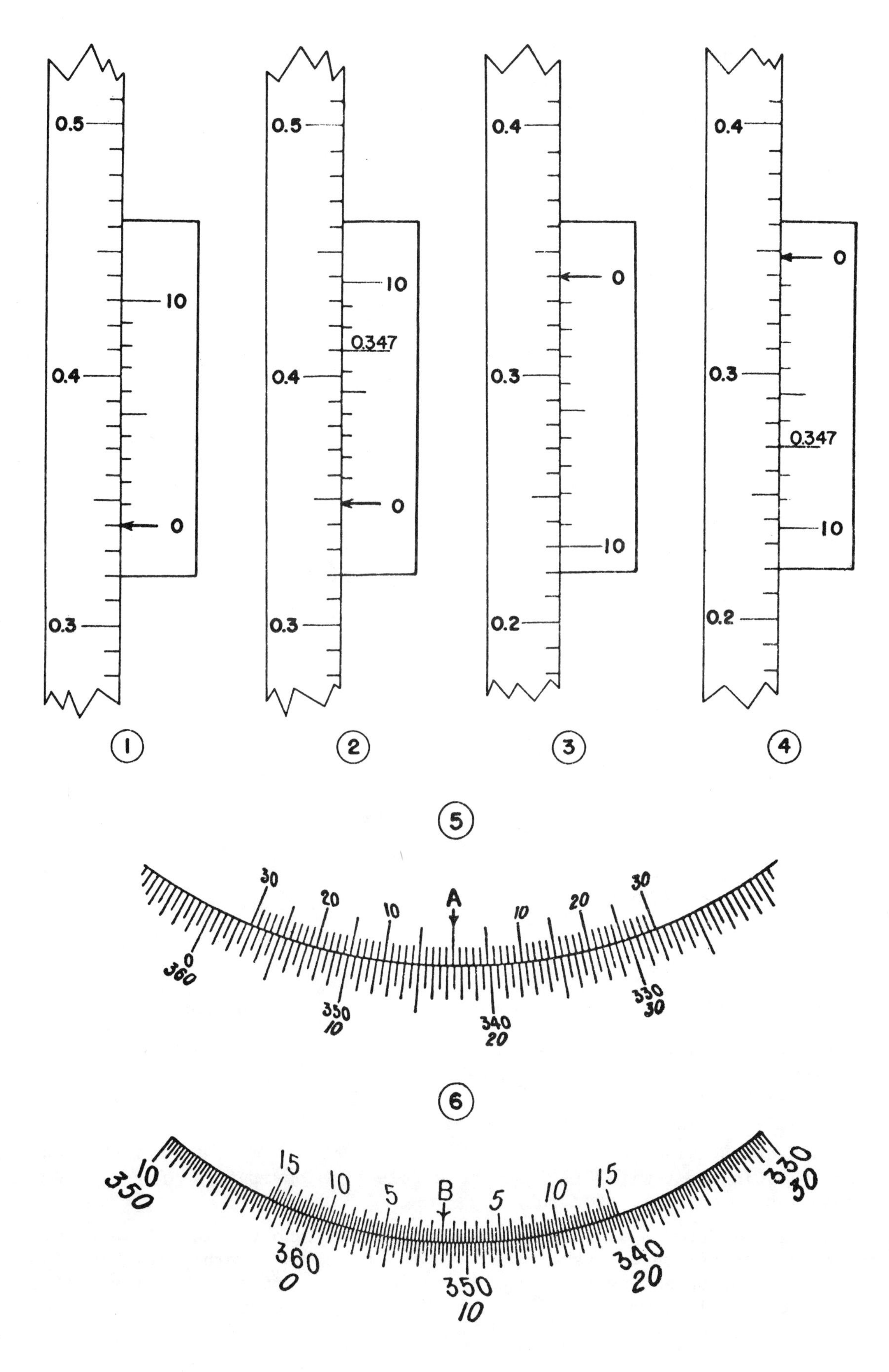

Figure 4-6. Verniers.

The combination of scale gradations and vernier divisions generally used on transits are shown in the table below.

Transit Scales and Verniers

Scale Graduations	Vernier Divisions	Least Count	Fig. No.
30 min	30	1 min	4-7a
20 min	40	30 sec	4-7b
30 min	60	30 sec	4-7c
15 min	45	20 sec	...
10 min	60	10 sec	4-7d

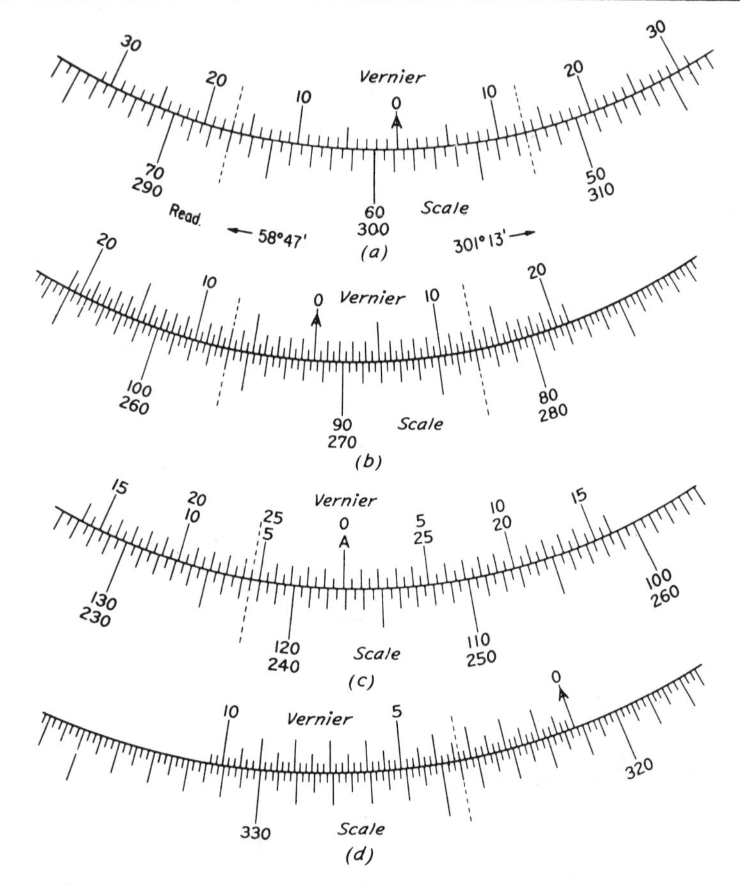

The three types of verniers shown in Fig. 4-7 are used on transits.

Direct, or single, vernier (Fig. 4-7d). This is read in only one direction and must therefore be set with the graduations ahead of the zero (index) mark in the direction to be turned.

Double vernier (Fig. 4-7a, b, and c). This vernier can be read either clockwise or counterclockwise, only one half being used at a time. Once the index mark has been set coincident with 0°00' on the circle, the observer is not limited to turning angles in one direction only.

Folded vernier (Fig. 4-7c). This type avoids the long vernier plate required by the normal double vernier. Its length is that of a direct vernier with half of the graduations placed on each side of the index mark. Except possibly for vertical arcs, the use of folded verniers is not justified by space or cost savings and is likely to cause reading errors.

Figure 4-7. Transit verniers.

uations decrease in both directions from the level line. The other scale, V, is graduated with 50 at level, to 10 as the telescope is depressed, and to 90 as it is elevated. For discussion of the stadia arc and its readings, refer to paragraph 7–3.

d. Reading the Vernier. The vernier, or vernier scale, is an auxiliary device by which a uniformly graduated main scale can be accurately read to a fractional part of a division. Both scales may be straight as on a level rod, or curved as on the circles of a transit. The vernier is uniformly divided, but each division is either slightly smaller (direct vernier) or slightly larger (retrograde vernier) than a division of the main scale (fig. 4–6). The amount a vernier division differs from a division of the main scale determines the smallest reading of the scale that can be made with the particular vernier. This smallest reading is called the *least count* of the vernier. It is determined by dividing the value of the smallest division on the scale by the number of divisions on the vernier.

(1) *Direct vernier.* ①, figure 4–6 shows a scale graduated in hundredths of a unit, and a direct vernier for reading it to thousandths of a unit. The length of 10 divisions on the vernier is equal to the length of 9 divisions on the main scale. The index, or zero of the vernier, is set at 0.340 unit. If the vernier were moved 0.001 unit toward the 0.400 reading, the first (1) graduation on the vernier would coincide with 0.35 on the scale, and the index would be at 0.341 unit. ②, figure 4–6 show the vernier moved to the seventh (7) graduation where it coincides with 0.41 on the scale. In this position, the correct scale reading is 0.347 unit (0.340 + 0.007). The index with the zero can be seen to point to this reading.

(2) *Retograde vernier.* ③, figure 4–6 shows a retograde vernier on which each division is 0.001 unit longer than the 0.01 unit divisions on the main scale. The 10 divisions on the vernier equal 11 on the scale. The retograde vernier extends from the index, backward along the scale. ④, figure 4–6 shows a scale reading of 0.347 unit, as read with the retograde vernier.

(3) *Circle verniers.* ⑤ and ⑥, figure 4–6 represent part of the horizontal circle of a transit and the direct reading vernier for the circle. The main circle graduations are numbered both clockwise and counterclockwise. A double vernier which extends to the right and left of the index, makes it possible to read the main circle in either direction. The vernier to the left of the index is used for reading clockwise angles, and the vernier to the right of the index is used for reading counterclockwise angles. The slope of the numerals in the vernier to be used correspond to the slope of the numerals in the circle being read. Care must be taken to use the correct vernier. In ⑤, figure 4–6, the circle is graduated to half degrees, or 30 minutes. On this vernier, 30 divisions are equal in length to 29 divisions on the circle, and the least reading of this vernier is the 30 minutes of the circle divided by the 30 divisions of the vernier, or 1 minute. The index (⑤, fig. 4–6) is seen to lie between 342° 30′ and 343°. In the left vernier the fifth (5) graduation is seen to coincide with a circle graduation. Then, the clockwise reading of this circle is 342° 30′ + 05′, or 342° 35′. Using the right vernier in the same way, the counterclockwise reading of the circle is 17° 00′ + 25′, or 17° 25′. In ⑥, figure 4–6, the circle is graduated in 15-minute divisions and each half of the double vernier contains 45 divisions. Therefore, 15′ (or 900″) divided by 45 equals a least reading of 20″ on this vernier. The clockwise reading of the circle and vernier is 351° 30′ + 05′ 40″, or 351° 35′ 40″. The counterclockwise reading is 8° 15′ + 09′ 20″, or 8° 24′ 20″.

4–4. Care

a. All surveying instruments are scientific instruments and must be handled with care. When an instrument is first received, it is packed in its carrying case, and in turn, usually in a larger wooden shipping case. The shipping case should be opened with care to avoid damage to the instrument and its carrying case. The carrying case should be placed on a firm surface for opening. Some carrying cases open at the top, while others have a door in the side of the case. When the case is opened, the exact position of the instrument and each accessory should be carefully noted. When the instrument is replaced, all circle screws are turned so that the instrument parts will give if the case is jarred. The screws should not be loose enough, however, to allow play in the various instrument parts. The carrying case should close without being forced. If the case does not close easily, the instrument or an accessory is misplaced. *Never use force to close the case.*

b. The transit is removed from the case by grasping the cross members of the standards, being careful not to grasp the vertical circle. With one hand, place the transit on the tripod by holding the transit by the standard *not* containing the vertical circle; with the other hand, loosen the lower motion clamping screw and aline the footplate on the tripod head. Rotate the footplate *counterclockwise* until a click is heard. This indicates a mesh of the footplate and the tripod head threads. Now, rotate the footplate *clockwise* until it bears firmly on the tripod fitting. *Do not overtighten.* After the instrument has been set up for use, do not allow anything to touch it or its tripod unnecessarily. The observer should not straddle a tripod leg, but should stand between two legs. Care must be exercised when moving around the instrument to avoid kicking or otherwise touching the tripod, or stepping close to where a tripod leg enters the ground. Tapepeople must be extra careful to prevent disturbing the instrument when measuring from the transit station. Never leave any instrument unguarded in the field. Those instruments used on a tripod should never be left standing on the tripod indoors, or on any other smooth surface unless the tripod is mounted in some type of holding device. Always keep the telescopic instruments in their carrying cases when not in use. In case of rain, telescopic instruments should be covered with their waterproof covers.

c. The transit must be kept clean and dry. During use, as necessary, and after use, the instrument should be cleaned as follows:

(1) Painted surfaces should be wiped with a clean cloth.

(2) The exterior of the vernier windows should be cleaned with a camel's-hair brush or wiped with a clean cloth or lens tissue.

(3) The exterior of the eyepiece and the objective lens should first be brushed gently with a camel's-hair brush to remove dust and then wiped with chamois or lens tissue to remove moisture. Care must be taken not to scratch the lenses and the coating on the lenses.

(4) The vertical circle and vernier should be brushed with a camel's-hair brush or wiped with a chamois across (perpendicular to) the graduations to avoid removal of the blacking.

(5) The leveling screws should be cleaned one at a time. Run the screw, which is diagonally opposite the screw to be cleaned, all the way up. At the same time, run the screw to be cleaned all the way down. The screw should be cleaned with a clean cloth. A string may be pulled through the grooves to aid in removing dust. Apply the proper lubricant immediately after the cleaning.

(6) The upper surface of the footplate should be wiped with a clean cloth. This surface must be kept clean at all times.

d. The transit should be carried under one arm, with the hand of the other arm under the head of the tripod or the footplate of the transit. In open areas, and where there is no danger of striking the head against any object, the transit may be carried over the shoulder. The screws should be clamped lightly to allow the parts to move if the instrument is struck accidentally. When the instrument is carried in a vehicle, it should be placed in its case with the shifting center in the center of the footplate, the leveling screws at approximately the same height, and the vertical motion clamped.

4–5. Adjustment

a. General. The transit must be kept in good adjustment to obtain correct and accurate results. There are six tests and adjustments of the transit that the surveyor must understand and be capable of performing. All these tests and adjustments are made with the instrument mounted on a tripod in a shaded area. These tests are made periodically and in the sequence in which they are discussed. When one of tests indicates that an adjustment is necessary, the adjustment is made and all previous tests must be repeated before proceeding with the next test.

b. Plate Levels. The test of the plate level vial bubbles should be made every time the instrument is set up for use. When an error in either plate level is indicated the adjustment should be made. However, this adjustment must be made before the other tests and adjustments can be accomplished. This adjustment (fig. 4–8) makes the axis of level bubbles perpendicular to the vertical axis and is performed as follows:

(1) Bring both level bubbles to the center of their vials using the leveling footscrews (①, fig. 4–8).

(2) Rotate the instrument about its vertical axis through 180° and note the amount the bubbles move away from their centered position (②, fig. 4–8).

(3) Bring the bubble of each level half the distance back to the center of its vial by turning the capstan adjusting screws under the end of each vial.

(4) Relevel the remaining distance with the footscrews and rotate 180°. Make a similar correction if the bubbles do not remain at the centered position.

(5) Check the final adjustment by noting that the bubbles remain in the centered position during an entire revolution about the vertical axis (③, fig. 4–8).

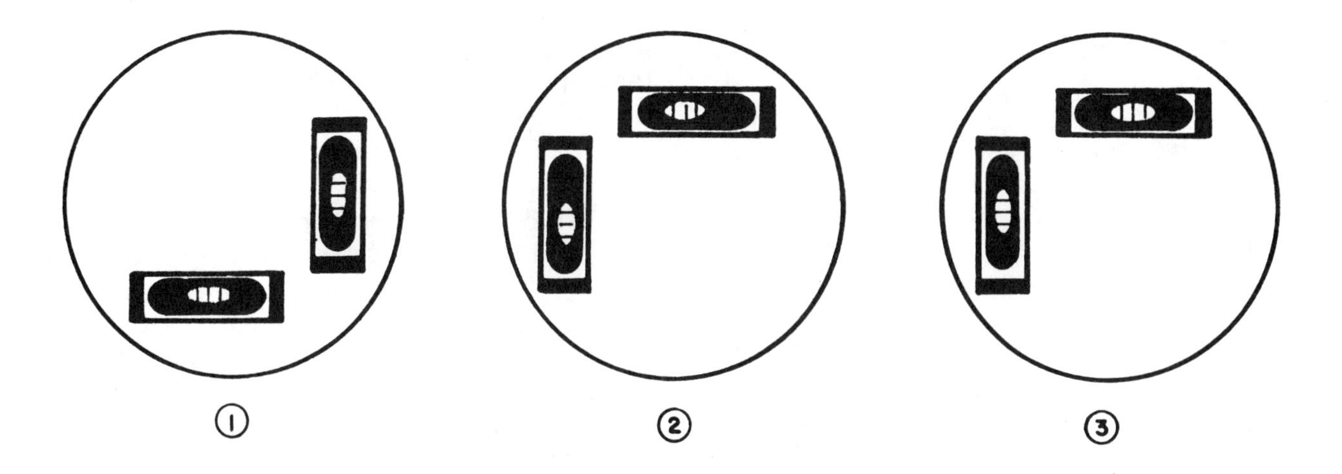

Figure 4-8. Adjustment of plate levels.

c. Vertical Wire. The vertical wire must lie in a plane perpendicular to the horizontal axis. The test and adjustment (if necessary) are performed as follows:

(1) Eliminate any parallax. Sight the vertical wire on a well-defined point, and with all motions clamped, move the telescope slightly up and down on its horizontal axis, using the vertical slow motion tangent screw. If the instrument is in adjustment, the vertical wire will appear to stay on (track) the point through its entire length.

(2) If it does not (fig. 4–9), loosen the screws holding the crosswires and rotate the ring slightly by tapping the screws lightly, tighten screws.

(3) Sight again on the point. If the vertical wire is still not tracking the point, repeat the procedure in (2) above. This procedure is repeated until the wire tracks the point through its entire length.

d. Crosswires. The line of sight must be perpendicular to the horizontal axis. The test and adjustment (if necessary) are performed (fig. 4–10) as follows:

(1) Sight on a point, A, not less than 100 meters from instrument and at about the same elevation, and clamp both horizontal plates.

(2) Plunge the telescope and set another point, B, at a distance equal to the first and at about the same elevation.

(3) Unclamp the lower motion, rotate the instrument about its vertical axis, sight on the first point and clamp.

(4) Plunge the telescope and observe toward the second point. If the instrument is in adjustment, point B will fall at position E and will be a straight line, AE. If the instrument is not in adjustment, the intersection of the crosswires will fall at a point C. Point C will be equidistant from position E as point B, only on the opposite side.

(5) Measure the distance B to C and place a

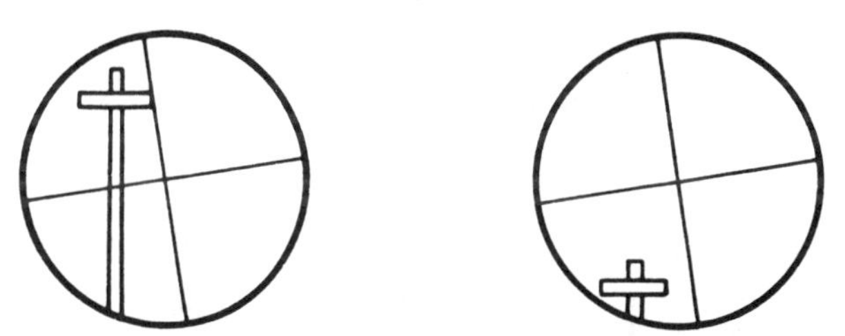

Figure 4-9. Vertical wire out of adjustment.

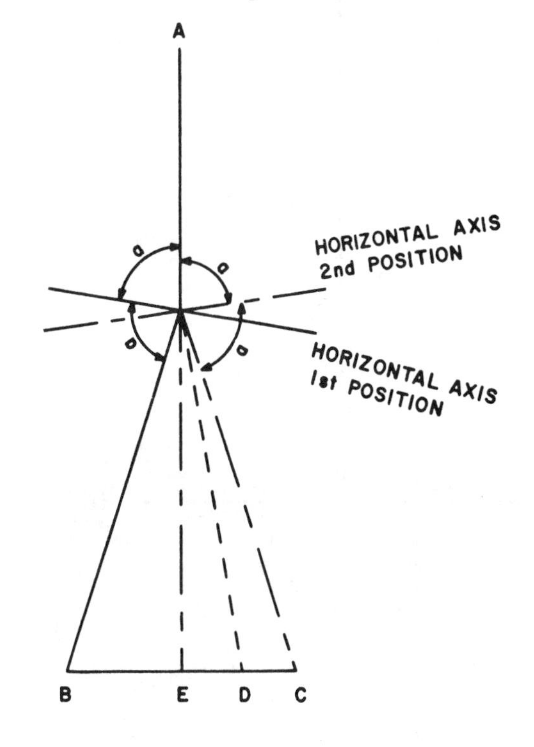

Figure 4-10. Crosswires out of adjustment.

point, D, at one-fourth this distance back from C toward B.

(6) Then, using the horizontal capstan screws, located on the side of the telescope, move the wire reticle until the vertical wire appears to have moved from point C to the corrected position, D. The reticle is moved by loosening one screw and tightening the opposing screw equal amounts.

(7) Repeat this procedure starting at (1) above, until point B falls at position E.

(8) If this adjustment is performed, the test described in *c* above must be repeated, because the vertical wire may have rotated.

e. Standards. The horizontal axis of the telescope must be perpendicular to the vertical axis of the instrument. The test and adjustment (if necessary are performed (fig. 4-11) as follows:

(1) Sight with the vertical wire on some high point, A, at least 30° above the horizontal, such as the end of a ridge of a stable building, and clamp the horizontal plates.

(2) Depress the telescope and mark a second point, B, at about the height of instrument.

(3) Plunge the telescope, unclamp the lower plate, and rotate the instrument about its vertical axis.

(4) Sight on the first point A.

(5) Clamp lower plate and depress the telescope. If the vertical wire falls on the lower point, B, the horizontal axis is in adjustment. In this case, point B is coincident with point D, in both the direct and reverse positions of the telescope.

(6) If not, mark the new point, D, at the same height as B. Measure the distance B to C.

(7) Mark a point, D, exactly halfway between B and C. Distance CD is the amount of correction.

(8) Adjust by turning the small capstan screw in the adjustable bearing at one end of the horizontal axis until point C appears to have moved to point D.

(9) This test is continued until the vertical wire passes through the high and low points in both the direct and reverse positions of the telescope.

(10) Repeat all previous tests if this adjustment if performed.

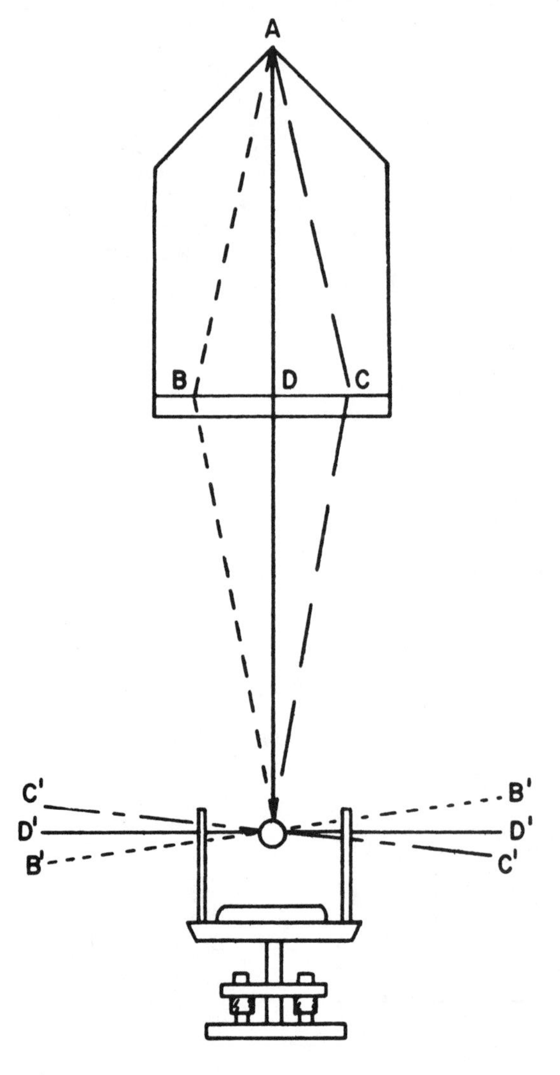

Figure 4-11. Adjustment of the standards.

f. Telescope Level. The line of sight must be parallel to the axis of the telescope level vial (fig. 4-12). This test is the same as the two-peg method described in paragraph 3-5*c*. The test and adjustment (if necessary) are performed as follows:

(1) Set the instrument up midway between two stakes about 100 meters apart.

(2) A reading is taken through the telescope on rods held on each of the stakes. The telescope must be carefully leveled before each reading. The difference between the reading is the difference in elevation between the stakes.

(3) Move, setup, and level the instrument and telescope so that the eyepiece is within 10 millimeters of one of the rods.

(4) The near rod is read through the objective lens, and the far rod in the normal manner, taking care that the telescope level is centered for both readings. The difference between the rod readings should equal the difference determined in (2) above, if the instrument is in good adjustment. If not, a correction must be made.

Figure 4-12. Adjustment of telescope level.

(5) To adjust, compute the reading that should be made on the far rod. This reading equals the near rod reading minus the difference from (2) above.

(6) Set the horizontal crosswire on the far rod for the computed reading using the vertical slow motion screw, and move one end of the level vial vertically using the adjusting nuts until the bubble is centered in the vial.

(7) Repeat the test to verify final adjustment.

g. Vertical Circle Vernier. The vertical circle vernier must read zero when the telescope is exactly level (fig. 4–13). The test and adjustment (if necessary) are performed as follows:

(1) Bring the telescope level vial bubble to the exact center of vial.

(2) Read the vertical circle vernier. If it does not read zero it must be adjusted.

(3) To adjust, loosen the capstan screws holding the vernier and move index until it reads zero on the vertical circle.

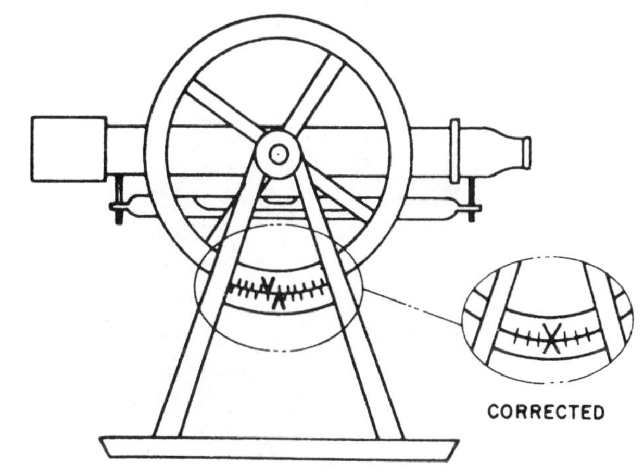

Figure 4-13. Adjustment of vertical circle vernier.

(4) Tighten the screws and read the vernier, check all level vials to make sure they have not moved during adjustment.

Section II. ONE-MINUTE THEODOLITE

4–6. General

The 1-minute directional theodolite is essentially a direction-type instrument. It can be used, however, to observe horizontal and vertical angles as with a transit. There are two versions of this theodolite: one version which reads directly to 1 minute; and another version which reads directly to 0.2 mil. There are many manufacturers of these theodolites and they may differ in appearance, but they are basically alike in their essential parts and operation.

4–7. Description

This theodolite is a compact, lightweight, dustproof, optical reading, direction-type instrument. The scales are readable directly to the nearest 1 minute or 0.2 mil and are illuminated by either natural or artificial light. The description given in this paragraph applies essentially to all models and makes of this theodolite. The theodolites shown in figures 4–14 and 4–15 are samplings of this theodolite.

a. Horizontal Motion. A horizontal motion clamp and tangent screw for moving the theodolite in azimuth are located adjacent to each other on the lower portion of the alidade. A horizontal circle clamp, which fastens the horizontal circle to the alidade, is located on the horizontal circle casting. When this horizontal (repeating) circle clamp is in the lever down position, the circle is attached to the rotating part of the instrument and turns with the telescope. With the circle clamp in the lever up position, the circle is unclamped and the telescope turns independent of the circle. This combination permits use of the theodolite as a repeating instrument. To use the theodolite as a direction-type instrument, the circle clamp is used only to set the initial reading. An initial reading of 0°30′ should be set on the plates when a direct and reverse (D/R) pointing is required. This will minimize the possibility of

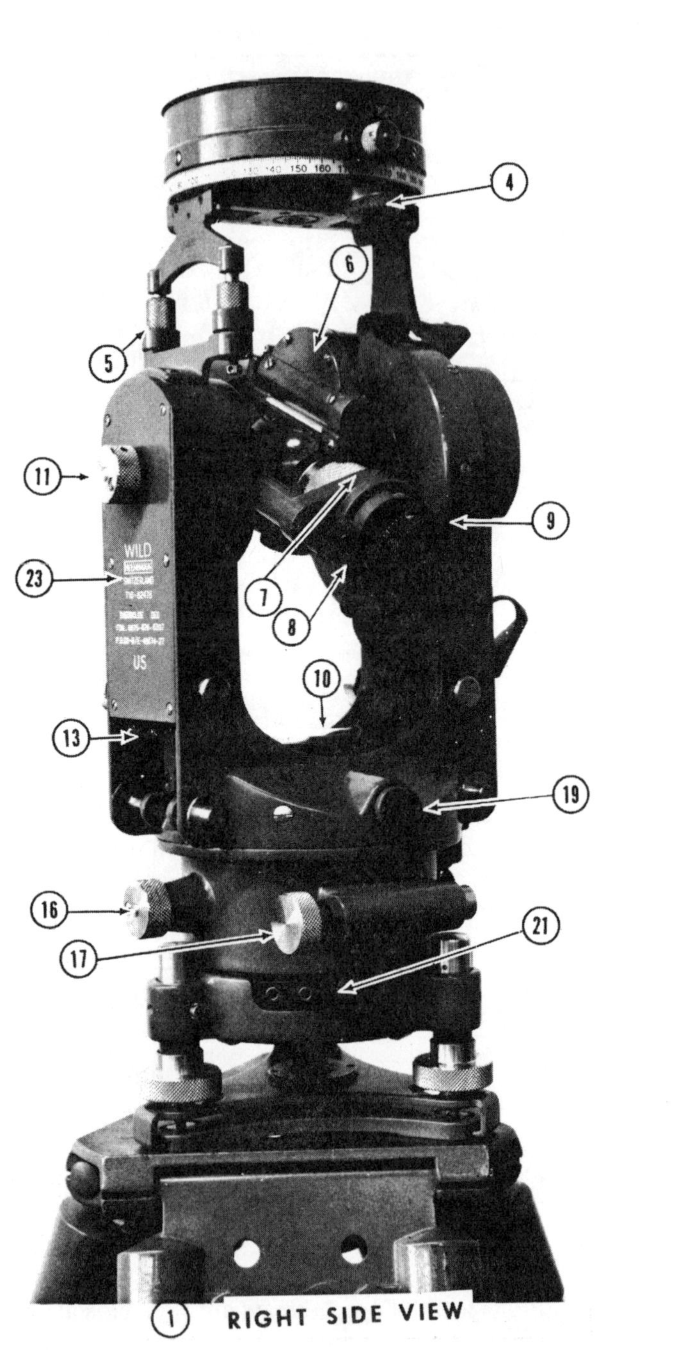

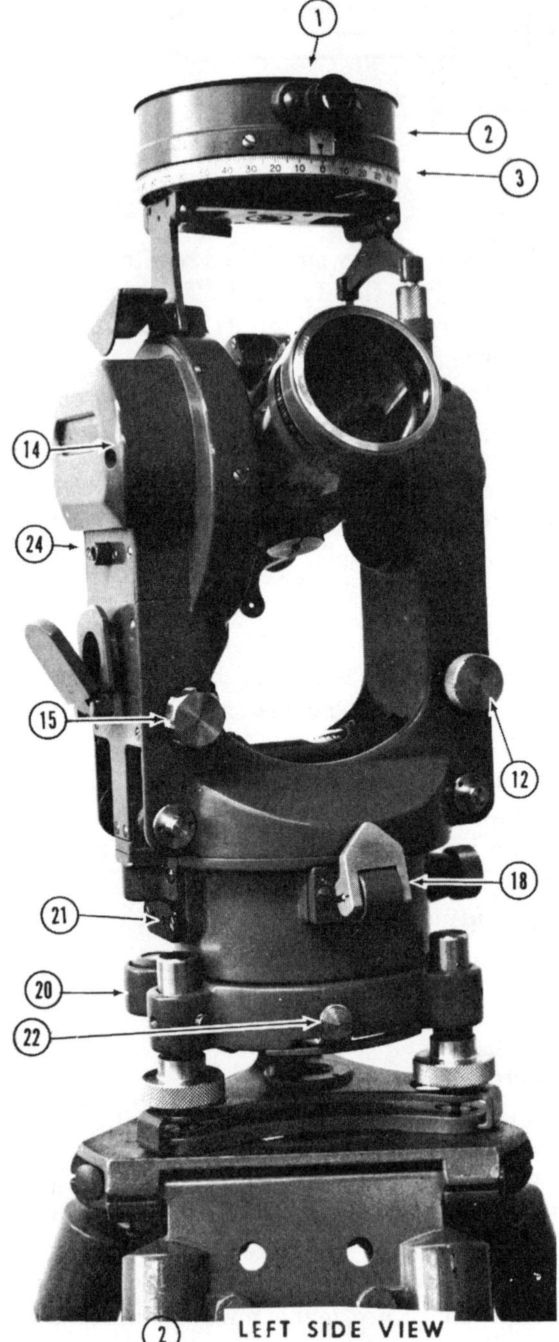

1 Compass eyepiece
2 Compass
3 Circle ring
4 Caging knob
5 Compass leveling screw
6 Telescope level
7 Focus drive
8 Microscope eyepiece
9 Telescope eyepiece
10 Plate level
11 Vertical clamp
12 Vertical slow-motion screw
13 Vertical slow-motion adjusting screw
14 Collimation level
15 Collimation slow-motion screw
16 Horizontal clamp
17 Horizontal slow-motion screw
18 Horizontal circle clamp
19 Optical plumb eyepiece
20 Circular level
21 Electric plugs
22 Tribrach clamp lever
23 Right cover
24 Left cover

Figure 4-14. One-minute theodolite (T-16).

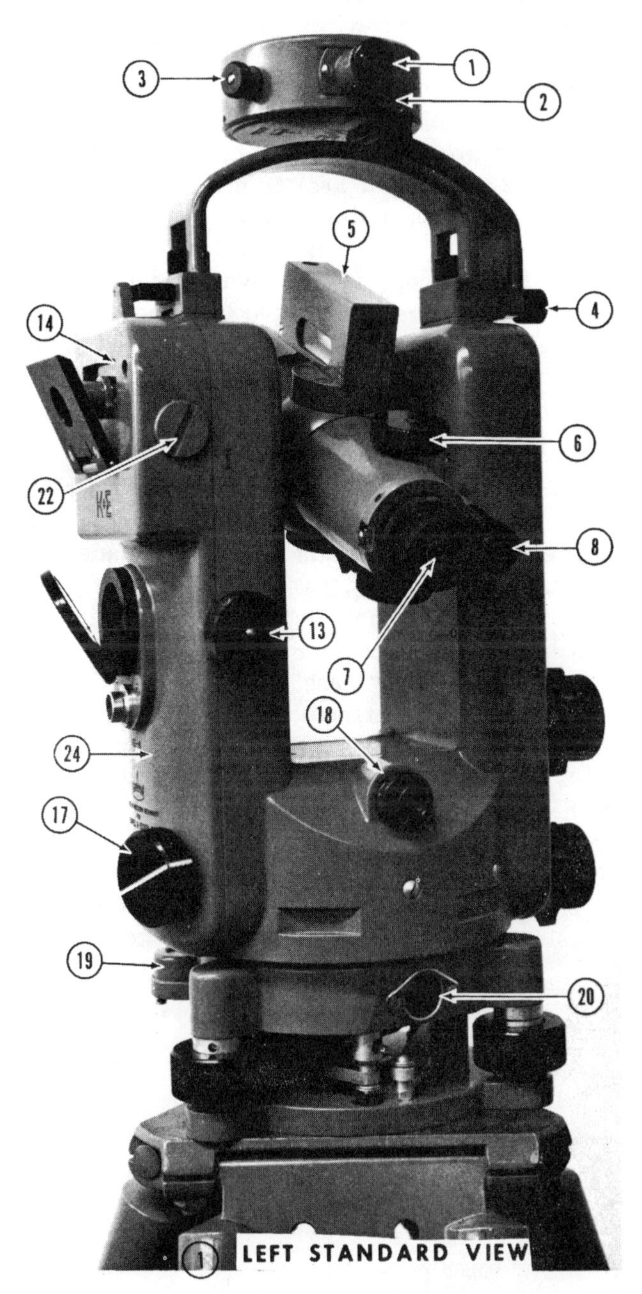

1 Compass eyepiece
2 Compass
3 Needle lifter knob
4 Compass mounting screw
5 Telescope level
6 Focusing knob
7 Telescope eyepiece
8 Microscope eyepiece
9 Plate level
10 Reticle illumination knob
11 Elevation clamp
12 Elevation tangent screw
13 Altitude level knob
14 Altitude level
15 Azimuth clamp
16 Azimuth tangent screw
17 Repetition clamp
18 Optical plummet eyepiece
19 Circular level
20 Electric plug
21 Clamp screw
22 Level access screw
23 Right standard
24 Left standard

Figure 4-15. One-minute theodolite(KE 73-0041 special).

ending the D/R pointing with a value below zero and of working with a negative value in determining the mean.

b. Vertical Motion. The vertical motion clamp and tangent screw are located on the standard opposite the vertical circle. The tangent screw is located to the lower left and at a right angle to the clamp. The telescope can be rotated in the vertical plane completely around the horizontal axis (360°).

c. Levels. There are usually three level vials on this instrument; the circular, the plate, and the vertical circle.

(1) *Circular level.* The circular level is located on the tribrach of the instrument and is used to roughly level the instrument.

(2) *Plate level.* The plate level is located on the lower part of the alidade between the two standards. This level is used for leveling the instrument in the horizontal plane.

(3) *Vertical circle level.* The vertical circle level (vertical collimation) vial is a coincidence type, referred to as a split bubble. The level vial is completely built in, adjacent to the vertical circle and viewed through a prism and 45° mirror system from the eyepiece end of the telescope. This results in the viewing of one-half of each end of the bubble at the same time. Leveling consists of bringing the two halves into exact coincidence using the collimation level tangent screw (fig. 4–16).

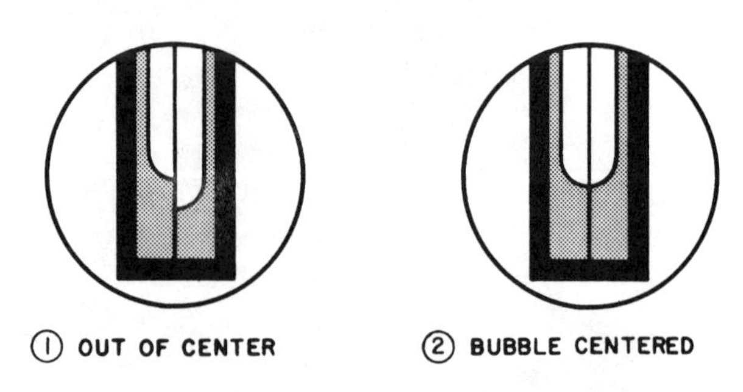

Figure 4-16. Coincidence-type level.

(4) *Telescope level.* Some models of this theodolite are furnished with a long telescope level. The telescope level vial assembly is mounted below the telescope in the direct position. For leveling operations, the telescope is plunged to the reverse position, bringing the level assembly to the top. A prism system and a 45° mirror permit viewing both ends of the split bubble from the eyepiece end of the telescope for leveling operations.

d. Telescope. The telescope of the theodolite is reversible; that is, it can be rotated around the horizontal axis for direct and reverse readings. It is usually a 28-power instrument with the shortest focusing distance of about 1.4 meters. The crosswires are focused (removing parallax) by turning the eyepiece; the image, by turning the focusing ring. The reticle (fig. 4–17) has horizontal and vertical crosswires and a set of horizontal and vertical ticks at a stadia ratio of 1:100. Some models have a solar circle on the reticle for making pointings on the sun. This circle covers 31 minutes of arc and can be imposed on the sun's image (32 minutes of arc) to make the pointings refer to the sun's center. One-half of the vertical wire is split for finer centering small distant objects. The telescope is the inverted image type. The crosswires are illuminated by either sunlight reflected by mirrors or by battery powered lights. The amount of illumination is adjusted through the use of the illuminating mirror for telescope diaphragm.

e. Accessories. The theodolite is issued with the following accessories: Canvas accessory kit, compass, eyepiece prisms, sun filters, sunshade, two jeweler's screwdrivers, two adjusting pins, camel's-hair brush, chamois, lubricant, plastic instrument head cover, operation and maintenance service manuals, battery case with lighting devices, and the universal tripod with plumb bob. The compass is mounted on the top of the standards and is used to check azimuths, to orient the sketch in the field notes, or as a means of assuming direction. To prevent breakage of the glass cover, always place the compass with the dial down in the pocket of the carrying case. The carrying case consists of a base assembly and a steel dome-shaped hood. When in the case, the instrument

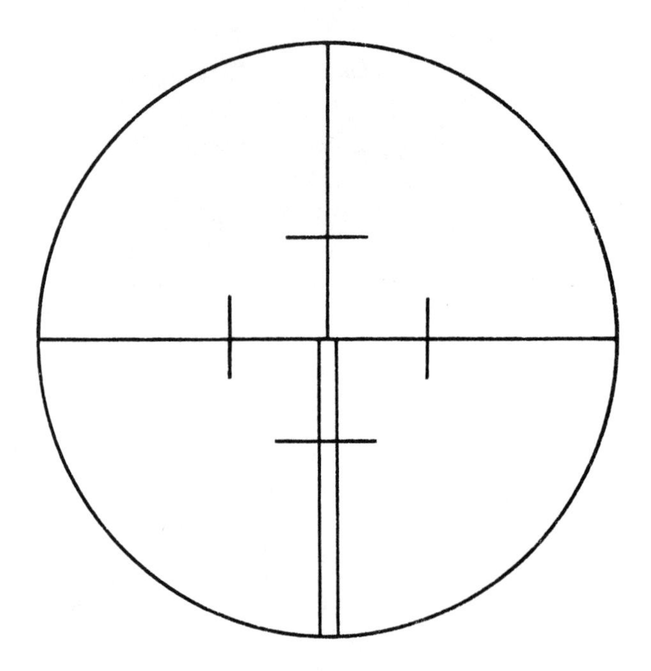

Figure 4-17. Theodolite reticle.

rests on two supports and is fixed to the supports with locking devices. A padded wooden shipping case is also furnished for transporting the theodolite in its carrying case. The tripod assembly consists of a tripod, plumb bob, plug-in sleeve and a key for the tripod leg nuts. The battery case with the light devices is used for the electric illumination of both circles, both main levels, and the telescope reticle. A rheostat is provided on the battery case for adjusting the intensity of the light.

f. Tribrach. The tribrach assembly (fig. 2–5) on many of the makes and models is a detachable part of the theodolite which contains the leveling screws, circular level, and the optical plumbing device. The leveling screws are completely enclosed and dustproof. A locking device holds the alidade and tribrach together. This assembly permits interchangeability of instruments without moving the tripod. After observations are completed, the instrument (alidade) is detached at its base and moved to the next station and another tribrach, with the target replacing the alidade. Used in this "leapfrog" method, the amount of time needed per instrument setup is reduced by half.

4–8. Reading

The circles of these theodolites are read through an optical microscope. The eyepiece of the microscope is located to the right of the telescope eyepiece in the direct position, and to the left in the reverse. The microscope consists of a series of lenses and prisms which bring both the horizontal and the vertical circle images into a single field of view. The horizontal circle, marked Az, appears below the vertical circle, marked V.

a. Degree Graduated. Figure 4–18 shows the image of both circles as they would appear through the reading microscope of the 1-minute theodolite. Both circles are graduated from 0° to 360° with a major or index graduation for each degree on the main scales. This scale's graduation appears to be superimposed over an auxiliary scale which is graduated in minutes to cover a span of 60′(1°). The position of the degree mark on the auxiliary scale is used as an index to get a direct reading in degrees and minutes. If necessary these scales can be interpolated or estimated to the nearest 0.2 minute of arc. The vertical circle of these theodolites reads 0° when the telescope is pointed at the zenith, and 180° pointed straight down. A level line reads 90° in the direct position, and 270° in the reverse. The values read

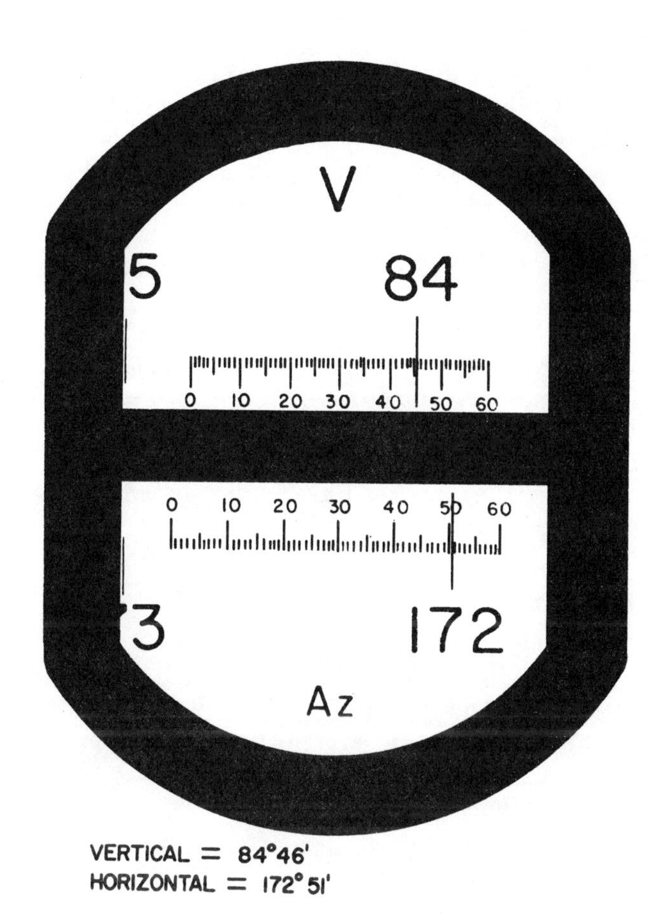

Figure 4-18. Degree graduated scales.

from the vertical circle are referred to as *zenith distances*, and not vertical angles. When necessary, these zenith distances can be converted to vertical angles as shown in figure 4–19.

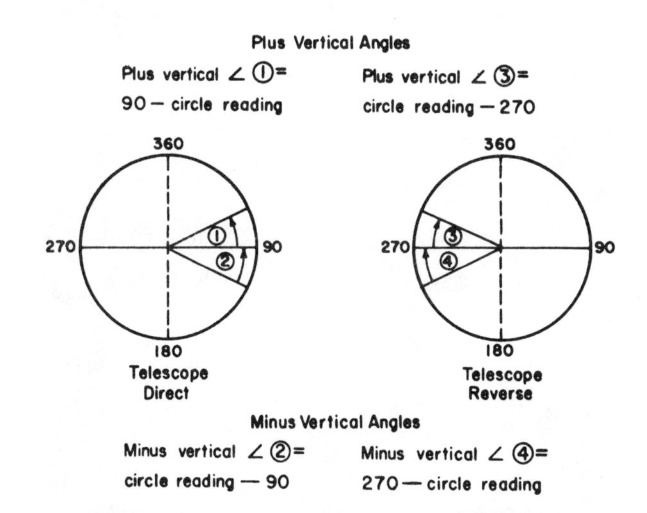

Figure 4-19. Vertical angles from zenith distances (degrees)

b. Mil Graduated. Figure 4–20 shows the image of both circles as they would appear through the reading microscope of the 0.2-mil theodolite. Both circles are graduated from 0 to 6400 mils. Therefore, 360° equals 6400 mils, or 1 mil equals 3.375 minutes. For conversion chart, refer to table D-1, appendix D. The main scales are marked and numbered every 10 mils, with the last zero dropped. The auxiliary scales are graduated from 0 to 10 mils in 0.2-mil increments. Readings on the auxiliary can be interpolated to 0.1 mil. The vertical circle of these theodolites reads 0 mil when the telescope is pointed at the zenith, and 3,200 mils straight down. A level line read 1,600 mils in the direct position, and 4,800 mils in the reverse. As with the degree graduated circle (*a* above), the values read are zenith distances (fig. 4-21).

4-9. Care

The theodolite is a delicate, scientific instrument and must be treated as such. It must be kept clean, dry, and properly oiled. Any accumulation of dirt and dust can scratch the machined or polished surfaces, or can cause friction and sticking in the motions.

a. Cleaning. Before, during, and after use the instrument should be cleaned as follows:

(1) Painted surfaces should be wiped with a clean cloth.

(2) The exterior of the eyepieces and the

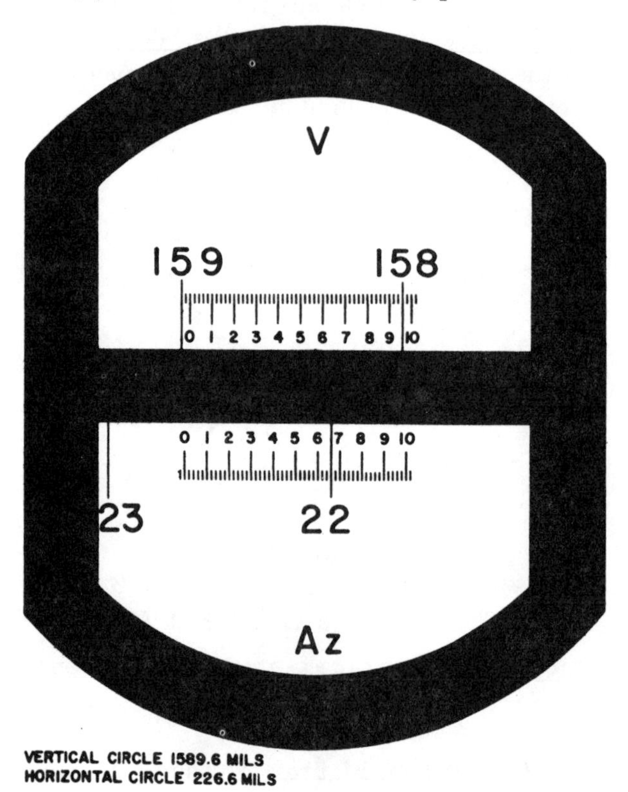

Figure 4-20. Mil graduated scales.

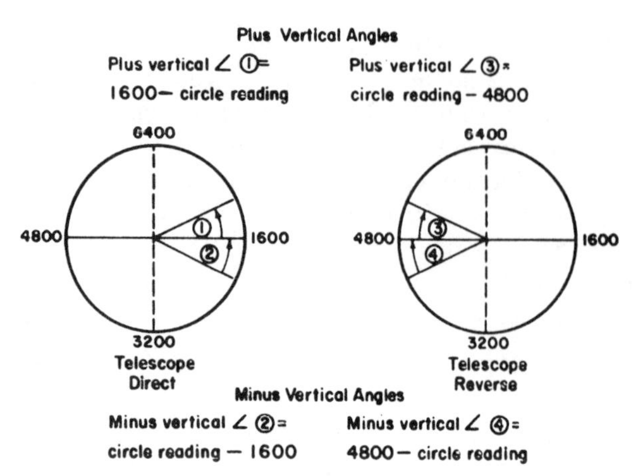

Figure 4-21. Vertical angles from zenith distances (Mils)

objective lens of the telescope should first be brushed with a camel's-hair brush to remove dust and then wiped with a chamois or lens tissue to remove moisture. *Care must be taken not to scratch the lenses or the coating on the lenses.*

(3) The tripod should be kept clean, and moving parts should be oiled lightly.

(4) If the instrument gets wet, remove outside moisture, and as soon as possible place instrument in a warm room or tent.

b. Inspection. Inspect the instrument for cracks, bumps, and dents. Inspect the machined surfaces and the polished faces of the lenses and mirrors. Try the clamps and motions for smooth operation without binding or grinding. Check the compass, diagonal eyepieces, sunglasses and other accessories. Try the lighting equipment for proper functioning.

c. Carrying. In moving the instrument from station to station, a person on foot may carry the instrument, mounted on its tripod, under one arm, with the hand of the other arm under the tribrach of the instrument. All motions should be loosely clamped with the telescope pointed upward. When the theodolite is carried over rough terrain, the instrument should be transported in its carrying case. When transported in a vehicle, the theodolite should be in the dome-shaped carrying case, and the case should be in the padded shipping crate. For short distances, the carrying case may be carried in an upright position on the lap of the instrument operator.

d. Shipping. The instrument is shipped in a dome-shaped carrying case inserted into a padded wooden shipping case or box. When removing and replacing the dome-shaped cover or hood, lift or lower it carefully and vertically to avoid hitting it against the theodolite. Release the clips which hold the instrument to the base. Note and remember the location of the circular bubble, the light plug, and leveling screws with respect to the clips, and the positions of the rods to which the clips are attached. Lift the instrument carefully out of its base and avoid hitting the clip stands. Hold the instrument by its base or its right standard when setting it on the tripod.

4–10. Adjustments

The theodolite must be in correct adjustment if accurate results are to be obtained. There are six tests and if necessary six adjustments that should be completed before using the instrument, periodically during use, and anytime the instrument receives a severe bump or jolt. When the test indicates a needed adjustment, *it must be made and checked before the next test* is started. All tests are made with the instrument on a tripod set in a shaded area, on firm ground, protected from the wind, and with the tripod head as nearly level as possible. Excessive manipulation of the adjusting screws will cause them to wear and can cause the instrument to go out of adjustment easily. The following is the sequence of tests and adjustments.

a. Plate Level. The vertical axis of the theodolite must be exactly vertical when the plate level vial bubble is centered.

(1) Using the circular level, approximately level the instrument.

(2) Bring the plate level parallel to any two foot screws and center the level bubble using those screws.

(3) Rotate the instrument 90° and center level bubble using only the third screw.

(4) Rotate the instrument back to the first position ((2) above). Re-center bubble if necessary.

(5) Rotate the instrument 180°. If bubble is off center, adjustment is necessary. To adjust, bring bubble halfway back to center using the capstan screws recessed in the right standard and the remaining distance using the two parallel leveling screws.

(6) Rotate the instrument 90° and center the bubble using the third screw.

(7) Rotate the instrument through 360°, the bubble must remain centered throughout (or within one division) or steps (5) and (6) above must be repeated.

b. Horizontal Collimation. The line of sight must be perpendicular to the telescope's horizontal axis of rotation.

(1) Select a well-defined point at least 100 meters away and at about the level of the telescope. Using the vertical crosswire sight on it carefully with telescope in the direct position.

(2) Read and record the horizontal circle reading.

(3) Plunge telescope and sight on the same point in the reverse position.

(4) Read and record the horizontal circle reading. The difference between the readings should be 180°. If the difference is 3′ or less, adjustment is not necessary since direct and reverse reading will compensate for this error. The error determined is equal to twice the collimation error.

(5) To make the adjustment, divide the error in half and apply half to the last reading. Using the tangent screws, set this corrected reading on the horizontal circle.

(6) Sight through the telescope. The amount that the crosswire is off the target is equal to the correction.

(7) The reticle screws are used to make this correction and they are the pull-action type. To move the crosswire to the right, loosen the left screw first a small amount and tighten the right an equal amount. To move the crosswire to the left, loosen the right screw and tighten the left. *Do not over-tighten these screws.* Do not try to make the entire correction in one step.

(8) Proceed in this manner, loosening and tightening in small amounts until the crosswire is exactly on the sighted point.

(9) Repeat the test as a check on the adjustment.

c. Vertical Collimation. The line of sight must be horizontal when the vertical circle reads 90° 00′ direct, or 270° 00′ reversed. This test and adjustment must be done after the horizontal collimation (*b* above).

(1) Select a well-defined point at least 100 meters away, and about the same level as the telescope.

(2) With the instrument direct, sight carefully on the point with the horizontal crosswire, bring the ends of the vertical collimation level bubble into coincidence with the vertical collimation slow motion screw. Read and record the vertical circle reading, check the level bubble after the reading to make sure it is still in coincidence.

(3) Plunge the telescope and sight on the same point in the reverse position. Bring the collimation level into coincidence, read and record vertical circle reading, and check the level after reading.

(4) The sum of the two readings should equal 360°. Any variation is equal to twice the error. As in the horizontal adjustment, a small error need not be adjusted because direct and reverse pointings will cancel each other.

(5) To make the adjustment, divide the error in half and apply with the proper sign (±) to the last vertical circle reading. Using the vertical tangent screw, set this corrected reading on the vertical circle and bring the circle back into coincidence with the vertical collimation slow motion screw.

(6) The collimation level bubble will now be out of coincidence. It is brought back into coincidence using the adjusting screw located under the vertical circle level cover. Remove the cover. The capstan head screw below the U-shaped level housing on the objective end of the telescope (direct position) is used for this adjustment.

(7) Repeat the test as a check on the adjustment.

d. Optical Plumb. The optical plumb must be alined and centered in the vertical axis of rotation of the instrument.

(1) Carefully level the instrument, hang a plumb bob under it, and then mark a point on the ground exactly under the point of the plumb bob.

(2) Remove the plumb bob, care must be taken so as not to disturb the instrument during its removal. View the marked point through the optical plummet, it should appear centered on the crosswire, or in the center of the circle, depending on which type is used. If it is not centered, the plummet needs adjusting.

(3) To adjust, remove the screw-type covers set about 30 millimeters either side of the plummet eyepiece. The adjusting screws are located under these cover screws and require a screwdriver about 25 millimeters long to reach them. The screws act against opposing springs.

(4) The screws are turned a small amount to bring the crosswire or circle in line with the ground point. *Always make the last movement clockwise* to compress the opposing spring.

(5) Check the adjustment by rotating the instrument through 360° and observing the ground point through the plummet eyepiece. If the plummet does not hold the point centered, repeat the adjustment. After final adjustment replace the cover-screws.

(6) The plummet may be tested and adjusted without using the plumb bob (during windy days or when observing from high towers). Level the instrument, and then a ground point is marked by an assistant while the observer sights through the plummet, directing the marking. The instrument is rotated through 180° while viewing the marked point through the plummet. If the mark does not remain centered, the plummet is adjusted by moving the crosswires (or circle) half the required distance, as described in (3) and (4) above and remaining distance by moving the instrument on the tripod. Then check as in (5) above and repeat adjustment, if necessary.

e. Telescope Level. When the instrument is equipt with this level, the axis of the level vial must be parallel to the line of sight. The test and adjustment of this level vial is similar to the two-peg test described in paragraph 4–5 and shown in figure 4–12. The only difference is that the theodolite level vial is of the coincidence type.

f. Verticality of the Vertical Crosswire. The 1-minute theodolite is designed so that the vertical crosswire remains vertical at all times. It does not need adjusting; but should be checked as described in paragraph 4–5*c*.

g. Screws. The collimation, vertical, and horizontal slow-motion (tangent) screws must turn easily and smoothly, without backlash, throughout their entire travel. The adjusting screws, located adjacent to the respective slow-motion screws, are used to obtain the proper tension for smooth operation.

h. Compass. The compass is mounted on the top of the alidade standards as shown in figures 4–14 and 4–15. After the compass is securely mounted, release compass circle by turning the caging knob clockwise and also release compass case by loosening the clamp screw. Rotate the compass case through 360°, if compass circle does not remain level throughout the rotation, and adjustment is necessary. To adjust, turn one or both compass leveling screws, located on the compass bracket below the locking nuts, until circle remains level throughout the rotation.

Section III. ONE-SECOND THEODOLITE

4–11. General

The 1-second theodolite is a precision direction-type instrument for observing horizontal and vertical directions. There are two versions of this theodolite: one version which reads directly to 1 second, and another version which reads directly to 0.002 mil. Both versions are identical except for the horizontal and vertical circle plates as stated above. There are many manufacturers of these theodolites and they may differ in appearance, but they are basically alike in essential parts and operation.

4–12. Description

The 1-second theodolite is a compact, lightweight, dustproof, optical reading, direction-type instrument similar to, but larger than the 1-minute theodolite. It is tripod-mounted and has one spindle, one plate level, a circular level, horizontal and vertical circles read by an optical microscope directly to 1 second (0.002 mil), clamping and tangent screws for controlling the motion, and a leveling head with three footscrews. The circles are read using the coincidence method rather than the direct method. There is an inverter knob for reading the horizontal and vertical circles independently. Some models are fitted with a spring plate to hold the three leveling footscrews to the tripod, while other models have a detachable tribrach arrangement. The description given in this paragraph applies essentially to all models and makes of this theodolite. The theodolite shown in figure 4–22 is a sampling of these theodolites.

a. Horizontal Motion. The horizontal motion system consists of two sections, the horizontal motion controls and the horizontal circle group. The horizontal motion is controlled by a horizontal clamp and a horizontal slow-motion (tangent) screw. The horizontal circle group consists of the circle and the apparatus necessary to set and read it. The circle is read through a microscope who's eyepiece is adjacent to the telescope eyepiece. The horizontal circle rotates about the vertical axis and may be set to desired reading with the horizontal circle drive knob. The knob is protected from accidental movement by a hinged cover which must be kept closed whenever the drive is not in use.

b. Vertical Motion. The vertical motion system consists of two sections, the vertical motion controls and the vertical circle assembly. The vertical motion is controlled by a telescope clamp and a vertical slow-motion (tangent) screw. The vertical circle assembly consists of a circle and a circle level. The level is a split level-type and the ends are brought into coincidence using the collimation slow-motion screw. The vertical circle is read through the same eyepiece as the horizontal circle, the instrument has an inverter knob, located on the right standard, that switches the microscope from one circle to the other.

c. Levels. There are three level vials on the instrument, the circular, the plate, and vertical circle. In addition there is an auxiliary level that can be used if necessary, the striding level.

(1) *Circular level.* The circular level is located on the lower part of the instrument just above the footscrews. On the instruments that have detachable tribrachs it is located on the tribrach. This level is used to roughly level the instrument.

(2) *Plate level.* The plate level is located in the center of the instrument between the two standards. This level is used for final leveling of the instrument in the horizontal plane. The sensitivity of this level is about 20 seconds of arc per 2 millimeters of the vial.

(3) *Vertical circle level.* The vertical circle level (vertical collimation) vial is a coincidence type, referred to as a split level. The level vial is completely built in, adjacent to the vertical circle and viewed through the collimation level prism. The prism is rotatable and can be viewed from either front or rear of telescope. The ends of the level bubble are brought into coincidence by using the collimation slow-motion screw. This level is used to level the vertical circle. The sensitivity of this level is about 30 seconds of arc per 2 millimeters of vial.

(4) *Striding.* The striding level is an auxiliary sensitive level attachment that can be mounted across the top of the horizontal axis of the telescope. All models and makes aren't equipped with this level. It is used to get a precise alinement of the horizontal axis when observing high altitude objects, such as stars. The striding level slips into a fitting on the vertical scale and rests on two rings concentric with the horizontal axis.

d. Telescope. The telescope of the theodolite can be used for direct and reverse readings. It is usually a 28-power instrument with the shortest

1 Diaphragm illuminating knob
2 Focusing ring
3 Telescope eyepiece
4 Microscope eyepiece
5 Telescope clamp
6 Vertical slow-motion screw
7 Collimation slow-motion screw
8 Collimation level
9 Inverter knob
10 Micrometer knob
11 Horizontal clamp knob
12 Horizontal slow-motion screw
13 Horizontal drive cover
14 Optical plumb eyepiece
15 Circular level
16 Vertical circle illumination
17 Horizontal circle illumination
18 Tribrach locking handle
19 Vertical circle housing
20 Plate level

Figure 4-22. One-second theodolite.

focus distance of about 1.4 meters. The crosswires are focused (removing parallax) by turning the eyepiece; the image, by turning the focusing ring. The reticle has horizontal and vertical crosswires and a set of horizontal and vertical ticks at a stadia ratio of 1:100. One-half of the vertical wire is split for finer pointings on small distant objects. The telescope is the inverting image type.

The crosswires, vertical plates, and horizontal plates are illuminated by either sunlight reflected by mirrors or by battery powered lights. The crosswire illumination is adjusted through the use of the illuminating mirror for telescope diaphragm.

e. Accessories. The theodolite comes with the following accessories: canvas pack accessory kit, containing diagonal eyepieces for the telescope and microscope, sun filter objective, dust cap, screwdriver, adjusting pins, camel's-hair brush, plastic instrument cover; a battery box; tripod assembly; and a carrying case, consisting of a base plate and a steel dome-shaped hood. Some models are equipped with an auto collimating eyepiece accessory. It is used for accurate collimation of the line of sight when taking short sights. It is a 20-power magnifier assembly that replaces the eyepiece lens and contains a prism assembly to project light from a plug-in illumination housing past the crosswires and through the telescope. The light is projected to a front-surfaced mirror of some distance ahead of the objective lens. The observer, looking through the telescope, sees the reflected and the directly viewed images of the crosswires. To collimate the instrument, the reticle ring is adjusted until both the direct and reflected images are superimposed.

f. Tribrach. The tribrach on some makes and models is a detachable assembly which mounts on the tripod head and contains the three leveling screws, circular level, and the optical plumbing assembly. The remainder of the instrument (alidade) head attaches to the tribrach by a locking assembly. The tribrach permits quick interchangeability of the theodolite heads, other instruments, and target sets without disturbing the tripod leveling setup or the position of the vertical axis with respect to the point on the ground.

4-13. Optical System

The horizontal and vertical circles of the 1-second theodolite have very fine lines etched on glass. A complex system of lenses and prisms (fig. 4-23) permits the observer to see small sections of diametrically opposite sides of either of the two circles.

4-14. Reading

The 1-second and 0.002-mil theodolite are direction-type instruments. The reading of the 1-second theodolite is discussed below.

a. General. The circle to be viewed is selected by turning the inverted knob on the right standard. The field of the circle-reading microscope shows the image of the circle, with lines spaced at 20-minute intervals, every third line numbered to indicate a degree, and the image of the microscope scale on which the unit minutes and seconds are read. The numbers increase in value clockwise, around the circle (0° to 360°). The coincidence knob, on the side of and near the top of the right standard, is used to read either of the circles. The circles are read by the *coincidence method,* in which the optical coincidence is obtained between diametrically opposite graduations of the circle by turning the *micrometer knob.* When this knob is turned, the images of the opposite sides of the circle appear to move in opposite directions across the field of the circle-reading microscope. The graduations can be brought into optical coincidence, that is, they appear to form continuous lines crossing the dividing line. The image is marked at the center by use of a single, vertical immovable wire that is *not* used in the reading, but as a reference index only. The upright degree value to the left of this index line is generally read.

b. Horizontal Circle. To read the horizontal circle, turn the *inverter knob* until its black line is horizontal. Adjust the illuminating mirror (or light) to give uniform lighting to both sections of the horizontal circle and the micrometer scale viewed through the circle-reading microscope. Focus the microscope eyepiece, by turning the eyepiece, so that the graduations are sharply defined. The view through the microscope should be similar to that shown in ①, figure 4-24.

(1) Turn the *micrometer knob* until the images of the opposite sides of the circle are moved into coincidence. The turning of this knob moves the micrometer scale in relationship to the circle. The view through the microscope now should appear similar to that shown in ②, figure 4-24.

(2) The degrees and tens of minutes are read from the image of the circle. The nearest upright number to the *left* of the index mark is the number of degrees (105). The diametrically opposite inverted number (the number ± 180) is 285. The number of divisions of the circle between the upright 105 and inverted 285 gives the number of *tens* of minutes. In ②, figure 4-24, there are five divisions between 105 and 285, the reading, therefore, is 105° 50′.

(3) The unit minutes and seconds are read from the image of the micrometer scale. This

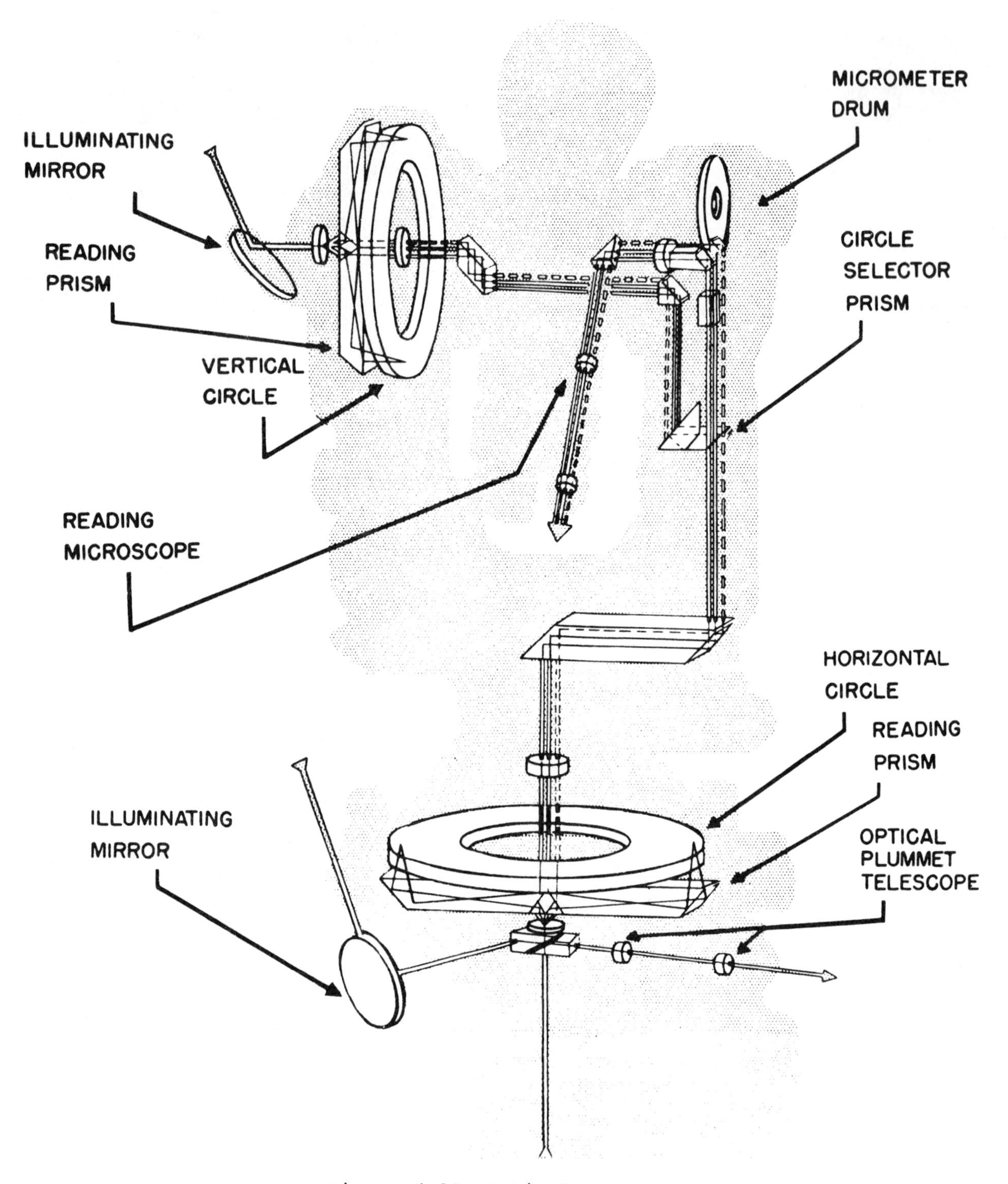

Figure 4-23. Optical systems.

scale has two rows of numbers below the graduations, the bottom row being the unit minutes and the top row seconds. In ②, figure 4–24, the unit minutes and seconds are read as 07′ 24″.

(4) Adding the values determined in (2) and (3) above, we have 105° 50′ plus 07′ 24″, or a final reading of 105° 57′ 24″.

(5) Although one can easily interpolate on the T–2 micrometer to the nearest half a second (or even to tenths), this practice adds very little if any to the precision of the results. One should either observe more positions or observe with a more precise instrument if more precision is desired.

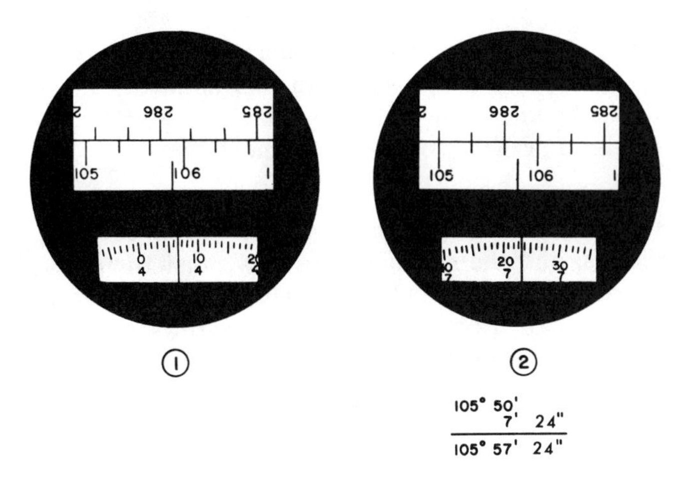

Figure 4-24. Horizontal circle reading.

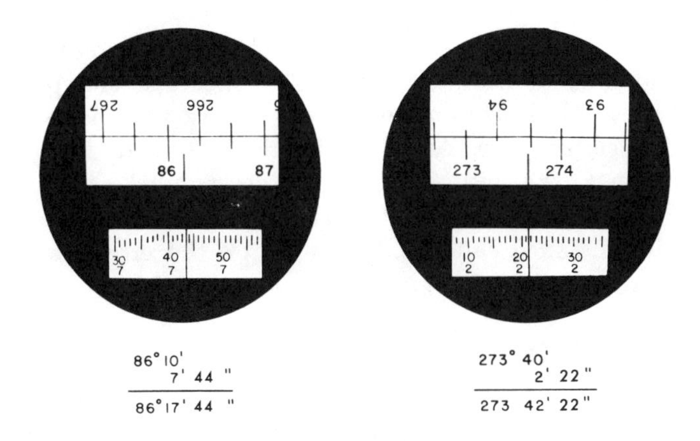

Figure 4-25. Vertical reading, direct and reverse.

c. Vertical Circle. To read the vertical circle, turn the *inverter knob* until its black line is vertical. Adjust the mirror (or light) on the left standard for illumination of plates. Focus the microscope eyepiece.

(1) Using the collimation slow-motion screw, bring the collimation level bubble ends into coincidence (fig. 4–16) as viewed in the collimation level prism.

(2) Read the vertical circle and micrometer scale as described in *b* (2) through (4) above.

(3) The vertical circle graduations are numbered to give a 0° reading with the telescope pointing to the zenith. Consequently the vertical circle reading will be 90° for a horizontal sight with the telescope direct, and 270° with the telescope reversed. Figure 4–25 shows the view in the circle-reading microscope for a direct and reverse pointing on a target.

(4) The readings obtained in figure 4–25 are referred to as zenith distances. These readings are used to determine the vertical angles, if necessary, as described below and shown in figure 4–19.

	Telescope direct	*Telescope reversed*
Circle reading	86° 17′ 44″	273° 42′ 22″
Zenith distance	86° 17′ 44″	86° 17′ 38″
Mean zenith distance	86° 17′ 41″	
Mean vertical angle	+ 03° 42′ 19″	

d. Predetermined Readings. There are occasions when the setting of the horizontal and/or vertical circles has been predetermined and the following procedure is used to set these values. In the first case, the horizontal circle is set to read a given value with the telescope pointed at a target. In the second case, the horizontal circle is set to a value of a given angle. In the third case, the vertical circle is set to a given zenith distance for locating a preselected star.

(1) With the theodolite pointed at the target, and with the horizontal clamp tightened, the circle is set as follows: Set micrometer scales to read the unit minutes and seconds of the given value. Then, using the horizontal circle drive knob, turn the circle until coincidence (or as near as possible) is obtained at the degree and tens of minutes value of the given reading. After setting the circle in this manner, the actual reading is determined as in *b*(2) through (4) above. This setting should normally be accurate to ± 5″.

(2) When measuring a predetermined angle, the instrument is first pointed along the initial line from which the angle is to be measured and the circle is read. The value of the angle is added to the reading to determine the circle reading for the second pointing. Set the micrometer scale to read the unit minutes and seconds of the value to be set on the circle. Then turn the instrument in azimuth and make coincidence (or as near as possible) at the degrees and tens of minutes value that is to be set. The horizontal slow-motion screw is used to get this coincidence. After setting the value in this manner, the actual reading is determined as in *b*(2) through (4) above. The predetermined value can usually be set on the circle in this way to ± 2″.

(3) It is seldom necessary to set a predetermined value on the vertical circle, except when adjusting the instrument, or for locating a preselected star for azimuth observation. Set the micrometer scale to the unit minutes and seconds of the given zenith distance. The vertical circle level bubble is then brought into optical coincidence. Then, the telescope is elevated (or depressed), to

make coincidence at the predetermined degrees and tens of minutes value. The telescope clamp and vertical slow-motion screw are used to make this coincidence. After setting the value in this manner, the actual reading is determined as in *b*(2) through (4) above. The predetermined zenith distance can usually be set on the circle in this way to $\pm$ 2″.

4–15. Care

This theodolite, like any other delicate scientific instrument must be handled and treated accordingly. It must be kept clean, dry, and properly oiled. If it gets wet, all exterior moisture should be wiped off immediately and should be brought into a warm, dry room or tent.

a. Cleaning. Before, during, and after use the instrument should be cleaned as follows:

(1) Painted surfaces should be wiped with a clean cloth.

(2) The exterior of the eyepieces and the objective lens of the telescope first should be brushed with a camel's-hair brush to remove dust and then wiped with a chamois or lens tissue to remove moisture. *Care must be taken not to scratch the lenses or the coating of the lenses.*

(3) The tripod should be kept clean, and moving parts oiled lightly.

b. Inspection. The instrument must be inspected when it is received and after any major shipment as detailed in its maintenance manuals.

c. Carrying. When moving from station to station on foot, the instrument may be carried while mounted on the tripod, theodolite in front, under the arm with the other hand under the tribrach or base. If the tribrach is detached and left on the tripod for other instruments, the theodolite head is held by the right standard with one hand, and the other hand is under the instrument to support it.

d. Shipping. The instrument is shipped in its dome-shaped carrying case inserted into a padded wooden shipping case or box. All motions should be loosely clamped.

4–16. Adjustment

The accuracy of surveys requiring the use of the 1-second theodolite also requires more attention to keep the instrument in good adjustment. The following tests must be made in the sequence that they are described with the adjustment, as necessary, of each step completed before proceeding to the next test. These tests and adjustments must be made with sufficient frequency to maintain the required accuracy of the instrument.

a. Screws. The collimation, vertical, and horizontal slow-motion screws, the three leveling screws, and the micrometer screw must turn easily and smoothly, without backlash, throughout their entire travel. The adjusting screws or nuts, located adjacent to the respective screws are used to obtain the proper tension for smooth operation.

b. Plate Level. The vertical axis of the theodolite must be exactly vertical when the plate level vial bubble is centered. The adjustment of the plate level should be tested every time the theodolite is set up for use. This test and adjustment is the same as for the 1-minute theodolite (para 4–10*a*), except that the plate level adjusting screw for this instrument is located on the left standard below the collimation slow-motion screw housing.

c. Optical Plummet. The optical plummet must be alined and centered in the vertical axis of rotation of the leveled instrument.

(1) Suspend the plumb bob from the leveled instrument and mark a point on the ground exactly under the point of the plumb bob. Remove the plumb bob from the instrument.

(2) Check the leveling of the instrument, and then look into the eyepiece of the optical plummet. If this plummet is correctly adjusted, the mark on the ground will be centered in the reticle circle or crosswires of the plummet.

(3) If the image of the mark is not centered in the reticle, bring the reticle to center by means of three adjusting screws located near the optical plummet eyepiece. Loosen the checknut. The vertical screw will raise or lower the crosswires in the direction that the screw moves. The side screws will move the crosswires opposite to the direction of screw travel. Always loosen the screw on the side opposite to the direction of travel and tighten the one in the desired direction.

(4) Retest and when finally correct, remember to tighten the checknut.

(5) For a more accurate adjustment, secure a piece of paper on the ground directly below the plummet. After carefully leveling the instrument, mark the position of the circle center on the paper. Release the spring plate, carefully lift and rotate the instrument through 120°, and reset it in the spring plate. Level the instrument and mark the new position of the circle center on the paper. Once again, lift, rotate 120°, reset, and level the instrument. Mark the third circle center

position on the paper. If the plummet is in adjustment, the three points will coincide. If not, the three points will appear as vertices of an equilateral triangle. Mark the center of the triangle and move the circle (as described in (3) above) until the circle center and triangle center coincide. Check the adjustment by rotating the instrument 120°.

d. Vertical Crosswire. The vertical crosswire must lie in a plane perpendicular to the horizontal axis of the instrument.

(1) Level the instrument carefully.

(2) Select a well-defined distant point as near the same horizontal plane as the instrument as possible.

(3) Center the vertical crosswire of the telescope on the selected point. With the vertical slow-motion screw, elevate and depress the telescope. If the vertical crossline continuously bisects the point, the adjustment is correct.

(4) If the vertical crosswire does not continuously bisect the sighted point, rotate the telescope reticle by turning the slant screws located between the focusing ring and eyepiece in opposite directions.

(5) Retest and readjust until the vertical crosswire remains on the point through the entire motion.

e. Horizontal Collimation. The line of sight must be perpendicular to the telescope's horizontal axis of rotation.

(1) Select a well-defined point about 100 meters away and at about the level of the telescope. Sight on it carefully with the telescope in the direct position.

(2) Read and record the horizontal circle reading.

(3) Plunge telescope and sight on the same point in the reverse position. Read and record the horizontal circle reading.

(4) The difference between the readings should be 180° ± 5″, if the instrument is in adjustment. If the deviation is greater than about ± 5″, the instrument should be adjusted.

(5) To make the adjustment, set the mean value of the two readings on the micrometer drum, and turn the slow-motion screw until the graduation lines coincide.

(6) The crosswire will now be off the sighted point, and the reticle must be moved horizontally until the point and wire coincide. The three adjusting screws are of the pull-action type set 120° apart with one of them horizontal. To move the reticle, the screw or screws opposite to the direction of movement are loosened first. Then the screws or screw in the direction that the crosswire is to move are tightened.

(7) Repeat these measurements and corrections until the collimation error is brought within the allowable limits.

(8) Recheck for verticality of crosswires as in *d* above.

f. Vertical Collimation. The line of sight must be horizontal when the vertical circle reads 90° 00′ 00″ direct, or 270° 00′ 00″ reversed. This test and adjustment is the same as that for the 1-minute theodolite (para 4–10*c*), except as noted. The sum of the two readings should be 360° ± 10″ for a well adjusted instrument. If the deviation is greater than ± 10″, the instrument should be adjusted. The adjusting screws for the vertical circle level vial are located on the left standard, at the same level and on both sides of the collimation level prism.

Section IV. TWO-TENTH-SECOND THEODOLITE

4–17. General

The two-tenth-second theodolite is a precision direction-type instrument that resembles the 1-minute and the 1-second theodolites. There are many models of this theodolite; with or without an optical plumbing device; with a 60 unit or a 2 minute micrometer scale; with or without target finding scopes; with or without camera attaching devices. All models are generally the same in their operation and location of the control levers and clamps. All models are capable of first order accuracies.

4–18. Description

The description given here is a general description and each model will vary a little, but all models are basically the same in their operation and location of the control levers, clamps, screws, and major components. Figure 4–28 shows a sampling of this theodolite.

a. This theodolite is similar to the two previously mentioned (secs. II and III) theodolites, except that it is larger, has a telescope 25.9 cm long, has its control knobs placed differently, and has levels and circles that can be read with higher precision. It uses the coincidence method of reading the circles like the 1-second theodolite, but it differs in the reading procedure. The horizontal and vertical circles and micrometer scales are read through the same microscope eyepiece, but are read independently through the use of an inverter knob.

b. The telescope's power can be changed from 24, to 30, or to 40 diameters by changing the eyepiece. The objective lens is 61 millimeter in diameter and has a greater light-gathering capacity than the 1-second theodolite. This is especially useful in star observations since it permits observing dimmer stars. The level vial sensitivity of the plate level is 5 to 7 seconds per 2-millimeters graduation, and of the vertical circle level vial, 12 seconds per 2-millimeters graduation. The telescope focusing ring, the eyepiece focusing, and the crosswire adjustment screws are the same as for the 1-second theodolite. The horizontal clamp and slow-motion screws are located diametrically opposite each other below the alidade. The clamp extends outward from a diameter of the circle, while the slow-motion screw is on the other side of the instrument and works tangent to the circle. The vertical clamp is inside of the standard opposite the vertical circle and faces upward, while the slow-motion screw is below the horizontal axis and is operated as a tangent screw.

c. Exterior adjustable mirrors or electrical lighting attachments permit illumination of the reticle and circles. The inverter knob is located on the outer face of the right standard. The collimation level is viewed through a prism system and is mounted on the left standard. The observer can see the level vial bubble from the eyepiece end of the instrument, without moving from the normal observing position. A knob atop the telescope at the horizontal axis permits the changing of the illumination of the crosswires through the turning of the telescope diaphragm.

d. The accessories for this theodolite are similar to the others, and include the diagonal eyepieces, filters, lighting attachments, and tripod. The carrying case is a similar dome-shaped type and the shipping case is made of wood and padded.

4–19. Reading

The coincidence method of reading the circle is similar to the one described with the 1-second theodolite. However, the circles and drum are arranged differently and the old models require double conicidence readings to obtain a final value.

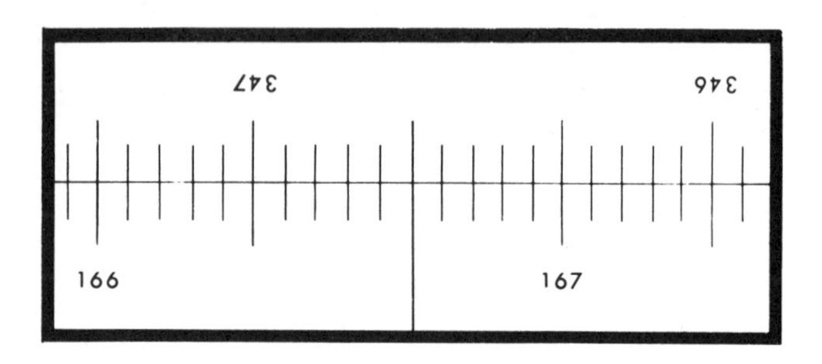

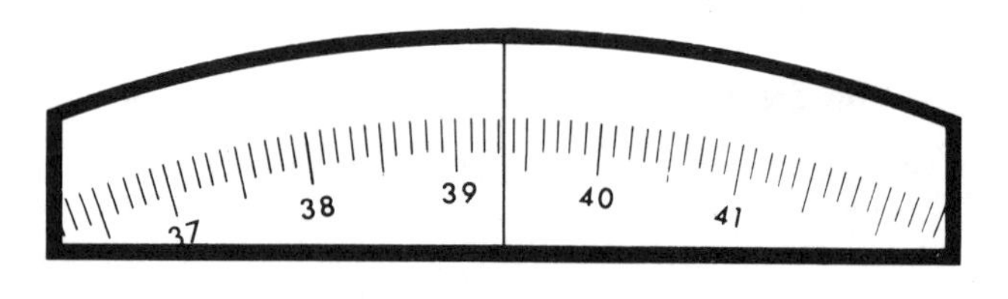

CIRCLE-READING 166° 40'

1st DRUM-READING 39".4

2nd DRUM-READING 39".3

166° 41' 18.7"

Figure 4-26. Horizontal circle and micrometer(old model).

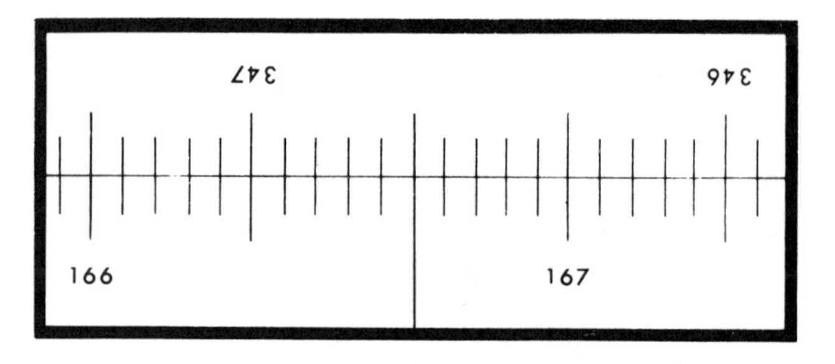

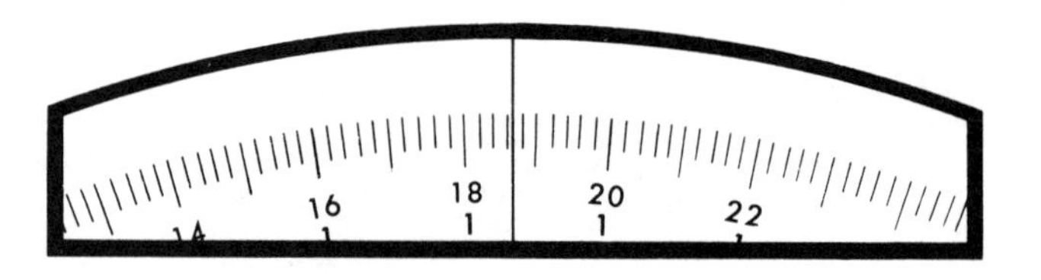

CIRCLE-READING 166° 40'

1st DRUM-READING 1' 18".6

2nd DRUM-READING 1' 18".8

166° 41' 18".7

Figure 4-27. Horizontal circle and micrometer(new model).

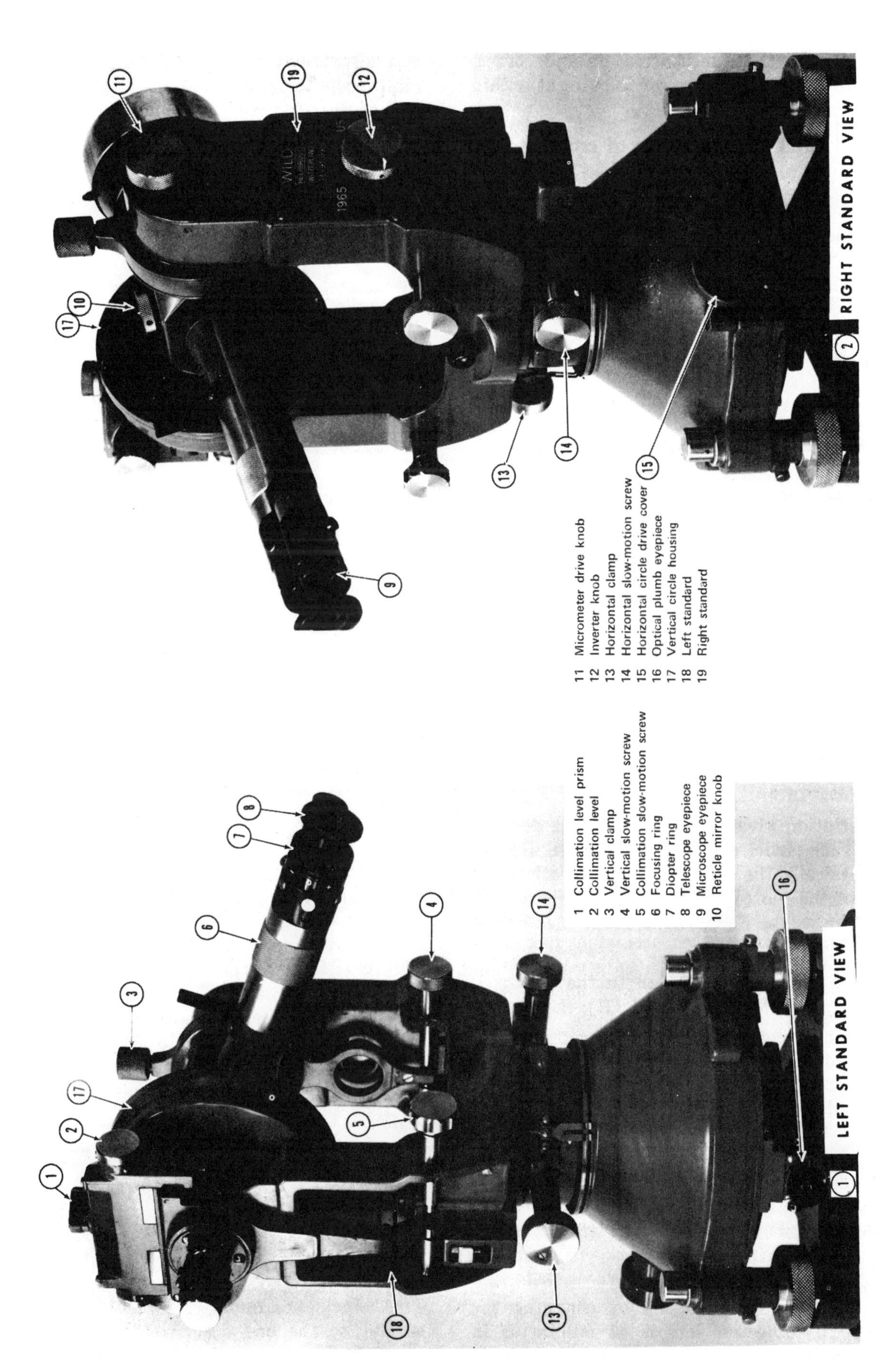

Figure 4-28. Two-tenth-second theodolite.

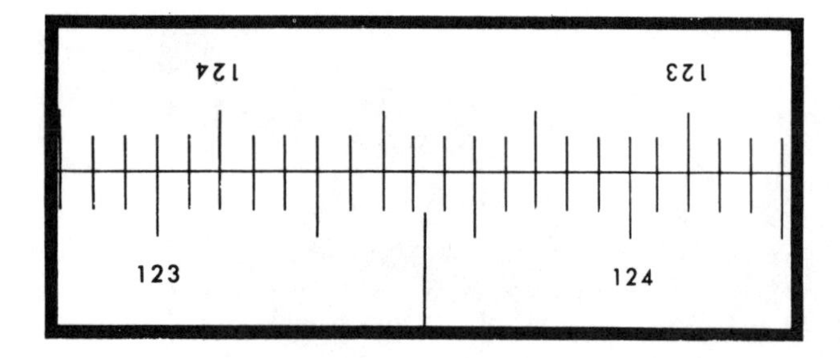

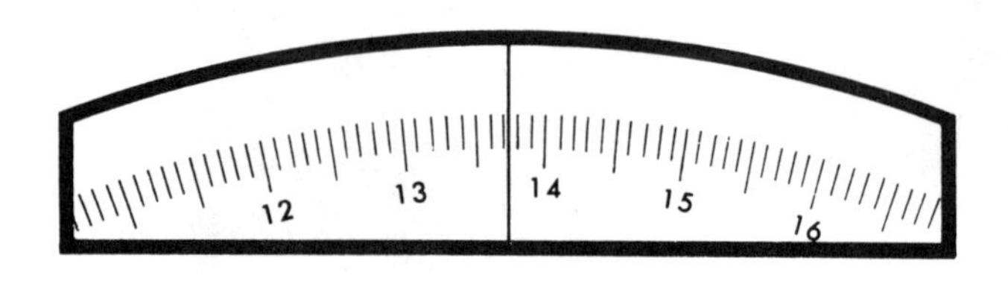

CIRCLE-READING	123° 34'	
1st DRUM-READING		$13^{u}.7$
2nd DRUM-READING		$13^{u}.7$
	123° 34' 27.4''	

Figure 4-29. Vertical circle and micrometer (old model).

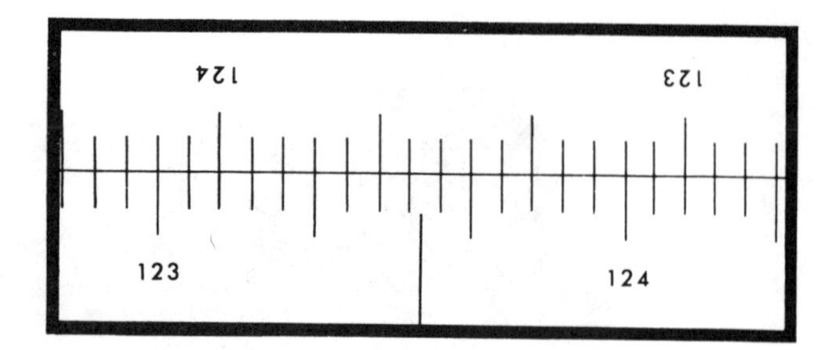

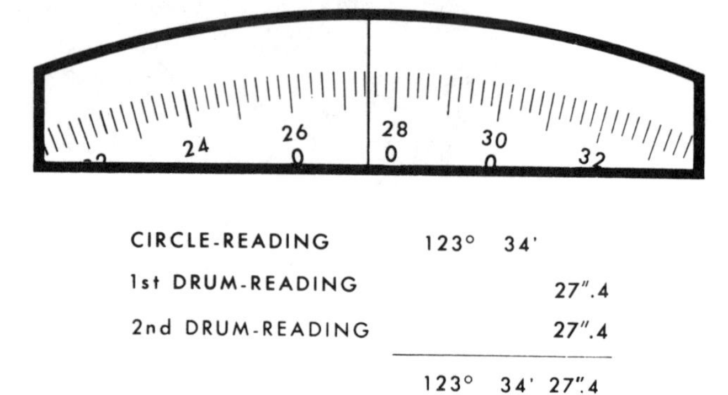

Figure 4-30. Vertical circle and micrometer (new model).

On the newer models the double coincidence is not needed to obtain the final value, but for accuracy and checking the double coincidence is usually required.

a. Horizontal Circle. The plate circle of a two-tenth-second theodolite is divided into 360°, and each degree into 15 graduations or 4 minutes per graduation. The coincidence method uses both the upright and inverted scales of the circle and allows readings to 2 minutes. On the older models, (fig. 4–26), the 2 minute micrometer scale is divided into 600 units and each tenth part is numbered. The reading on this micrometer scale equals one-half of the true value, but instead of doubling this reading, procedure requires making a double coincidence and adding both coincidence readings for the final micrometer reading. On the newer models (fig. 4–27), the micrometer scale is divided in 0.2 second segments, and the micrometer is read directly without adding the two coincidences, but procedure requires making a double coincidence and meaning the two coincidence for the final value. The procedure for reading the horizontal circle and micrometer for both models is described below and shown in figures 4–26 and 4–27.

(1) Turn the inverter knob all the way to the right to bring the horizontal circle into view in the reading microscope.

(2) Bring the graduations of the upright and inverted images on the upper scale into coincidence. There will be only one position where the graduations will coincide. The image is marked at the center by use of a single vertical immovable wire that is *not* used in the reading, but as a reference index only. The upright degree value to the left of this index line is generally read.

(3) The degrees and *even minutes* are read from the image of the circle. The nearest upright number to the *left* of the index is the number of degrees (166). The diametrically opposite inverted number is 346. The number of divisions of the circle between the upright 166 and inverted 346 is multiplied by two to obtain the number of even minutes. For both models there are 20 divisions between 166 and 346, the reading, therefore, is 166° 40′.

(4) The remainder of the reading is taken from the micrometer scale. The value is read, estimated, and recorded to the nearest 0.1 unit or 0.1 second. The old model is $39^{u}4$ and the newer model is 01′ 18″6.

(5) The circle is moved off coincidence slightly, and then reset. Read, estimate and record this second value. These readings are $39^{u}3$ (old model) and 01′ 18″8 (new model).

(6) On the older models, add the two values,

$39^{u}4 + 39^{u}3$ = 01′ 18″7. On the new models, mean the two values, 01′ 18″6 and 01′ 18″ = 01′ 18″7.

(7) Add the totals, (3) and (6) above, for the final value (166° 40′ + 01′ 18″7 = 166° 41′ 18″7).

(8) Remember it is the *sum* of the two scale readings that gives the final value on older models, and the *mean* of the two scale readings that gives the final value on newer models.

(9) Normally, the survey requires direct and reverse readings, and the sum (mean) of the two coincidences in direct position and the sum (mean) of two coincidences in reverse must be made. When making the coincidences, the micrometer drive knob should be turned in the same direction both times. If the coincidence readings in either the direct or reverse differ by more than 0.3 unit (old model) or 0.5 second (newer model), a third coincidence should be taken (or until two agree with $0^{u}3$ or 0″5).

b. Vertical Circle. The vertical circle and micrometer scale are read exactly the same way as the horizontal circle on the old and new models, respectively (figs. 4–29 and 4–30).

(1) Turn the inverter knob all the way to the left to bring the vertical circle into view in the reading microscope.

(2) A coincidence of the graduations is made as described in *a*(2) above. Due to the method of graduation of the vertical circle on this theodolite, the same figures will appear upright and inverted.

(3) The collimation level is brought into exact coincidence using the collimation slow-motion screw. Then, the degrees and *even* minutes are read from the image of the circle as described in *a*(3) above. The number of divisions are counted between upright left degree value to the *same figure* inverted on the right. The reading for both models is therefore 123° 34′.

(4) The double coincidence micrometer scale values are read as described in *a*(4) through (8) above.

(5) The total value for both models is 123° 34′ 27″4 in the direct position. or a reverse reading, the telescope is plunged, the instrument reversed, and the circle read as above.

(6) Remember, the vertical circle collimation level bubble must be in coincidence before the vertical circle and micrometer scale are read.

c. Vertical Angle Computation. This computation is unique to the Wild T–3 theodolite. The vertical circle is set so that a perfectly level line reads 90° both in the direct and in the reverse telescope positions.

(1) To differentiate between direct and reverse, the nomenclature, *circle left* and *circle right* respectively, is normally used. This refers to the location of the vertical circle on the standard which is on the observer's left when the instrument is direct, and on his right, when in reverse.

(2) Elevating the telescope in a circle left (direct) position increases the circle reading, and in a circle right (reverse), decreases it.

(3 The value read on the vertical circle equals *only one-half* of the vertical angle added to or subtracted from 90°, and is *not* the vertical angle added to or subtracted from 90°. However, to compute the vertical angle, the value of one reading is not doubled, but a circle left and a circle right reading is taken. The smaller one is subtracted from the larger and the sign of the angle is determined from the circle left reading. If the circle left reading is over 90°, the vertical angle is positive (+), if less than 90°, the vertical angle is negative (—).

(4) *Example:*

	Old Model		*New Model*		
	92° 32′		92° 32′		
D(Circle left)					
1st coincidence		$33^{u}3$		01′	06″6
2d coincidence		$33^{u}2$		01′	06″4
Final	92° 32′	66″5	92° 33′		06″5
R(Circle right)	87° 26′		87° 26′		
1st coincidence		$26^{u}4$		00′	52″8
2d coincidence		$26^{u}5$		00′	53″0
Final	87° 26′	52″9	87° 26′		52″9
Vertical angle =	+ 05° 06′	13″6	+ 05° 06′		13″6

(5) To compute the zenith distance, the vertical angle is subtracted from 90° if plus, and added to 90° if minus. A more direct method uses "(circle right + 90°)—circle left." The zenith distance is the angle from the zenith (point directly overhead) to the point sighted. In topographic, geodetic, and astronomic surveys the zenith distances are used in the computation of differences in elevation.

4–20. Care

The care of the two-tenth-second theodolite is similar to that for the 1-second theodolite (para 4–15). Protection from accidental damage and proper maintenance should be routine and continuous. The instrument must be kept clean and in proper operating condition in order to obtain the high precision of which it is capable.

4–21. Adjustment

The testing and adjusting of this theodolite follows the procedure set forth for the 1-second theodolite, paragraph 4–16. The same sequence is followed. The location of the adjusting screws and nuts may be different. For instance, the vertical collimation level bubble is mounted atop the left standard and the adjusting screw is toward the observer in the circle left position. In making the collimation tests, remember; that the total vertical circle readings add up to 180°, not 360° as in the 1-second theodolite; and that on the old models the coincidences have to be added to obtain the correct circle readings.

4–22. Calibration

When using this theodolite for high order accuracies (1st and 2d order), it is necessary that the level vials be calibrated to determine their exact sensitivity, which is expressed in seconds of arc per 2 mm division of the level vials.

Section V. ONE-TENTH-SECOND THEODOLITE

4–23. Description

The one-tenth-second theodolite (fig. 4–31) is an instrument generally used for high precision astronomic position determinations. However, this instrument may be used to observe horizontal directions if the occasion demands. This theodolite, although large and heavy, is nevertheless basically the same as the other theodolites previously discussed. Since the observations must be of great accuracy, the size and weight are an advantage rather than a handicap. The primary assemblies of the theodolite are the base and yoke, the telescope with vertical circle, the Horrebow-Talcott levels, the eyepiece micrometer, and the hanging level.

a. This theodolite is sometimes referred to as a "broken" telescope transit, so-called because its eyepiece is mounted in the horizontal axis of rotation at a right angle to the objective axis. A 45° mirror is mounted at the intersection of the two axis to turn the line of sight from the objective to the eyepiece. This allows the use of 540 millimeter focal length objective and 70 power magnification.

b. The eyepiece micrometer is designed with a reticle system of a fixed plate and a movable plate whose lateral displacement can be measured. The fixed plate has a series of precisely spaced cross-wires. The moveable plate uses one wire which is always parallel to the split wires as it moves across the field of view. The eyepiece can be rotated through a 90° angle, so as the micrometer screw is turned, the movable wire is also turned. A drum on the micrometer has a series of 12 platinum contact strips, 10 are accurately spaced around the circumference, and 2 are placed on each side of the zero mark. As the micrometer is turned, the strips make electrical contact, and when connected to a chronograph, they make a record on a tape at each contact.

c. The hanging level is used for the fine leveling of the instrument. It is a long level vial suspended or hung from the horizontal axis of the telescope. The sensitivity of this level vial is usually about 1.0 second of arc per 2mm division. The vertical circle is equipt with a vertical collimation level vial with a sensitivity of about 2 seconds of arc per 2 mm division. In addition, a pair of Horrebow-Talcott levels are attached to the horizontal axis for precise measurement of tilt changes in the telescope.

d. The circles are read by the coincidence method, similar to 1-second and two-tenth-second theodolites. The horizontal circle is readable directly to one-tenth of a second. The vertical circle is readable directly to two-tenths of a second.

e. The terms, direct and reverse, have no mean-

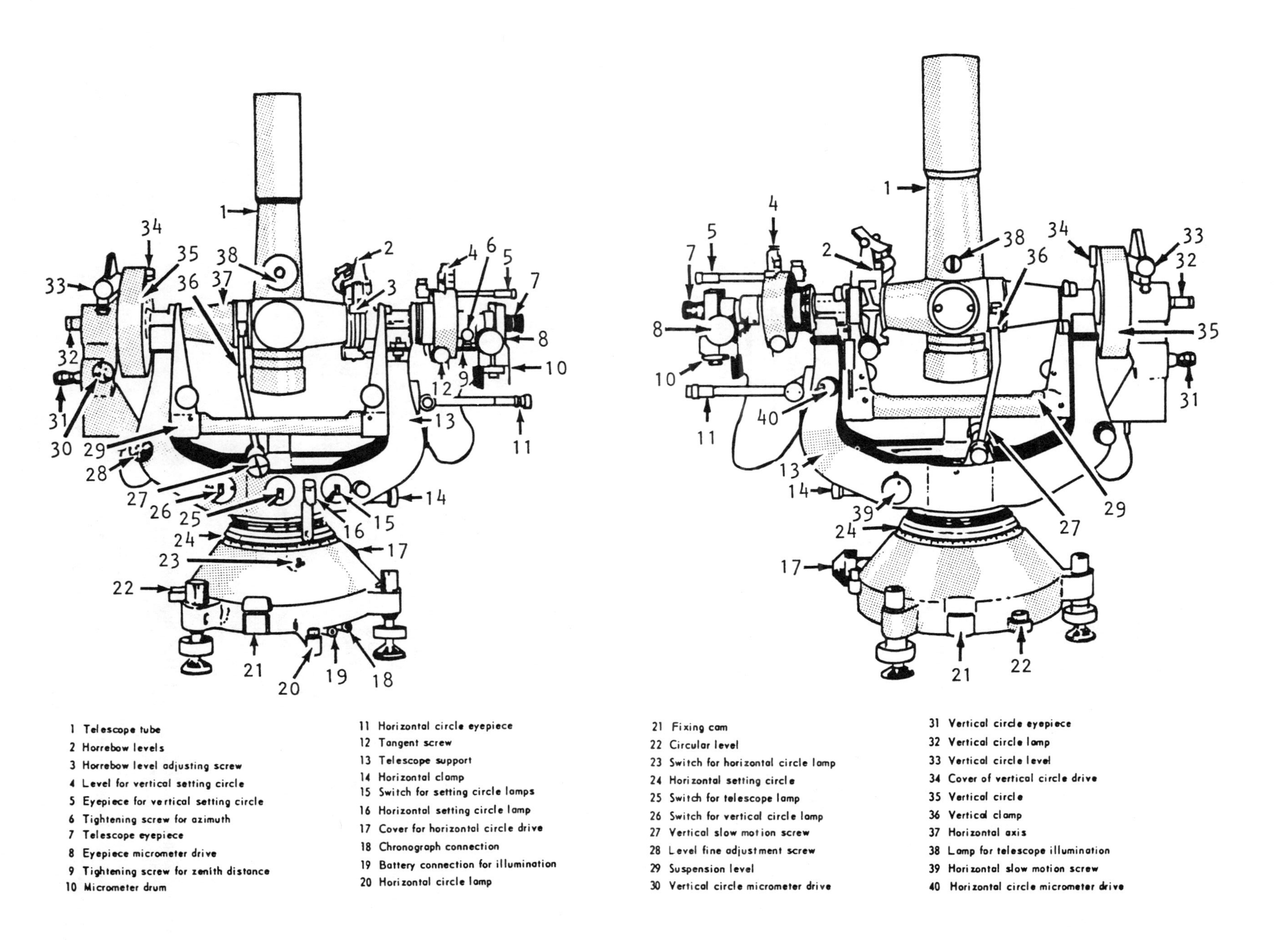

Figure 4-31. One-tenth-second theodolite.

ing when using this instrument. Instead, the positions are described as *eyepiece right* or *left, circle right* or *left,* or *ocular east* or *west.* The terms, ocular and eyepiece, refer to the viewing eyepiece of the telescope; circle is the vertical circle. Left and right are defined by the location of the vertical circle when looking at the observed object over the telescope tube.

f. This instrument is also equipped with a zenith distance setting circle that is used to set predetermined zenith distances. The circle has every degree numbered and a scale graduated in 10-minute intervals. This circle can be estimated to the nearest minute.

g. A lighting system is furnished to illuminate the circles and crosswires for working at night.

4–24. Operation and Maintenance

This instrument is usually used for the observation of astronomic latitude and longitude, however it can be used to observe astronomic azimuth and horizontal directions.

Section VI. GYRO-AZIMUTH SURVEYING INSTRUMENT

4–25. Description

The gyro-azimuth surveying instrument (fig. 4–32) is a portable gyrocompass used to determine directions from a fixed position in conjunction with a 0.002-mil theodolite. This instrument comes in two models, designated ABLE (Autonetics Base Line Equipment) Models C2A and C2B. Both models are basically the same. In operations, the gyro is placed in a fixed relation to the earth. Its orientation is determined by observing the effect of the rotation of the earth on the gyroscope and applying appropriate corrections. It can establish a true azimuth reference in any weather, day or night, without the aid of stars, landmarks, or other visible stations.

a. Sensing Element. The sensing element consists of the alinement head with a highly sensitive single axis rate gyroscope, and the theodolite. Mounted to the gyroscope assembly is a mirror assembly to check the collimation of the instrument. The mirror may be seen through the circular window installed in the side of the alinement head. A 0.002-mil theodolite is mounted on the top of the alinement head in such a manner that the horizontal circle of the theodolite is locked to the movement of the alinement head. This allows the azimuth of any line to be determined with reference to true north after orientation has been completed. The base of the instrument contains the leveling and alinement assembly. The instrument is leveled in the same manner as a theodolite. The alinement controls consist of an azimuth lock and an azimuth vernier knob with an odometer-type scale.

b. Control-Indicator Box. The electronic package converts the power to operate the instrument. The control panel (fig. 4–33) includes the controls and indicators necessary for the operation and to determine the directional alinement of the instrument. An integral voltmeter circuit is furnished to check various operations of the instrument and serves as an aid to troubleshooting. The control indicator contains six removable subassemblies called modules. This type of component packaging facilitates the replacement of faulty components.

c. Power Source. The gyro-azimuth surveying instrument is designed to operate from a power source of 24 volts direct current (dc). However, a 24-volt dc output is provided in the control indicator box to convert available 115 volts alternating current (ac) to 24 volts direct current.

d. Interconnection Cables. The instrument is provided with several cables. One cable is used to carry power from the electronic package to the alinement head. When the instrument is operated on direct current, either a 5-foot power cable for separate batteries or a 25-foot power cable for batteries installed in a vehicle is used to carry the power from the source to the electronic package. When the instrument is operated on alternating current, two powerpack cables are used, one connects the power source to the powerpack and the other connects the powerpack to the electronic package.

e. Tripod. The tripod is heavy duty, specially designed with wooden legs or shorter metal legs, which are interchangeable to allow operation under a wider scope of situations.

4–26. Operation

The gyro-azimuth instrument is a precise and sensitive piece of equipment. It should be handled with the same care as any other precise survey

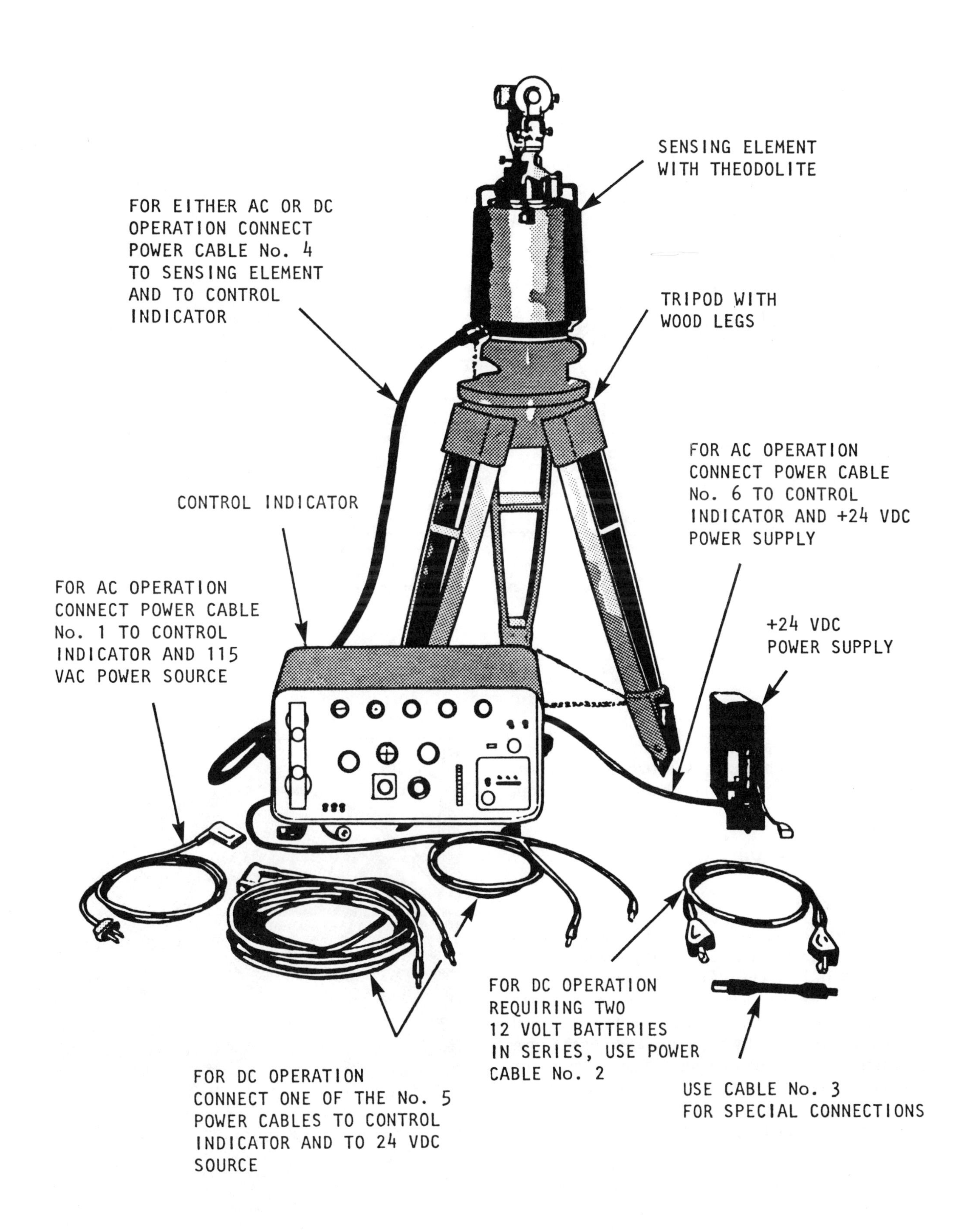

Figure 4-32. Gyro-azimuth surveying instrument.

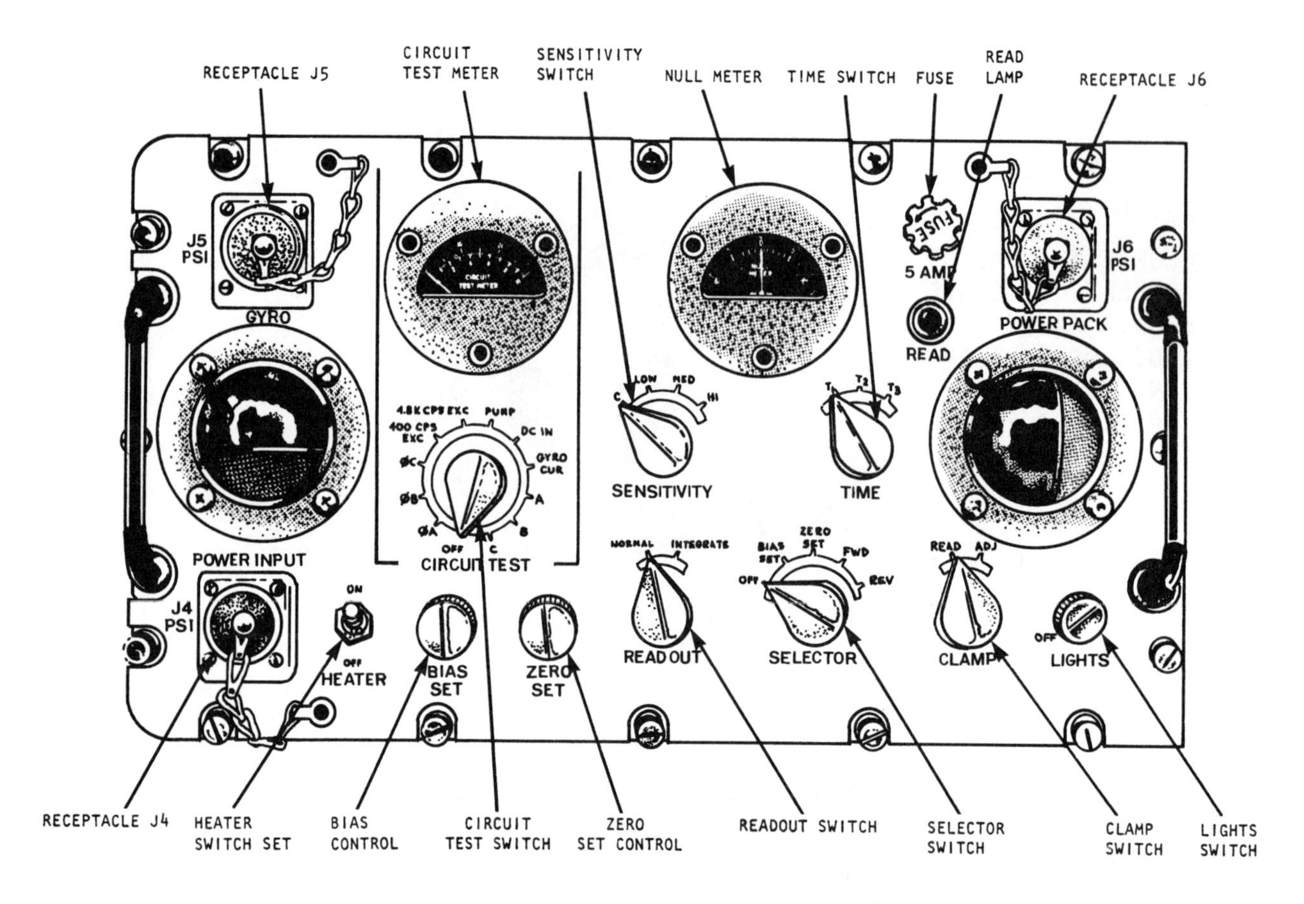

Figure 4-33. Control-indicator panel.

instrument. For operation, the instrument is set up, the proper cables are selected and attached, the instrument alined, and the readings taken.

Section VII. COMPASSES

4–27. General

The magnetic compass is the most commonly used and simplest instrument for measuring directions and angles in the field. While most compasses are not considered as delicate survey instruments there are some that are used in surveying. All compasses are basically the same and consist of a magnetic needle mounted on a fine bearing for easy pivoting. A balance or counterweight is added to adjust for any dip in the needle and to keep it in a horizontal plane.

a. Lensatic compass. The lensatic compass is the most commonly used compass. The surveyor uses it for orienting himself on maps, for sketching and describing control points, and to find approximate north lines.

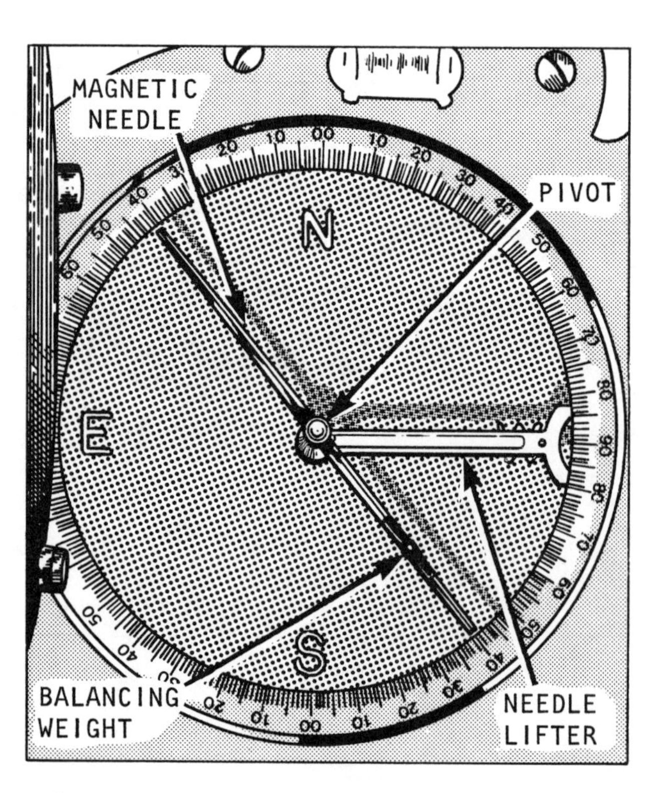

Figure 4-34. Engineer transit compass.

4–28. Surveyor's Compasses

The following are the compasses most generally used by the field surveyors.

a. Engineer Transit Compass. The compass on the engineer transit (fig. 4–34) is located between the standards on the upper plate. The needle swings in a circle graduated from 0° at north and south to 90° at east and west. The cardinal directions are marked on the dial face. However, the east and west markings are reversed. This permits direct reading of the magnetic direction. In figure 4–34, the needle points to N40°E. This means that the telescope (which is in line with the north-south zeros) is pointed N40°E, even though the needle itself is pointed at magnetic north. The compass is used on the transit for approximate measurements of magnetic bearings or azimuths and as a check on horizontal angles measured with the plates.

b. 1-Minute Theodolite Compass. The 1-minute theodolite compass (fig. 4–14) is similar to the compass described in *a* above, and is used for the same purposes.

c. Plane Table Compass. The planetable compass is attached to the alidade for orienting the planetable. This compass is mounted in a trough or box, and has a lever and cam arrangement to lift the needle off the bearing for travel. The compass box permits the needle only a limited angle of motion either side of north. The alidade must be facing approximately north before releasing the needle. This compass and its uses is discussed in chapter 7, Planetabling.

d. Brunton Pocket Transit. This instrument (fig. 4–35) is a combination compass and clinometer. It may be mounted on a light tripod or staff or it may be cradled in the hand. The cover of the compass box is hinged perpendicular to the sighting axis. It has a mirror on the inside to permit readings of the graduated circles or arcs while a sight is being taken, and has two peephole sights. A center line is cut on the mirror between the sight holes. The opposite side of the compass box

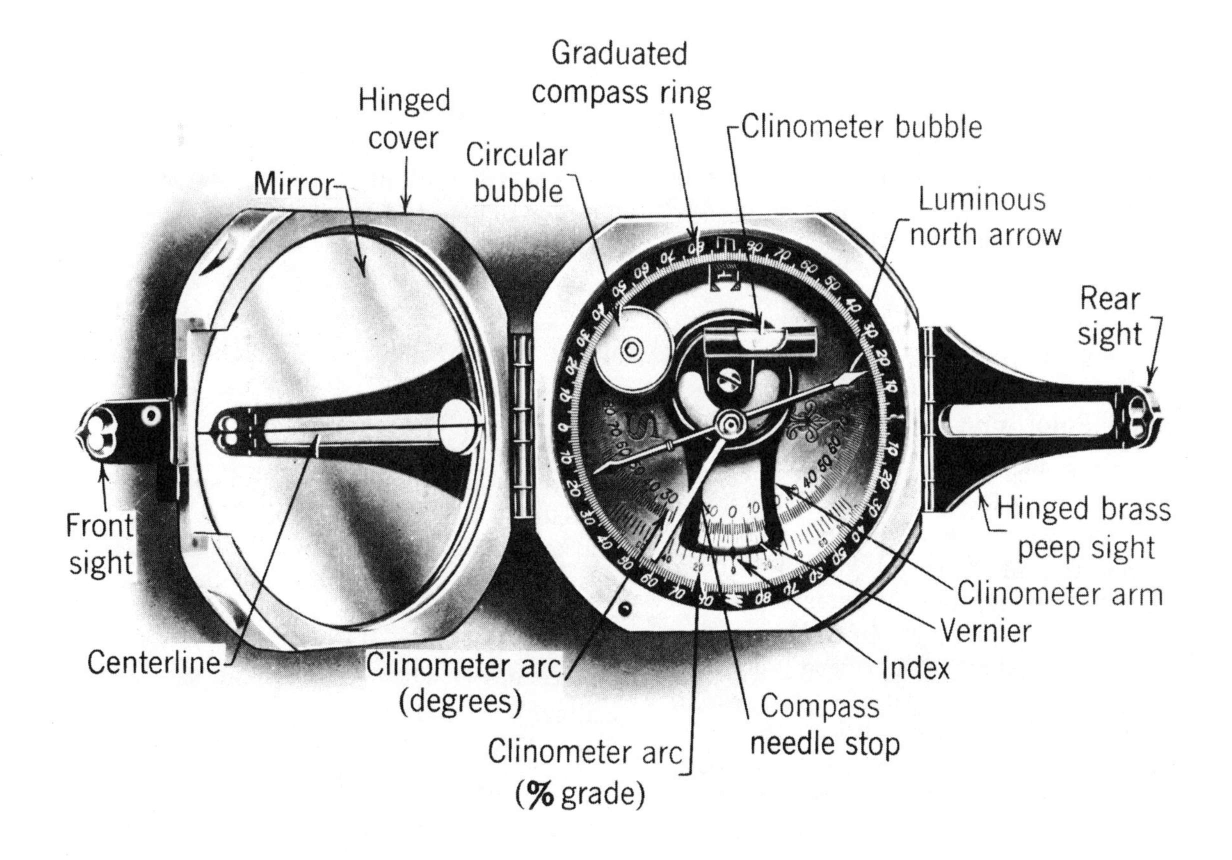

Figure 4-35. Brunton pocket transit.

carries a hinged brass peepsight and sighting point. The compass ring is graduated in degrees running either counterclockwise from 0° to 90° in each quadrant or, clockwise from the north, from 0° to 360°. When used as a compass, it is held or mounted with the compass ring horizontal, a circular bubble being used for leveling. The hinged cover and peepsight are raised to such a position that the observer can see both the point sighted and the reflected image of the compass reading. Used as a clinometer, the compass box is placed in a vertical plane with the clinometer bubble on top and the clinometer arcs below. The peepsight holes are used for sighting along the slope to the station ahead and, when the clinometer bubble is level, the reading of the index of the clinometer arm is taken. One clinometer arc is graduated in degrees and reading to 5 minutes arc obtained from a vernier on the arm. The lower arc is graduated in percent of grade and is read opposite the index of the arm. The instrument weighs about one-half pound and is easily carried in a pocket. When it is to be mounted on a tripod (fig. 4–36), the pocket transit is clamped in a yoke which is supported, through a ball-and-socket joint, by a head fitting on a small tripod. Knobs are provided on clamps so that the instrument, when leveled, can be clamped in position on the desired line of sight.

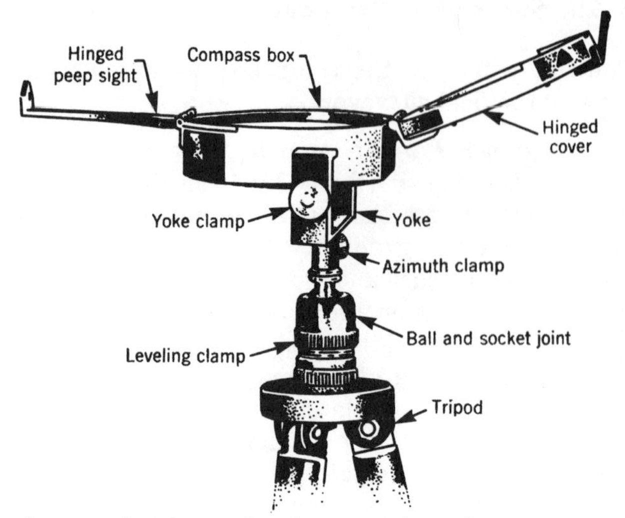

Figure 4-36. Tripod mounting for Brunton pocket transit.

4–29. Care and Handling

The magnetic compass needle must swing freely on its bearing pivot. Any obstruction or undue friction on the bearing will create a drag on the needle and give false readings. To maintain the free swing, the needle must always be raised off its pivot when the instrument is moved. It is good practice to lower the needle only during actual use, and immediately thereafter to raise the needle. A magnetic needle is easily affected by any nearby iron or steel and by electro-magnetic waves created by nearby powerlines. When using a compass, always make sure that no party member holding some steel or iron object is standing close to the instrument, that no vehicle is close enough to pull the compass off line, and that the setup is not directly under, above, or close to a high tension powerline, telephone line or underground water supply or sewer lines.

Section VIII. DIRECTION AND ANGLE MEASURING ACCESSORIES

4–30. Range Poles and Targets

a. Range Poles. The range pole is also called a line rod, flag, or flag pole. It is a wooden or metal rod about 2 meters long with a pointed tip at one end. It is painted in alternate red and white bands, either 0.15 meter or 0.3 meter wide. The range pole is useful as a target in angle measuring. It can be mounted in a tripod vertically and sighted upon. Its coloring makes it highly visible against many backgrounds. At times, small sections may be seen through trees or underbrush to line up a sight on some smaller target. Range poles are also used in pairs by taping parties to mark the direction of a tape line from station to station.

b. Survey Targets. The crosswires in survey instruments obstruct a portion of the field of view. As the distance increases, the target should subtend 4 to 6 seconds of arc to be visible and permit accurate centering. At one kilometer, 1 second of arc equals 0.5 centimeters, 6 seconds equals 3 centimeters, while at 10 kilometers, the 6 seconds equal 30 centimeters. These widths are the minimums during good visibility. To make the target more easily visible, it should be constructed in alternate bands of red and white, or yellow and red. A triangular-shaped target is easy to see and easy to bisect. For visibility from one direction, a flat triangle is sufficient. But to be seen from all directions, a triangular pyramid should be constructed from wood and covered with a colored cloth. The tripod gives the greatest stability and durability. The issue target, or surveying beacon,

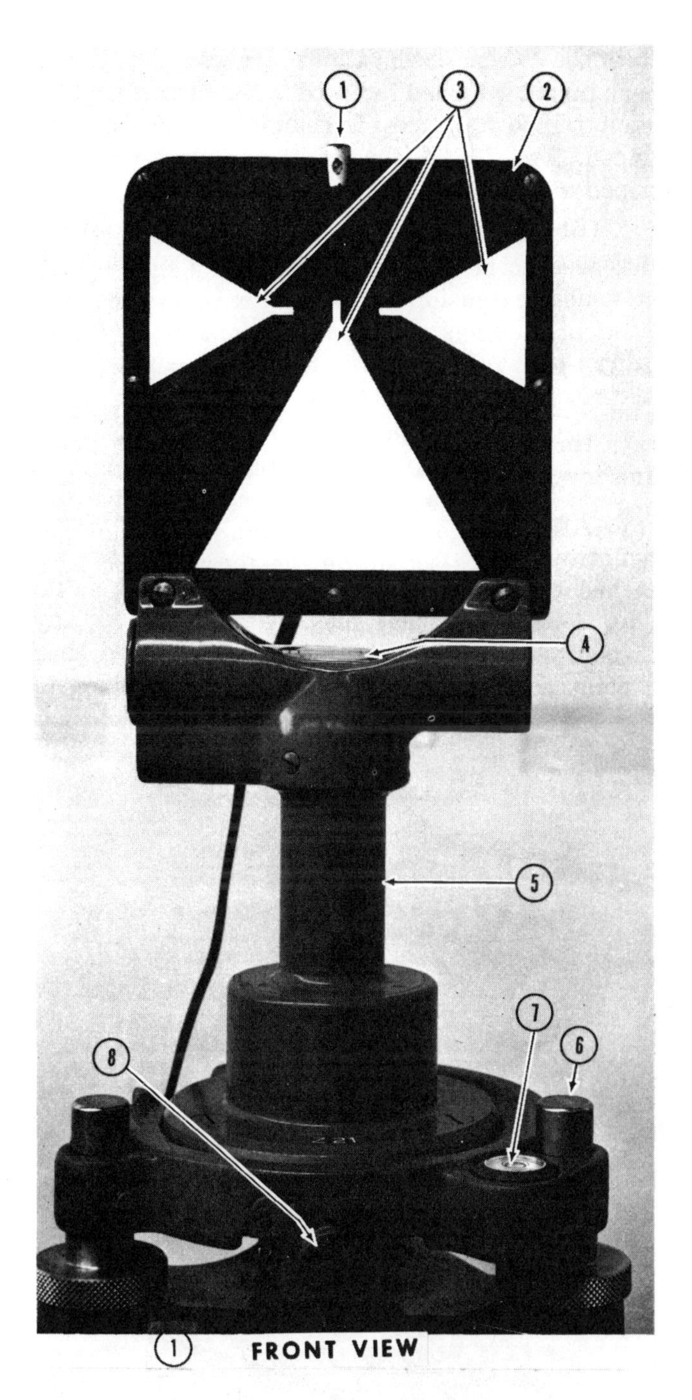

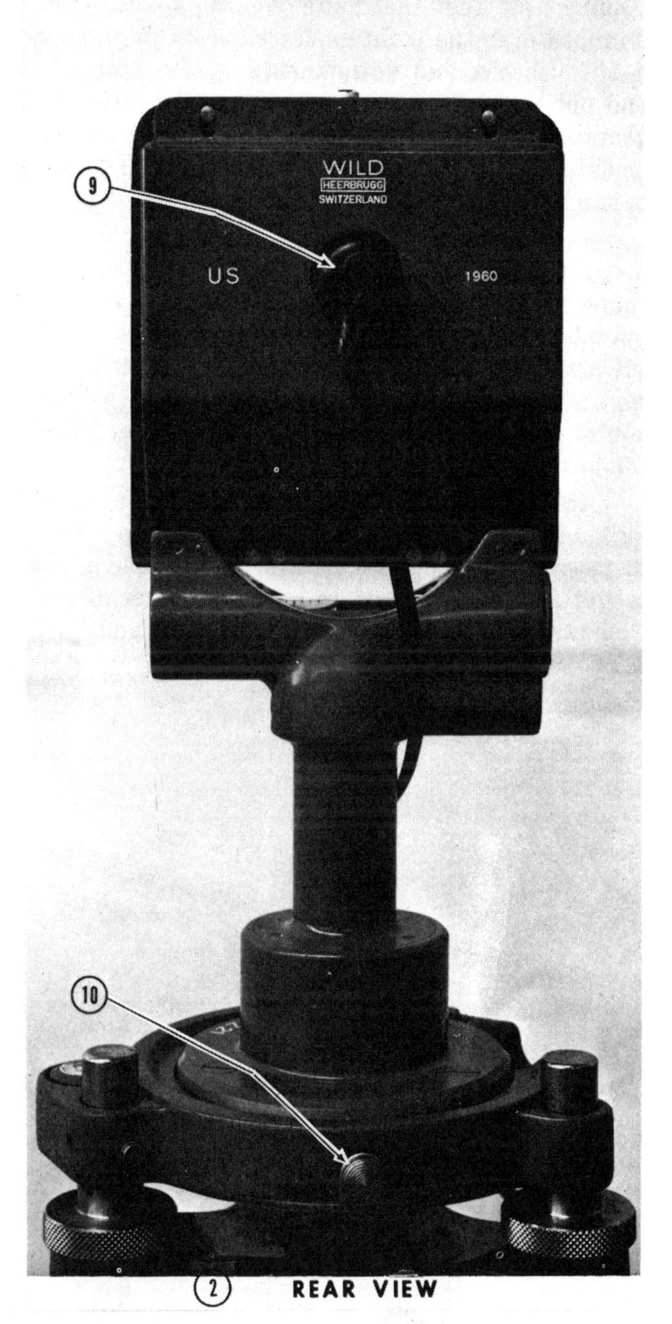

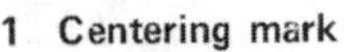

1 Centering mark
2 Target frame
3 Illuminating triangles
4 Target level
5 Target base
6 Tribrach assembly
7 Circular level
8 Optical plumb eyepiece
9 Light holder
10 Tribrach lockscrew

Figure 4-37. Target assembly.

is of the bipod type and is carried from station to station dismantled. It must be guyed or braced when being set up.

c. Target Sets. The target set is used to obtain accurate alinement when establishing lines and directions. It is designed to mount on the same type of tripod as the 1-minute and 1-second theodolites.

(1) The target set is furnished in a wooden top-opening box which holds two targets, two tri-

brachs, and two night-light attachments. In addition, two tripods with plumb bobs and adjusting wrenches, and two battery cases are issued with the set.

(2) The target assembly (fig. 4–37) consists of a tribrach with a three-screw leveling head, circular bubble, and optical plumbing assembly; an upper group containing a plate with three triangles, a long level vial, and a lighting attachment. The tribrach arrangement makes the target set interchangeable with the theodolites and the tripods do not have to be disturbed once set up. The lighting attachment makes the set useful during night observations and when the observing conditions are poor.

(3) The target is leveled in exactly the same manner as the theodolite (para 4–10*a*) and plumbed over the point as described in paragraph 4–10*d*. Checks and adjustments of the long vial and optical plummet are described in these paragraphs. After leveling and plumbing, the target is oriented so that it faces the line of sight from the instrument.

(4) To sight on the target for a horizontal direction, the large vertical triangle is bisected by the vertical crosswire. For shorter sights, only the apex has to be bisected, and on very short sights the small extension on the vertex is bisected. For vertical readings, the smaller triangles as a horizontal pair are bisected in the same manner using the horizontal vertices as the sighting points.

(5) In observations requiring continual backsights and foresights, such as traverse, and where distances between stations are not excessive, the target sets can be used in a leap-frog technique. The targets are set up on the forward and back stations. When observations are completed, the back target is moved forward to the former instrument tripod and the instrument to the forward tripod. The front target is set up on the back

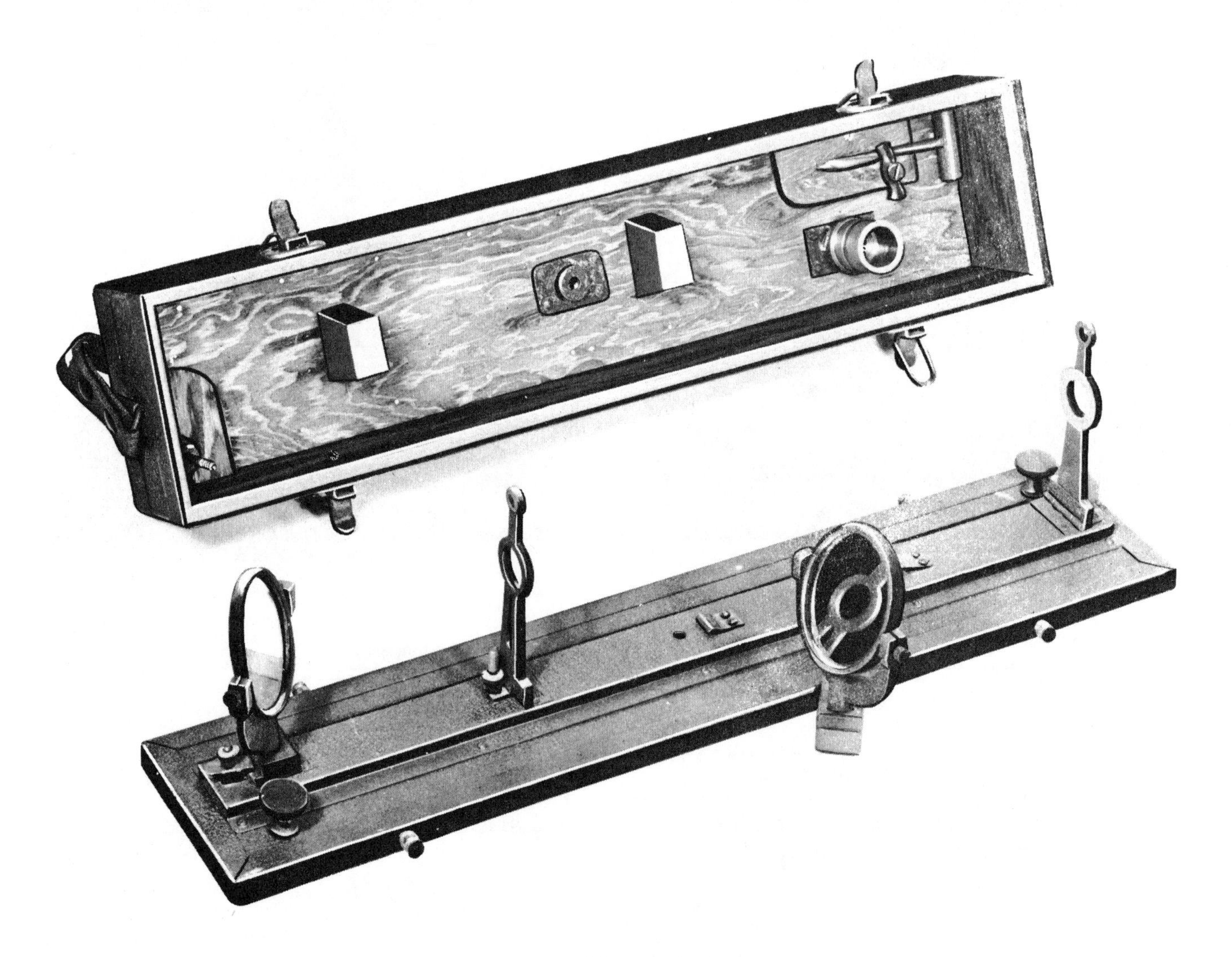

Figure 4-38. Heliotrope.

tripod moved to the new forward station.

(6) The target set is a precise instrument just as any theodolite and should receive the same care and handling.

4–31. Heliotrope

The heliotrope is a device which reflects the sun's rays through a pair of mirrors set over a point and toward an observer on another station.

a. After the box cover is removed, the issued heliotrope consists of a flat base to which are attached two plane mirrors and the pair of sights (fig. 4–38). The base may be mounted on any theodolite or range pole tripod and plumbed over a point with a plumb bob. One mirror is fixed to the base, though it may be rotated and tilted, and directs the reflected rays of the sun through the sights to the observing instrument. If the sun is in the direction of the observer, the fixed mirror is all that is needed. The second mirror is on a movable arm for use when the sun is shining toward the observer. It serves to direct the light into the fixed mirror, and thence to the observer. No provision is made for adjusting the pointing in a vertical plane. Flat wood chips or folded cardboard shims between the base and the tripod head will solve this. The main point to remember is that all the fittings must be tight, as any play or looseness will cause trouble. Whichever mirror is receiving the direct rays of the sun must be constantly attended and moved with the movement of the sun. With experience, the operator becomes adept at this. The sun mirror should be just tight enough so that light taps with the fingernail or pencil will maintain the correct angle. The peep sights are seldom alined perfectly with the light apertures, and tests should be made prior to use so that the proper allowances can be made.

b. The heliotrope's use is limited by, first necessity for sunlight; second, difficulty to use it as a target while making simultaneous observations from the same ground station; and, third, difficulty of lining in without radio communications. The second difficulty can be solved by placing the heliotrope to one side of the station and making careful measurements of its eccentricity. It can also be used in this position for simultaneous vertical angles by measuring the difference of elevation from the station. The third difficulty is not often present as most parties are equipped with radio, but faulty technique can cause much fumbling. After communications and identifications have been established, the observer, with his eye at the telescope and the speak button of the handset depressed, gives a running stream of one-way directions which might go as follows:

"NO LIGHT ________ NO LIGHT ________ NO LIGHT ________ NO LIGHT________ FLASH NO LIGHT________ FLASH, FLASH ________ NO LIGHT ________ FLASH, FLASH ________

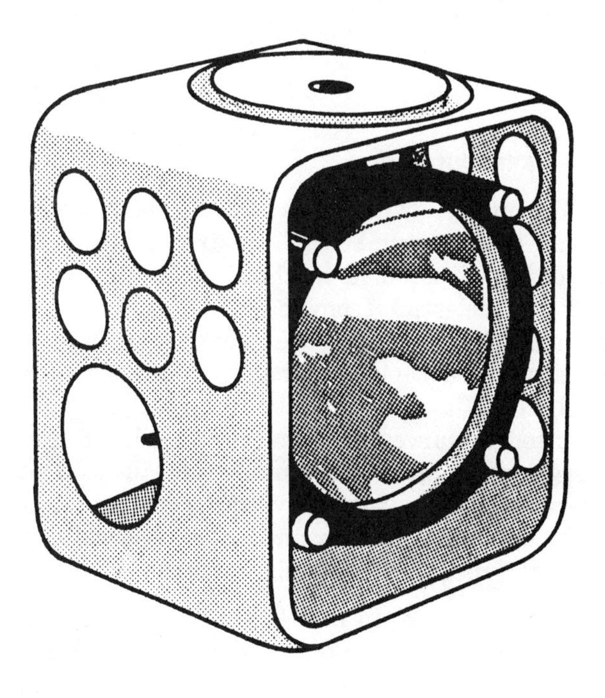

Figure 4-39. Signal lights(5 inch).

FLASHING, FLASHING, HOLD IT! ________ STEADY LIGHT."

At this time, the helio attendant needs only to note the position of the sun's image on the sights and periodically adjust the sun mirror. Many surveyors tend to shy away from using the heliotrope because they have not been able to get good results due to faulty techniques. There are times when use of this instrument is the only way to solve a line-of-sight problem, as the reflected rays of the sun will cut through hazy, smoky atmosphere and can be seen when nothing else can be.

4–32. Signal Lights

A signal light is useful at night and whenever the distance between stations is too great to use the target set. The signal lamps are made in two sizes; one with a 2-inch diameter reflector, and a larger size, which is most common, with a 5-inch reflector (fig. 4–39).

a. Both lamps contain five basic parts—a frame, reflector, sighting device, focusing device, and a light bulb. The top and bottom of the frame have threaded holes which permit stacking of lamps one above the other. Each lamp can thus be pointed at a different observing station. The reflector pivots around a horizontal axis and is rotated up or down to point exactly at any station. The size of the bulb depends on the desired brilliance of light. For short sights, a low power bulb is sufficent, whereas for long sights and for daylight observations, a high powered bulb is necessary. To power these bulbs, it may be necessary to connect the batteries in series or parallel combinations, depending on the rated voltage of the bulb and the length of time the bulb is to remain lighted.

b. Since different sizes of bulbs are used, each must be focused by the screw at the rear of the reflector. This is done by aiming the light at a flat surface perpendicular to the beam and at least 50 meters away. This aiming surface may be a building, tent, or a piece of cloth suspended vertically from a tree or bush. The bulb is focused (moved in or out) until the bright part of the image is slightly larger than the lens of the lamp. When no distant object is available, a field expedient is to hold your hand about 6 inches in front of the light and adjust until a *dark* spot the size of a quarter appears in the center of the beam.

c. The horizontal and vertical pointing of the lamp at the observer can be critical in the more accurate surveys. Observations on a mispointed light will be on an eccentric point and not on one directly over the station. The light should be pointed while the observer at the distant station is viewing it through his instrument. Radio instructions to the light-tender to pivot the lamp until the best image is visible, will define the exact pointing.

Section IX. THE SOLAR TRANSIT

4–33. Introduction

Beginning with the Burt solar compass in 1836, a number of instruments have been designed to solve mechanically the pole-zenith-sun celestial triangle by means of an attached solar unit. Because such instruments can be oriented rapidly without reference to a backsight or a new direct observation, they are suited to surveying through timber, dense undergrowth, and mountainous terrain. The modern solar transit is fully equipped for making the necessary stellar and solar observations by direct means as well (fig. 4–40).

The solar unit is in fact a second instrument, operating in its own right independently. Latitude and declination arcs remain clamped and set to their proper values. The polar axis conforms with the line of collimation of the solar telescope. The mounting is designed to bring the vertical plane of the polar axis into parallel with the vertical plane of the transit.

When oriented, the vertical plane of the polar axis is in the plane of the great circle of the meridian. When turned in hour angle at the moment of an observation, the plane normal to the axis of the reflector is in the plane of the great circle that passes through the pole and the sun. The sun's hour angle at that moment is the angle measured along the plane of the Equator, intercepted between the plane of the meridian and the plane of the great circle that passes through the pole and the sun. This angle reads "apparent time" on the hour circle of the solar unit.

The vertical angle inclination of the polar axis equals the latitude of the station; this angle is set on the latitude arc. The angle on the plane

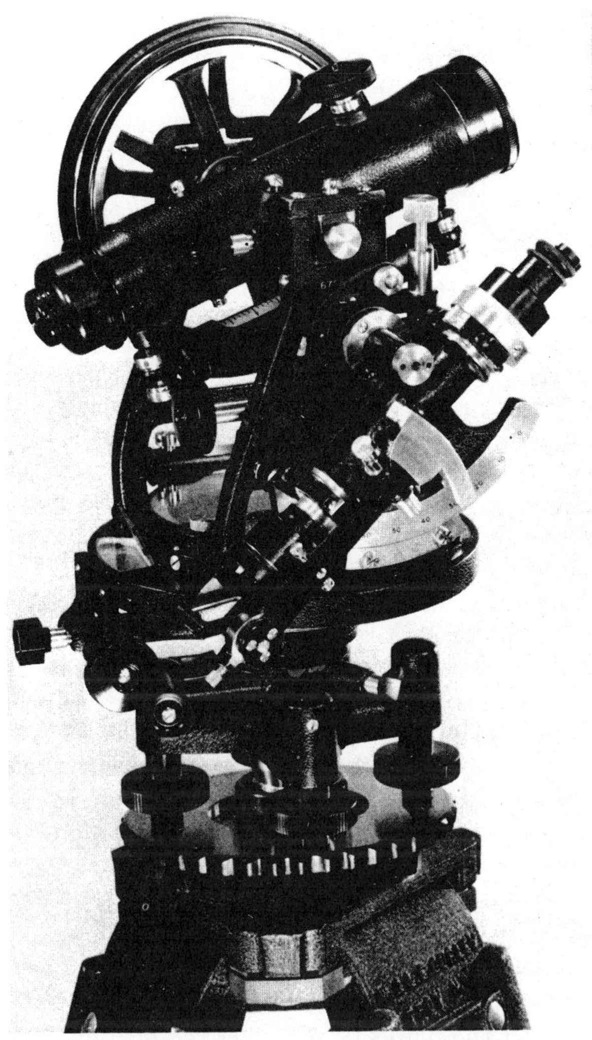

Figure 4-40. Standard model solar transit.

of the great circle that passes through the pole and the sun, counting between them equals 90° minus the sun's north declination, or 90° plus the sun's south declination, corrected by an increment equivalent to the refraction in polar distance. The settings for this angle are computed for each day in advance; it is set on the declination arc to agree with the apparent time of observation. The correct position of the sun's zenith distance measured on the vertical plane of the great circle that passes through the sun is secured by the careful leveling of the transit.

After setup and careful leveling, the solar transit may be instrumentally oriented by an experienced surveyor in less than two minutes. The accuracy or acceptable "tolerance" is equal to that of any single, unverified, average direct altitude observation on the sun.

In line running, through timber and undergrowth, there may be 20, 30, or more setups to the mile, each by solar orientation without cutting or opening the line to secure an exact backsight. In this practice, the net result for the mile is the mean of the whole number of the observations, in which many of the smaller differences are compensated, and in which the azimuth of the line between the monuments should normally be brought well inside of the tolerance of 1′ 30″.

What is more, each azimuth determination gives the angular value referred to the true north at that station. This of course is the only method by which a true parallel of latitude can be run by instrumental orientation. The determinations of the true parallel by the "tangent" or "secant" method require the careful running of a "back-and-foresight" line with measured offsets.

Use of the solar unit avoids the cumulative error normally encountered in long "back-and-foresight" lines and in traverse lines where there are many turns. A traverse line may be run *by occupying each alternate station,* cutting in half the time required for the instrumental work. Heavy winds or insecure ground, windfalls, timber, undergrowth, and obstructions that require offset are not in themselves any preventative to rapid and accurate solar orientation.

a. The instrumental orientation of the solar unit is made possible through five elements in the construction, as follows:

(1) A telescope whose line of collimation is the polar axis; the polar axis corresponds to an element of the more elaborate observatory "equatorial instrument mounting," which is designed for the telescope to follow a star's travel in diurnal circle. The solar telescope is mounted in collar bearings whose bases are attached to a vertical limb; the telescope may be revolved or turned 12 hours in hour angle.

(2) The vertical limb is an arc that is graduated to read in latitude; a vernier mounted on the base frame gives the reading in latitude; the center of the limb is called the latitude axis, and is horizontal.

(3) A reflector at the objective end of the telescope picks up the light rays of the sun; its axis is normal to the line of collimation. An arm controls the angle of the reflector in the plane of the great circle that passes through the pole and the sun; a vernier on the arm gives the

reading on a graduated declination arc.

(4) A small graduated circle on the telescope, normal to the line of collimation, reads in hour angle from 6 a.m., 7, 8, up to 12, and 1, 2, 3, up to 6 p.m.; this reads directly in *apparent time.*

(5) The plan of the reticle includes three "equatorial wires" that are set parallel to the axis of the reflector. One is in the line of collimation, the others parallel, spaced at 15′ 45″ to conform with the sun's July 1 diameter (the smallest for the year). A fourth cross wire, normal to the others, passes through the line of collimation, indicating the center of the field in *time.*

Orientation of Sun's Image on Reticle

The frame of the solar unit is attached to one standard of the transit, controlled in position by three foot posts. This is the standard on the east when the transit is oriented in the meridian. The vertical plane of the polar axis may be adjusted to true parallel with the vertical plane of the transit telescope. By first setting the plate reading at zero, when preparing to orient, all horizontal angles will count from the meridian.

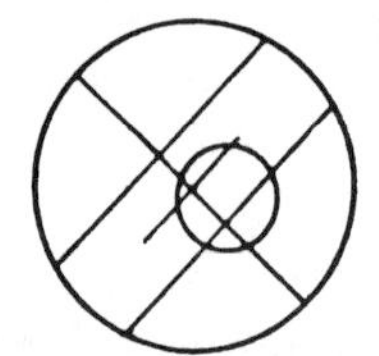

Not oriented
When turning the transit in horizontal angle, the image of the sun cuts across the equatorial wires.

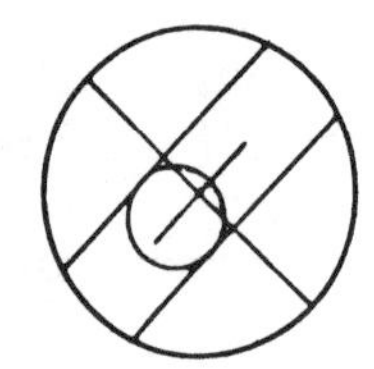

Oriented
The travel of the sun's image is along the path of the equatorial wires.

Figure 4-41. Reticle orientation of sun's image.

When oriented, and turned in hour angle to agree with the apparent time, the image of the sun will travel across the field of the solar telescope along the path of the equatorial wires. If the transit is turned away from the meridian to the right or to the left, the sun's image will cut across the equatorial wires. Thus to bring the solar unit into proper orientation, all that is needed is to see that the image of the sun is centered anywhere along the length of the equatorial wires. This centering is done with the lower tangent motion, the plates clamped at zero (fig. 4–41).

On the declination arc, an actual arc or segment of 5° is graduated for reading 10°; this is because a movement of 5° in the reflector position makes an angle of 10° between the light rays of incidence and those reflected. At zero declination the plane of the reflector is at 45° to the line of collimation. The declination arc is graduated north and south from zero for the range of the sun's position during the year.

One important element of the mounting is that the three points of control at the foot posts are placed to form a right-angle, one side of which is vertical, the other horizontal. In adjustment, the foot post at the 90° angle remains fixed. One of the foot posts controls the position of the latitude axis in horizontal. One foot post controls the direction of pointing of the line of collimation when in horizontal sighting, to bring that into parallel with the vertical plane of the transit. The foot-post controls are secured with capstan or hexagon nuts.

At one end of the frame that supports the collar bearings, there is a mechanism (corresponding to that of a telescope level) for adjusting the polar axis so that it is normal to the latitude axis. This is needed when the solar telescope is changed in latitude setting.

b. With the transit and solar unit in satisfactory adjustment, the simple steps in solar orientation at any setup are these: carefully level, with the solar unit on the west; reverse the instrument and correct *half* of any discrepancy in the centering of the plate bubbles; set the plates at zero; set the latitude and declination, or check the previous setting; turn the solar telescope to the reading in approximate apparent time; move the whole instrument in horizontal angle for position near the meridian, at this time bringing the sun's image into the field of the solar telescope, then tighten the lower clamp; use the lower tangent motion for final orientation, in which step the sun's image should be centered on the equatorial wires.

The solar transit is equipped for making any type of stellar or solar observation that may be employed profitably in land surveying practice as adapted to the one-minute transit. This calls for accuracy to a tolerance of ±15″ in the direction of lines, where that may be required. This accuracy is greater than that called for by the precision of distance measurement unless the character of the survey is such as to justify the greatly increased cost of exactness in measurement.

The use of the solar unit may be almost continuous as when running the line through timber or tall undergrowth, or it may be more or less incidental as when running in an open country. It is important, too, in the open country, and on almost any type of survey, to have the use of the solar unit in making the start in the line running or observing. Even on the work that requires the greatest refinement in the important lines, there are many off-line stations to be occupied for collateral data, mapping, or traversing, where the direction from true north should be employed.

For these reasons, the preparation of each day's work requires that the data shall be at hand, in the field tablet, for the sun's declination for the day, reading for value in the apparent time of the local meridian, and to which has been applied the correct refractions in polar distance for that position of the sun and that latitude. Additionally, the true latitude and the *instrumental latitude* should be known.

Proper accuracy in solar orientation becomes attainable as soon as the sun is high enough to reduce the refraction correction to not over 4′ or 5′, and continuing until 10:30 a.m. or a little later with care and suitable checks; a corresponding period applies in the afternoon. Thus from 10:30 a.m. to 1:30 p.m., or for about that period, the line running should be by back-and-foresight. If the sun becomes obscured, the usual transit methods are employed. In stopping for the day, an azimuth mark should be set for use the next morning.

c. Computing Hourly Declinations of the Sun.

For use with the solar unit, hourly declinations of the sun for each date may be prepared in tabular or graphic form, the graphic form being most advantageous.

Example of a table of hourly declinations of the sun, combined with refraction in polar distance, for March 14, 1972, at station in latitude 33°10′ N., longitude 116°45′ W. (7^h-47^m):

Declination of the sun at Greenwich apparent noon, March 14, 1972 .... 2°22′30.8″ S.

Difference in time from Greenwich apparent noon to 7 a.m., app. time, longitude 116°45′ W.

For longitude .... = 7^h47^m

For time, a.m., $12^h - 7^h00^m$ .... = −5 00

2.78^h = 2^h47^m

Hourly difference in declination .... = 59.21″

Difference in declination from Greenwich apparent noon to 7 a.m., app. time, longitude 116°45′ W.,
2.78 × 59.21 = 164.6″ .... = 2′44.6″ N.

Declination of the sun, 7 a.m., app. time = 2°19′46.2″ S.

Apparent time	True declination	Refraction	Declination setting
7 a.m.	2°19′46″ S.	2′39″ N.	2°17′07″ S.
7½	2 19 17	1 47	2 17 30
8	2 18 47	1 21	2 17 26
9	2 17 48	0 57	2 16 51
10	2 16 49	0 47	2 16 02
11 a.m.	2 15 49	0 44	2 15 05
Noon	2 14 50	0 41	2 14 09
1 p.m.	2 13 51	0 44	2 13 07
2	2 12 52	0 47	2 12 05
3	2 11 53	0 57	2 10 56
4	2 10 53	1 21	2 09 32
4½	2 10 24	1 47	2 08 37
5 p.m.	2 09 54	2 39	2 07 15

Example of a table of hourly declinations of the sun, combined with refraction in polar distance, for August 11, 1972, at a station in latitude 47°10′ N., longitude 111°00′ W. (7^h-24^m).

Declination of the sun at Greenwich apparent noon, August 11, 1972 ...... 15°09′38.1″ N.

Difference in time from Greenwich apparent noon to 6 a.m., app. time, longitude 111°00 W.

For longitude .... = 7^h24^m

For time, a.m., $12^h - 6^h00^m$ .... = −6 00

1.40^h = 1^h24^m

Hourly difference in declination .... = −44.73″

Difference in declination from Greenwich apparent noon to 6 a.m., app. time, longitude 111°00′ W.,
1.40 × 44.73 = 62.6″ .... = 1′02.6″ S.

Declination of the sun, 6 a.m., apparent time .... = 15°08′35.5″ N.

Apparent time	True declination	Refraction	Declination setting
6 a.m.	15°08′35″ N.	3′28″	15°12′03″ N.
6½	15 08 13	2 21	15 10 34
7	15 07 51	1 45	15 09 36
8	15 07 06	1 08	15 08 14
9	15 06 21	0 51	15 07 12
10	15 05 37	0 41	15 06 18
11 a.m.	15 04 52	0 38	15 05 30
Noon	15 04 07	0 36	15 04 43
1 p.m.	15 03 22	0 38	15 04 00
2	15 02 38	0 41	15 03 19
3	15 01 53	0 51	15 02 44

Apparent time	True declination	Refraction	Declination setting
4	15 01 08	1 08	15 02 16
5	15 00 23	1 45	15 02 08
5½	15 00 01	2 21	15 02 22
6 p.m.	14 59 39	3 28	15 03 07

Examples of diagrams showing declinations of the sun for given dates, combined with refractions in polar distance, are given in figure 4–42. The horizontal lines represent each hour of the day; the vertical lines represent intervals of one minute in declination. It is convenient to use the right-hand side of the sheet to represent north, the left-hand side to represent south. North declinations increase numercially to the right-hand side of the sheet, south declinations to the left-hand side. The vertical lines are numbered to suit the range of declination for the date.

The advantage of the diagram method is found in the avoidance of errors of computation and the ease with which it is checked, together with the fact that in the use of the diagram actual values are obtained at any time instead of by a linear interpretation.

Two points are marked on the diagram to agree with the true declination of the sun; the first point is marked with the argument of declination agreeing with the declination of the sun taken from the Ephemeris for Greenwich apparent noon, with the argument of time agreeing with the apparent time at the longitude of the station, corresponding to Greenwich noon; the second point is marked agreeing with the proper declination and time 10 hours later. The straight line determined by the two points agrees with the sun's true declination for the apparent time at the longitude of the station. The proper refractions in polar distance are then scaled from the straight line to the N. for each tabulated refraction, a. m. and p. m., taken from table 23, Standard Field Tables, appropriate to the latitude of the station, and declination of the sun. The latter points are then connected to form a smooth curve representing the declinations of the sun, corrected for refraction in polar distance, for use with the solar unit. The scale of the refractions must equal the scale of the intervals of 1′ in declination; the refractions are laid off along or parallel to the horizontal lines, and *not* normal to the line of true declination. At any time throughout the day the proper declination for use with the solar unit is obtained by reference to the curve at the point corresponding to the time of observation. To obtain any true value of the sun's declination for use in the reduction of *altitude observations* reference may be made to the

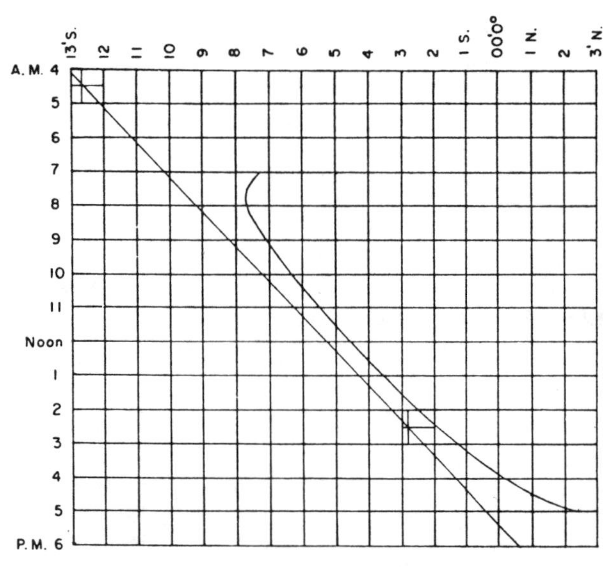

Date, March 20, 1970.
Station: Latitude 37°30′ N. Longitude 112°30′ W. (7^h30^m)
Declination at Greenwich app. noon
(4^h30^m a.m., app. time) = 0°12′39.3″ S.
Difference in declination for 10^h = 592.5″ = 9′52.5″ N.
Declination at 2^h30^m p.m., app. time = 0°02′46.8″ S.

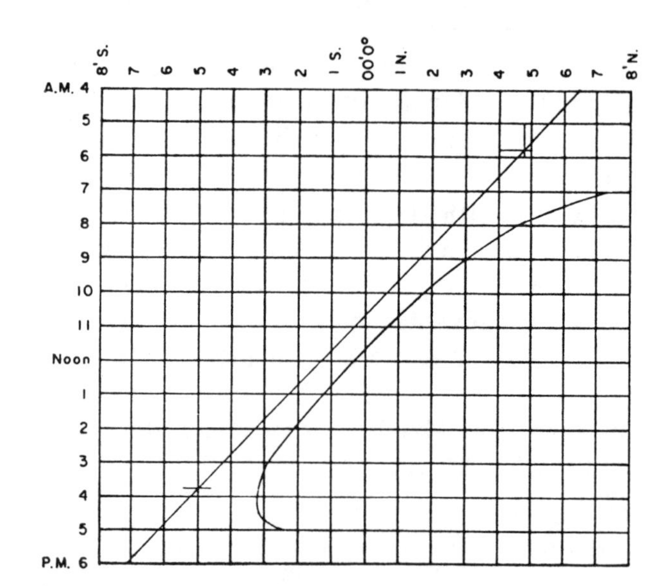

Date, September 23, 1971.
Station: Latitude 47°30′ N. Longitude 94°30′ W. (6^h18^m)
Declination at Greenwich app. noon
(5^h42^m a.m., app. time) = 0°04′44.0″ N.
Difference in declination for 10^h = 584.1″ = 9′44.1″ S.
Declination at 3^h42^m p.m., app. time = 0°05′00.1″ S.

Figure 4-42. Diagram of the sun's declinations.

straight line of true declination at the point corresponding to the time of observation.

4–34. Apparent Time from the Solar Unit

The solar unit of the solar transit has a graduated hour circle mounted normal to the polar axis. The readings are indicated at intervals of 10 minutes from 6 a.m. to 6 p.m., apparent time; the readings may be estimated to about ±1 or 2 minutes. This accuracy is sufficient for taking out the sun's declinations, which are calculated in terms of apparent time. It can be an approximate check upon the altitude and meridian observations on the sun for apparent time.

By applying the equation of time to the reading of the hour circle, the watch may be set to approximate local mean time, with the tolerance indicated above. This will be accurate enough for the finding positions for the stellar observations. It is also sufficiently accurate for the observations on Polaris for azimuth at elongation or latitude at culmination.

An exact method for time determination should be followed for the watch correction in local mean time necessary for hour angle observations on Polaris.

4–35. Checks of the Solar Unit

The following checks of the solar unit are all that are ordinarily required at the beginning of a survey:

(1) The reading of the latitudinal vernier with the solar telescope in true horizontal position.

(2) The reading of the latitude arc at noon with the solar telescope oriented on the meridian and the correct declination setting. This is the "instrumental latitude."

(3) The reading of the declination vernier when set in true zero declination, 15° north declination, and 15° south declination.

(4) The check for parallelism when the solar telescope is set and clamped in the latitude of the station.

(5) The checks for orientation when compared with a true meridian.

The solar unit should be checked on a true meridian at least weekly in normal use and whenever it has been subject to unusually hard bumps or jars.

Reference should be made to the maker's bulletin or a surveying textbook for care and adjustment of the transit.

4–36. Errors in Azimuth Due to Small Errors in Declination or Latitude

It may frequently happen with a solar transit, especially at the beginning of a survey or with an instrument insufficiently tested, that the first meridional trials are made with slight errors in the settings of the latitude and declination arcs, resulting in small errors in azimuth. This may be particularly true prior to a determination of the instrumental latitude. The discrepancies in azimuth due to such errors have been tabulated in the Standard Field Tables, which may be applied to results of single observations with considerable certainty. The corrections are not applicable to a series of observations as in ordinary line work owing to the

Latitude	Hours from noon			
	2^h0^m	2^h20^m	3^h0^m	
45° 00′	2.83	2.55	2.00	
46 21.5	....	2.62	....	Declination coefficient.
50 00	3.11	2.81	2.20	
45 00	2.45	2.10	1.41	
46 21.5	....	2.16	....	Latitude coefficient.
50 00	2.69	2.81	1.56	

The corrections are then applied as follows:

Indication of solar in test	= S. 0°	05′.0 W.
Correction for declination	= 0	01 .3 E. = (2.62 × 0.5)
Correction for latitude	= 0	03 .2 E. = (2.16 × 1.5)
Corrected indication of solar	= S. 0°	00′.5 W.

Figure 4-43. By reference to the Standard Field Tables.

changing values (for hours from noon) of the correction coefficients. The explanation with the table gives a key to the direction of the azimuth errors on account of small incorrect values in setting the latitude and declination arcs.

For example, at 9^h40^m a.m., app. t., at a station in latitude assumed to be 46° 20′ N., a test was made with a solar transit whereby the trial indication was found to be S. 0° 05′ W., or 0° 05′ west of the true meridian. Subsequent determinations of the true latitude of the station and of the correctness of the vernier of the declination arc showed that the actual latitude of the station was 46° 21′.5 N., and that the vernier of the declination arc had an index error which gave readings 0° 00′.5 S. of the calculated declination (i.e. reading 15° 19′.5 N. for a calculated declination of 15° 20′ N.) Thus in the test the latitude arc was set 1′.5 S. of the correct latitude of the station, and the declination arc was actually set 0′.5 N. of the value that would have been set had the index error been known.

The above corrections will often serve to explain the apparent errors of the solar transit, but are not intended for use in line work, and cannot be accepted in lieu of subsequent tests based on correct values.

CHAPTER 5

DISTANCE MEASURING EQUIPMENT

Section I. INTRODUCTION

5–1. General

This chapter covers only distance measuring equipment. For a description of the operation, use, and computation of distance measuring, reference is made to chapter 6.

Section II. TAPES

5–2. Description

Tapes are used in surveying to measure horizontal, vertical, and slope distances. This process of measuring distances is called "taping" or "chaining." The common survey tapes are made of a ribbon or band of steel, an alloy of steel, or a "cloth" of metal and cotton or linen. Tapes (fig. 5–1) are issued in various lengths and widths and graduated in a variety of ways.

a. Metallic Tapes. The metallic tape (ⓔ, fig. 5–1) is made by weaving fine metal wires and cotton or linen threads into a band or tape, ⅝ to ⅞ of an inch wide and 50 (sometimes 100) feet long. The tape is impregnated with a paint-like material for protection and graduations are applied to the surface. A leather-covered metal case with a built-in reel is furnished to wind up the tape when not in use. The materials in the tape are susceptible to temperature and humidity changes and to being stretched. For this reason, metallic tapes are used for low accuracy work. To measure a distance, the tape is pulled out of the reel the required amount. Just enough tension is applied to straighten the tape and keep it from sagging while the measurement is being made. Do *not* stretch the tape by applying a measurable tension to it. If the tape must be carried out of its reel for any distance, it should be lifted above the ground since the coating will wear away if it is dragged over a rough surface such as rocks. With the coating removed, the tape threads will absorb moisture and cause inaccuracies in the tape.

b. Steel Tapes. The steel tape (ⓐ and ⓕ, fig. 5–1), is the most common tape used in surveying. It is a ribbon of steel varying in width from ¼ to ½ inch and in thickness from 0.020 to 0.025 inch. The tape lengths that are commonly used are 30 meters, 50 meters, 100 feet, and 300 feet long. Some of these have foot graduations on one side and metric on the other. Loops are riveted to each end of the tape and rawhide thongs may be attached to each loop. Steel tapes are more accurately graduated than the metallics and are used to measure distances up to and including second-order accuracy under special conditions. Although steel tape is more rugged than the metallic for normal use, it must never be dragged over any surface. This will wear away sufficient material to cause the tape to yield when tension is applied. About 22 to 28 pounds (approximately 10–13 kilograms) of tension is used to offset the sag due to the added weight of a steel tape. A steel tape should never be jerked, pulled around corners, or run over by vehicles since it can kink or break. A steel tape must be carried by the tapemen, and if necessary, a third man should support it at the center point during movement. The tape should be wiped clean after use and oiled lightly. When not in use, the tape should be stored in its reel. If a reel is not readily available, it should be rolled into a circle (about 20″ diameter) and made into a *figure eight* (fig. 5–2) and carried this way.

c. Low Coefficient Tapes. Nickel-steel alloy tapes, known as Invar, Nilvar, or Lovar, have a coefficient of thermal expansion of about one-tenth that of steel and are used in high precision taping. These tapes must be handled exactly in the same manner as other precise surveying instruments. The alloy metal is relatively soft

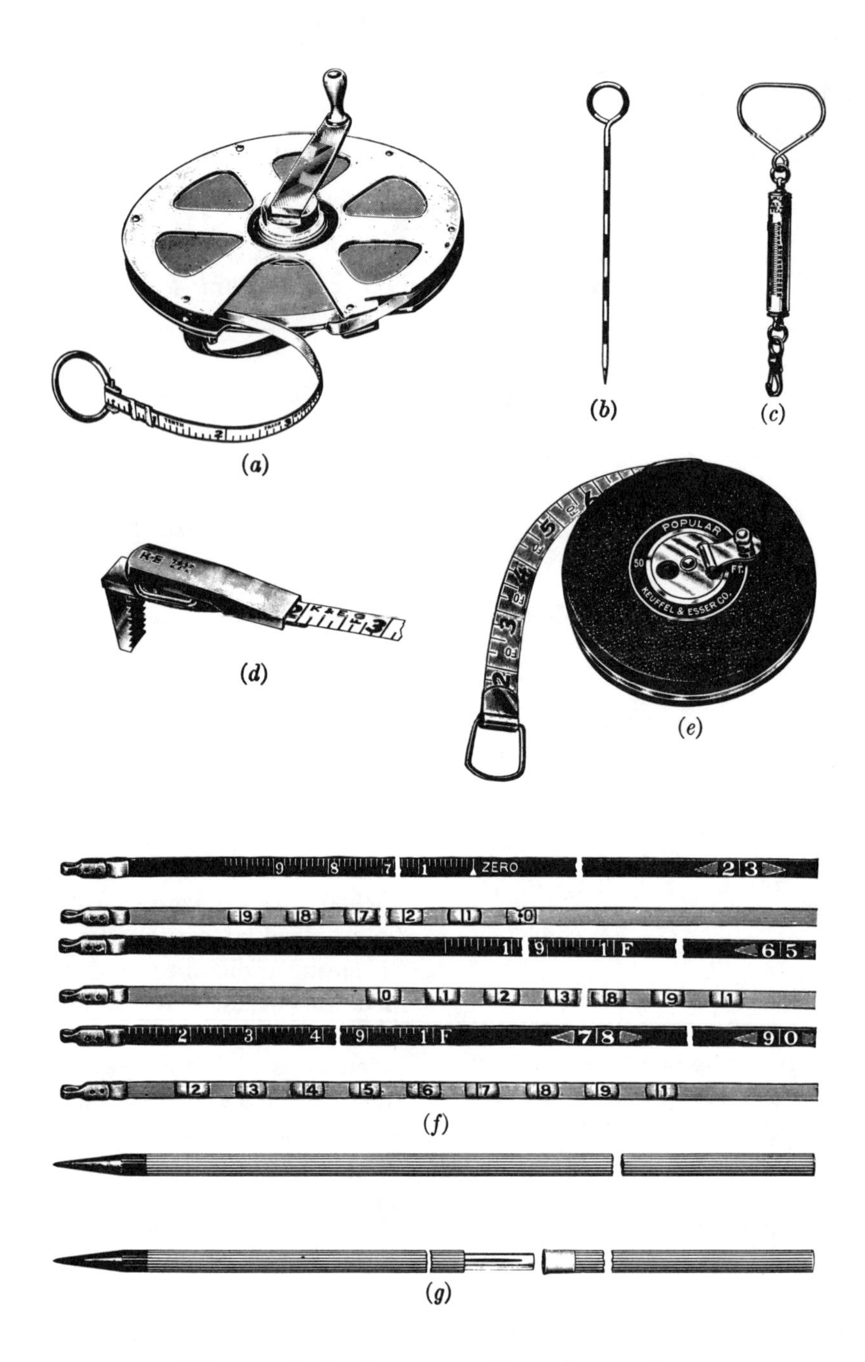

Figure 5-1. Tapes and accessories.

(a) Steel Tape (100 ft) on Reel; (b) Steel Marking Pin; (c) Tension Handle; (d) Clamping Handle; (e) Woven Tape; (f) End Arrangements for Steel Tapes; (g) Steel Range Poles. *Keuffel & Esser Co.*

and can be easily broken or kinked if mishandled. Low coefficient tapes should not ordinarily be used when a steel tape can get the desired accuracy under the same operating conditions. The tape is used to measure full tape lengths or evenly divided sections of a full tape length and a short steel tape is used to complete the measurement. When not in use, the tape should be stored in its reel. Except for special locations where the ground surface is hard and flat, such as roadways or railroad beds, the low coefficient tape is used over special supports or stools and is not permit-

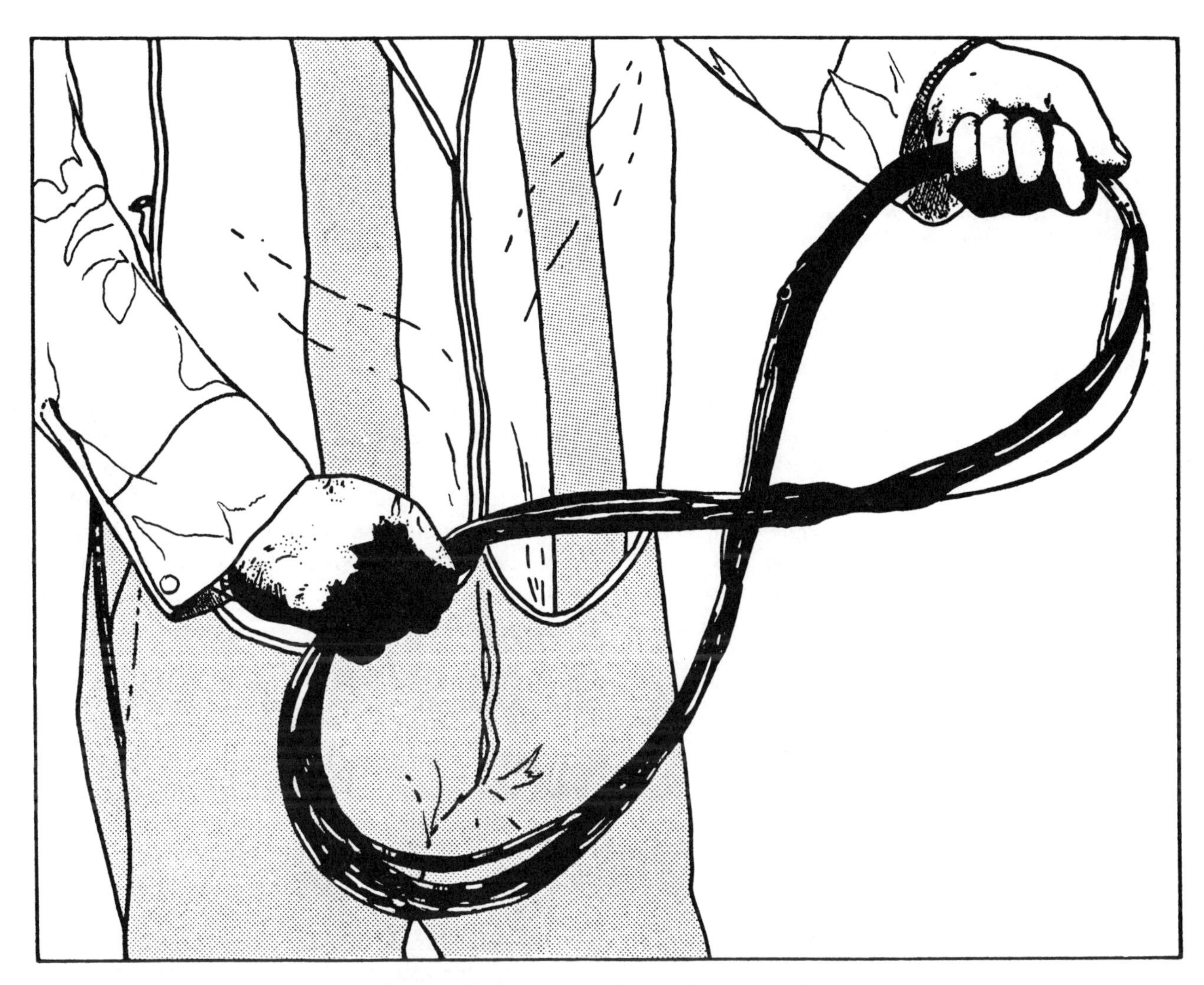

Figure 5-2. Tape in a figure eight.

ted to touch the ground.

d. Pocket Tapes. Small steel tapes (6-foot long) are issued with some survey sets. They are used for rough measurements of lumber, clearances, and distances, when preparing station descriptions and when erecting towers and targets. These tapes are the most convenient to use to measure the height of an instrument on a tripod. They may be used to measure any short distance when an accuracy of no more than a couple of millimeters is needed. Where accurate short distances must be measured, the steel surveying tapes (*a* above) must be used. These measurements include offsets, eccentricities in target and station setting, and target heights.

5–3. Graduations and Methods of Reading

Survey tapes are graduated in several ways. Care must be taken to prevent the loss of diligent taping by making a mistake or blunder in reading.

a. Location of the Zero Mark. The zero mark may be located in different places on the different tapes. On some, zero is the end of the tape ribbon and the attached loop is not included in the graduated portion. On others, the zero refers to the end of the loop and the graduations begin a couple of hundredths of a foot later. On still others, the zero is marked at some distance from the end of the ribbon and there is a blank piece of tape between the zero and the loop. In addition, some tapes have an extra foot or an extra decimeter graduated from the zero toward the loop. The tapeman must make sure of the location of the zero on his tape and be constantly aware of it during measurement.

b. Tape Graduations. Tapes are graduated in the following ways:

(1) 50-foot metallic in feet, tenths, and half-tenths (0.05) of a foot.

(2) 100-foot steel in feet, tenths, and hundredths throughout its length.

(3) 100-foot steel (as in (2) above) and the reverse side graduated for 30 meters in meters, decimeters, and centimeters. The first decimeter is also subdivided into millimeters.

(4) 100-foot steel in feet only, except the last foot (99 to 100) and 1 extra foot from zero toward the loop graduated in tenths and hundredths.

(5) 30-meter steel in meters, decimeters, and centimeters throughout its entire length with the first and last decimeter in millimeters also.

(6) 50-meter steel in meters, decimeters, and centimeters with an extra decimeter from zero toward the loop in millimeters.

(7) 50-meter low coefficient alloy in 5-meter intervals with additional graduations at the 12.5 and 37.5 meters. (A steel tape is used for any distance between the 5-meter marks.)

(8) 300-foot steel in feet with the last foot and 1 extra foot from zero towards the loop in tenths and hundredths of a foot.

(9) 300-foot steel in feet with the first and last foot in tenths.

(10) 6-foot steel in feet, inches, and sixteenths of an inch.

c. Method of Reading. Measuring or laying off distances requires using full tape lengths, partial tape lengths, or a combination of both.

(1) When measuring or laying off distances greater than the length of the tape, it is necessary to mark off one or more successive full lengths of the tape. These distances normally are marked off from the last graduation (100 ft, 30 m, 50 m, or 300 ft). When the tape has an extra unit outside of the zero graduation, special care must be exercised to avoid the unintentional reading of this extra unit. Otherwise, this will give an overall distance which will be one unit shorter per measured tape length than the actual distance.

(2) When measuring a distance less than the length of the tape, the method of reading the distance on the tape depends on the way the tape is graduated.

(*a*) With a tape graduated throughout its length either to hundredths of a foot or to millimeters, the measurement is made from the zero graduation. The distance is read directly on the tape.

(*b*) With a tape having an extra subdivided unit outside of the zero graduation, the measurement is made from one of the foot, meter, or decimeter graduations on the main part of the tape to one of the graduations on the extension. The distance is then the sum of the two distances indicated on the tape both ways from the zero graduation.

(*c*) With a tape having the first unit of the main graduations subdivided into smaller units, the measurement is made from one of the main graduations to one of the graduations on the subdivided unit. The distance is then the value of the main graduation minus the value of the subdivision graduation. For example, if the 68-foot graduation is read at one of two points, and the 0.60-foot graduation is at the other point, the distance between the two points is 68.00 minus 0.60, or 67.40 feet.

5-4. Standardization and Calibration

All tapes are graduated under controlled conditions of temperature and tension. When they are taken to the field, the conditions change. The tape, regardless of the material from which it is made, will be either too short or too long. For low accuracy surveys, the amount of error is too small to be considered. As accuracy requirements increase, variations due to temperature and sag must be computed and applied as corrections to the measured distance (ch. 6). In the higher orders of accuracy, the original graduation is checked for accuracy or calibrated at intervals, against a standard distance. This standard may be two points a tape length apart which have been set and marked using a more precise or a previously checked tape. The standard may be just the precise or checked tape (known as the "master" tape) which is kept in a safe location and not used for making field measurements, but only to check the field tapes. For the highest orders of accuracy, the tapes are sent to the National Bureau of Standards (or equivalent agency) for standardization under exact conditions of tension, temperature, and points of support. A tape standardization certificate is issued for each tape showing the amount of error under the different support conditions and the coefficient of expansion. The certificate (or a copy) is kept with each tape. For field operations, the tapes are combined in sets and one is selected as the master tape while the others are used as field tapes.

5–5. Tape Repair

Regardless of how well tapes are protected and cared for, at one time or another a tape must be repaired in the field to avoid unnecessary delay in the work. For this reason, every taping party, especially when operating some distance from the base, should be equipped with a tape repair kit similar to the one shown in figure 5–3. A tape repair kit usually contains a pair of small snips, the tape sections of proper size and graduations, a hand punch or bench punch with block, an assortment of small rivets, a pair of tweezers, a small hammer, and a small file. When repairing a broken tape, first square the broken ends. Next, arrange the broken tape with both parts alined over the spare tape section so that graduations are the correct distance apart. Select the spare tape part long enough to span distance between the squared ends and to overlap each end not less than 1 inch or 2 centimeters. Each overlap is then riveted so that the edges of all parts of the joints are snug. This is done by placing the rivets close to the ends, about $\frac{1}{16}$ to $\frac{3}{32}$ of an inch or 1.5 to 2 millimeters. Before being reused, a repaired tape must be compared with other tapes for accuracy of the repair.

5–6. Accessories

Some types of taping require special accessories, these include the taping stool or buck, taping arrows or pins, spring balance or tension handle, scissor clamp, and cutting tools, as well as range poles (para 4–30*a)* and the Abney hand level (para 3–2). The transit or theodolites (ch. 4) are used at times to measure vertical angles to convert taped slope distances to correct horizontal distances. Each accessory's use will be more fully discussed under the operation in which it is normally required (ch. 6).

a. Taping Stools. The taping stool or buck (fig. 5–4) is a metal three-legged stand with an adjustable sliding head, and a handwheel operated device for locking the plate in any desired position. A line is scribed on a plate attached to the sliding head. During taping operations, the head is moved until the scribed line is directly under the required graduation, and the handwheel used to lock the head. When the tape is moved forward, the graduation is held exactly over the line until the next stool is adjusted and locked. The setting is checked after which the rear stool is picked up and carried forward for further use. Taping stools furnish stable, elevated surfaces on which taped distances can be marked accurately.

b. Taping Arrows. Steel pins with a ring at one end and pointed on the other are called taping arrows (fig. 5–5). These pins are used for marking measured tape lengths on the ground and for checking the number of tape lengths measured since the last station. The ring and part of the pin are painted red, while the remainder of the pin is white. Taping arrows are issued in sets of 11, 10 of which are given to the front or head tapeman and one to the rear tapeman at the start of a measurement. When a tape length is measured, the front tapeman inserts the pin to mark the position of the tape end. After the following tape length is measured, the rear tapeman picks up the pin or arrow and carries it with him. As the tap-

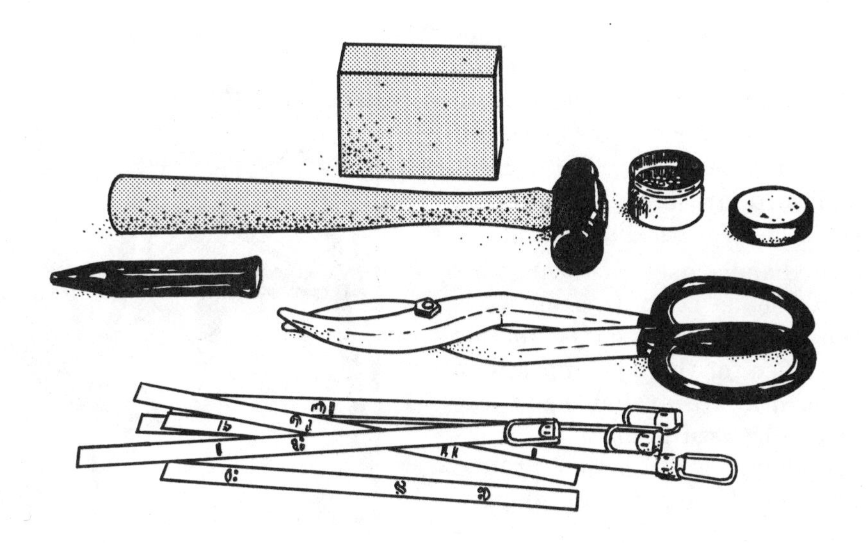

Figure 5–3. Tape repair kit.

Figure 5-4. Taping stool or buck.

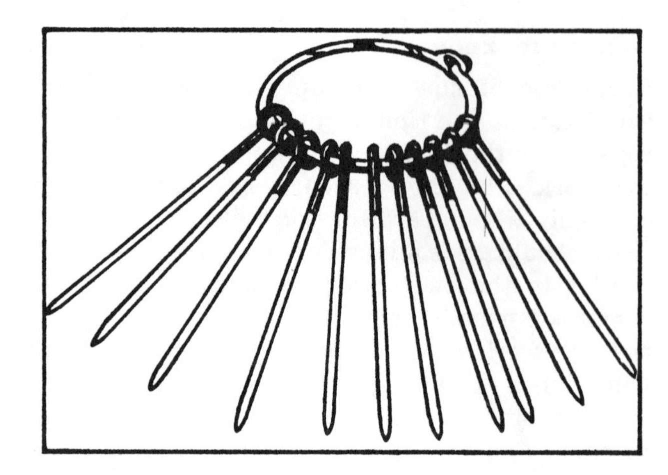

Figure 5-5. Taping arrows or pins.

ing progresses, the number of pins which the rear tapeman holds at any time denotes the number of full tape lengths measured since the last station. Properly used, taping arrows will prevent "dropped" tape lengths, which are a common mistake in distance taping.

c. Tension Scale and Spring Balance. Two types of tension devices are issued for taping. The tension handle, (①, fig. 5–7) with an integral scale graduated in pounds, is used for the low orders of accuracy. The spring balance (②, fig. 5–7), graduated in kilograms, issued for the higher orders is part of the stretcher issued for baseline taping.

(1) The tension handle uses a linear scale and is graduated in pounds from 1 to 30. It is clipped to the tape loop and tension is applied until the reading specified for the tape appears on the scale. If only two men are available for taping, a pair of poles can be used to make the work easier. The rawhide thongs are wrapped around the poles at a convenient height and gripped firmly (fig. 5–6). The bottom end of the pole is braced against the outside of the foot, while the upper end is tucked under the arm. Tension is applied by using the shoulder and leaning against the poles. The tapemen are in position to read the scale and hold the mark, or to move the taping buck plate to the desired graduation.

(2) The spring balance uses a circular dial graduated up to 5 kilograms per revolution and can be read up to three revolutions (15-kg) maximum. A drawbar extends from the scale. A hook at the end of the drawbar is attached to the tape loop during operation. A counterpoise is used to aline the scale so that the pull is completely axial, and the drawbar does not create friction as it is pulled out. A staff is mounted on a pivoted fitting between the counterpoise and the scale. During use, the staff is firmly anchored at its contact with the ground and the top pivoted around that point until the correct tension is read on the scale.

d. Scissor Clamp. When measuring less than a

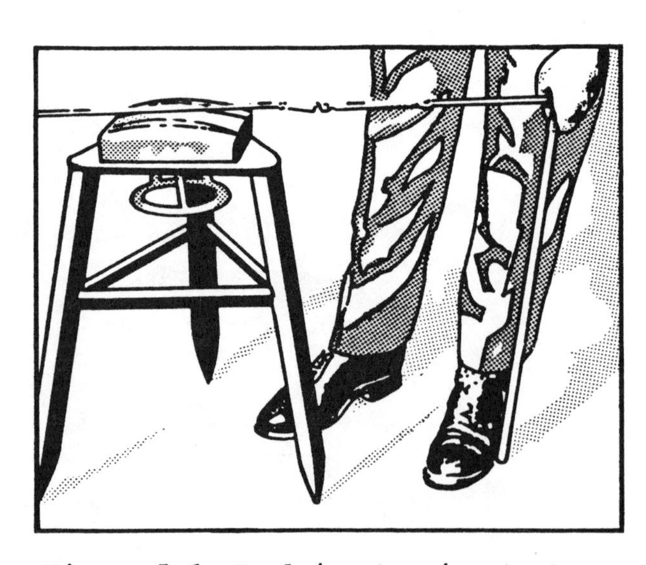

Figure 5-6. Applying tension to tape.

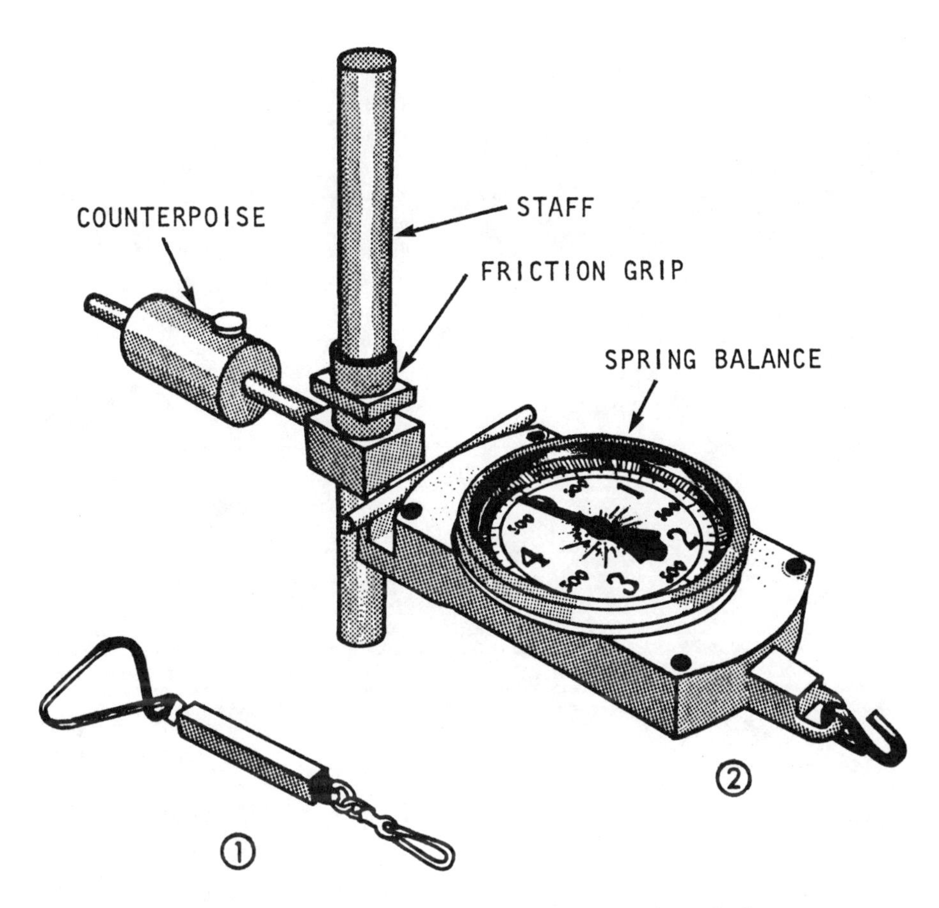

Figure 5-7. Tension scale and spring balance.

full tape length, tension is applied by using a steel tape clamp handle or scissor clamp. This is a mechanical device which grips the flat ribbon of steel without kinking it. For use with poles, the thong attaches to the scissor clamp.

e. Thermometers. Tape thermometers are used in accurate taping to determine the temperature of the tape at the time of the measurement. The temperature is used to determine a correction to the measured distance. Two types of thermometers are issued. One is graduated to 1.0° celsius (centigrade), while the second is a precision type which is graduated to 0.5° celsius, and is tested by the National Bureau of Standards for accuracy to within 0.3°. The thermometers are mounted in holders and can be attached near the ends of the tape for reading during the taping operation.

f. Other Equipment. Taping will frequently require the use of other equipment, some of which has been previously described. To get straight line clearance between taping points, cutting tools, such as axes, hatchets, or machetes, may be required to clear the vegetation. For alining the tape from station to station, range poles (para 4–30*a*) are used. The Abney hand level (para. 3–2) or engineer level (para 3–3) and leveling rod (para 3–13) are needed during slope taping to convert the measured distance to a horizontal distance. In horizontal or "break" taping, a plumb bob (para 2–9) is needed to transfer tape readings to ground position or to read the tape directly above a marked position on the ground. The use of the different equipment during taping operations is discussed in chapter 6.

Section III. MICROWAVE INSTRUMENTS

5–7. General

The microwave measuring systems use high frequency radio waves to make precise distance measurements. A master unit transmits a modulated radio wave in a narrow conical beam to a remote unit, where the wave is changed into a pulse, superimposed on another modulated wave and transmitted back to the master. At the master, the information is converted into a trace on a

Figure 5-8. Micro-chain(MC-8).

cathode ray tube (CRT) on the old systems, or a mechanical readout on the new systems, which is used to compute the distance. All makes and models can be operated from tripods, stands or towers.. An additional feature is the internal radio system which permits communications between units during setting up and measuring procedures. The newer models may be used as either a master or remote unit depending upon which operation is required. For a discussion on the use of this equipment refer to chapter 6.

5–8. Theory

The microwave measuring system measures distances by comparing the phase of a transmitted wave from a master unit to the phase of a returning wave from a remote unit. Both units contain independent oscillators which emit carrier waves that are modulated by pattern frequencies controlled by quartz crystals. The master unit's transmission is received at the remote unit, detected, amplified, and superimposed on the remote's transmission, and returned to the master. At the master, the phase of the returning wave is compared with the phase of the transmitted wave. This comparison shows that some number of full waves plus a portion of another wave have traveled to the remote and back. The master measures the portion of the wave and records it as a percentage of one full wave. The next frequency is then transmitted and measured and so on.

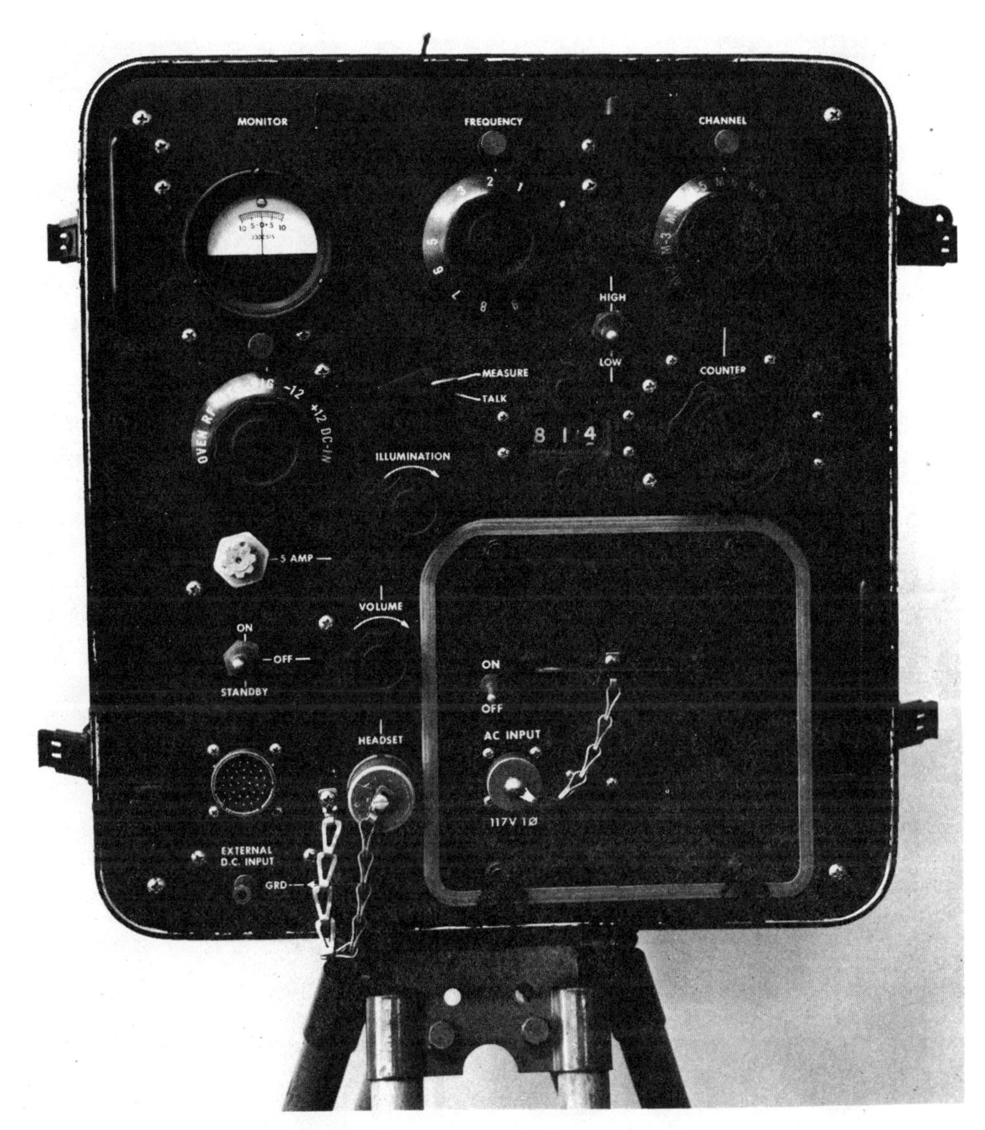

Figure 5-9. Surveying instrument, electronic(Model 99).

5–9. Types of Microwave Equipment

There are many makes, types, and models of microwave measuring equipment that are in use today. The three instruments listed below are compatible with each other and they may be used in combinations, but the units must be calibrated to each other before use.

a. Fairchild Micro-Chain (MC–8). The MC–8 is a portable, transistorized, electronic distance measuring instrument (fig. 5–8). The rated range is from 200 to 50,000 meters with an accuracy of 4 parts per million plus or minus 1.5 centimeters. The operating temperature range is —25° to 125°. The instrument package is weatherproof and designed to operate over a wide range of environments utilizing power from a self-contained 12-volt nickel-cadmium battery or from an external 12- or 28-volt dc or 115-volt ac, 50 to 60-hertz power source. Each unit may be operated as a measurer (master or responder (remote). These instruments must be calibrated before starting and upon completion of a project, or every 6 months, whichever is sooner. Upon replacement of any major component, such as the klystron, the instrument must also be calibrated.

b. Applied Devices, Surveying Instrument, Electronic (Model 99). The model 99 is a portable, transistorized, electronic, distance measuring instrument (fig. 5–9). This instrument is almost identical to the MC–8. The rated range is 200 to 50,000 meters with an accuracy of 4 parts per million plus or minus 1.5 centimeters. The operating temperature range is —25°F. to 105°F. with the internal battery and —65°F to 105°F with external power source. The

Figure 5-10. Tellurometer(MRA 301).

instrument package is weatherproof and designed to operate over a wide range of environments utilizing power from a self-contained 12-volt nickle-cadmium battery or from an external 22- to 28-volt dc or a 105- to 126-volt, 60-hertz, ac power source. Each unit may be operated as a measurer (master) or responder (remote). This instrument must be calibrated in the same way and at the same time interval as the MC–8.

c. Tellurometer (MRA 301). The MRA301 is a portable, transistorized, electronic distance measuring instrument (fig. 5–10). The rated range and accuracy are about the same as those of the MC–8 and Model 99. The MRA 301 is generally the same as the MC–8 and Model 99 in all features and operation except as noted. The major difference between this instrument and those in *a* and *b* is that it has a wider tuning range. The frequency control dial on this instrument is numbered from 1 to 12 with nine dots appearing between the numbers 3 to 10. The nine dots correspond to the numbered positions (1 through 9) on the MC–8 and Model 99 frequency control dial. When using this instrument in pairs or with a MRA3, the 12 number positions of the frequency control dial are used.

5–10. Care

The care and precautions noted in this paragraph pertain to all three instruments described in paragraph 5–9. As with all delicate scientific equipment, these instruments must be treated with the proper care and operator maintenance to insure their continuous operation.

a. Inspection. Upon receiving the instrument, it should be checked for damage and missing parts.

b. Cleaning. The instrument must be kept clean to insure proper operation of all parts. For cleaning metal parts, use only an approved cleaning solvent. For nonmetal parts, use a mild solution of warm water and soap. When drying the parts which have been cleaned, use clean, dry, lint free cloth or *low-pressure* compressed air. The use of wire brushes, sandpaper, or abrasive cloth should be avoided, because the use of these may render a part dimensionally unserviceable.

c. Short Distance Travel. When using these instruments in the field and travel between stations is short, the instrument should be backpacked. Before backpacking disconnect all cables and headset, secure front panel and antenna covers, and pack accessory in the proper packs.

d. Long Distance Travel. When traveling long distances between stations or shipping the instrument, it must be packed as in *c* above and then packed into the transport case. If traveling by vehicle, secure the transport to prevent movement and possible damage.

e. Special Cautions. The following special precautions must be taken when using these instruments.

(1) *The issued battery is an alkaline type. Do not add acid or use filling or testing equipment that has been contaminated with acid. Use only distilled or deionized water to refill cells.*

(2) *Do not add water to battery until it is charged.*

(3) *Always observe correct polarity in connecting leads to power source. Incorrectly connected leads will result in blown fuses and possible damage to the instrument.*

(4) *When an external 12- or 24-volt dc power source is used, insure that the proper power cable is used and connect to the external dc input.*

(5) *After setting up the instrument, set the ON-OFF-STANDBY switch to STANDBY. Allow instrument to remain in standby condition until monitor meter begins to fluctuate between "0" and an on-scale reading (approximately 2 minutes). Failure to do this usually results in a blown fuse.*

(6) *Extreme heat and solar radiation affects the accuracy of the instrument. When operating in extreme heat, shield the instrument from the direct rays of the sun.*

(7) *When servicing any part of the instrument, make sure the power is switched OFF. High voltages are present and electrocution is a possibility if electrical system is operating.*

(8) *When charging the battery a gas (hydrogen and oxygen) is produced and it is highly flammable. A proper ventilated area must be used when charging.*

5–11. Adjustments

There are many adjustments and alinements that must be done on the microwave instruments, but these cannot be done by the operator. These adjustments and alinements require the skill of a trained electronic repairman with access to a fully equipped electronic repair shop. The adjustment and alinement procedure is detailed in the respective maintenance manuals.

5–12. Calibration

Each combination of master (measurer) and remote (responder) units to be used in a survey must be calibrated before, periodically during and after any extended survey (usually 6 months). Upon replacement of any major component, such as the klyston or a crystal, the instrument should also be calibrated. Variations in crystals, frequencies, the aging of the instruments, and other considerations will cause the distances measured to vary from the true distance. Much of this discrepancy can be eliminated by determining instrument constants for each pair of instruments. Calibration can be accomplished by one of the following methods. The first method should be used prior to the use in a field survey of any combination of master and remote units or after any major overhaul or repair. The second method is less elaborate and is used for the periodic checks of the instruments. Before the calibration is started with either method, all crystals must be checked with a frequency counter and adjusted to the proper operating frequencies. A record of all frequency checks and calibration data will be kept with each instrument and in the project files. The calibration is for one combination of master and remote unit and cannot and will not apply to any other combination of master and remote unit. The calibration constant determined applies only when the instruments are used in the same operational modes that were used during the calibration. If the instruments are to be used in both operational modes during the survey, then the paired instruments must be calibrated in both modes. The accuracy requirements for the calibration depends primarily upon the required accuracy of the survey being accomplished. Normally, a pair of instruments will be calibrated only for the accuracy of the highest order that is to be performed.

a. First Method. Calibration is accomplished by comparing microwave distance measurements to accurately tape measured distances. Calibrating distances shall be measured with a tape and shall be a series of distances approximately 200, 250, 400, and 500 meters from a common point. Base line accuracy is required in these tape measurements. The microwave distance measurements made over these taped courses shall be made and recorded for each combination of master and remote unit to be used in the survey. Any difference in the slope of the tape measured distance and the microwave distance will require a reduction of the microwave distance to the tape measured distance. The calibration constants, as determined by the taped distance minus the microwave distance, shall be computed for each separate distance. The constant for the instrument shall be the mean of three or more individually computed constants. Any individually determined constants should not differ by more than 2 centimeters from the mean. If this difference is exceeded, additional measurement should be performed until the criterion is met.

b. Second Method. In this method all distances are measured by microwave and no tape measurements are involved. With each combination of master and remote unit, measure a series of distances approximately 400, 500, and 600 meters. Then divide each of these distances approximately in half (the shortest being no less than 200 meters) and measure each line in two segments. The calibration constant will be the full measured length of each line minus the sum of the two partial lengths. The accuracy of the above distances is not critical except that no microwave measured distance should be less than 200 meters. The order in which these measurements are performed is not critical so long as the record is properly coordinated. Any offset of the microwave unit, either horizontally or vertically, requires a reduction of all measurements to a common line. A mean constant is determined as in *a* above. If, when using this method, the constant thus determined differs by more than 2 centimeters from that determined by using the first method, additional measurements should be taken until the criterion is met or a complete calibration using the first method must be accomplished.

5–13. Accessories

The following accessories are issued with each instrument.

a. Transport Case. The transport case is a large wooden box that contains the instrument and all accessories, except the tripod. This case should be used to transport the instrument over long distances during a survey. The case must be used when shipping the instrument.

b. Accessory Case. The accessory case contains the following items:

(1) One 12-volt dc power cable (6 feet).

(2) One 24-volt dc power cable (25 feet).

(3) One 115-volt ac power cable (25 feet).

(4) Two 12-volt batteries (alkaline type).

(5) One headset with cable and connector.

(6) One screwdriver.

(7) One wrench.

(8) One socket-head wrench.

(9) One spare parts box, with extra fuse and bulbs.

c. Instrument Back Pack. A canvas back pack for carrying the instrument over short distances during a survey.

d. Accessory Case Back Pack. A canvas back pack for carrying the accessory case over short distances during a survey.

e. Tripod. The tripod is identical to the one issued with the 1-minute and 1-second theodolites. The microwave instrument and the theodolites can be interchanged on this tripod, thus permitting both distance and direction (angle) measurements to be made from the same setup. The tripod is issued in a separate packing case and it should always be shipped in this case.

5–14. Auxiliary Equipment

Atmospheric conditions affect all microwave distance measurements. With the microwave instruments the distance obtained is the distance the microwave would travel in a standard atmosphere plus the differences caused by the different atmospheric conditions. Atmosphere variations change the speed of these waves. To reduce the effect caused by the variation in atmospheric conditions along the length of the line to be measured, temperature (both wet and dry bulb) and atmospheric pressure must be determined. These determinations are made at the terminal points using a aneroid barometer or an altimeter (para 3–21) and psychrometer (para 3–22). Chapter 6 discusses the use of these instruments in distance measuring.

Section IV. LIGHTWAVE INSTRUMENTS

5–15. General

The lightwave measuring system uses electro-optical instruments to measure distances accurately. The system consists basically of two units; the measuring unit (transmitter/receiver), and the reflector unit. The distance is measued by precisely timing the amount of time it takes a modulated lightwave to travel from the transmitter to the reflector and return to the receiver. There are many types, makes, and models of electro-optical measuring instruments, but they all function generally on the same principles and theory. For the many uses of these instruments refer to chapter 6.

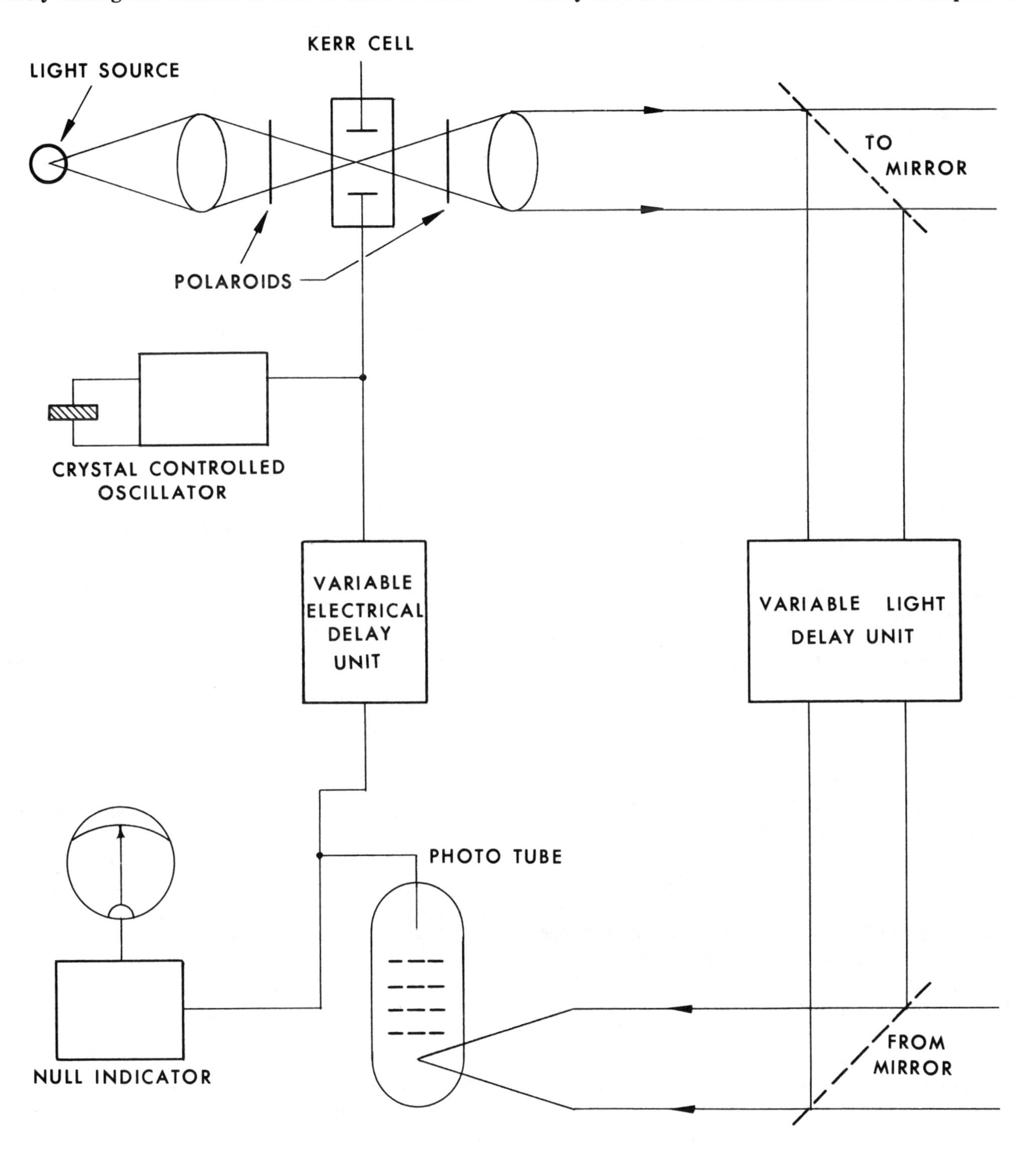

Figure 5-11. Electro-optical system.

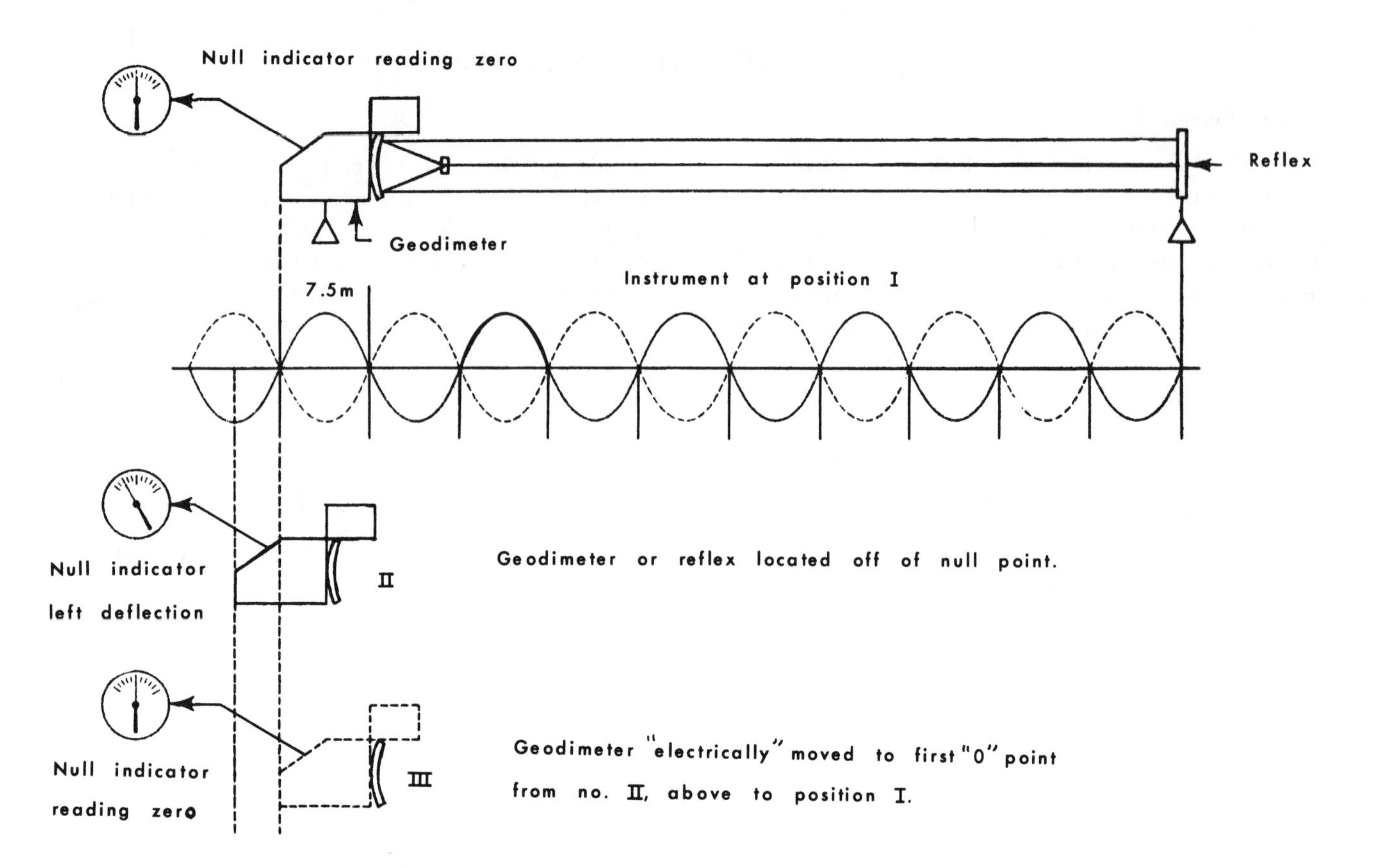

Figure 5-12. Schematic diagram of measurement.

5-16. Theory

These electro-optical instruments use the fundamental constant, the velocity of light, to precisely measure distances. A beam of light is modulated and transmitted on a controlled frequency in a narrow beam to a reflector and is returned to the receiver (fig. 5-11). By means of a phase detector, the returning light is instantaneously compared in phase with that of the light leaving the transmitter.

a. The units are setup to measure a line in schematic form in figure 5-12. Assume that the length of the line is an exact multiple of 7.5 meters, which is a quarter wave length of the accurately controlled modulated frequency. The null indicator reads zero in this case. If the measuring unit is moved away from the reflector a short distance the null indicator is now to the left or right of zero depending upon the direction moved. Instead of physically moving the measurer, an electrical delay on the instrument can be changed in order to zero the null indicator. In effect, the measurer has been moved electrically. A line to be measured will always contain an integral number of quarter wave lengths plus a fraction part of a quarter wave length.

b. The model 1 and 2 geodimeters are provided with a built in variable light path which is calibrated in millimeters. The light conductor provides means for calibrating the electrical circuits against a variable light path of known distance. Briefly, the light sent through the transmitter objective is deflected up to the light conductor proper. Here the parallel beam of light is sent back and forth between the rows of right angle prisms. Internally plated metal tubes take care of scattered light. These direct the beam towards the continuously variable port, consisting of two prisms mounted on a slide maneuvered by means of a chain and turning knob. The distance is read on a scale divided in centimeters. The continuously variable delay covers the interval between two coarse steps with some overlapping. It is thus possible to adjust the light path to any distance between about 1.15 and 10 meters reflected light path (actual light path is twice these figures).

5-17. Types of Lightwave Equipment

There are many types and models of lightwave measuring instruments that are commercially available today. The specifications listed with each instrument are those of the manufacturer's extracted from their brochures.

a. Geodimeter, Model 4D.

(1) *Range.* The range depends upon visibility, light conditions and the type of lamp in the instrument. With the standard lamp the daytime distance is 1.5 km (1 mile) and nighttime is 15 km (10 miles). With the mercury arc lamp daytime distance is 4.5 km (3 miles) and nighttime 25 km (15 mile).

(2) *Accuracy.* Standard deviation less the 10 mm + $2 \cdot 10^{-6}$ times the distance.

(3) *Power requirement.* Approximately 60 watts with standard lamp and 300 watts with mercury lamp, 110 (220) volts, ac, 50/60 hertz.

(4) *Weight.* Total unit less than 50 pounds.

(5) *Climatic limits.* Working temperature of —40°C. to + 30°C. (up to + 50°C. with slightly decreased crystal stability).

b. Geodimeter, Model 6.

(1) *Range.* Minimum 48 feet, maximum daytime 1 to 2 miles and nighttime 10 miles.

(2) *Accuracy.* Average error of 1.0 CM ± 2 millionths of the distance.

(3) *Power requirement.* Approximately 30 watts, standard 12 volt battery.

(4) *Weighy.* Approximately 68 pounds with transport case and tripod.

(5) *Climatic limits.* Not listed.

c. Geodimeter, Model 8, Laser.

(1) *Range.* Minimum 15 meters and maximum 60 kilometers.

(2) *Accuracy.* 6mm + 1 ppm.

(3) *Power requirement.* 75 watts, 12 volt, dc, laser power output of 5mW.

(4) *Weight.* 51 pounds.

(5) *Climatic limits.* Not listed

d. Laser Ranger (LSE, Inc).

(1) *Range.* 0 to 6 kilometers.

(2) *Accuracy.* ± 5mm + 1 ppm

(3) *Power requirement.* 12 volt battery, 3 mW laser power output.

(4) *Weight.* 25 pounds.

(5) *Climatic limits.* —18°C. to + 54°C. (0° to 130°F.).

e. Range Master (LSE, Inc).

(1) *Range.* 10 yards to greater than 40 miles.

(2) *Accuracy.* 10 ppm (10^{-6}) or 1 cm, whichever is greater.

(3) *Power requirement.* 100 watts, 100 volt, ac, 50/60 hertz. (Being modified to operate on a vehicle battery using 90 watts, 12-volt dc.)

(4) *Weight.* 50 pounds

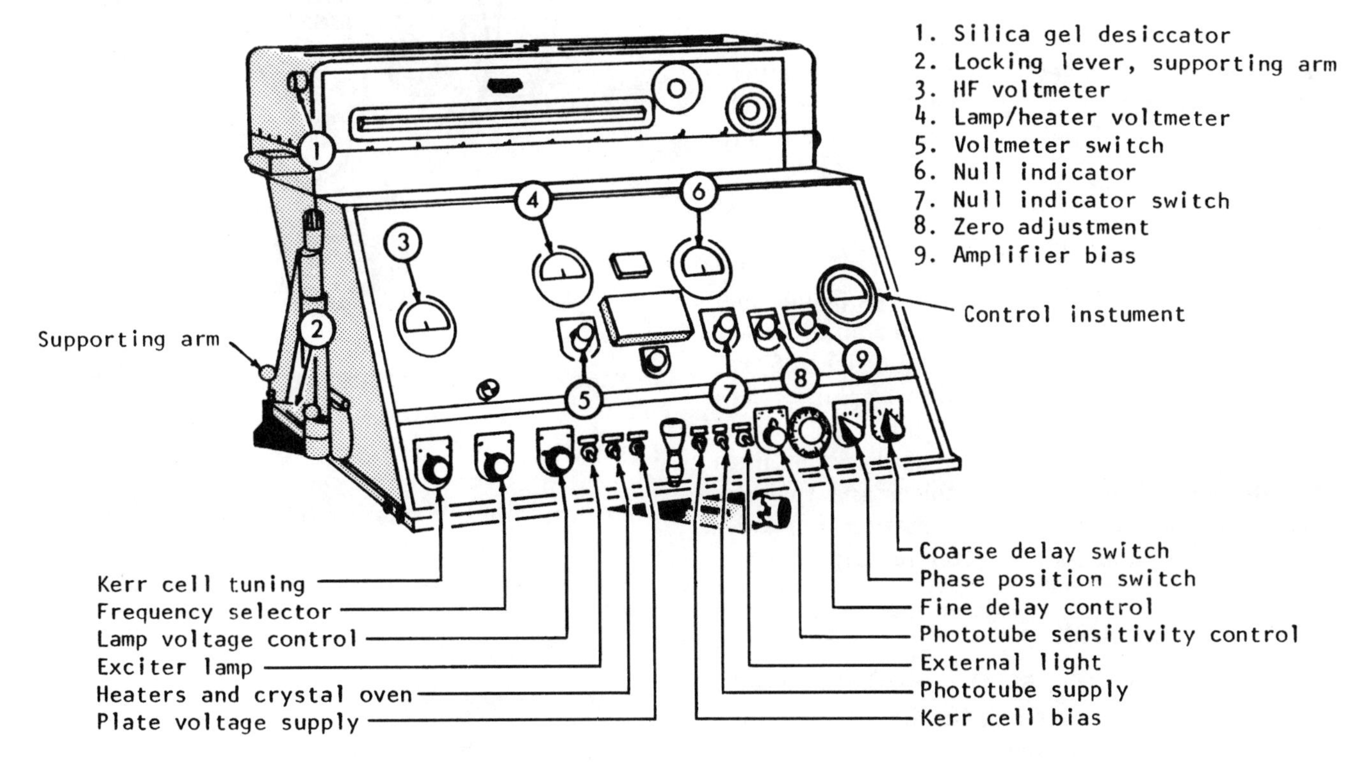

Figure 5-13. Geodimeter(Model 2A)

(5) *Climatic limits.* Not listed.

f. Geodolite.

(1) *Range.* 2 meters to 60 kilometers

(2) *Accuracy.* 1 mm ± 1 part in 10^{-6}

(3) *Power requirement.* 400 watts

(4) *Weight.* 150 pounds

(5) *Climatic limits.* —20° F. to 120° F.

5–18. Geodimeter Model 2A

The model 2A (fig. 5–13) is considered to be one of the most accurate electronic distance measuring instruments used today. The instrument's accuracy is primarily limited by the accuracy of the modulated frequency, as this directly determines the course length. Aging effects of the crystals make this accuracy difficult to predict, as these effects vary among individual crystals. Experience has shown, however, that after a few months running time, the frequency error will be smaller than $1 \cdot 10^{-6}$. All lightwave measuring instruments consist functionally of two systems; the electronic and optical.

a. Electronic. The electronic system contains the crystal oscillator with power amplifier for the light modulator, the photo multiplier, and the power supply. The unit weighs about 108 pounds.

b. Optical. The optical system consists of a projector for transmitting the modulated light to the reflector, a receiver for receiving the returning light, and a built-in light conductor for internal calibration. This unit weighs about 113 pounds.

5–19. Use of Geodimeter Model 2A

The rated range of this instrument is 50,000 meters. This instrument can be used for all orders of base lines in triangulation and all orders of traverse distance measurements. However, it is principally used for first order base lines. Because of the size and weight, it should be used only on lines that do not require a survey tower and that are easily accessible by land transportation. For a detailed discussion of its uses refer to chapter 6.

5–20. Reflector Unit

The reflector unit (fig. 5–14) generally used with the electro-optical system is the Reflex, Geodimeter, Retrodirective Prisms, 360° rotation, tripod mounted, Type A. However, the reflector unit may contain any number of prism mirrors from 1 to 7. If more reflection power is needed, the reflector units can be stacked as shown in figure 5–14.

Figure 5-14. Reflector units, stacked.

5–21. Accessories

When using the model 2A, a generator and atmospheric determining equipment is needed.

a. Power Supply. The primary power requirement is about 140 watts. A built-in autotransformer transforms the main voltage from 230, 220, 130, 120 or 110 volts 50/60 hertz to 220 volts, the voltage for which the power supply is designed. In the power supply unit the main voltage is converted to the specific AC and DC voltages needed for operation of the various electronic circuits.

b. Atmospheric Equipment. Atmospheric conditions affect all electronic distance measurements. The distance obtained in lightwave measurements is based on the velocity of light in a vacuum. atmosphere changes the speed of these waves. Therefore, to eliminate or reduce the effects caused by variations in the atmosphere, atmospheric measuring equipment is needed. This equipment consists of a psychrometer, altimeter or barometer, and a thermometer. Refer to section VI, chapter 3 for a discussion of this equipment.

5–22. Care

The lightwave instrument combines electronic and optical subassemblies which are precisely adjusted. Handling must be careful enough at all times not to damage the components, nor disturb the adjustments which will affect the operation.

a. The first inspection must be for physical damage to any piece of equipment. A thorough examination should be made of the prisms, lenses, and mirrors to see that their surfaces are clean and undamaged. These surfaces are highly polished and covered with an invisible protective coating. Special techniques are used in cleaning or wiping so as not to scratch nor remove the coating. Avoid touching the surfaces since the oil on the skin will leave smudges and require cleaning.

b. Some of the internal circuits in the instrument are susceptible to high humidity. One or more silica gel desiccators are inserted in the unit to maintain low humidity. These should be checked and replaced, as needed, depending upon the prevailing humidity conditions.

c. The electronic system is designed to operate in the optimum range well below the maximum for any electronic component and should not require much servicing. If the instrument does not function properly, the units must be checked by using proper electronic test equipment.

d. One precaution must be observed. The internal circuits carry high voltage. To avoid bad shocks, all switches must be OFF, or better still the power plug disconnected before starting work inside the measuring unit.

e. To move the instrument to another site, separate it into its components, and pack each one in its proper case.

5–23. Calibration

The model 2A should be calibrated at the start and conclusion of a project or every 6 months, whichever is sooner. The replacement of any major component within the geodimeter will require a calibration. The purpose of this calibration is to determine the light conductor constants, i.e., (1) the distance from the reference point to the zero point of the light conductor where fine and coarse scales are set at zero, and (2) the length of one coarse interval. This is done by measuring at least two but preferably several known distances. The focusing adjustments of the model 2A do not permit operation on shorter distances than about 25 meters and a good calibration should be made over a distance interval of at least 10 to 15 meters, a minimum distance of 35 to 40 meters is required.

CHAPTER 6

DISTANCE MEASURING

Section I. TAPING

6-1. Introduction

There are two basic methods for measuring distance with a tape. These methods are slope taping, and horizontal taping. In the slope taping method, the tape is held as required by the slope of the ground, the slope of the tape is measured, and the horizontal distance is computed. In the horizontal taping method, the tape is held horizontally, and the position of the required graduation is projected to the ground with a plumb bob. Under certain conditions, each of the methods of taping has its advantages. The slope taping method is the most precise and is always used for taping base lines and second and higher order traverse distances. Third order tape traverse measurements are usually done using slope taping, horizontal taping can be used, but it is not recommended. The horizontal taping method is generally used for lower order tape traverses in mapping projects and construction.

6-2. Party Organization

A taping party can consist of a minimum of two people for horizontal taping to as many as nine people for precise slope taping. When working in areas of heavy underbrush or jungle type environment an additional clearing party is sometimes required to clear the path of the survey.

a. Horizontal Taping. The minimum taping party consists of two people, one at each end of the tape. However, this is not recommended except for very low order surveys. The standard taping party should consist of at least 4 people; a recorder, stretcherperson, rear and front tapepeople.

(1) *Recorder.* The recorder keeps a complete record of all measurements made by the taping party, makes all necessary sketches, writes descriptions of stations and reference points, records all other pertinent supplementary data, and usually completes all necessary field computations. The recorder is usually the chief of the taping party.

(4) *Front tapeperson.* The front or head tapeperson is responsible for making the measurement on the tape during the taping.

(3) *Rear tapeperson.* The rear tapeperson holds the tape on or directly over the rear point and also maintains the alinement of the tape between stations.

(2) *Stretcherperson.* The duty of the stretcherperson is to apply and maintain the correct tension and marking the new point on the ground.

b. Slope Taping. A slope taping party will usually consist of a minimum of 4 to as many as nine people depending upon the accuracy requirements. As in all surveys, some of the duties are absorbed by other members of the party if sufficient personnel are not available. The following is a generalization of the duties assigned to the party members.

(1) *Chief of party.* The chief of the taping party; directs all measurements, the establishing of stations, and other activities of the party. When traversing, the chief of the traverse party will usually be directly in charge of the taping party.

(2) *Recorder.* The duties of the recorder of the slope taping party are the same as those in *a*(1) above.

(3) *Front and rear tapeperson.* The tapepeople's duties are that of making the actual measurement.

(4) *Front and rear stretcherperson.* The stretcherpeople's duty is to apply and maintain the proper tension on the tape while the tapepeople are making the measurements.

(5) *Level party.* For precise tape measurements the slope of the tape is usually determined by direct (differential) leveling. This type of leveling requires an instrumentperson, rodperson and a

recorder. For third and lower order taping, the slope of the tape is usually determined by using an Abney topographic hand level. When using the hand level, the tapemen or the stretchermen are responsible for the slope measurement.

(6) *Instrumentman.* One of the requirements of taping is that the taping be done on a straight line between stations. The order of the work being done defines the permissible limits of misalinement between stations. For precise taping the allowable misalinement is very small and a theodolite is used to keep the alinement of the tape within the limits. For third and lower order taping, the alinement of the tape can be done by "eyeing in" the tape by the rear tapeperson.

6–3. Slope Taping Procedure

In the slope taping method, the tape may lie on smooth ground, on a paved road or other fairly smooth and uniformly sloping surface, or its ends may be supported by taping stools or stakes. The following procedures are generally used when taping third or lower order traverse distances. Normally, the equipment consists of a range pole, one set of taping pins, one steel tape, one tape thermometer and one Abney hand level.

a. The range pole is set on line slightly behind the point toward which the taping will proceed.

b. The rear tapeman, with one taping pin, stations himself at the point from which the taping will start. The head tapeman, with the zero end of the tape and 10 taping pins, moves toward the distance point. When he has gone nearly the full tape length, he is halted by a signal from the rear tapeman. The rear tapeman holds the last graduation at the initial point and directs the head tapeman into alinement. The head tapeman then pulls the tape taut, bringing it onto the line. The tapemen exert and maintain the specific tension. When the exact graduation is on the initial point, the rear tapeman calls "mark" or "stick" and the head tapeman marks the distance by sticking the arrow into the ground at the zero graduation of the tape. The head tapeman then calls "marked" or "stuck", both tapemen release the tension, and repeat the measurement as a check. The distance is now recorded along with the tape temperature. An Abney level reading is now taken along the line coincident with or parallel to the slope of the tape and recorded.

c. Before moving forward, the rear tapeman pulls out the taping pin at his position (not used for initial tape length) and carries it with him. The front tapeman leaves his where it is set. Thus, one taping pin always remains in the ground, and the number of pins held by the rear tapeman indicates the number of full tape lengths from the initial point to the remaining arrow.

d. Both tapemen move forward and make another complete measurement starting from the taping pin that marks the end of the previous measurement.

e. When the taped distance is over 10 tape lengths, the head tapeman signals the rear tapeman to come forward at the end of the tenth length, and to bring the 10 taping pins he is holding. Both tapemen check the number of pins to see that none have been lost. The head tapeman checks the recording book to see that 10 taped distances, temperatures readings, and slopes have been recorded. The rear tapeman gives 10 pins to the head tapeman and the measurement is then continued.

f. At the end of the line, the head tapeman holds the zero graduation on the terminal point. The rear tapeman pulls the tape taut and holds the appropriate even graduation at the taping pin. The fractional unit is now read by the head tapeperson.

6–4. Horizontal Tape Procedure

In horizontal taping, the tape is supported only at its ends and held in a horizontal position. Plumb bobs are used to project the end graduations of the tape to the ground or the ground to the tape. Care must be exercised in the use of the plumb bobs, in exerting a steady pull on the tape ends, and in determining when the tape is nearly horizontal.

a. Plumbing. Plumbing is accomplished when the tape is alined horizontally, and is under the proper tension.

(1) The rear tapeman holds his plumb bob cord at the proper graduation of the tape, and the point of the plumb bob about 1/8 inch above the marker from which the measurement is being made. When their plumb bob is directly over the marker, they call "mark."

(2) The head tapeman holds his plumb bob cord at the correct graduation of the tape, with the point of the plumb bob about 1 inch above the ground. He allows his plumb bob to come to rest;

sees that the tape is approximately horizontal; checks its alinement and tension; and, when the rear tapeman calls "mark," allows his plumb bob to fall and stick in the ground. This spot is then marked with a taping arrow or pin.

(3) Any swinging in the plumb bob is eliminated by gently lowering the tape and plumb bob until the tip touches the ground, and then slowly raising the two back to the horizontal position. This procedure may have to be repeated once or twice to stop the swinging entirely.

(4) After the point has been marked, the measurement must be repeated for a check. The tapeman must be certain that the point on the ground is a good transfer from the elevated tape. This check measurement is very important in horizontal taping.

b. Leveling the Tape. The head tapeman can easily keep his tape horizontal by using the level bubble on the hand level. On the Abney hand level, he must first set the arc for a level reading. He holds the level against the tape while the downhill man raises his end of the tape until the bubble is centered. This will mark the height at which to hold the tape for both tapemen. Once the height is established, the level is removed and the tape returned to this same height for measuring.

c. Measuring Full Tape Lenghts. Taping consists of measuring some number of full tape lengths, and a partial tape length at the end of the line.

(1) To measure the first full tape length, the front tapeman gives 1 taping pin to the rear tapeman, keeping 10 pins in his possession. The pin given to the rear tapeman represents the first full tape length. The front tapeman moves toward the forward station with the zero end of the tape. As the end of the tape reaches the rear station, the front tapeman stops, either on the count of paces or on the command given by the rear tapeman. The rear tapeman sights along the tape toward the forward station and signals the direction that the front tapeman should move to aline the tape, first with the forward station and then with an estimated horizontal plane. Each tapeman places the leather thong on his wrist and the plumb bob cord on the proper graduation on the end of the tape. At the command "pull," each tapeman exerts a pull on the tape. After the front tapeman has properly alined the tape, the rear tapeman places his plumb bob exactly over the rear station and calls "mark." At this command the front tapeman drops his plumb bob and then marks the point of impact with a taping pin. When the pin has been placed in the ground, the front tapeman reports "marked." The reading is checked and recorded in the notebook, and the rear tapeman moves forward to measure the next tape length. When a team is taping on sloping ground void of brush and tall grass, the plumb bob need *not* be used at the uphill end of the tape. The end of the tape may be held on the ground immediately adjacent to the taping pin.

(2) To measure succeeding full tape lengths, the tapemen use the procedure discussed above except that the rear tapeman should place his plumb bob exactly over the point at which the taping pin enters the ground. When the measurement is completed, the rear tapeman pulls the taping pin or arrow from the ground before moving forward to the next pin position. If a taping pin is lost during the measurement of the distance, the tapemen must make the entire measurement again, rather than complete the measurement from a recovered pin hole. The number of arrows in the rear tapeman's possession indicates the number of full tape lengths which have been measured.

d. Breaking Tape. To measure tape lengths when the tape cannot be alined with a horizontal plane within ½ meter (1½ ft) because of the slope of the ground, the tapemen use a special procedure known as *breaking tape.*

(1) The front tapeman pulls the tape forward a full tape length and drops it approximately on line. He then comes back along the tape until he reaches a point at which a partial tape length, when held level, is below the armpits of the downslope tapeman. At this point, the front tapeman selects any convenient full meter (or foot) graduation. The tapemen then measure the partial tape length, applying the proper tension to the tape.

(2) After he has placed the taping pin, the front tapeman waits until the rear tapeman has come forward. The front tapeman tells the rear tapeman which full graduation was used, e.g., "Holding 25," which is repeated by the rear tapeman. He receives a pin from the rear tapeman and moves forward, repeating this procedure until the zero mark is reached.

(3) When holding a point on the tape other than the zero graduation, the front tapeman must receive a pin from the rear tapeman before moving forward. At the zero point, the rear tapeman picks up the pin as in any full tape length measurement to maintain his count.

e. Measuring Distances in Excess of 10 Tape Lengths. To measure a distance longer than 10 full tape lengths, the tapemen use the procedures discussed above and instructions outlined in paragraph 6–3*e*.

f. Measuring Partial Tape Lengths. To measure the partial tape length between the forward station and the taping pin representing the last full tape length, the tapemen use the following procedure:

(1) The front tapeman moves to the forward station and places the plumb bob cord on the zero graduation of the tape. The rear tapeman moves forward along the tape to the taping pin.

(2) If slack is needed, the rear tapeman allows the tape to move forward. When the front tapeman is ready, he calls "pull" and the rear tapeman exerts a pull on the tape, using the clamping handle to hold the tape. As he applies tension to the tape, the rear tapeman slides his plumb bob cord along the tape until he reaches a full decimeter or foot division. He moves the tape until the graduation is exactly over the pin and calls "mark." The front tapeman slides his plumb bob cord along the tape until his plumb bob is over the station. He reads the value on the tape and asks the rear tapemen for the even value he is holding.

(3) The front tapeman records the reading of the final increment counting the number of pins in the rear tapeman's possession. Both tapemen should now check the taping record.

g. Determining the Taped Distance. The notebook contains a record of the number of exchanges of pins (10 tape lengths) ; plus a number of tape lengths, determined by the number of pins held by the rear tapeman; plus the final partial increment measured last. These figures should be checked by both tapemen and initialed as correct.

h. Laying Off a Given Distance. Frequently, a taping party is required to "lay off" a given distance and establish a new point on the ground. This is measuring using a known distance on the tape and transferring it to the ground.If the distance is greater than a tape length, the method under *c* above, is used until the required number of full tape lengths is measured. The remaining partial tape length is measured as described under *f* above, except that the plumb bob is moved along the tape until the required graduation is reached. The distance is then transferred to the ground. Other considerations in laying off distances are discussed under practical problems, appendix A.

6-5. Use of Taping

Present day practices are incorporating electronic distance-measuring devices in traverse and base line distance measuring. Frequently, however, taping is the only practical way of making these measurements. Construction taping usually covers comparatively short distances, and electronic measuring is not practical.

a. Triangulation Taping. A base line is the measured side of a triangle in a triangulation net. It is used to start the survey, to check its accuracy along the way, or to complete the survey. Triangulation is discussed in more detail in chapter 10.

(1) Taping of base lines must be precise enough to maintain the accuracy of the survey. The base line measurement results in a horizontal distance between two control stations in the triangulation net selected by reconnaissance. The taping party must measure the distance between these stations.

(2) Triangulation associated with lower accuracy surveying normally does not extend beyond 15 to 20 kilometers, and the base lines are less than 3 kilometers in length. Where possible, the base line is selected to obtain a straight line measurement between stations, and preferably, an even slope or a horizontal line. If horizontal taping is not practical, slope taping is used. The slope is determined by a level line run over the tape ends.

(3) The tape is either standardized or calibrated against a standardized tape. The measurements are made carefully, using full tape support or center and quarter-point supports. Each measurement is marked by the scribed line on the taping stool or by using a fine mark on a stake. After each measurement is marked, it is repeated as a check.

(4) After taping in one direction, the base line is taped in the opposite direction independently, resetting the stools or the stakes.

(5) When both measurements are completed, they are converted to the horizontal distances, and both values are meaned. The final required distance is the result.

b. Traverse Taping. Traverse is the measurement of a series of distances and directions from one known control station to another whose position may be known or may have to be established by the measurement. The distance between these stations may be one straight measurable line, or as normally is the case, a series of straight lines which follow the ground features. The traverse party chief will select the angle-measuring points in locations to keep instrument setups to a minimum. The survey accuracy will determine whether the distances require taping or can be measured trigonometrically. If taping is required or more practical, the party chief will also select the points which present the least number of obstacles for the tapepeople. Traverse is more fully described in chapter 10.

(1) Traverse may require horizontal or slope taping. The party chief will decide which method will be used and will also assign the extra taping party members when necessary.

(2) Traverse accuracy will determine whether single taping or double-run taping is needed. Single-run taping is done in one direction only. Double-run or double taping requires two independent tape measurements between stations. After one run is completed, the taping party remeasures the line in the opposite direction.

(3) Conversion of slope distances to horizontal distances is a computing function and will include all tape corrections where necessary. However, if horizontal distances are required during the survey, conversion tables are in appendix D. The tables show the conversion based on the slope angle, but can also be used when the difference in elevation between tape ends is known.

c. Construction Taping. Construction surveys for road, railroad, and pipeline alinement require taping of comparatively long distances as described under traverse taping (*b* above). In most other construction surveys, corrections for sag, temperature, and tension are too small to affect the required accuracy and are not considered. Horizontal taping is the usual method. The tape is hand held and tension is applied to remove any excessive sag, but tension scales are not used. The tape is read and recorded to the nearest graduation as required by the survey accuracy.

6-6. Taping Corrections

Every measured distance has to be corrected to reduce it to the true length. These computations are based on known errors, such as errors due to tape length (standardization error), slope, temperature, supports, and tension. Correction computations for support and tape length can be made from data supplied by the standardization agency, or by comparing the tape against a calibrated or standardized tape. The comparison is made by applying the standard tension and by using the different methods of support that are required bv the survey accuracy. For example, the tape may be fully supported, supported at the ends only, or supported at the ends and at the midpoint. Once the correction is determined, the amount is applied to each tape length as it is measured. Temperature and slope corrections cannot be predetermined. Slope and temperature readings must be made with each tape measurement and the correction applied to each one. The effect of the correction, that is, whether it is plus or minus, depends upon the type of correction. Tape length and temperature variations may be in either direction from a normal or standard. However, the horizontal and vertical misalinement in horizontal taping results in a value that is longer than the actual distance. These errors are called negative and are deducted from the measured value on the tape to get the true distance. In slope taping, vertical misalinement may give a longer or shorter value and is a random error.

a. Length of Tape. Tape length varies with temperature, tension, and manner of support. The original graduations were placed on the tape under conditions which are not the same as those that are encountered in the field. Thus, the standard length is referred to a definite set of conditions. The conditions encountered in the field are measured and compared to the standard. The differences are converted into a correction.

(1) The precise tapes are sent to the National Bureau of Standards or its foreign equivalent where the length is determined using a standard tension and several different support conditions. In addition, a specified temperature is maintained during the standardization and a coefficient of thermal expansion is computed for each tape. A tape standardization certificate is issued for the tape, listing all this information. This tape can be used to calibrate others which in turn are used to compare field tapes.

(2) The comparison establishes a "standard" length for each tape under specific conditions. Any variation from these conditions is measured and the correction can be computed.

(3) Any tape which is to be used for measuring distances should be checked for length against a standardized or a calibrated tape. The value of the comparison is recorded in the taping notebook as a plus or minus correction for each tape length as it is recorded. Since these comparisons are made under a definite applied tension, the same tension must be used in the field to get standard readings.

b. Method of Support. Even under standard tension, a tape supported or held only at the ends will "sag" in the center an amount based on its weight per unit length. This sag will cause the recorded distance to be greater than the length being measured. When the tape is supported at its midpoint, the effect of sag in the two sections is considerably less than when the tape is supported only at its ends. As the number of intermediate supports, equally spaced, is increased, the distance between the end graduations will approach the length of the tape when supported throughout its length. When more than two supports are used, the tops of all supports must be at the same level. The correction for the error due to the sag between the two supports for any section can be determined by the following equation:

$$C_s = \frac{-W^2 l^3}{24t^2}$$

in which C is the correction for sag, *w* is the weight per unit length of the tape, *l* is the length of the suspended section of tape, and *t* is the tension applied to the tape. In this equation, *w* and *t* must be in the same units of weight measure, then C will be in the same unit of linear measure as *l*. The weight (*w*) per unit length can be determined accurately enough for most purposes by weighing the tape and dividing the result by the number of units in the tape. A strong wind acting on an extended tape will create a condition similar to sag. A correction for this condition is almost impossible to determine and the more precise taping should not be attempted in high winds. For lower accuracy surveys, the wind effect can be minimized by increasing the number of support points or by measuring with a fully supported tape.

c. Atmospheric Effects. When measuring with a tape, the only atmospheric condition that has to be considered is temperature. The temperature that is considered is only that of the tape and not the air. This temperature is used to determine the thermal expansion or contraction of the tape because of temperature changes. The tape is standardized at a specific temperature, and the temperatures which are recorded during the taping are used to correct the measured length. The order of accuracy of the survey and the type of tape being used will determine the number of thermometers used, the placement of the thermometer(s) on the tape, and the required accuracy of the reading of the temperature. The standard length of a tape is usually determined at a temperature of 68° F. (20° C.). When the temperature of a steel tape is less than 68 F., the length of the tape will be less than its standard length; and conversely, when the temperature of the tape is greater than 68° F., the length of the tape will be greater than its standard length. For example, at temperatures higher than the standard, a tape length will measure a distance actually greater than the tape value. Therefore, any correction must be added. For colder temperatures, the tape is short and the measured ground distance is less than the value on the tape, and correction is subtracted (minus). The change in the length of a steel tape because of a difference between the actual temperature of the tape and the standard temperature can be found in table D–10, appendix D, or by the following equation:

$$C_t = 0.000\,006\,45\ (T-68)\ l$$
$$C_t = 0.000\,011\,61\ (t-20)\ l$$

in which C is the change in length because of temperature, *T* is the temperature of the tape in degrees Fahrenheit, *t* is the temperature of the tape in degrees celsius, *l* is the length of tape used. The value of C is in the same unit of measure as *l*. The sign of the correction will disclose the way the correction is applied.

d. Slope of Tape. When a measurement is made with the tape inclined, the slope distance is always greater than the horizontal distance. The slope of the tape can be measured in two ways: difference in elevation between taping stations, or value of vertical angle. The percent of slope is used as a criteria to determine which formula is used with the difference in elevation method. The small errors that normally occur in the alinement of the tape will have no appreciable effect on the measured distance when taping on the lower order surveys.

(1) *Vertical angle.* When the slope is measured as a vertical angle, the correction to be subtracted from the slope distance is determined by the following equation:

$$C_h = s \text{ vers } a$$
$$= s\ (1 - \cos a)$$

in which C_h is the correction to the slope distance, *s* is the measured slope distance (usually one tape length), and *a* is the vertical angle of slope. For lower order surveys the correction for percent of slope may be extracted from table D–11, appendix D.

(2) *Difference in elevation.*

(*a*) The difference in elevation between supports is measured in the same units as the slope distance measurement, and a correction (to be subtracted from the slope distance to obtain the horizontal distance) is computed by the following equation:

$$C_h = \frac{-h^2}{2s}$$

in which C_h is the correction, h is the difference in elevation, and s is the slope distance. This equation gives an approximate value for C_h to be used for slopes of 5 percent or less.

(*b*) For slopes greater than 5 percent, closer approximation of the true value of C_h can be determined by use of the following equations:

$$C_h = \frac{-h^2}{\frac{2s - h^2}{2s}}, \text{ or}$$

$$C_h = \frac{-h^2}{2s} - \frac{h^4}{8s^3}$$

in which C_h, h, and s are defined in (*a*) above.

6–7. Taping Mistakes and Errors

Taping, like all survey measurements, is susceptible to mistakes and errors. Most mistakes can be avoided and errors minimized as described below.

a. Mistakes. Mistakes may result from poor work habits, lack of judgment, or confusion. They are often costly, time consuming, and difficult to detect. The easiest way to avoid them is to establish a definite procedure and follow it, being constantly alert during the operations where mistakes are possible. Some of the more common ones are as follows:

(1) Failing to hold the zero graduation of the tape over the point.

(2) Transposing figures, such as recording 48.26 for 48.62.

(3) Reading the figure upside down and obtaining, for example 9 for 6.

(4) Measuring from the wrong end of the tape and misreading fractional tape lengths, for example, 28 instead of 22.

(5) Reading the wrong foot mark, such as 38.32 for 37.32.

(6) Subtracting incorrectly when reading a tape.

(7) Omitting an entire tape length.

b. Errors. Errors in surveying can be divided into two classes, systematic and accidental (app. B).

(1) *Accidental.* The effect of accidental errors may be reduced by computing the artithmetical means of several measurements. Accidental errors are likely to be plus or minus and have a tendency to compensate during a series of observations. Examples of these types of errors in taping are: variation of the tension applied to the tape, inaccurate sticking of the taping pin, and inaccurate determination of the slope of the tape during slope distance measurements.

NOTE

Vertical misalinement in slope taping may be considered an accidental error, but horizontal misalinement for either horizontal or slope taping are accidental errors, but have a systematic effect on the measured distance—making it always too long. Accidental errors can be minimized by the carefulness of the field people during the taping operations.

(2) *Systematic.* Systematic errors, or also called constant errors, are those which occur from well-understood causes and can be greatly reduced by taking proper precautions to eliminate the causes. Examples of these types of errors in taping are—Calibration errors of tape, the thermometer, and the tension scale; constant side wind during the measurements (measures distances would always be too large) ; and adjustment error of the hand level vial (if slope is measured in one direction only).

6–8. Recording Procedures

The field notebook for taping is prepared in the same manner as all survey notebooks (para 1–7). The general purpose notebook is used for taping notes. In addition to this preparation, one page of the notebook should contain a record of the comparison or calibration of the tapes (①, fig. 6–1). It should show the points of support and the actual measured distance for the tape length. The rest of the notes (sample shown in ②, fig. 6–1) contain the field recorded values in the first five columns. The other information is computed to reduce all distances to horizontal.

a. The top of the page shows where the line was run, to what accuracy, by whom, when, and with what equipment. The tension used is also recorded.

b. The column headings are entered as shown and are self-explanatory.

Project DESIGNATION 537/2S/3TT/41 DATE ______ 19__

10 July 19 through 30 July 19

Tapes and Calibration Data:

Tape	Support	Value
Tape #16	20 lbs Tension – K =	0.009
	Supported throughout (T/o) =	100.028
	" at 0, 50, & 100 =	100.026
	" at 0 & 100 =	100.019
Tape #2	20 lbs Tension – K =	0.011
	Supported throughout (T/o) =	100.002
	" at 0, 50, & 100 =	99.999
	at 0 & 100 =	99.991
Tape #5	20 lbs Tension – K =	0.012
	Supported throughout (T/o) =	100.034
	" at 0, 50, & 100 =	100.031
	" at 0 & 100 =	100.022

Remarks

Tape #16 accidently broken on 15 July 1970. Returned to Co. Hqs. for repair. Tape #5 received as a replacement on 16 July 1982

(1)

DESIGNATION 3rd Order taping DATE 11 July 19

From: Δ Fox To: Wolf Head Tapeman: J. Doe
Rear tapeman: W. Roe Recorder: R. Coe

Station	Support	Temp.	Slope %	Measured Distance	Temp Corr.	Slope Corr.	Tape Corr.	Total Corr.	Horiz. Distance	Remarks
Δ Fox	2	72° F	2.5	100.00	+0.003	-0.031	+0.019	-0.009	99.991	Tape #16
	2	72	2.8	100.00	+0.003	-0.039	+0.019	-0.017	99.983	Tension: 20 lbs
	2	73	1.5	100.00	+0.003	-0.011	+0.019	+0.011	100.011	
	2	73	1.0	100.00	+0.003	-0.005	+0.019	+0.017	100.017	
	2	74	1.3	100.00	+0.004	-0.008	+0.019	+0.015	100.015	
	2	74	1.8	100.00	+0.004	-0.016	+0.019	+0.007	100.007	
⊙1	2	74	0.5	83.72	+0.003	-0.001	+0.018	+0.020	83.740	
Fox to ⊙1	Totals			683.72				+0.044	683.764	
⊙1	2	74° F	2.3	100.00	+0.004	-0.026	+0.019	-0.003	99.997	
	2	75	1.8	62.50	+0.003	-0.010	+0.016	+0.009	62.509	
	T/o	75	1.5	100.00	+0.005	-0.011	+0.028	+0.022	100.022	
	T/o	74	1.5	100.00	+0.004	-0.011	+0.028	+0.021	100.021	
	T/o	75	1.5	100.00	+0.005	-0.011	+0.028	+0.022	100.022	
⊙2	2	75	0.0	59.77	+0.003	0.000	+0.015	+0.018	59.788	
⊙1 to ⊙2	Totals			522.27				+0.089	522.359	

✓ [illegible]

(2)

Figure 6-1. Taping field notes and computations.

c. The stations are entered in column 1.

d. The method of support, in column 2, explains whether the tape was supported at two points, three points, or in its entirety. Remember that the support points are equally distributed along the tape.

e. The temperature value, column 3, is the tape temperature.

f. Column 4 lists the percent of slope. Note that the value could be the difference in elevation between tape ends. The DE would be divided by 100 to get the percent of slope. Tape values less than 100 feet long use a proportioned amount of the DE to get percent of slope.

g. The measured distance, column 5, is the actual tape reading.

h. The temperature corrections (TEMP CORR) were determined for each tape length as described in paragraph 6–6*c*. Then, as a check, the average of all the temperature readings was used to apply a temperature correction to the total line distance.

i. The slope corrections (SLOPE CORR) are determined as described in paragraph 6–6*d*. In this instance, the percent of slope is the difference in elevation in feet for a slope distance of 100 feet. The slope correction must be determined for each measured value.

j. Values in the column headed TAPE CORR depend on the manner of support of the tape during the measurement, and as described, the tape correction differs between the two-point support and the totally supported tape.

k. Values in the column headed TOTAL CORR are the algebraic sum of the values on the same line, in the three preceding columns.

l. Values in the column headed HORIZ DIST (horizontal distance) are equal to the MEAS DIST (measured distance) corrected by the value of the total correction.

6-9. Recording Precautions

The field notes must contain all the data needed to convert the taping information into horizontal distances. This responsibility rests with the recorder and the chief of the party. The order of accuracy of the survey will establish the type of taping to be done, the precision of the measurements, and the readings other than tape measurements that must be taken. Thus, if the survey calls for slope taping, the slope or the difference in elevation of the tape ends must be measured. If survey precision further requires the temperature correction, the tape thermometer reading must be noted for each tape length. When taping with a specified tension, not only must the tension be recorded, but a tape calibration against a standard (①, fig. 6–1) must also be shown. The described techniques of taping pin exchange and checking of notes by both tapemen should be a standard procedure. The head tapeman should make a practice of pacing the distance as he moves forward with the tape. This will call attention to possible mistakes if the paced and measured distance do not agree. Tape measurements are subject to many mistakes which, if not discovered in time, will require reruns and waste valuable time. Every attempt should be made to eliminate the practices which cause mistakes as soon as possible. This will minimize the number of reruns.

Section II. MICROWAVE

6–10. Introduction

There are several makes and models of microwave distance measuring instruments. On the older models the distances were determined by reading the time lapse in millimicroseconds that occurred between the outgoing signal and its return. On newer models, the distances appear as direct readings on a dial, as described in this section. The readings are uncorrected meters, and corrections must be applied to the reading.

6–11. Electronic Theory

Knowledge of electronic theory is not essential to operate the microwave instrument. However, a review of the fundamentals of radio wave phenomena may prove helpful in increasing the operator's familiarity with its application. Radio waves, like lightwaves, are oscillatory electromagnetic waves of definite velocities which are the same in a vacuum. For measurement purposes, this velocity is accepted as 299,792.5 kilometers per second ± 0.4 kilometer per second, and is the standard adopted by the International Union of Geodesy and Geophysics.

a. Radio waves are bent by the atmospheric strata due to the varying densities. This bending or refraction of radio waves is of primary importance in electronic surveying. *Reflection* is also an important consideration in electronic surveying. Radio emissions from both master (measurer) and remote (responder) units are reflected from various surfaces, such as water and undisturbed ground. Errors from reflections must be corrected in computation of distance. The radio *wave form* is a sine curve. One complete section of the curve, starting at any point and progressing to the next identical position on the curve, is called a hertz *(cycle)*. *Wave frequency* is the number of hertz (cycles) that occur in a given unit of time. A frequency of 10 megahertz (megacycles) per second indicates that the wave completes 10 million hertz (cycles) per second. *Wave period* is the unit of time required to complete one hertz (cycle); thus, the greater the frequency, the smaller the period. The *wave length* is the distance traveled during the period of one hertz (cycle).

b. Phase is a portion of a complete hertz (cycle) of a wave. If one complete hertz (cycle) is represented by 360°, one-quarter hertz (cycle) would be 90° phase. *Phase comparison* uses the algebraic difference between the phase of one wave and the phase of another at the same instant of time. There are methods of altering the phase of a waveform and then measuring this change, but a reference waveform must be used for a phase comparison. In microwave distance measuring the signal from the master (measurer) unit is used as a basis for phase comparison against a similar return signal from the remote (responder) unit, the difference is called *phase shift* and is the actual measure of the distance. In the new microwave models the phase of the returning signal is changed to match that of the sending signal. The change is done by turning the counter control knob on the instrument until the null meter is zeroed. The phase difference is then read from the counter dial. Since the frequencies (and thus, the wavelenghts) of the radio waves are controlled by crystals of known values, the phase difference readout represents a distance difference. By using several frequencies and integrating the results, the distance between units can be determined. Table 6–1 shows the crystal relationship and how frequency differences results in wave length differences for the MC–8 and Model 99. The crystal relationship for the MRA 301 is slightly different, but the resulting distance measurement is the same.

Table 6–1. Crystal Relationship

Channel	Measurer frequency (mh_z)	Responder frequency (mh_z)	Channel difference	Frequency difference	Distance (meters)
1	7.492377	7.493875			
2	7.492377	7.490879	2–1	14.984754 hz	10
3	5.993902	5.995400	3–1	1.498475 hz	100
4	7.342530	7.344028	4–1	149.847 hz	1,000
5	7.477393	7.478891	5–1	14.984 hz	10,000
6	7.490879	7.492377	6–1	1.498 hz	100,000

6–12. Instrument Accuracies

The microwave instruments are designed to give accuracies which are expressed in terms of two factors, a number of parts per million plus or minus a number of centimeters. Each of these factors is based on a possible area of error.

a. The parts per million (ppm) figure considers the effects of atmospheric variation between the measuring units. Even though the atmospheric variables can be measured with some accuracy at each unit, the atmosphere between the units is not constant and its effect on the microwave cannot be measured. The figure in ppm is assigned to indicate the possible loss in accuracy due to this variation. This atmospheric variation error cannot be offset nor eliminated until some new method of interpreting readings is devised. At the present time there are several government agencies and civilian companies testing new methods of determining this atmospheric variation, such as the use of balloons and airplanes to measure intermediate atmospheric conditions along the measuring line.

b. The centimeter figure is assigned to offset

internal errors within the units. The tubes, capacitors, and resistors can change their electronic value due to age and temperature. These changes in turn create small variations in the measurements. However, these internal errors can be minimized by calibrating the instruments at regular intervals (para 5–12).

6–13. Conditions Affecting Measurements

Microwave measurements are made between two separated units by a radio wave that must travel between these units in a direct beam. The accuracy of the measurements is affected seriously by meteorological and topographic conditions which exist between the units and they must be compensated for or controlled.

a. Meteorological Effects. Although the lightwave and radio wave velocities are equal in a vacuum, they differ considerably under other conditions. For instance, water slows light waves to about three-quarters of their vacuum velocity, and radio waves to about one-ninth. The water vapor in the atmosphere has a similar effect in retarding the waves. For accurate distance measuring, those velocity changes must be corrected by computing the vapor pressure in the air from the temperature, relative humidity, and barometric readings taken and recorded during the measuring operations.

b. Topographic Effects. The emitted beam is not a cylinder or pencil of radio waves, but is a cone. Any surface struck by part of this beam will create reflections. The intensity of these reflections varies with the type of surface. Smooth surfaces, such as still water, ice fields, sand, paved or graded areas, and flat arid plains, will reflect the wave strongly. On the other hand, woodland, plowed fields, choppy water, and grass or brush covered land will absorb or scatter the reflection into numerous paths. The reflections that reach the instrument will affect the readings, since the microwave receiver cannot distinguish between a direct and reflected wave. The result of such reflections is referred to as *ground swing.* To minimize the effects, lines over highly reflective surfaces should be avoided. Observations over expanses of water, if unavoidable, should be made when the surface is wind roughened. Observations over cliffs which may cause reflection anomalies can be shielded from the reflected rays by the use of ground features (fig. 6–2). Line of sight should be selected over terrain with diffusing reflection characteristics, such as plowed fields, broken ground, or terrain well covered by vegetation. A desirable sight is one over a ground line falling away from the station of about 5° for a few hundred feet to a floor between 15 and 100 meters below.

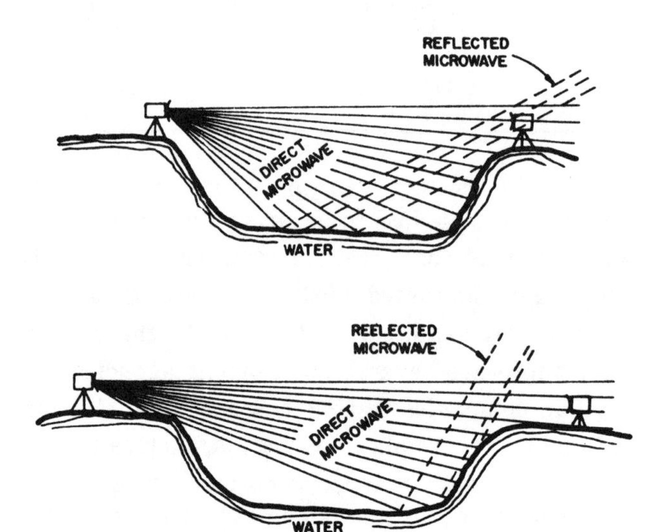

Figure 6-2. Ground reflections.

6–14. Operational Procedure

The operational procedure for making a distance measurement is classified into four functional groups—setting-up, for making the initial contact; *operating,* for making the measurement; *monitoring,* for checking the circuit adjustments; and *internal,* for present adjustments of the instrumental circuits. The operations are divided into those performed at the measurer (master) and responder (remote).

6–15. Use of Microwave Instruments

The microwave instruments are used by the surveyor in geodetic, topographic, and construction surveys as outlined below.

a. Conventional Triangulation. The use of microwave instruments in conventional triangulation is very limited. They are only used for the measurement of class II, second order and third order base lines. The specification requirements of all first order and class I, second order base lines do not allow for the use of microwave instruments.

b. Braced Triangulation. In braced triangulation the microwave instruments are used extensively. The measurement of the distances within the figures of braced triangulation for all orders of accuracy are done with these instruments.

c. Traverse. The distance measuring of traverses of all conventional orders of accuracy can be accomplished with microwave instruments, providing the minimum distance criteria of the individual instruments are met.

d. Trilateration. The distance measuring for trilateration of all conventional orders of accuracy can be accomplished with microwave equipment, providing the minimum distance criteria of the individual instruments are met.

6–16. Recording Procedures

The observing party will complete the field records to the extent that the correct slope distance is realized. The forms are generally self-explanatory and are available through publication suppliers. All recording and computations will be done in a dark, reproducible ink, preferably black or blue-black. All corrections will be made by crossing out the erroneous value and entering the correct value above. No erasures or recopying of the field records will be done.

6–17. Computation of Distance

The computation of the slope distance is done by the field observing party on station. The horizontal and geodetic distance is computed by the computer at the survey office.

Section III. LIGHTWAVE

6–18. Introduction

The lightwave measuring system uses electro-optical instruments to measure distance by timing pulsed light beams after they travel both ways over the course to be measured. Generally all electro-optical instruments function on the same principles and theory.

a. The transmitter/receiver unit is set up over one terminal station, and the reflector unit is set up over the other. A light beam whose intensity varies sinusoidally is transmitted toward the reflector. After reflection, the returned beam is received by a light sensitive device and converted into electrical pulses or *photo current.* This current is amplified and sent to a *null detector* which is provided with an indicating galvanometer, referred to as a *null indicator.* The sensitivity of the light receiving device is varied in synchronization with the transmitted light pulses and can be "moved" electronically to coincide into the proper phase relationship (para. 6–11*b*) with the returning light.

b. To visualize the next few steps, the shape of the light pulse is represented by the shape of a sine wave (fig. 6–3). During one cycle of this wave or 360 electrical degrees (using the mean value as a starting point), the pulse passes from a mean value to minimum, back through mean to maximum and back to mean again. With the pulse at maximum or minimum, a 180° change will reverse it or change it to a minimum or maximum respectively. A $\pm$ 90° change from either position will bring the pulse to a mean value. This procedure is called a *phase change.* If the returning pulse is exactly in phase with the transmitted pulse, the course would be an even number of half wave-lenghts. If, however, the two pulses are exactly 180° out of phase, the course would be an odd number of wavelengths. The electro-optical instrument will detect these phase conditions. Then it can also detect half wavelength changes in the distance which will take place for each quarter wavelength change in the distance between the transmitter/receiver and reflector units.

c. Therefore, the quarter wavelength is the constant for the above mentioned instruments and is referred to as a *unit distance (u).* The distance to be measured is made up of some number of unit distances, plus a partial value of the unit distance. This partial value (1) is determined by the delay line setting. The only unknown value needed to determine the distance is the number of unit lengths. This is resolved by measurements on at least two known modulating frequencies. For a very comprehensive discussion of lightwave measurements reference is made to "Surveyor's Guide to Electromagnetic Distance Measurement" edited by J.J. Saastamoinen, University of Toronto Press, Canada.

6–19. Instrument Accuracies

The electro-optical instruments are designed to give accuracies similar to those of the microwave instrument (para 6–12), that is, they are expressed in two factors. The first factor is a number of parts per million (ppm) and the second is plus or minus a number of centimeters.

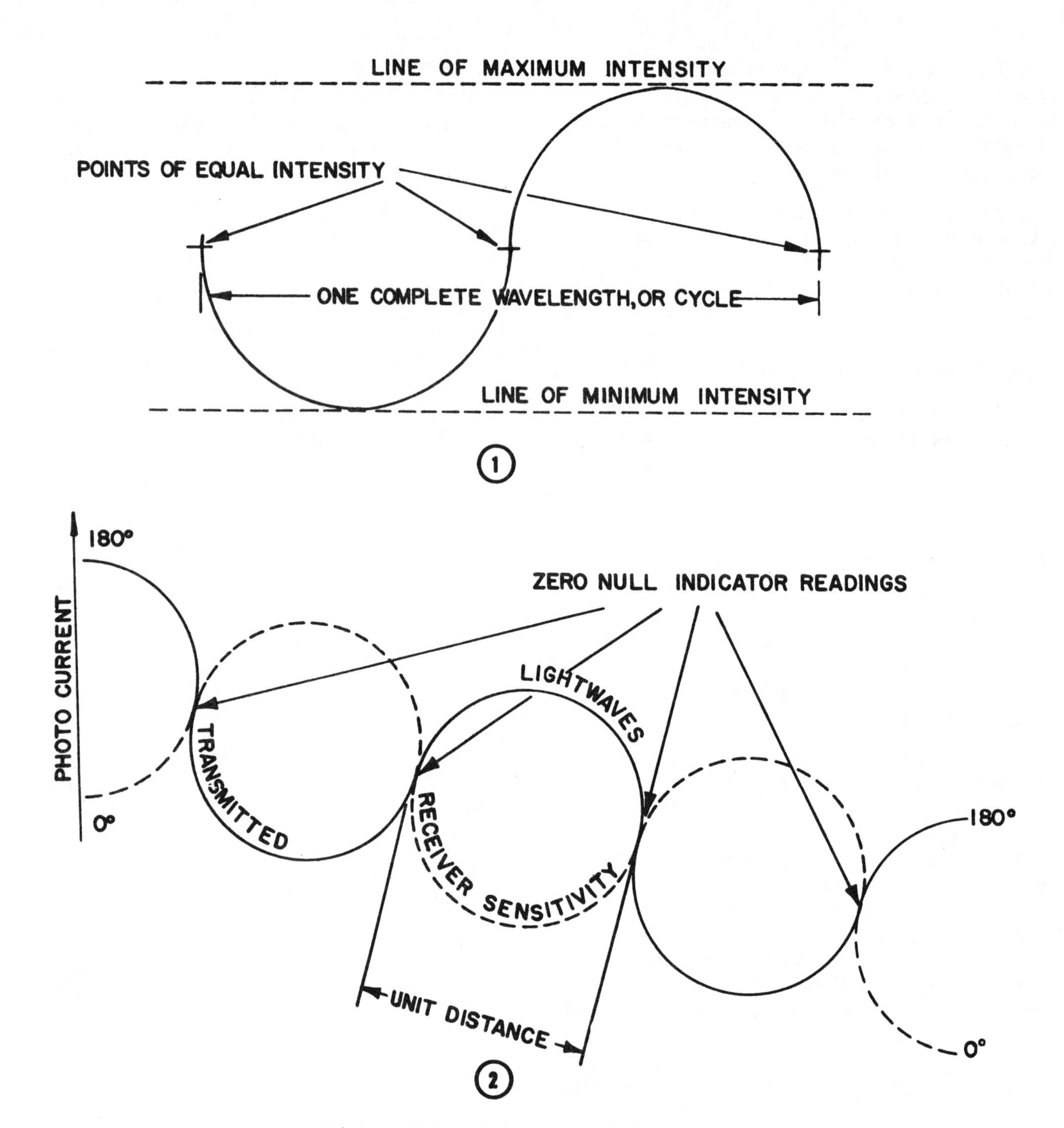

Figure 6-3. Light pulse (sine wave).

a. The ppm factor is the possible error due to external variations which affect the light beam. Local disturbances between the transmitter/receiver unit and the reflector unit cause small changes in atmospheric density and are not reflected in the meteorological readings taken at the terminal points. On the most accurate of these instruments the amount of variation is estimated as one-half millionth of the distance or usually about 2 ppm.

b. The plus or minus number of centimeters is the factor based on internal variables due to components changing values due to heat, humidity and age. This error can be minimized by calibrating the instruments at regular intervals (para 5-23). In the most accurate of these instruments this error will usually be about $\pm$ 1 centimeter.

c. Measurements for high accuracy surveys are usually done at night when using the electro-optical instruments. One reason for this is the greater atmospheric stability during hours of darkness. Another is the absence of extraneous light. Bright ambient light during daylight hours has a tendency to "drown" the light beam, and thus shorten the range of the instrument. However, with the introduction of the mercury arc lamp and the laser beam into the new models of these instruments the requirement for darkness has been minimized.

d. The range of these instruments is usually given as the nominal maximum distance that can normally be measured within the accuracy capabilities of each instrument used. Distances greater than those specified as maximum have been measured, but these were done under ideal external conditions. The minimum distance that can be measured depend mostly on the internal variables of the instrument being used. For example, the internal variable for the Model 2A Geodimeter is a constant of ± 1 centimeter (cm) for all distances, and resolves to a 1:1,000,000 error at a 1 kilometer (km) distance. If the error is greater, say ± 3 cm at 1 km, it amounts to 1:33,333. Thus, depending on the accuracy required by the survey, the minimum distance is more or less determined by the capability of the instrument being used.

6–20. Conditions Affect Measurement

Lightwave measurements are made between a measuring unit at one end of the line to be measured and a group of reflectors at the other end with the beam of light traveling from the measuring unit to the reflectors and back to the measuring unit. A condition that affects the measurement in regards to limiting the range is bright light. However, this instrument has a unique ability in that it will not register light that is received by the photo sensitive tube unless it is in phase with the emitted light. The accuracy of the measurements is affected by two conditions, meteorological and topographic.

a. Meteorological Effects. The meteorological effects on the electro-optical instruments are not as great as those on the microwave instruments (para 5–12*a*), but for high accuracy surveys they must be corrected. The velocity of light in vacuum is by definition a constant, but as soon as measurements are made in ordinary atmosphere the influence of atmospheric conditions must be considered. These conditions, namely temperature, pressure, and humidity, each has its effect on the refractive index of air. For a standard atmosphere of 0° celsius, 760 millimeters of mercury, and 0.03 percent CO_2 the refractive index of modulated light for group velocity is 1.0003036. Variations in temperature (t°C), pressure (p mm Hg) and humidity (e mm Hg) will change this refractive index. Therefore, temperature, pressure, and humidity are recorded simultaneously at both ends of the distance being measured. The temperature readings must be made with special care because of the greater influence of temperature on the velocity of light. This is not to say that pressure and humidity are unimportant, because they are, it's just that temperature is the most important. All atmospheric conditions should be measured with calibrated instruments (sec. VI, ch. 3).

b. Topographic Effects. The effects of topography on lightwave measuring are very minor in comparison to those of microwave measuring (para 5–12*b*). The only requirements are; (1) that the line of sight is clear of all obstructions, such as tree branches, clusters of wires, etc and (2) that the transmitter/receiver station be a ground station when using the model 2A Geodimeter, because of its size and weight.

6–21. Operating Procedure

The operating procedure for each type, make, and model of electro-optical instrument is similar, but there are differences and the manufacturer's operating manual must be consulted before attempting any measurements.

6–22. Use of Lightwave Instruments

The lightwave instruments are most used in geodetic and topographic surveys. They can be used for all orders of baselines, conventional triangulation, braced triangulation, traverse, and trilateration. These instruments can be used in construction surveys, but usually the accuracy requirements do not justify their use in place of microwave instruments.

6–23. Recording Procedure

The recording of the geodimeter observations are completed by the observing party on station site prior to dismantling equipment. All recording will be done with a dark, reproducible ink. All computations and values will be checked to insure accuracy. The checking will be done by one who did not do the computation. The atmospheric measurements are made and recorded at both the geodimeter and reflector units.

a. Model 2A. One set of measurements is recorded to include all three frequencies. The recording and computations done are self explanatory.

b. Model 4. One set of measurements is recorded to include all three frequencies. The recordings are generally self-explanatory.

6–24. Computation of Distance

The computation of the final distance is usually done by the field observing party at a base camp or in the field office.

Section IV. INDIRECT METHODS

6–25. Introduction

Distance measuring is normally performed with distance measuring equipment. However, there are some indirect methods that employ the use of angle measuring equipment to determine distances. These methods are usually used for fourth and lower order surveys when microwave instruments are not available or inoperative and the terrain is prohibitive to taping. These methods are stadia, short base, and subtense.

6–26. Theory of Stadia

Stadia is a method which uses the geometeric principle of similar triangles to measure distances. When stadia is used to measure distances the difference in elevation is also determined and this is referred to as trigonometric leveling.

a. Two horizontal lines or hairs are placed at a fixed distance above and below the center hair on the telescope reticle. When viewing through the telescope at a rod some distance away (①, fig. 6–4), these two lines (U and L) will appear on the rod at U_3 and L_3. This distance is called the *stadia intercept* or *stadia interval.* If the rod is moved to position 2, the stadia interval will be U_2 L_2; and at position 1, U_1 L_1. The angle UOL is fixed and remains constant regardless of the distance. The stadia hairs are spaced so that the distance, UL, is some even proportionate value of the distance from 0 to the line through U and L. By similar triangles, U_1 L_1, is the same proportion of the distance from 0 to U_1 L_1, and similarly U_2 L_2 and U_3 L_3 for their respective distances. No matter where the rod is held, the stadia intercept will equal the same proportion of the distance from the instrument to the rod. In surveying, the principle is stated as: The distance to any point, equals so many times the stadia intercept on a rod held on that point. There are three common stadia intervals in use—1:100, 1:200, and 1:333.

b. Distances can be measured to any point that is 200 times the length of the rod and visible from an instrument setup. Only the rod has to be moved from point to point for the distance to be determined. To establish a position, the direction from the instrument to the point must be known. This is accomplished by measuring an angle when using the transit or theodolite. Another common method uses a planetable and alidade (para 7–9). After orientation, the direction is drawn on the planetable and the distance is plotted to a convenient scale.

c. Stadia, as described above, has assumed a horizontal line of sight. When the line of sight is inclined (②, fig. 6–4), the stadia intercept on a rod held vertically is not a true proportion of the distance since the rod is not perpendicular to the line of sight. To make the rod perpendicular, it would have to be titled from the X position to a postion like Y. This tilt would be almost impossible to maintain. In practice, the rod is held vertically, and *not* tilted. Note in the figure, that the extension from X is perpendicular to the horizontal distance, H. The angle between X and Y is equal to angle a, and the horizontal distance, H, and difference in elevation, DE, can be computed from the intercept read on the vertical rod as follows:

$$H = \frac{f}{i} I \cos^2 a + C \cos a$$

$$DE = \frac{f}{i} I \tfrac{1}{2} \sin 2a + C \sin a$$

where $\frac{f}{i}$ is the stadia interval factor,

I is the stadia intercept on the rod held *vertically* at X, and C is the stadia constant of the instrument.

d. The stadia constant (C) is equal to f + c (the focal length of the telescope's objective lens (f) plus the distance from the objective lens to the center of the instrument (c)). Ordinarily, this value is determined by the manufacturer and is noted on the inside of the instrument carrying

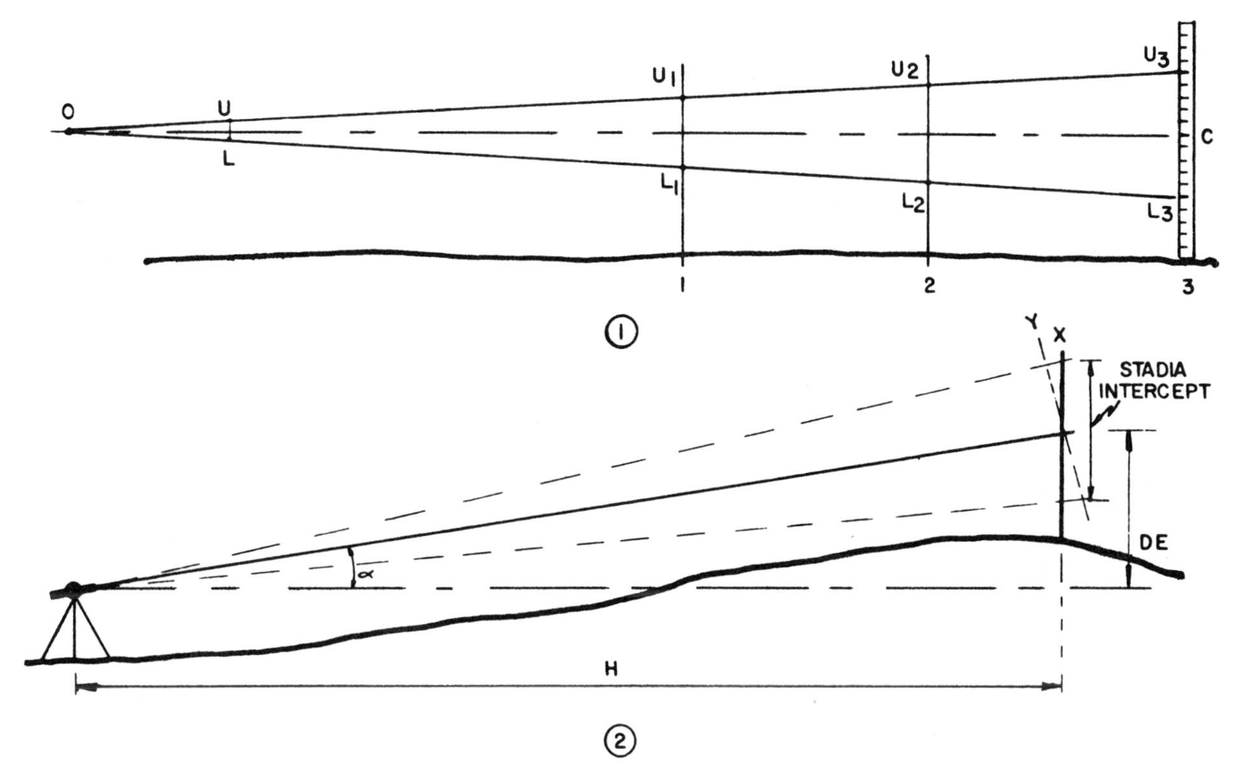

Figure 6-4. Theory of stadia.

case. For normal stadia work, the constant (C) is assumed to be zero for the internal focusing type instrument and 1.0 foot for the external focusing type instrument. The 1.0-foot value is not universal. A C-factor is determined by the manufacturer for each instrument and marked on a tag mounted in the carrying case. It must be checked for each instrument.

(1) For internal focusing instruments, C = 0, and the second part of both formulas above becomes zero. Thus, both H and DE are dependent upon the interval and angle, and the stadia interval factor.

(2) For external focusing instruments, C equals 1 foot. In the first formula, the cos a is multiplied by one. However, when the stadia angles are small (below 10°), the cos a would be near 1, also (cos 10° = 0.9848). The combined effect would approximately equal the value, 1 foot. For practical purposes, the stadia constant, C, is added to the horizontal distance, especially when the angle does not exceed about 5°. In the second formula, C is multipled by sin a. For small angles, the sin a is close to zero, and the product (C sin a) can be considered zero. The DE would depend only on the first part of the formula.

e. The stadia interval factor is the focal length of the telescope's objective lens (f) divided by the actual distance between the stadia hairs (i). When the stadia interval factor in the field must be checked, the following method is used:

(1) Select an area where a horizontal sight of approximately 300 meters (800 to 1,000 feet) can be made.

(2) By taping, mark a series of points at 50-meter (100-foot) intervals from the instrument. For external focusing instruments, only the *first* 50-meter (or 100-foot) point should include the stadia constant (C). The remaining points are set at the normal 50-meter (100-foot) interval.

(3) Read and record the stadia intercept on a rod held on each of the points in turn.

(4) Divide the measured distance to each point by the stadia intercept reading for that point. The result is the stadia interval for each distance.

(5) Take a mean of all the values as the stadia interval factor $\frac{f}{i}$ for the instrument.

f. The frequency of the interval factor check depends on the type and accuracy of the survey and the instrument associated with it. For higher orders of leveling, the factor determination is performed before the start of work and at anytime the instrument is bumped or jolted. On the lower order accuracy surveys such as plane-table-alidade stadia, a check when the alidade is received, a recheck about once a week and when the instrument is jolted, may be sufficient.

g. The formulas used in the reduction of stadia readings can be solved by table D–6, appendix D, or by the use of a calculator. The various stadia reduction devices give a tabular or mechanical solution of the stadia formulas in which a standard value is used for $\frac{f}{i}$ (usually 100) and the value for C is disregarded.

6–27. Stadia Measurement

The stadia measurement consists of two basic steps; one, to determine the stadia intercept for the distance, and the other to obtain a rod reading and vertical angle or scale reading for the difference in elevation.

a. The readings are made on a Philadelphia rod (para 3–13), or on a specially graduated stadia rod (para 3–15). The foot-graduated stadia rod has painted graduations in feet, tenths, and half tenths (0.5) of a foot. Both the stadia rod and the Philadelphia rod are normally read to the hundredths of a foot for the distance and to the tenths of a foot for elevation.

b. For a complete discussion of the stadia arc refer to paragraph 7–3. To take a reading, the telescope is moved vertically until all three wires fall on the rod. The instrument is clamped. With the slow motion screw, the motion is continued in the same direction until the vertical index on the arc is exactly on a division. The value of the arc division and the center wire reading are recorded for the vertical reading. The top and bottom wire values (stadia intercept) are also read at this time, and the value opposite the horizontal index recorded to determine the horizontal distance. If greater accuracy is required, the vertical angle must be read and the formulas in paragraph 6–26*c* used for computing.

c. During stadia surveys, the rod is sometimes held in a location where only two hairs can be seen on the rod; the thrid hair being obscured. Another possibility, especially on a level sight, is that only one hair intercepts the rod; the other two will be visible but will be above or below the rod. The geometry of stadia can be used to obtain a complete reading.

(1) The two stadia hairs or wires are set at equal intervals each side of the center wire. When two of the three wires can be read on the rod, the value of the third can be computed. If the center and either outside wire are readable, the difference between their readings (the half-intervals) is read, and the center wire is moved to the outer wire position. This brings the off-the-rod wire onto the rod. The second half interval can be read and added to the first to obtain the full value. If only the two outer (stadia) wires are visible, their difference divided by two is the distance from either wire to the center one.

(2) Another geometric relationship is available in the stadia method to aid in reading. The distance between adjoining graduations on the vertical scale is equal to one stadia interval, which is also equal to the distance between the two outer stadia wires. The distance from the center wire to either outer wire is one-half of the interval (one-half of the graduation). If only one wire falls on the rod, the position is noted and the center wire is brought to this point. The telescope has been moved an amount equal to one-half of a division,

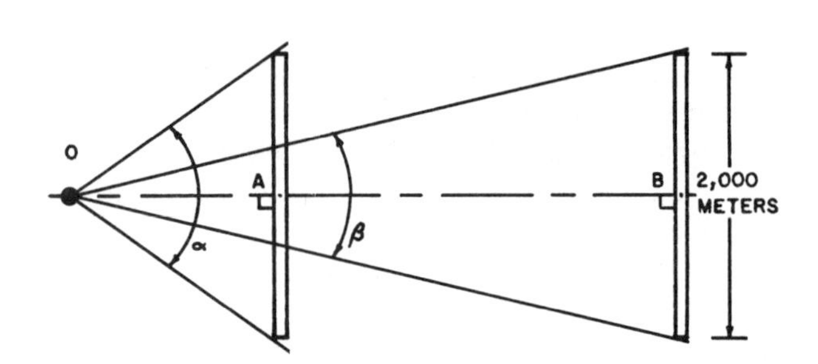

Figure 6-5. Subtense method.

and the readings can be taken and the computations made. Note that the center wire is one-half of the interval away from its original position. If the instrument was leveled before the move, the outer hair will now coincide with the level line of sight. Applying the readable one-half interval to the center wire reading will result in a rod value for a level reading. Some telescopes have a reticle with the stadia interval subdivided into 10 equal parts. This permits estimating one-tenth of the interval in these circumstances and may not require moving the telescope at all.

6–28. Stadia Accuracy

Stadia measurements are affected by numerous factors that may cause large errors in the horizontal distances and differences in elevation. The

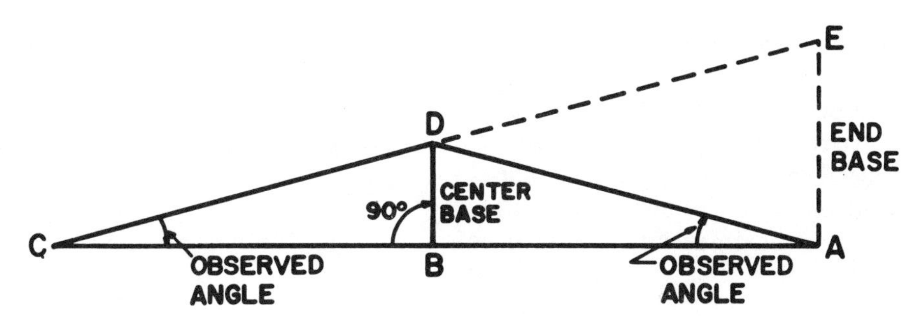

Computation

END BASE (AE)

AC = AE X cot angle ACE

CENTER BASE (BD)

AC = BD X cot angle BAD + BD X cot angle BCD

① END AND CENTER SHORT BASE

Computation

PERPENDICULAR BASE (BD)

AC = BD X cot angle BAD + BD X cot angle BCD (See center base above)

DIRECTIONS OBSERVED: 4

OBLIQUE BASE (BF)

$$AC = BF \times \frac{\sin AFB}{\sin FAB} + BF \times \frac{\sin BFC}{\sin FCB}$$

DIRECTIONS OBSERVED: 10

② PERPENDICULAR AND OBLIQUE SHORT BASES

Figure 6-6. Short base geometry.

use of incorrect stadia constants will cause large errors in measurements. Errors in the adjustments of the instrument, and incorrect plumbing of the rod will cause errors especially when sights are steeply inclined. Inaccuracies in the rod graduations will cause erroneous intercept readings. Sights passing close to the ground may be affected by differential refraction. Long sights on sunny days over terrain which reflects the heat will cause heat waves in the sights that make accurate rod reading impossible. With reasonably careful stadia measurements over terrain that does not require steep sights and the use of the correct values for the stadia constants, horizontal distances can be determined to a degree of accuracy of 1:500. Under the same conditions, differences in elevation can be determined with errors not exceeding 0.2 foot times the square root of the distance in miles (or approximately 0.048 $\sqrt{km}$).

6-29. Subtense Method

The subtense method measures distance by using a bar of fixed length (usually 2.000 meters), and the measured angle subtended by this length (fig. 6-5).

6-30. Short Base Method

The short base method is similar to subtense, except that an accurately measured base line is used instead of the two-meter bar.

a. The base line or short base is placed along

the line to be measured and at a 90° angle to the line (①, fig. 6–10). It is usually placed at the center (BD = center base) and since it is at a right angle to the traverse line, it is called a perpendicular base. Occasionally, it cannot be placed at a 90° angle, and must be at some other angle (②, fig. 6–10). It then becomes an oblique base (BF). If the terrain does not permit a center base, a perpendicular end base (AE) may have to be substituted. However, every attempt should be made to establish a perpendicular center base.

b. The base must be long enough to minimize the effects of instrumental error on the distance to be measured. This requires it to be larger than one-twentieth, and preferably about one-tenth, of the distance to be measured. Table 6–2 lists the error in the computed length caused by a 2-second angular error. As the base to distance ratio gets smaller, the error in distance increases.

c. One end of the base is placed exactly on line between the stations. The base is then measured away from the line using a distance determined by the expansion ratio and the accuracy desired. The advantage in using a perpendicular base instead of an oblique base is that it requires shorter distance measurements, fewer angular observations and simpler computations. Due to the geometry involved, if an oblique base must be used, the smaller of the two angles at the base terminals should be greater than 60°.

d. The horizontal directions should be measured with the appropriate number of circle positions (para 4–14) as required by the specifications for the order of traverse. The total error for each station is determined through horizontal closures and by appropriate adjustments. Speed in the observations can also be improved by leaving the tribrachs with the optical plummets attached to the tripods and interchanging the theodolite and target set.

Table 6–2. Short Base Expansion Ratio

Ratio base to distance	Angle subtended	Error in computed length due to 2-second error in angle
1/10	5°42′40″	1/10,300
1/20	2°51′45″	1/5,152
1/25	2°17′25″	1/4,122
1/30	1°54′35″	1/3,438
1/35	1°38′12″	1/2,946
1/40	1°25′55″	1/2,578
1/45	1°16′22″	1/2,291
1/50	1°08′45″	1/2,062

CHAPTER 7

PLANETABLE TOPOGRAPHY

Section I. EQUIPMENT

7–1. Introduction

Planetabling is a method of surveying by which the topographer can make a map in the field while viewing the area being mapped. The planetable method eliminates the necessity for lengthy field notes; the only field record is the finished map. The planetable map can be tied into existing control or can be an uncontrolled sheet. Planetable surveys are only conducted for the lowest orders of accuracy. Very seldom, if ever will the accuracy of a planetable survey be greater than 1 part in 1000 for horizontal position and about 1 ft $\sqrt{\text{miles}}$ for elevations.

7–2. Description

The planetable, alidade, and stadia board or Philadelphia rod are used in combination in a planetable survey. With these instruments, the direction, the distance, and difference in elevation can be measured and plotted in the field. The planetable operation produces a completed sketch or map without need for further plotting or computing.

a. Planetable. The planetable (fig. 7-1) consists of a drawing board mounted on a tripod with a leveling device designed as part of the board and tripod. The leveling head used by the surveyor is of the ball-and-socket-type (fig. 7–2). The board (G), either 46 x 61 centimeters, or 61 x 79 centimeters, has an attached recessed fitting which screws onto the top of the spindle (A). Wingnut (B) controls the grip of parts C and D on the cup E. By releasing wingnut (B), the drawing board can be tilted in any direction to level it. Wingnut (F) acts only on the spindle (A) and when released, permits the leveled board to be rotated in azimuth for orientation. The tripod is shorter than the theodolite or level tripods, and when set up, brings the planetable board about waist high for easy plotting. One precaution must be observed in attaching the planetable to the tripod head. A paper gasket should be placed between the fittings to prevent sticking or "freezing" of the threads.

b. Alidade. The alidade (fig. 7–3) is a straightedge with a sighting device parallel to the edge. The more precise types have telescopes for sighting, special reticles for distance measuring, and graduated arcs for measuring vertical angles. A new version also includes a self-leveling, optical-reading system with enclosed graduated arcs.

(1) The open-sight alidade (①, fig. 7–3) which is very useful in sketching small areas, has collapsible open sights attached to the straightedge. A level vial is mounted on the straightedge for keeping the alidade level. A trough compass is also furnished for attaching to the board. By sighting through the rear peep sight, it is possible to determine a level line and the slope from the sighting point. No magnification is provided, so the sight lines are kept relatively short. The distance can be measured by pacing or can be measured with a tape if more accuracy is required. A 10-mil graduated scale, number every fifth line from 0 to 40, runs up the right edge, and down the left edge of the front sight for determining slopes.

(2) The telescopic alidades (② and ③, fig. 7–3) consists of a straightedge with a rigidly mounted telescope which can be rotated through a vertical angle of about ± 30°. One type has a telescope set on a high standard or post to raise it above the table. This permits direct viewing through the telescope which is at a comfortable height. The other type has the telescope mounted on a short standard, close to the straightedge. A right-angle prism is attached to the eyepiece and permits viewing through the telescope by looking down into the eyepiece.

(3) The telescope on the high standard usually is 16 power and on the low standard 12 power. Both are the inverting type with internal focusing. The prismatic eyepiece inverts the image top to bottom, so that it appears erect but

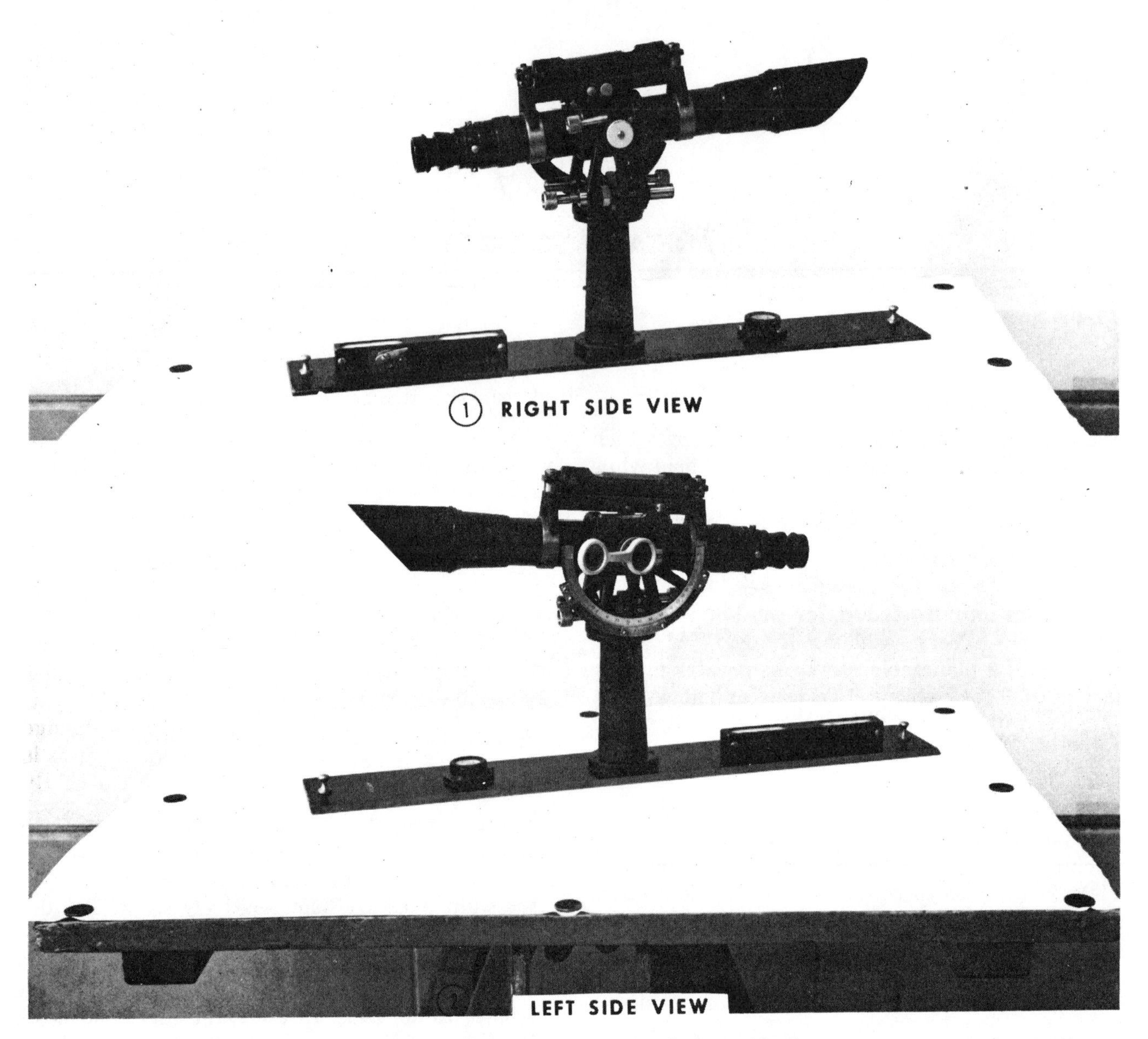

Figure 7-1. Planetable with tripod.

reversed side to side. The line of sight through the telescope in a level position is parallel to the straightedge on the base. The telescope reticle has horizontal and vertical crosswires and a set of stadia wires (para 2–12). The stadia wires are used to measure distance (para 6–26). The vertical distance between the upper and lower stadia wires is carefully read and multiplied by the stadia interval factor. This value is the straight line distance between the center of instrument and the rod.

(4) A circular level vial and a magnetic compass are attached to the base. These are used to level the planetable board and orient it in the proper position. Since the tripod head does not permit as fine a movement as a leveling screw, the bubble is centered as accurately as possible. Then the wingnut (B, fig. 7–2) is set firmly but not tight. Light tapping on the proper corner of the board is used to refine the leveling after which the wingnut is properly tightened. To orient the planetable board, wingnut (F, fig. 7–2) is loosened and the board rotated. It is good practice to draw a magnetic north line on the cover sheet or on two pieces of tape attached near the edges of the board. The straightedge is set on this line during orientation. When the board is rotated and the alidade is faced approximately north, the magnetic needle is released and will have room to swing in its case without hitting the sides.

(5) The telescope alidades have two other important features used in planetable surveying.

These are the detachable striding level and the stadia arc (para 7–3). The striding level contains a long bubble, and when attached, permits accurate leveling of the line of sight. The bubble is mounted in a metal tube with V-fittings on each end. These fittings are placed astride the telescope and bear on built-in polished brass rings on each side of the center standard. A spring clip on the level grips a center pin on the top of the telescope and keeps the level from falling or being knocked off during operation. A button on the side of the level releases the clip for removing the level. For checking and adjusting, the level is reversible. The striding level is normally used to establish a

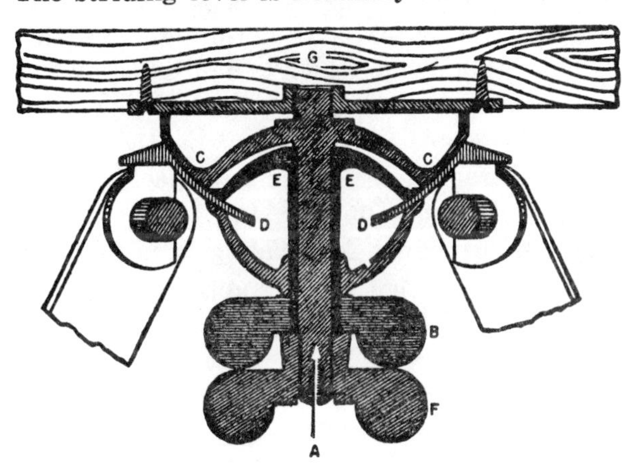

A—Spindle
B—Wingnut for tilting clamp
C—Orientation clamp
D—Socket
E—Tilting clamp
F—Wingnut for orientation clamp
G—Planetable

Figure 7-2. Cross section of tripod head.

horizontal line of sight and to use the alidade as a level.

(6) A new model telescopic alidade is currently in use and it is a self-leveling, optical-reading instrument. Instead of the exterior arc and level bubble, a prism system with a suspended element, and enclosed arcs is used. As long as the alidade base is leveled to within one-half degree of horizontal, the suspended element (or pendulum) will swing into position and the vertical arc index which is attached to it will assume a leveled position. The scales are read directly through an optical train. This combination permits faster operation and there is no chance of forgetting to index the arc bubble and introduce errors into the readings.

c. Rods or Boards. The stadia board (para 3–15) is most generally used in planetable surveys. However, any rod or board may be used as long as it is graduated. When the stadia board is not available, the standard leveling rod (para 3–13) is usually used. This rod is read directly without the use of a target.

d. Auxiliary Equipment. The auxiliary equipment consists of a coated plastic or paper planetable sheet on which the map or sketch is drawn, drawing materials (scribing tools for coated plastic or pencils for the paper), scales for plotting distances, triangles, waterproof table covers, umbrella, and notebook. The planetable sheet is attached to the board by flat-headed, threaded studs which fit into recesses in the board and do not obstruct the alidade's movement. The planetable sheets are of scale-stable plastic with a roughened

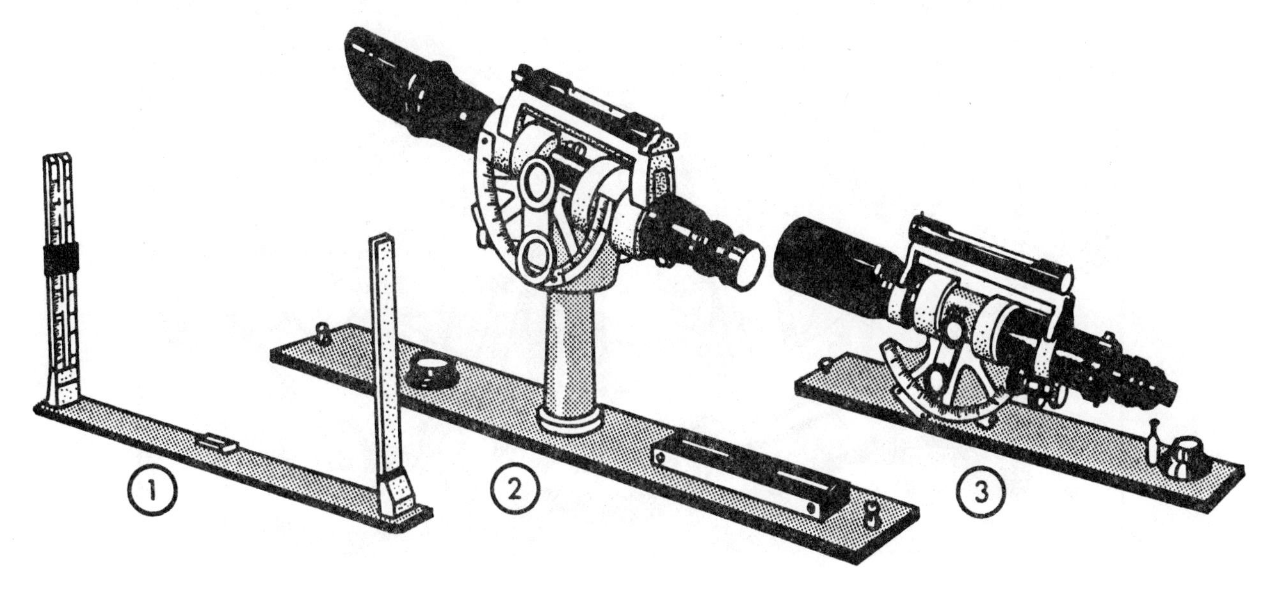

Figure 7-3. Alidades.

surface to retain ink or pencil lines, or a paint-coated surface which can be scribed with steel points. The painted surface will accept pencil or ink for drawing the information just as on the paper, but when the paint is scribed (cut away) with the steel points, a negative copy is produced.

7–3. Stadia Arc

The stadia arc assembly (fig. 7–4) consists of a vertical arc mounted on the end of the left trunnion and a vernier attached to the left bearings by means of an arm. A level vial is attached to the upper end of the arm and a tangent screw controls the movement of the vial. Once adjusted this vial establishes a reference from which vertical angles can be measured even if the planetable is not exactly level. The stadia arc is a group of three scales; one to measure vertical angles; the second, to compute the difference in elevation when the line of sight is above or below a level line; and the third, to obtain a correction factor to convert the measured slope distance to the horizontal distance.

a. Vertical Scale. To obtain difference in elevation (DE) between the instrument and rod, the scale marked VERT. or V is used. The index point of this scale is marked 50, so a reading of less than 50 indicates that the telescope (line of sight) is depressed, while a reading greater than 50 shows that the telescope is elevated. The number of graduation above or below 50 is the factor by which the stadia intercept is multiplied to obtain the difference in elevation.

(1) The telescope is moved vertically first, until the three wires fall on the rod, and then until the vertical scale index is exactly on a graduation. The whole number of graduations which results is easier to multiply by the intercept to get the DE.

(2) The value of the intercept between the upper and lower wires is read through the telescope. This is the value which is multiplied by the vertical scale reading minus 50 to get the DE.

(3) The center wire reading on the rod is added or subtracted to get the total backsight or foresight for leveling (para 8–4).

(4) The procedure for reading is as follows:

(*a*) Assume that a foot rod is used, the arc reading is 43, and the three wires read 4.10, 7.30, and 10.50.

(*b*) The multiplying factor is the arc reading minus 50 (43 — 50 = 7).

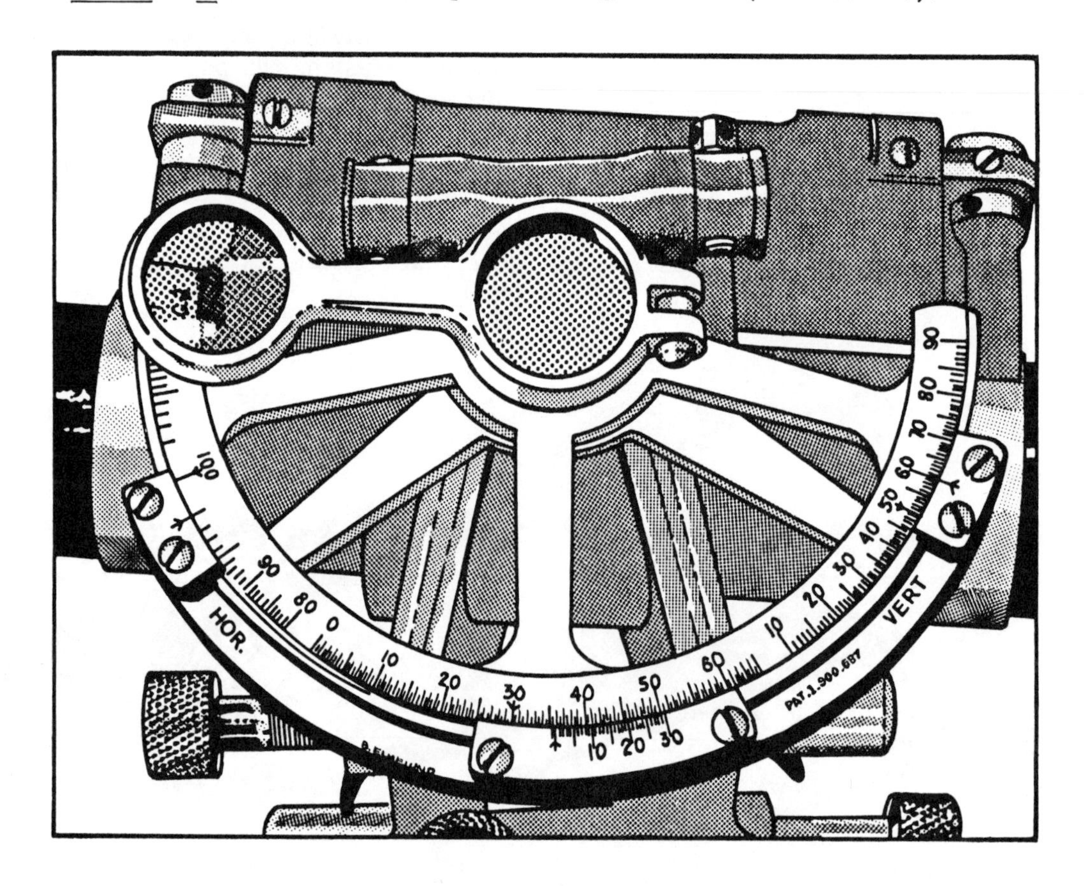

Figure 7-4. Stadia arc.

(*c*) The stadia intercept is the upper wire minus the lower wire (10.50 — 4.10 = 6.4).

(*d*) The DE is the multiplying factor times stadia intercept (—7 x 6.40 = —44.8 ft). The minus sign indicates the object is below the horizontal plane of the instrument.

(*e*) The DE from the point on which the rod is held to the height of instrument is equal to the DE ((*d*) above), minus the rod reading (— 44.8 — 7.3 = —52.1 ft). The minus indicates the point is below the HI.

(*f*) This total distance is added to the station elevation if the reading is a backsight, and subtracted from the height of instrument if a foresight. A detailed discussion of rod readings and DE in backsights and foresights is given in paragraph 8–4 and figure 8–2.

b. Horizontal Scale. The horizontal distance is found by means of the scale marked HOR or H. The numbers refer to the percent difference between the slope intercept value and the horizontal distance. When the reading is 100, the telescope is level and the stadia intercept multiplied by the 100 equals the horizontal distance. When the reading is 90, the horizontal distance is 90 times the stadia intercept reading. Thus, with an intercept of 6.40 feet and a horizontal scale reading of 90, the horizontal distance would equal 90 x 6.40, or 576 feet. When the instrument has a stadia constant affecting the accuracy of the work, the C value must be added before computing the distance. The 6.40 intercept would be equivalent to a 640-foot slope distance, but a 1-foot value for C would make the total distance equal 641 feet. The 90 percent of 641 feet would equal 576.9 feet.

c. Vertical Angle Scale. The third scale on the arc is a vertical angle scale with an index value of 30°, and a vernier to read to the nearest 1 minute in one direction only. After the angle is read, 30° is subtracted from the reading. If the result is a plus angle, the telescope was elevated; if minus, the telescope was depressed. This is an actual vertical angle, and can be used in trigonometric leveling (para 8–17) or in slope taping (para 6–3).

7–4. Care

The alidade's telescope and stadia arc, the planetable's tripod, and the stadia rod are similar to equipment previously described. Their care is identical to the other equipment. The drawing board and straightedge are unique to planetable surveys and their care is described below.

a. Tripod care is discussed in paragraph 2–4. The telescope and stadia arc are similar to the transit's telescope and vertical circle and the care of both is the same (para 4–4). The stadia rod's care is similar to other leveling rods (para 3–13*c*).

b. The planetable is used for careful drawing of map features and for sketching. Dents or grooves in the board will hinder good drawing. The surface should be protected from injury. A coating of wax will protect the finish and prevent moisture from causing warpage or unglueing of the joints. When the manuscript sheet is in place, it should be covered with a protective plastic or paper sheet to prevent injury to the drawn or scribed copy. The straightedge is a drafting instrument and must be free of nicks and dents. The blade must be flat since it will affect the leveling of the planetable (para 7–5*e*) as well as the drawing of straight lines.

c. All equipment should be kept clean of dirt and dust. Any that accumulated must be wiped off. Lenses should be brushed and wiped with chamois or lens tissue; wooden parts wiped with damp cloths and dried; and metal parts wiped and lightly coated with oil where it will not act as a dust catcher and cause trouble.

d. In operation, the greatest chance of damaging the instrument is when placing the alidade on the planetable for leveling the board. The board can tip and the instrument can slide off. Do *not* release the alidade until the board is leveled and locked in place by the wingnut. When setting up the tripod, hold the alidade until the board is in place and leveled by eye. Place the alidade on the table, but do not let go of it. Level the board and set the wingnut before releasing the alidade. The planetable operator must never lean on the board. To carry the alidade, the standard is hooked in the crook of the elbow, the blade or standard held against the body, and the forearm crossed in front of the body. For long distances, the alidade is placed into its case and the planetable removed from the tripod. From station to station travel, the planetable is carried under the arm or over the shoulder, depending on the terrain and vegetation cover. Special care must be taken when traveling through brushy or wooded areas. The planetable extends beyond the body limits and can easily catch on a tree or be scraped by the underbrush. It is advisable to carry the planetable under the arm, board forward, to avoid hitting obstructions and damaging the leveling joint or splitting the board. A waterproof cover is provided for use when a sudden shower comes up. The umbrella is extremely useful in planetable operations. The glare from the drawing surface when plotting in sunny weather can tire the eyes. The umbrella will shade the board and permit faster and more accurate plotting.

7–5. Adjustments

The telescopic alidade requires six adjustments. They should be made in the listed sequence. A seventh adjustment given under *g* below, is the only one required for the self-indexing alidade. Prior to the alidade adjustment, the planetable is set up and carefully leveled.

a. Collimation of the Crosswires. To make the line of sight through the crosswire intersection coincide with the axis of the telescope—

(1) Point the alidade at a distant well-defined point.

(2) Rotate the telescope in its sleeve. The intersection of the crosswires should remain on the distant point. If the distant point appears to move away from either or both the crosswires, they should be adjusted.

(3) Adjust each crosswire separately until the intersection of the crosswires continually bisects the distant point as the telescope is rotated through 180°.

b. Verticality of the Vertical Crosswire. To make the vertical crosswire perpendicular to the horizontal axis of the telescope—

(1) See that parallax is eliminated and that the alidade is leveled. Sight the vertical wire on a well defined point, and move the telescope slightly up and down on its horizontal axis with the slow motion screw. If the instrument is in adjustment, the vertical wire will appear to follow the point through its entire length.

(2) If it does not, loosen the screws holding the crosswires and slightly rotate the ring by tapping the screws lightly.

(3) Sight again on the point and if the vertical wire does not follow the point through its entire length as telescope is moved up and down, rotate the ring again.

(4) Repeat this process until the condition is satisfied.

(5) Repeat the collimation check (*a* above).

c. Striding Level. To make the axis of the striding level parallel to the line of sight—

(1) Clip the striding level into place on the telescope.

(2) Center the level bubble using the tangent screw.

(3) Unclip, reverse, and reclip the striding level.

(4) If the bubble is off center, bring it halfway back using the tangent screw.

(5) Complete the centering, using the pair of capstan screws at *one* end of the bubble tube.

(6) Repeat the test and adjustment until a reversal of the striding level does not move the bubble off center.

d. Vertical Arc Control Level. To make the vertical arc read true vertical angles—

(1) Place the alidade on a stable, flat, approximately level surface.

(2) Place the striding level on the telescope.

(3) Center the bubble of the striding level in its vial.

(4) Move the zero graduation of the vernier into coincidence with the 30° graduation of the vertical arc. If the bubble of the vertical arc control level comes to rest off center, use the adjusting screws near one end of the vertical arc control level to move the bubble until it is centered in the vial.

e. Circular Level.

(1) Set up and approximately level the planetable.

(2) Place the alidade near the center of the drawing board.

(3) Draw a line along the length of the alidade blade.

(4) Turn the alidade 180° and replace the edge of the blade on the line previously drawn on the board. The bubble of the circular level should now come to rest at the center of the circle.

(5) If the bubble comes to rest off center, the blade must be checked for flatness. When the test indicates the blade is warped, the blade must be flattened. If a test of the level still indicates an error, the bubble should be adjusted by placing small shims under the edge of the bubble holder.

f. Stadia Arc. To make the stadia arc read the true stadia factors for horizontal and vertical corrections—

(1) Test and adjust the vertical arc control level (*d* above).

(2) Inspect the stadia arc index mark or marks. The index for horizontal corrections

should be in exact coincidence with the arc graduation numbered 100. The index for the vertical corrections should be in exact coincidence with the arc graduation numbered 50.

(3) When the bubbles of the striding level and the vertical arc control level are both centered in their vials and the stadia arc is not properly positioned, loosen the index plate holding screws with a screwdriver, move the plate to its proper position, and clamp in place by retightening the screws.

g. Self-Indexing Alidades. To set the scales of the self-indexing alidade at their correct values when the line of sight is horizontal—

(1) Setup and level the planetable over one or two selected points at about the same elevation and about 75 meters apart.

(2) Place the rod against the planetable, slide the left side of the alidade up to the rod, read, and record the exact height of the friction adjusting screw on the end of the telescope axle. A pencil mark at this point on the rod will be helpful.

(3) Move the rod to the other selected point, sight upon the marked point, read, and record the vertical angle scale.

(4) Move and set up the planetable at the second position.

(5) Check the height of the adjusting screw at this point ((2) above) and move the rod to the first point.

(6) Sight on the second marked ((5) above) point (if not the same as the first point), read, and record the vertical angle scale.

(7) If the instrument is in adjustment, the sum of the two readings ((3) and (6) above) will equal 180°. If the sum is not 180°, the instrument needs adjustment.

(8) To adjust, loosen the capstan locknut to the right of the tangent screw and move the reading an amount equal to one-half the difference between the sum and 180°. *For example:*

Reading at position (1) = 89° 48′
Reading at position (2) = 90° 20′
Sum = 180° 08′
Amount of correction = 08′/2 = — 04′

(The sum was greater than 180°, so the correction is minus.)

(9) With the instrument still set up at the second position, the value is changed, 90° 20′ — 04′ = 90° 16′.

(10) After making the adjustment repeat the procedure (2) through (7) above as a check.

Section II. FIELD PROCEDURE

7-6. Description

Planetable surveys combine the stadia method (para 6–26) and field sketching on a drawing board, or planetable (para 7–2*a*). Stadia determines the distance and elevation, while the planetable is used to obtain the direction to selected points. If the planetable is set up over or near a station of known location and elevation, the relative positions and elevations of unknown points can be determined and plotted on the drawing board.

a. The primary considerations in planetable surveys are the area that the survey will include, and the purpose for which it will be used. These will determine the size of the sketch that must be drawn on the board. The sketch must be drawn to fit on the drawing board and to serve its intended purpose, by assigning a scale to be used. The scale is the equivalent distance on the drawing that corresponds to the distance on the ground. Thus, 1:1000, called a *representative fraction* or *scale,* signifies that 1 unit on the sketch equals 1,000 similar units on the ground. This value could be stated as 1 meter on the sketch equals 1000 meters on the ground or 1 centimeter equals 10 meters. In most construction surveys feet and inches are used to express scale; however, the representative fraction is always stated in the same units. *For example,* 1″ = 400′ is generally used in construction sketches. This states that 1 inch equals 400 feet, but the representative fraction for this scale, would be 1 inch equals 4,800 inches (same units) or 1:4800. For mapping large areas, modern photogrammetric methods have almost entirely replaced field mapping methods. Photogrammetric topographic maps are usually made for reproduction in large quantities at scales of 1:50,000 or smaller, with contour intervals of 10 meters or more. Regardless of modern advancements, small areas must often be mapped to a larger scale, with a smaller contour interval, by the use of the field mapping methods discussed in this section. The maps made by these methods usually have a scale between 1:600 and 1:12000, and a contour interval of ½ foot to 10 feet.

b. Once the sketch size and scale have been established, the location of the area to be sketched must be referenced to some definite position or place. This is done by plotting the existing control (horizontal and/or vertical) on the prepared planetable sheet. This existing control along with the control to be established in the area forms the skeleton which is to be filled in with the necessary detail. The positions of all horizontal control points are computed on a single rectangular grid system for ease of plotting. The grid system is plotted at the correct scale and then the horizontal control is added. Vertical control that is established on the horizontal control stations should also be noted on the planetable sheet. On some large scale construction surveys, previously established control may not be available or may be at some prohibitive distance away. Under these conditions, the planetable party should assume and establish their own starting stations and mark them on the ground for future tie-in to the control network. When more than one planetable sheet is required for the area at the required scale, neat (border) lines should be drawn on the sheets to act as match lines for the data plotted on the adjacent sheets. The sheets should have an overlap area for ease of matching.

c. After the control and grid is plotted, the planetable, alidade, and rods are used to perform the survey operations.

7–7. Party Organization

The planetable party consists of three or four people — topographer or planetable operator, computer, and rodpeople.

a. *Topographer*. The topographer is the chief of the party. He or she sets up, levels, and orients the planetable, makes the necessary readings for determination of horizontal distances and elevations, and plots the detail on the planetable sheet as the work proceeds. The topographer also directs the other members of the party.

b. *Computer*. The computer, reduces stadia readings to horizontal and vertical distances, and computes the ground elevation for rod observations. They carry and position the umbrella to shade the planetable, and perform other duties as directed by the topographer. At times, the computer may be used as an additional rodperson, especially when the terrain is relatively flat, and the computations are mostly for leveling alone.

c. ***Rodpeople***. The number of rodpeople will depend on the terrain and the number of personnel available. As stated in *b* above, the computer may be used as a rodperson. The rodperson carries a stadia board (or rod) and holds it vertically at detail points and at critical terrain points. An inexperienced rodperson must be directed by the topographer to each point where the board is to be held. An experienced rodperson will expedite the work of the party by selecting the proper board positions themselves, and by returning at times to the planetable to draw in special detail that he or she may have noticed.

7–8. Orienting the Planetable

Before the plotting of positions and elevations can begin, the planetable must be oriented. Orientation consists of rotating the planetable around its vertical axis until the plotted information is in exactly the same relationship as the data on the ground. There are several methods of orienting the board. Some require the planetable to be set up over a point whose position is plotted on the board. Others do not have this requirement. One method uses the magnetic compass for orientation.

a. *Backsighting*. The usual method of orienting the planetable is to place the alidade straightedge along a line between two plotted points; one representing the occupied station, and the other, a known station visible from the occupied station. Then, with the orienting clamp loose, the drawing board is turned until the alidade sights on the selected known station. The line between the two stations on the ground is now parallel to the line between the plotted points. The clamp is tightened. The orientation is checked by sighting on another visible and plotted point. The direction to any other visible point can be plotted as a ray from the plotted position of the occupied station.

b. *Use of the Magnetic Compass*. To use the magnetic needle, a magnetic north line must be drawn on the planetable sheet. The planetable must first be set up and oriented by backsighting as described above. The alidade is moved toward one edge of the board, and the magnetic needle lowered onto its pivot. With the board clamped, the alidade is carefully positioned in a direction where the needle lines up with the index marks in the compass box, and the alidade is facing north. This alinement is maintained while a north line is drawn along the straightedge blade. Once this line is drawn, orientation consists of placing the straightedge blade along the drawn north line, and rotating the table until the needle lines up with the index marks. Wherever the planetable is oriented with the magnetic needle, it will always be alined in the proper direction. The great disadvantage of magnetic orientation is the inconsistency of the earth's magnetic field (para 9–6) which causes unpredictable shifts in the pointing. For this reason, the magnetic needle should not be used for orienting when the orientation can be accomplished by backsighting. In addition, areas where local attraction exists will cause deviations in the needle. The operator must constantly be aware of this possibility and use other methods.

c. *Resection.* The methods (described in *a* and *b* above) require occupying one of the stations whose position is plotted. However, the planetable may also be oriented without occupying any plotted stations. A technique, called resection, uses two or more visible points whose positions are plotted on the planetable. Rays are drawn from the plotted points back toward the occupied, but unplotted point. In some methods of resection, orientation is simultaneous with the point location; in others, the newly located position is used in conjunction with the previously plotted points to orient the planetable.

(1) *Fundamental methods.* The basic method of resection uses a direction or ray previously drawn to a prominent or well defined feature. Occasionally, this direction, if not previously drawn, is established by setting up on the extension of a line through two previously plotted and visible points. In areas without local magnetic attraction, a magentic orientation can furnish a starting point. Another very effective method of resection uses three visible points and their plotted positions in a graphic construction on tracing paper. These four methods are described below.

(*a*) A previously drawn ray is used in this basic method of resection. For example, while occupying a high top during a survey, the planetable operator sees a lone tree on a ridge some distance away. He draws a ray toward this tree and labels it. At some later time while in the vicinity of the lone tree, the operator sets the planetable on line between the tree and the previously occupied top. Placing the alidade along the drawn ray, the operator rotates the planetable until the line of sight intersects the position on the previously occupied top. He clamps the planetable. In this manner, the planetable is oriented without occupying a plotted position. Sighting at one or more visible and plotted points, the operator can draw resection rays, and thus plot and check the position of the planetable.

(*b*) A special case of the method described in (*a*) above, is shown in ①, figure 7–5, where the occupied station (D) is ranged in on the line through two known stations (A and B). In this case, a previously drawn ray is not required. Instead, the operator must locate the planetable setup so that it is exactly on line with points A and B. When the alidade is set in position so that the straightedge passes through the plotted positions of A and B, and the planetable rotated so that the line of sight intersects A and B, a ray drawn along the straightedge will also pass through the setup position (D). The planetable is oriented by backsighting along this known line and clamped. Then a ray is drawn through another plotted and visible station (C), locating the position of the unknown station (D) at the the intersection of the two lines. The accuracy with which point D can be located depends on the precision with which line ABD can be established and the angle of intersection CDA. The nearer the angle to 90°, the stronger the intersection. Also, if the length of AB is greater than BD, any small error in ranging on AB will create a still smaller error in establishing point D. Thus, a point located at D′ will have a stronger location than D.

(*c*) Occasionally, in field mapping, a planetable operator will find himself in a location from which he can only see two plotted control stations, and the method described in (*b*) above, is not practical. If the area is relatively free of local magnetic attraction, the table can be oriented by magnetic needle, and resected using the known plotted points (a′ b′, ②, fig. 7–5), and the equivalent visible stations (A and B). c′ is the location of the planetable and the board is oriented.

(*d*) The graphic construction on tracing paper plots the position of an unknown occupied location, but requires three visible and plotted points. The accuracy is better than with any of the three methods described above, but is not as great as with the three-point method ((3) below). Tracing paper or transparent plastic on which lines can be drawn is required. The procedure is as follows:

1. Set up, level, and clamp the planetable. A general, even though rough, orientation is very helpful at this point.

2. Locate three well defined, visible points at some distance from the planetable, whose positions are plotted on the planetable. If a choice is available, select the points around the horizon to give strong intersections.

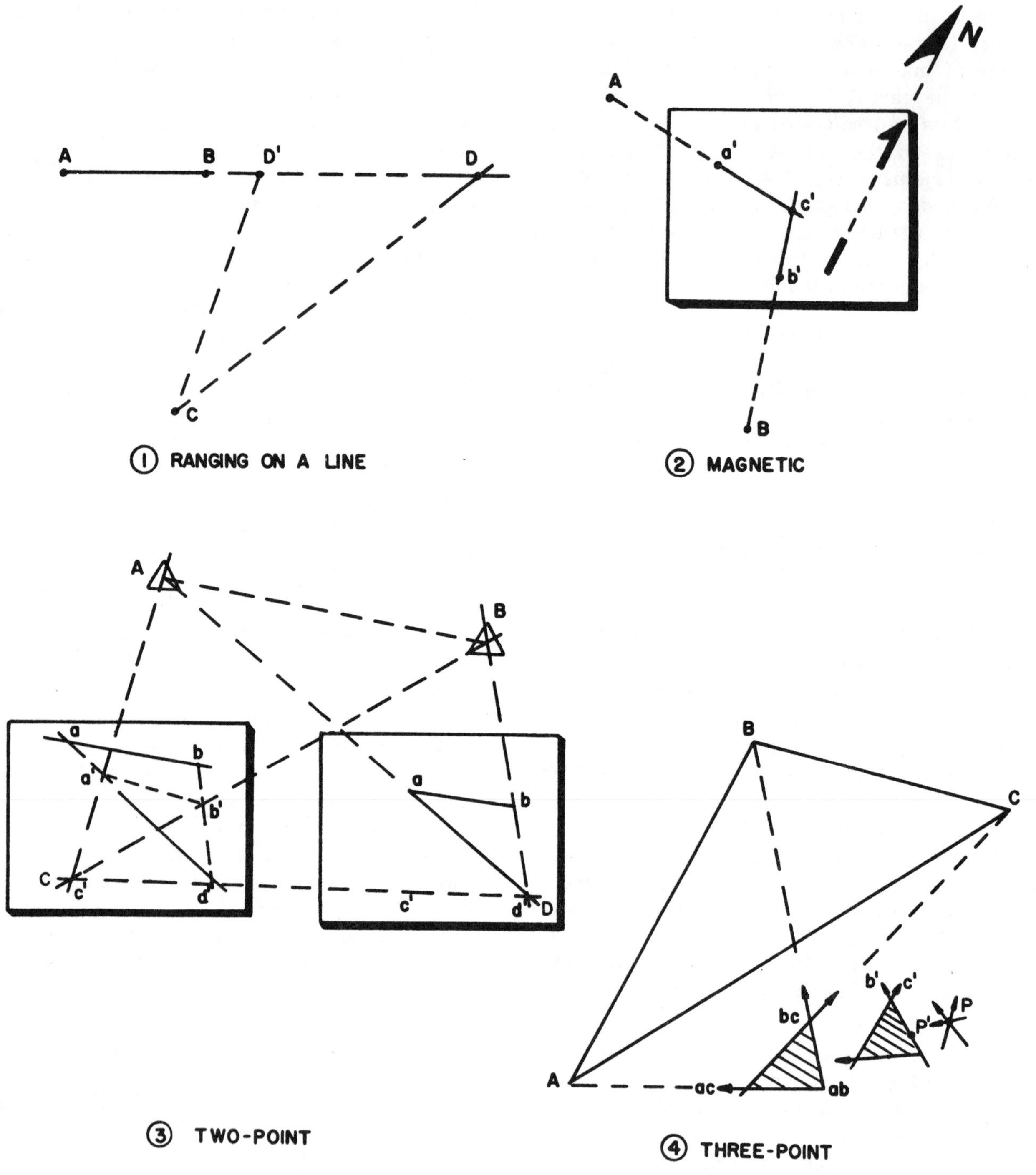

Figure 7-5. Planetable orientation.

3. Attach the tracing paper or plastic directly over the planetable drawing.

4. Estimate a tentative location for the setup and mark the point carefully on the tracing paper.

5. Moving only the alidade, sight on each of the three selected points, in turn, using the estimated location as a pivot point.

6. Draw a ray on the tracing paper in the direction of each of the sighted points using fine lines. When drawing these rays, disregard the plotted positions of the three points. Use only the plotted tentative location and the three visible points on the ground.

7. Remove the alidade from the planetable and release the tracing.

8. Shift the tracing until all of the three drawn rays pass through the respective plotted positions of the points sighted upon.

9. The pivot (tentative location) point is now exactly over the required location on the planetable. It can be transferred to the planetable by marking through the tracing.

10. The planetable is oriented finally by using the newly marked point and one or more of the three visible points.

11. The orientation and location must be checked by sighting upon another visible and plotted point and drawing a ray from it toward the marked location. If the ray does not intersect the marked point, the orientation is not correct and the tracing and shifting procedure should be checked.

(2) *Two-point method.* When two points are visible, but the local magnetic attraction is too great for needle orientation, orientation is still possible on an unplotted point. The two-point method is used to orient the planetable and establish a station position when the two control stations cannot be conveniently occupied.

(*a*) In ③ figure 7–5, a and b are the plotted positions of the visible, inaccessible control stations A and B, and unplotted point C, is approximately estimated and marked (c′).

(*b*) Select an auxiliary point D where the resection lines from A and B will give a strong intersection (greater than 30°). D must be located with respect to C so that rays to A and B from both C and D will also give strong intersections.

(*c*) Set up the planetable at point D and orient by eye, approxiately. Resect on A and B. The intersection of these lines is d′, the tentative position of D. Draw a ray from d′ toward C. Plot the point c′ on this ray at the estimated distance from D to C.

(*d*) Set up the planetable at point C and orient by backsight on D. The error in this orientation is the same in magnitude and direction as it was at D. Sight on A and draw the ray through c′ intersecting the line ad′ at a′. In a like manner, sight on B to obtain b′. Quadrilateral a′b′d′c′ is similar to ABCD. The line a′b′ will always be parallel to the line AB, and the error in orientation will be indicated by the angle between line ab and a′b′.

(*e*) To correct the orientation, place the alidade on the line a′b′ and sight on a distinctive, distant point. Then, move the alidade to line ab. Rotate the planetable to sight again on this point. The planetable is now oriented, and a resection from A and B through a and b established the position of the desired point c.

(3) *Three-point method.* The three-point method involves orienting the planetable and plotting a station when three known plotted stations can be seen but not conveniently occupied.

(*a*) Set up the planetable at the unknown point P. (④, fig. 7–5), and orient by eye or compass, approximately. Draw rays to the known points A, B, and C. The point ab denotes the intersection of the ray to A with the ray to B. Points bc and ac are similar in their notation. If the planetable is oriented properly, the rays to the three known points will intersect at a single point. Usually, the first orientation is not accurate, and these rays intersect at three points (ab, bc, and ac) forming a triangle, known as the *triangle of error.*

(*b*) From the geometry involved, the location of the desired point, P, must fulfill three conditions with respect to the triangle.

1. It will fall to the same side of all three rays, that is, either to the right or to the left of all three rays.

2. It will be proportionately as far from each ray as the distance from the triangle to the respective plotted point.

3. It will be inside of the triangle of error, if the triangle of error is inside of the main plotted triangle; and outside of the triangle of error, if it is outside of the main triangle.

(*c*) In the figure, note that the triangle of error is outside of the main triangle, and almost twice as far from B as from A, and about equally as far from C as from B. The desired point, P, must be about equidistant from the rays to B and to C, and about one-half as far from the ray to A, and the three measurements must be made to the same side of the respective rays. As drawn, only one location will fulfill all these conditions and that is near P′. This is assumed as the desired location.

(*d*) The planetable is reoriented using P′ and backsighting on one of the farther points (B). The new rays (a′,b′, and c′) are drawn. Another (smaller) triangle of error results. This means that the selected position, P′, was not quite far enough. Another point, P, is selected using the conditions above, the table is reoriented, and the new rays drawn. If the triangle had become larger, a mistake was made and the selected point was on the wrong side of one of the rays. The directions should be rechecked and the point reselected in the proper direction.

(*e*) The new point, P, shows no triangle of error when the rays are drawn. It can be assumed to be the desired location of the point over which the planetable is set. In addition, the orientation is correct. Using a fourth known and plotted point as a check, a ray drawn from that point should also pass through P. If not, an error has been made and the process must be repeated.

(*f*) The third try should, normally, bring the triangle of error down to a point, and in many cases, only two are needed. If after three tries, the triangle has not decreased to a point, a circular arc should be drawn through one set of equivalent intersections (ab, a′b′) and another arc through either of the other sets (bc, b′c′, or ac, a′c′). The intersections of the two arcs will locate the desired point, P. This intersection is used to orient the planetable. A check sight on a fourth point will prove the location.

(*g*) There is one possible combination of three points which can present a problem, and the planetable operator should be aware of this possibility. When the three known points and the unknown point are on the circumference of a circle, the orientation will always result in a point intersection (not a triangle of error). No matter which way the table is oriented, the rays will always intersect at a point. These points, if sufficient number is selected, will form a circle with the known points on the circumference. When the first orientation results in a point, the operator must use a fourth known point to prove the location. This fourth point should be selected so that it cannot possibly be on a circumference with the other three known points. If the fourth ray intersects the drawn point, the orientation and location may be assumed correct.

7–9. Point Location

Point location determines horizontal positions. The planetable triangulation uses a graphical determination. Any two plotted points on the planetable sheet can act as a base for triangulation. A ray drawn from each of these points to some unknown point will form a triangle, with the distance between the two known plotted points as the third side. The newly plotted position of the third point will be at the crossing of the rays. The rays to the unknown point may be drawn while occupying the known stations. This is called intersection. The rays may also be drawn while occupying the unknown point, and this is called resection.

a. Resection. The methods of resection are described in paragraph 7–8*c*. Using resection, it is unnecessary for the operator to occupy the known stations, which at times may be inconvenient or impractical. However, resection should include more than two known points to determine the location to a higher degree of precision.

b. Intersection. Intersection is accomplished by setting up and orienting the planetable at each of two or more known stations in turn. At each of these stations, the alidade is pointed toward the unknown point, and a ray is drawn from the plotted position of the occupied station toward the point being located. As such rays are drawn from two or more known stations, their point of intersection is the plotted position for the required station. Two known points are the minimum requirement to establish a location. For more accuracy, three or more known positions must be occupied.

c. Radiation. In planetable surveys when intersection is used, a series of radiating rays are drawn and marked. This technique is called radiation since all rays radiate from the known station. Radiating rays drawn from other known stations will intersect the first rays to locate the points. When drawing the rays, it is necessary to identify to what object each one was drawn. The object may not look the same when viewed from another direction and the intersecting ray can be drawn to some other point resulting in an incorrect point location.

d. Combined Method. The planetable can combine resection with intersection to establish positions. This method is used all through the planetable survey. Let us assume that while on a known point, and after completing the primary observations, the topographer draws another series of radiating rays toward obvious or prominent features visible from his station. These rays may not be needed immediately, but at any time in the future that the prominent feature is occupied, the ray can be used as an orienting line. A resection (para 7–8*c*) from other visible and plotted points will locate the feature. This technique is very useful in areas where the magnetic declination is erratic.

7–10. Planetable Traverse

Traverse starts from a known position and uses a continuous series of direction and distances to establish positions.

a. In planetable traverse after orientation, the direction is drawn to the next point on the survey with a radiating ray. The distance between the occupied point and the new point is measured and plotted along the ray. The new plotted position is now considered a known point, and can be occupied and used as the next station on the line. The planetable is set up over this station and oriented, and another radiating ray is drawn to the next point. The traverse, radiating ray and distance, is continued until the next plotted position is reached.

b. Orientation plays a very important part in planetable traverse. Slight errors in direction at each setup can accumulate rapidly and become large in a short time. Long traverses should be avoided except in reconnaissance surveys. If a traverse must be extended, it is a good practice to use intersection methods first (para 7–9*b*) and establish tie points so that no more than three traverse setups have to be made without a position check. The survey's direction can be maintained within acceptable limits in this manner, and with carefully measured and plotted distances, the accuracy of the survey can be maintained.

c. After a traverse is closed, any accumulated errors, both in distance and direction are equally divided among the stations. Small closure errors are divided by the number of traverse setups. One line drawn from the final surveyed point to the previously plotted tie-in point will define the direction in which the corrections must be made. Lines are drawn through each of the plotted traverse stations parallel to the correction line. A proportioned amount of correction is applied at each of the stations along the parallel lines, and results in a corrected traverse line. The error can also be adjusted by using a piece of tracing paper. The traverse points are carefully traced on the tracing paper. Using the starting point as a pivot, the tracing is swung until the final surveyed point and the plotted tie-in point coincide, or both are in line with the starting point. The stations along the route are transferred to the planetable in their adjusted positions. Small amounts of error in the direction of the line from the starting point to the tie-in point are prorated and applied at each station. Any additional information located during the survey can now be transferred to the adjusted traverse line.

d. For greater accuracy, the traverse can be run using a transit or theodolite. The traverse station values are computed and adjusted, and the adjusted positions are plotted on the planetable to form a control skeleton from which the sketching can start.

7–11. Plotting Detail

A topographic survey is made to determine the shape or relief of part of the earth's surface, and the location of natural and artificial objects thereon. Detail is the term applied to information which is added to the planetable drawing to make a completed sketch or map.

a. The detail that is included is divided into several categories, which in a general way follow the colors printed on the map. The subdivision is as follows:

(1) Works of man or culture, such as buildings, roads, railroads, bridges, and dams.

(2) Drainage features, such as streams, rivers, and lakes.

(3) Natural features, such as woods, and brushwood.

(4) Relief, including contours and elevations.

b. Detail positions are plotted to scale with respect to the plotted position of the occupied station. They are normally located by radiating rays and distances measured by stadia, and plotted directly on the planetable sheet. The ground elevation at each point is determined and shown on the sheet at the plotted position, using the decimal point of the figure as the plotted point. A further discussion of locating detail will be found in chapter 13. On larger scale maps, it is often possible to represent the true shape of features to scale. On small scale maps, buildings and other features must often be symbolized with the symbol centered on the true position, but drawn larger than the scale of the map. Such detail is portrayed on the map by means of standardized topographic symbols.

7–12. Contouring

Detail points and elevations for contouring are usually located at *key points,* distinct changes in ground slope, or in the direction of a contour. After a number of key points have been located (usually from one occupied station) and plotted, sketching of the contour lines is started.

a. Contour lines are drawn on the map by logical contouring. Ground elevations are determined at key points plotted on the planetable sheet. Such key points are located at the following positions:

(1) Hill or mountain tops.

(2) Ridge lines.

(3) Top and foot of steep slopes.

(4) Valleys and streams.

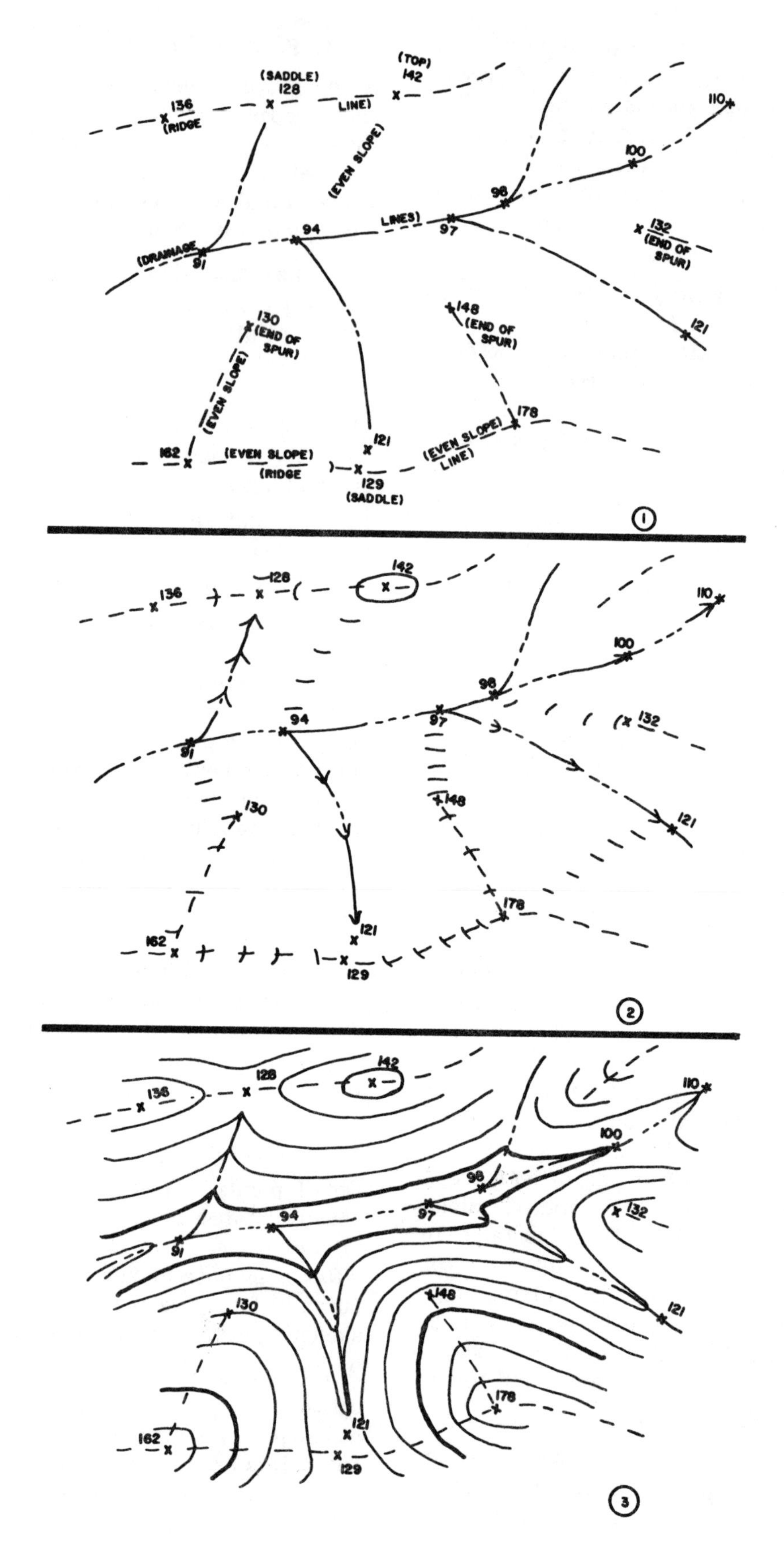

Figure 7-6. Plotting detail and contouring.

(5) Saddles between hills.

b. Figure 7–6 shows a portion of a planetable sheet and the manner in which contour lines are drawn. Key points with description and elevation are shown as in ①, figure 7–6. To aid in drawing the contours, the drainage and the ridge lines are sketched. The contours crossing these lines are marked by interpolation. Along the slopes, the contours are interpolated between the marked key points using short dashes (②, fig. 7–6). The final step is to join the equivalent contour lines (③, fig. 7–6). When joining contours, the topographer must utilize his view of the terrain. If the terrain along the slopes is curved smoothly, the topographer should join his contours with smooth convex or concave curves to depict the terrain. Relatively straight and angular slopes are reflected by more or less straight contour lines. Slopes with numerous small drains should be contoured using lines with indentations at the proper locations to show this condition. This method of contouring should continue as the survey progresses. In addition to the sketching, the topographer should constantly be alert for any additional key points around him that may be used for future contouring. These points are included in the survey and marked to be used at some future setup.

c. When contouring, it must be remembered that stream and ridge lines have a primary influence on the direction of the contour lines, and that the slope of the terrain controls the spacing of the contour lines. Contour lines crossing a stream follow the general direction of the stream on both sides, then cross the stream in a fairly sharp V that points upstream. Also, contour lines curve around the nose of ridges in the form of a U pointing downhill and cross ridge lines at approximately right angles.

d. Every fifth contour line should be drawn heavier than the other, and the elevations of these heavier lines should be shown at frequent intervals. These heavy, numbered contour lines are those representing multiples of the 5-, 10-, 25-, 50-, 100-foot or 5-, 10-, 20-, or 50-meter elevations. For example, with a 10-foot or meter contour interval, the 50-foot or meter contour lines would be heavy and numbered.

7–13. Practical Hints

The topographer sketches the terrain using the measured and plotted information as a skeleton. His ability to see and interpret the ground features, and to transfer them to the planetable sketch is important. Some of the problems have been discussed in the different phases above. Below are listed some practical hints.

a. The topographer should not remain continuously at the planetable, but should move around to acquaint himself thoroughly with the terrain.

b. The magnetic needle can often be used to advantage to orient the planetable approximately at the start of a resection problem.

c. When drawing a ray toward a detail point, draw only a short line near the estimated position for the point, and label the ray.

d. Extend the scope of the work by drawing radiating rays to easily distinguished points at some distance from the planetable, even though no distance measurement is made. A future intersecting ray from a different angle will locate the point.

e. Accurate plotting of distances is critical.

f. Planetable traverses should be kept to a minimum.

g. The topographer should face the area he is contouring.

h. Contour lines should not be drawn beyond the determined and plotted elevations.

7–14. Compilation

The field measurements necessary for planetable compilation use the basic method described. However, sometimes it is more convenient to complete the drawing in the office from field data, while at other times, it is better to compile in the field as the measurements are made.

a. Office Compilation. The measurements are made in the field, recorded in a notebook, and later used in the office either by the field man or by an office man who may be familiar with the terrain.

(1) *Transit and tape.* In this method, the engineer transit and a tape are used to determine the horizontal positions of the required points, and the engineer level or transit is used to determine their elevations.

(2) *Transit and stadia.* In this method, the transit and stadia are used to determine both the horizontal and vertical measurements to the required points.

b. Field Compilation. The measurements are made in the field and converted and plotted directly onto a sketch or drawing before or while the next measurement is being made.

(1) *Transit and planetable.* In this method, the transit and tape or transit and stadia are used to locate the required points. They are plotted on the planetable as the measurements are made.

(2) *Planetable, telescopic alidade, and stadia.* This method is similar to that of (1) above, except that all measurements for the location of the detail points are made with the telescopic alidade by the stadia method.

c. Comparison. At times there is doubt as to the advantage of some method of field measurement. Accuracy requirements will dictate the precision, and in most cases, the proper instrument and method. However, if a choice is possible, the following list will help to make the decision:

(1) *Advantages.* Field sketching has the following advantages:

(*a*) The terrain, being in view, reduces the possibility of missing important detail.

(*b*) Detail can be sketched in its proper position on the map with a minimum of measurements. This applies especially to the representation of relief, where the contour lines can be sketched between plotted key detail points.

(*c*) A greater area can be mapped in a given period of time.

(*d*) Field notes are required, but only until the measurements are plotted. There is less chance for an error to be overlooked until it is too late to correct.

(*e*) Fewer detail points will have to be located, because the topographer can sketch in considerable detail by eye.

(*f*) Triangulation can be accomplished graphically and computations avoided.

(*g*) Office work is reduced to a minimum.

(2) *Disadvantages.* The disadvantages of the field sketching system are not as many numerically, but may offset the advantages. This does not include the accuracy comparison of each method.

(*a*) The planetable and alidade are rather bulky, and cumbersome to handle. In rough terrain or brushy country, the planetable is awkward to carry from point to point.

(*b*) Sketching under field conditions can be uncomfortable and even difficult especially in cold weather.

(*c*) Field measurements are recorded in the notebook, but not identified exactly, so the methods of checking designed to uncover mistakes cannot be used effectively once field work is completed.

(*d*) Numerous items of equipment such as drafting scales, triangles, and dividers must be used, and therefore carried in the field during the survey.

Section III. STRIP TOPOGRAPHY AND PROFILES

7–15. Method

When a road, railroad, or canal is to be constructed, a *preliminary traverse* usually is run in the general direction of the final alinement to aid in the final selection of the route. To determine the amount of earthmoving that will be required (cut or fill) to bring the natural grade to the final grade, profiles are run over the traverse line. Additional measurements can then be made from which a topographic map can be drawn. This map normally represents the topography along the traverse line, and for 300 to 400 feet (approximately 100 meters) on each side of the traverse. Such a topographic map is a long narrow strip on which the location engineer can lay out the tangents and curves for the final alinement, and compute the volume of earth to be added or removed to establish the final grade.

a. When the preliminary traverse is run in the field, stakes normally are set at 100-foot, 50-foot, or 20-meter intervals along the traverse lines depending on the survey. Profile levels (para 8–14) are run over the staked lines to determine the ground elevation at each station and at any other point along the line where there is a definite change in the slope of the ground.

b. By use of the Abney hand level or a clinometer, approximate ground slopes are determined along cross section lines at each of the marked stations. These cross sections extend for distances of 300 to 400 feet on both sides of the traverse line.

c. The strip can be readily compiled in the drafting room from the traverse, profile, and cross slope measurements.

7–16. Field Procedures

This is the process carried out in the field for final office compilation.

a. The Abney hand level or clinometer is used for the approximate measurement of slopes. For use with this instrument, two rodpeople are often equipped with short rods marked at the same height as the instrumentperson's eye. These rods are about 6 feet long with 1-inch adhesive tape wrapped horizontally at the instrumentperson's eye level.

b. The topographer, with the clinometer, stands over a traverse stake and estimates the direction of a line perpendicular to the traverse line (app. A). A rodperson who is out on one side of the line an estimated distance of 300 to 400 feet, is waved onto the perpendicular line. Then the topographer measures and records the percent of slope between his or her eye and the mark on the rod. The topographer then turns around and repeats the procedure with the second rodperson. The three people advance and measure the slopes from the next marked station.

c. When there is a definite change in the ground slope between marked stations, a metallic tape may be used to measure the distance to the point of change, and the ground slope is determined (as in *b* above) using the change point as a marked station.

d. An experienced topographer can often operate without the rodpeople. In this case, he or she estimates the distance above the ground at which they will sight when measuring the cross slopes.

7–17. Plotting

Strip topographic maps are usually plotted to a scale of 1 inch equals 400 feet, but a scale of 1 inch equals 200 feet may sometimes be required. The normal contour interval is 5 feet.

a. Since the preliminary traverse forms the basic horizontal control for the strip topographic map, it must be plotted to scale before the topography is plotted. Every 100-foot station is marked on the plot, and the ground elevation at each of these stations is marked alongside the station mark. Lines are drawn through each station mark, perpendicular to the traverse line, and distances are marked on these crosslines to represent 300 and 400 feet each side of the traverse line. This forms a network of points marking the corners of three rows.

b. The measured ground slopes are used with the ground elevation at each traverse station to determine the approximate ground elevation at the points 300 to 400 feet out from the traverse line, and at intermediate points as required. Percent slope indicates the change in elevation of one unit for each 100 units of horizontal distance. The elevation of each corner of the squares is determined. The contours can be drawn easily between these elevations.

c. In heavily wooded terrain, the method of strip topography can often be used to advantage for the construction of a topographic map of a large area. In this case, straight parallel transit and tape traverse lines are run through the area at intervals of 100 and 200 meters. Side slopes are measured for distances of 50 to 100 meters each side of the parallel lines. A topographic map of the area is compiled from the measurements using the technique described above and filling between the strips with the same technique.

CHAPTER 8

LEVELING

Section I. INTRODUCTION

8-1. Description

Leveling is the operation of determining the difference in elevation between points on the earth's surface. A level reference surface or datum is established and an elevation assigned to it. Differences in the determined elevation are subtracted from or added to this assigned value and result in the elevations of the points. A level surface is one on which every point is perpendicular to the direction of the plumb line. It differs from a plane surface which is flat and is perpendicular to a plumb line at any one point. A body of still water will assume a level surface. If the changes in the surface of the ocean, caused by such influences as tides, currents, winds, atmospheric pressure, and the rotation of earth could be eliminated, the resulting surface would be level. The ocean's level surface is determined by averaging a series of tidal height observations over a Metonic Cycle (approximately 19 calendar years). This average, called mean sea level, is the most common datum for leveling and is usually assigned an elevation of zero. This datum remains in effect until continuing observations show a significant difference, and it becomes worthwhile to change to the new datum.

8-2. Types of Leveling

Leveling is subdivided into two major categories: direct and indirect.

a. Direct Leveling. Direct leveling is usually referred to as differential or spirit leveling. In this method the difference in elevation between a known elevation and the height of instrument, and then the difference in elevation from the height of instrument to an unknown point, are determined by measuring the vertical distance with a precise or semiprecise level and leveling rods. This is the only method that will yield accuracies of third or higher order. Refer to section II of this chapter for a description of this method.

b. Indirect Leveling. There are two methods of doing indirect leveling: trigonometric and barometric.

(1) *Trigonometric.* This method applies the fundamentals of trigonometry to determine the differences in elevation by observing vertical angles (above or below a horizontal plane) and a horizontal distance (measured or computed) to compute the vertical distance between points. This method is generally used for lower order leveling where the terrain is prohibitive to direct leveling. Refer to section III of this chapter for a description of this method.

(2) *Barometric.* This method uses the differences in atmospheric pressure as observed with a barometer or altimeter to determine the differences in elevation between points. This is the least used and least accurate method of determined differences in elevation. This method should only be used in surveys when one of the other methods is unfeasible or would involve great expense in time or money. Generally in surveying this method is used for small scale mapping projects as necessary. This method can and is used extensively where the primary interest is the "lay of land" and not accurate differences in elevation. Refer to section IV of this chapter for a description of this method.

8-3. Orders of Accuracy

Leveling, like all other methods of survey is divided into degrees (orders) of precision. The type and purpose of the survey being accomplished will dictate the order of precision required, and which in turn will dictate the method and instruments needed to achieve that precision. In engineer surveys, leveling is divided into first, second, third, and lower orders of precision. Lower order is considered to be any degree of precision less than third order in engineer surveys.

Section II. DIFFERENTIAL LEVELING

8-4. General

In direct leveling a horizontal line of sight is established using a sensitive level bubble in a level vial. The instrument is leveled and the line of sight adjusted to be parallel to the level vial axis. When leveled, the line of sight of the instrument describes a horizontal plane (fig. 8–1). This procedure is referred to as differential or spirit leveling as described below.

a. The leveling operation (fig. 8–2) consists of holding a rod vertically on a point of known elevation. A level reading is then made through the telescope on the rod, known as a *backsight* (BS), which gives the vertical distance from the ground elevation to the line of sight. By adding this backsight reading to the known elevation, the line of sight elevation, called *height of instrument* (HI), is determined. Another rod is placed on a point of unknown elevation, and a *foresight* (FS) reading is taken. By subtracting the foresight reading from the height of instrument, the elevation of the new point is established. After the foresight is completed the rod remains on that point and the instrument and back rod are moved to forward positions. The instrument is set up approximately midway between the old and new rod positions. The new sighting on the back rod is a backsight for a new height of instrument and the sighting on the front rod is a foresight for a new elevation. The points on which the rods are held for the foresights and backsights are called *turning points* (TP's). Other foresights made to points not along the main line are known as *sideshots*. This procedure is used as many times as necessary to transfer a point of known elevation to another distant point of unknown elevation.

b. Normally, for third and higher orders, sight distances are kept below 75 meters, except when necessary to pass or cross an obstacle. For lower order lines, the length of sight depends upon the optical qualities of the instrument and atmospheric conditions, with the maximum being about 600 meters under ideal conditions. Before starting leveling, a reconnaissance of the terrain must be made. Probable locations of turning points and instrument setups can be noted. The scope of the terrain is a prime consideration in leveling. The normal instrument height at any setup is about 1.5 meters. On even downhill slopes, the ground, where the instrument is set up, must not be more than 1 to 1.5 meters below the turning point for a level backsight. On the foresight, the extended 4-meter rod can be held on the ground about 2.5 meters below the instrument ground level, and still permit a reading to be taken. This means that going downhill, the tendency will be to make foresight distances longer. The backsights tend to be longer in running uphill. During the reconnaissance, line of sight can be estimated by sighting through a hand level. This can determine possible instrument and rod setups. The distances between are paced remembering to balance the foresights and backsights. The procedure is to sight at the uphill point with the hand level making sure that the line of sight is above ground level. The distance from the proposed turning point to the proposed instrument position is paced and the same amount paced to establish the next turning point. Once the distance between points and instrument

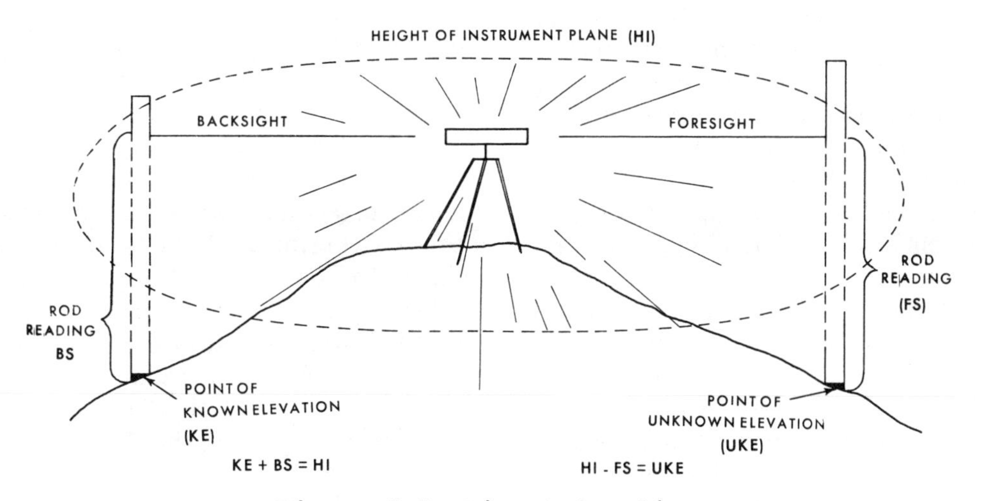

Figure 8-1. Direct leveling.

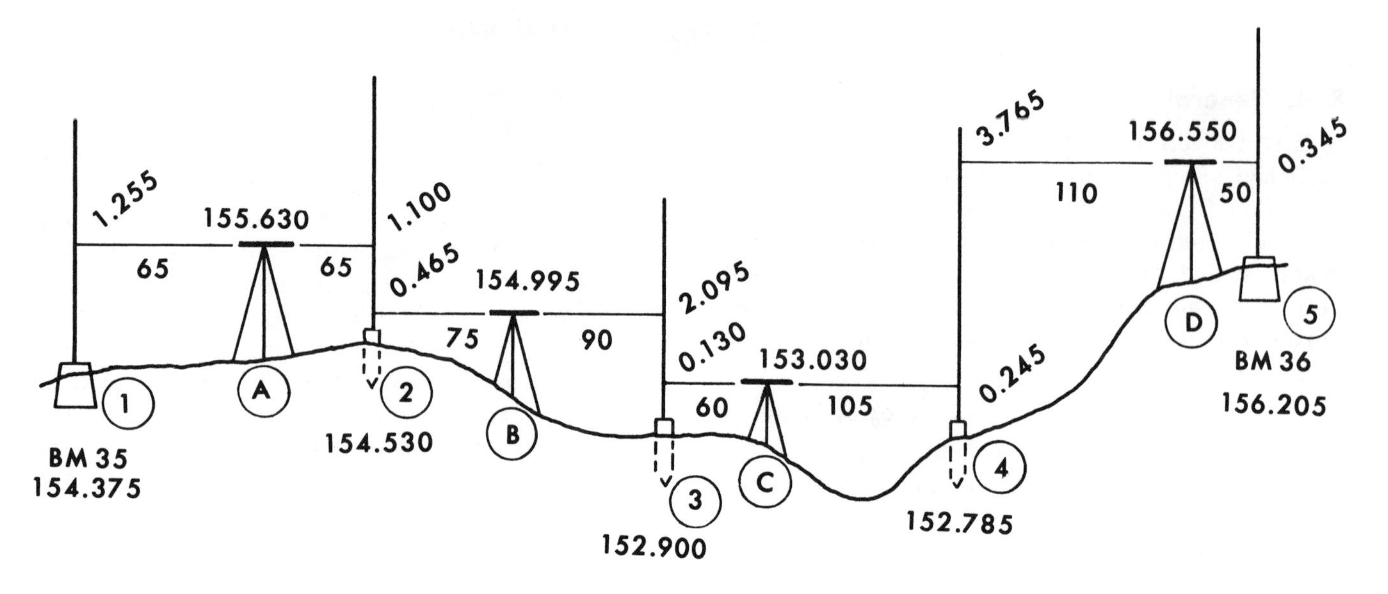

Figure 8-2. Leveling operation.

is determined, this same amount can be paced as long as the slope remains about the same. This procedure balances the distances and makes sure that a level line will fall on the rod. The balancing of the foresight and backsight distances is very important in leveling.

c. Slight errors in line of sight adjusting will distort the horizontal plane into a conical plane above or below the horizontal. Unequal distances between backsight and foresight rod positions will cause an error which will increase in proportion to the distance (fig. 8–3). If the same sight (back or fore) are consistently longer, the error will accumulate. To eliminate this source of error, the level should be set up midway between the TP's. This is not always possible, and the next best method is to balance backsights and foresights at every opportunity. In practice, the distance from the back rod to instrument and instrument to the front rod is measured at each setup by stadia and then recorded. A separate running total of backsight and foresight distances is kept and the two totals should be continually balanced and not left until the last few setups before closing the line or made up by combining one very long and one very short sight. However, if the inequality is introduced by one long sight, then it should be compensated for in one sight before a change in refraction takes place. Balancing of backsides and foresights is used to minimize errors caused by the line of sight not being horizontal. Balancing the sums of the backsights and foresights does not correct curvature and refraction errors, which depend on the square of the distance.

d. A sample differential leveling run is shown in figure 8–2. The rod is held on a point of known elevation, Bench Mark 35, ①. The level is set up at Ⓐ midway between ① and ②. The backsight reading is added to the BM elevation and gives the resulting height of instrument. The distance is read and recorded for balancing. The rod is moved from ① to ②. The foresight reading on ② is subtracted from the HI to get the elevation of point ②. The distance is read and recorded at this point, also. The rod is held at ② which becomes a turning point, and the level is moved to Ⓑ. Here the slope is becoming a factor. The setup at Ⓑ will require a longer foreight than backsight, and point ③ must be selected to get across the valley on one setup. The rod still at ② is rotated to face the instrument at Ⓑ and the backsight and distance are read. Since balancing had been achieved between ① and ②, the new balancing starts again with ②. When the HI at Ⓑ is computed, the rod moves to ③ for a new foresight and elevation. The process continues (level to Ⓒ, rod to ④, level to Ⓓ, and rod to ⑤) and appropriate readings are taken until BM 36 is reached. After point ④ is reached and the valley crossed, instrument position Ⓓ is selected to balance the distances before reaching the next BM. If BM 36 has a previously established elevation, the elevation computed from the level run is compared to it to determine the amount of error in the line. If BM 36 is being established, the level line must continue until a known elevation is reached and the level run value can be checked against it. This known elevation must have been established to an order of accuracy equal to or greater than the accuracy required for the level line.

8-5. Curvature and Refraction

The level surface defined in paragraph 8-1, will follow the curvature of earth.

a. A direct line between two points on this level surface will also follow the curvature of the earth and is called a level line. A horizontal line of sight through the telescope is perpendicular to the plumb line only at the telescope, and is therefore a straight line, not a level line (fig. 8-4). The line

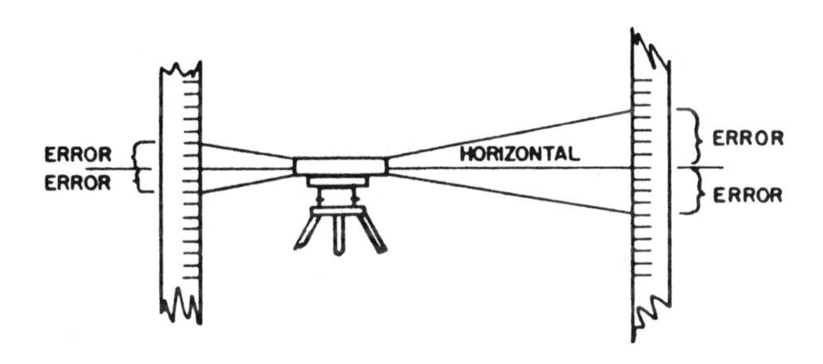

Figure 8-3. Error due to unbalanced sights.

OH is a horizontal line perpendicular to the plumb line at point A. Line OL is a level line and parallels the surface of the earth. At each point, OL is perpendicular to a plumb line. In leveling, as the distance between points increases, the correction for curvature of earth must be applied to account for the difference between a level line and a horizontal line of sight (fig. 8-4).

b. Another correction which must be considered is the atmospheric refraction. The earth's atmosphere will refract or bend a ray of light due to differences in the density of the air between the instrument and the point being viewed. Figure 8-4 illustrates this. Due to the atmospheric density difference the ray of light will follow the path OR. When viewed through the telescope point R will appear to be at point H.

c. To make the correction for curvature and fraction, the computations must first locate point R from point H, and second determine the amount to bring point R down to point L and establish a level line. In practice, the two corrections are combined. RH is about one-eighth of LH, and the value is given by the formula:

$h = M^2 \times 0.0000676$

where:

h is the correction in millimeters

M is the distance between points in meters.

Table 8-1 lists the values for h and M for distances up to 300 meters.

8-6. Party Organization

The organization of the differential level party will depend upon variables such as the order of accuracy required and the number of experienced personnel available. The smallest party consists of two people , an instrumentperson and rodperson. The instrumentperson records the notes. To improve

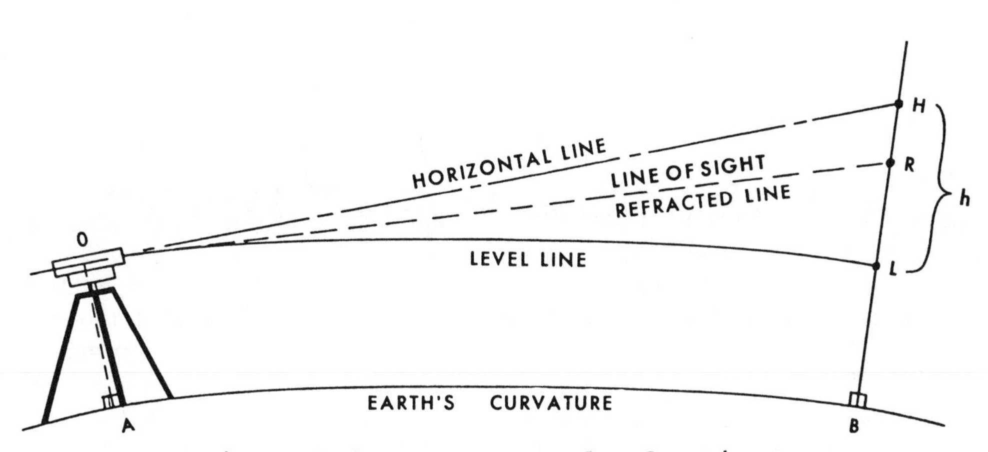

Figure 8-4. Curvature and refraction.

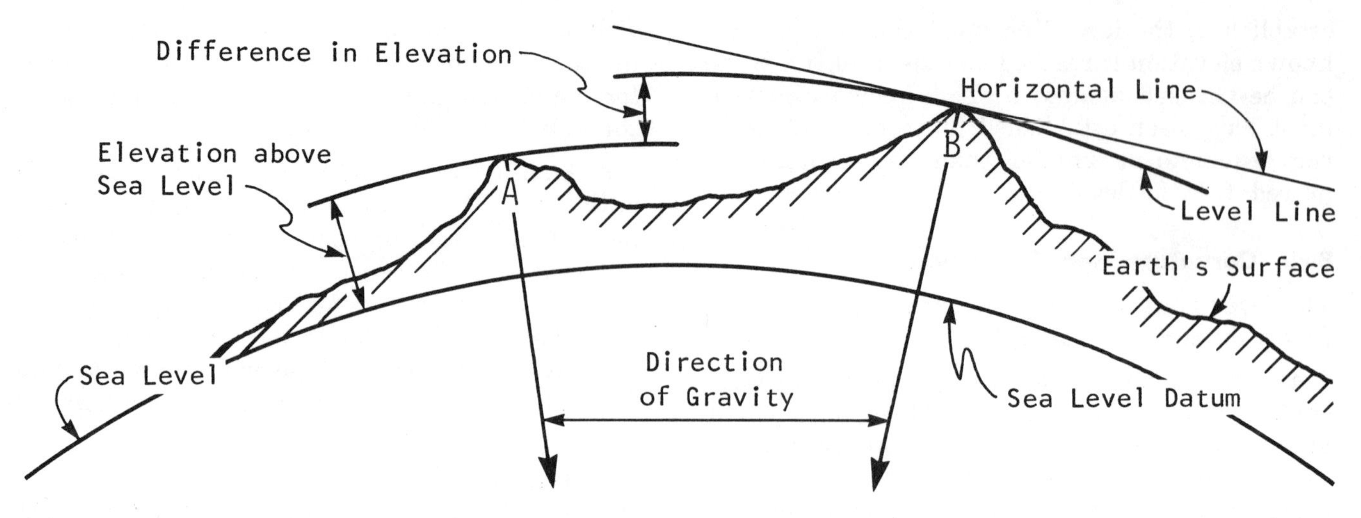

Figure 8-5. Earth's curvature and elevation.

the efficiency of the different leveling operations, additional personnel are required. A second rodperson to alternate on backsights and foresights will speed up leveling. A recorder will allow the instrumentperson to take readings as soon as the rodpeople are in position. In surveys requiring a shaded instrument, an umbrella person is necessary.

Table 8-1. Correction for Curvature and Refraction

Distance meters	Correction millimeters	Distance meters	Correction millimeters
0 to 27	0.0	160	1.8
28 to 47	0.1	170	2.1
48 to 60	0.2	180	2.3
61 to 72	0.3	190	2.6
73 to 81	0.4	200	2.8
82 to 90	0.5	210	3.0
91 to 98	0.6	220	3.3
99 to 105	0.7	230	3.7
106 to 112	0.8	240	4.0
113 to 118	0.9	250	4.3
119 to 124	1.0	260	4.7
125 to 130	1.1	270	5.0
131 to 136	1.2	280	5.4
137 to 141	1.3	290	5.8
142 to 146	1.4	300	6.2
147 to 150	1.5		

8-7. Equipment

The type of equipment needed to run a differential level line depends upon the order of accuracy requirements. The various types of leveling instruments and accessories are discussed in chapter 3.

8-8. Field Procedure

The leveling operation requires the teamwork of both the level and the rodpeople at the moment of reading to achieve consistent results. The survey accuracy depends upon the refinement with which the line of sight can be made horizontal, the ability of the rodperson to hold the rod vertical, and the precision to which the rod is read. Accuracy with instruments using spirit level bubbles must also consider the adjustment of the level vial and the precision with which the bubble axis and the line of sight are made parallel. Self-leveling instruments use a pendulum-prism combination to level the line of sight automatically when the instrument is first brought near level with a circular bubble. Due to the geometry of the prism assembly, the greatest accuracy in leveling the line of sight is achieved when the circular bubble is most accurately centered and the pendulum movement is at a minimum.

a. Instrument. The level must be in adjustment before starting the leveling operation (para 3–5, 3–8, or 3–12). For precise leveling, the adjustment should be checked every day before the start and again at the end of a day's work to verify the adjustment. Other types of leveling require one check a day. Once the instrument is adjusted, operation consists of setting up and leveling it, and taking the readings to the specified accuracy. Taking a reading consists of determining the position where the crosswire apparently intersects the rod, and of recording this value. Each instrument setup requires one backsight reading to establish the height of instrument, and at least one foresight reading to establish the elevation of the forward point (either a turning point or elevation station). Additional foresights may be required to other points visible from the instrument setup for which elevations are also required. Depending on the type of survey and instrument used, either the center wire, all three crosswires, or the micrometer method may be used.

(1) In the one-wire method, only the middle crosswire is used. The levelperson, looking through the telescope, reads the value on the rod that the center wire apparently intersects. If the survey accuracy requires more precise readings, the target with its vernier is used. The levelperson, while sighting through the instrument, signals the rodperson to move the target up or down until the crosswire bisects the horizontal line between the alternate red and white quadrants on the target. When the bisection is achieved, the levelperson signals "OK", and the rodperson locks the target in position until the reading is complete. After locking the target, a check reading should be made to see that the target did not slip during the locking operation. The recorder should be near the level to record direct readings or near the rodperson to record target setting operations. Immediately before taking any reading, the levelperson should check the level bubble and bring it back to center if necessary. A small amount of tripod settling may be encountered, especially if the tripod legs are not set firmly or if the ground is soft. The practice of checking the bubble just before taking each reading will minimize the errors.

(2) The sequence for taking a level reading is as follows:

(*a*) The level is set up and leveled (para 2–8).

(*b*) The telescope is pointed so that its vertical crosswire is just off to one side of the rod and the instrument clamped.

(*c*) The objective is focused and the parallax is checked (para 2–12*c*).

(*d*) The level bubble is checked for centering and improved if necessary.

(*e*) The rod is read and the value recorded.

(*f*) The bubble is rechecked for centering. If it is off center, it must be recentered and the reading repeated.

(*g*) Once the levelperson is satisfied that the bubble has remained centered while the reading was taken, the intercept between the upper and lower wires is read to measure the distance from the level to the rod. This distance is used for balancing foresights and backsights and does not have to be read closer than the nearest centimeter.

(*h*) The levelperson signals "OK" so that the rodperson can proceed to the next position.

(*i*) The telescope is unclamped, revolved, pointed at the next rod position, focused, parallax checked, rod read, and bubble centering rechecked.

(*j*) This method continues until the desired number of foresights are taken and a turning point established. The distance to the rod at the turning point is read and recorded. The rodperson holds the position on the turning point.

(*k*) The level is moved to its next setup position and the procedure repeated.

(3) The target setting procedure requires the levelperson to use hand or voice signals to the rodperson to move the target until the crosswire bisects the target. If the rod must be extended, the target is set and the upper section of the rod is moved up or down until the reading is set. The rod is clamped. After the levelperson has completed the operation and signals "OK", the rodperson reads the rod and vernier if necessary.

(4) The three-wire method requires the same preparation as the one-wire method. It is good practice to check and clear the parallax very carefully for this operation. The position of each horizontal wire on the rod is read separately. They are recorded as the upper, middle, and lower wires. The differences between the upper and middle, and the middle and lower wires are known as the half stadia intervals or thread intervals. The two differences should agree within an allowable amount or the readings must be repeated more carefully. An example of this allowable amount is 0.003 meter for second and third order accuracy. A separate distance reading is unnecessary in this three-wire method of operation, since the difference between the upper and lower wires multiplied by the stadia factor is the distance.

(5) The micrometer method requires the same preparation as the one-wire method. After establishing a horizontal line of sight, the micrometer knob is used to set middle wire (or wedge-shaped lines) of the reticle to the near graduated line (usually centimeter) on the rod. The graduated value of this line read directly from the rod and decimal portion of the reading is read from the micrometer. The distance is found by reading the top and bottom stadia wires as in the one-wire method. The Wild N3 level is equipped with this type of micrometer, and for a detailed description of the method of reading it refer to the manufacturer's operators manual.

b. Rod. The action of the rodperson, and the care with which the rod is positioned and held will affect the speed and the accuracy of the leveling operation.

(1) Before setting the rod on any point, the

rodman should clean the top of the point and the rod shoe for a good contact.

(2) The rod is placed firmly on the point, and the rodman stands facing the instrument and slightly behind the rod, holding it in front of him with both hands. His feet should be spaced about ½ meter apart for a comfortable stance.

(3) The rodman holds the rod as nearly vertical as possible, places a rod level (para 3–17) against the rod, and moves the top end of the rod until the bubbles are centered. Without a rod level, the rod is balanced, using the finger tips to prevent it from falling over. A properly balanced rod will stand for several seconds before starting to fall. This process of balancing the rod vertically is known as *plumbing the rod.*

(4) During a strong wind, it is difficult to plumb the rod and to hold it steady for any length of time. Under this condition, the levelman may call for the rodman to *wave the rod.* The rod is waved by pivoting it on its base and swinging it in a slow arc toward the instrument and away. The shoe should be kept firmly seated during this operation. The motion of the rod permits the levelman to read the rod when it reaches a vertical position at the top of the arc, and the smallest value appears on the rod. Before or after this vertical position, the rod reading will increase.

(5) When setting a turning point, the rodman should set the turning pin or pedestal (para 3–18) firmly in contact with the ground. Any unfirm footing can sag under the weight of the rod and result in erroneous readings between foresight and backsight. During freezing and thawing weather, the ground surface can heave in a comparatively short time. Pins and pedestals can be affected by the heave between a foresight and the following backsight. For higher order of accuracy surveys the instrument man should be aware of this possibility and select firm locations.

8-9. Recording

The field notebook or single sheet recording form is the permanent record of the survey, the notes must be clear, legible, and recorded with a reproducible ink, preferably black or blue-black. Leveling notes are kept in the Level, Transit, and General Survey Record Book or on any single sheet recording form. No survey recording is considered complete until the notes and computations have been checked and initialed by the chief of party or the designated representative.

a. One-Wire Method Recording. A form is used for the one-wire method and is completed as shown in figure 8–6. The recording illustrated in this figure is based on the leveling operation shown in figure 8–2.

(1) The leveling notes start with a known elevation or bench mark which is generally described from previous surveys. The identification of this point and its elevation are entered in the proper columns on the left-hand page. The right-hand page must show a reference to the source of the elevation and description. At times a level survey is run to establish grade and the exact elevation above a datum plane is not necessary nor readily available. A more or less permanent point is selected as a starting point and a fictitious value assigned to it for use on the survey. This elevation and all elevations determined from it can be tied in to a known elevation at a later date. The notebook must describe this point and state that the elevation was assumed.

(2) The first reading is a *backsight* (BS = 1.255) and is *added* to the elevation (154.375) to obtain the height of instrument (HI = 155.630).

(3) The next reading is a *foresight* (FS = 1.100) and is *subtracted* from the HI to get the elevation (Elev. = 154.530) of the next point (TP ② or turning point in the sample notes).

(4) The first instrument setup Ⓐ was selected midway between the bench mark ① and the turning point ②. This appears in the distance column as 65 meters on the left-hand side to show the distance from BM to instrument, and 65 meters on the right-hand side for the distance from instrument to TP.

(5) The instrument is moved to the next setup Ⓑ while the rod remains at the turning point ②. The backsight (0.465) and the distance to the rod (75) are read and recorded, and the HI computed (154.995). The instrument is pointed at the next turning point ③, and the foresight (2.095) and distance (90) are read and recorded. The elevation (152.900) is computed.

(6) This method continues until the survey is tied to the next BM. The distance balancing (sum of foresight distances to equal the sum of backsight distances) is done as the survey progresses, and should be completed before the bench mark is reached.

(7) The recorder's computations can be verified by adding the backsight rod readings and the foresight rod readings separately. The difference between the two totals is the difference in elevation between the starting and final elevation.

Project 537/2S/LL/A/

DESIGNATION BM35 to BM36 DATE 15 July 19__

Party Chief: J. Doehn
Instr.: W. Cael
Recorder: R. Roe
Rodmen: C. Day & B. Rayht
Instr: Mil-10x #2312
Weather: Clear, Calm, & Hot

Station	B.S.	H.I.	F.S.	Elev.	Dist.	Remarks
	+		−			
① BM35				154.375 ✓		✓Elevation and Description from third order line 2, Pg. 18, Project 537/2/S/3L/2/70-Book #3. - Recovered as described.
					65	
Ⓐ	1.255	155.630 ✓				
					65	
② TP			1.100	154.530 ✓	65✓ 65✓	
					75	
Ⓑ	0.465	154.995 ✓				
					90	
③ TP (PPM14-3-1382A(V))			2.095	152.900 ✓	140✓ 155✓	✓ PPM14-3-1382 A (V) described on page 15, this book.
					60	
Ⓒ	0.130	153.030 ✓				
					105	
④ TP			0.245	152.785 ✓	200✓ 260✓	
					110	
Ⓓ	3.765	156.550 ✓				
					50	
⑤ BM36			0.345	156.205 ✓	310✓ 310✓	✓Elevation and Description from third order line 2, Pg. 18, Project 537/4S/3L/2, Book #5 - Recovered as described.
	+5.615		−3.785			
	−3.785 ✓					
	+1.830 ✓					
	154.375 ✓					
	156.205 ✓ check					✓ JEB

Figure 8-6. Lower order level notes.

Applying this total to the starting elevation should result in the final elevation. Any disagreement is the result of an error in the computations which must be rechecked.

(8) In the example, the curvature and refraction correction (para 8–5) would not be applied since the correction for the longest distance would only be 0.8 millimeter (table 8–1), and would not affect any of the readings.

8–10. Reciprocal Leveling

This procedure is used for either differential or trigonometric leveling when a long sight across a wide river, ravine, or similar obstacle must be made. This long sight will be affected by curvature and refraction, and by any small error in alining the line of sight with the bubble axis. The alinement error can be minimized by balancing the long sight, and the curvature can be computed. The atmospheric conditions will vary so much over an open expanse that the refraction correction will be very erratic. Reciprocal leveling is designed to minimize the effect of the atmosphere as well as the line of sight and curvature corrections.

a. In reciprocal leveling (fig. 8–7), the backsights and foresights are balanced as carefully as possible before reaching the obstacle. A turning point (N) is selected close to the edge of the obstruction, so that it is visible from a proposed instrument location (B) on the other side. A second rod is held on the other side of the obstruction at F. Point F should be selected so that the equivalent distances AN and FB, and AF and BN are approximately equal. The instrument is set up at point A and leveled very carefully. A backsight reading is taken on the N rod and a foresight on the F rod. These readings are repeated several times. The instrument is moved to point B, set up, and carefully leveled. The rods remain at their stations. Once again, a backsight is taken on the N rod, and a foresight on the F rod, and repeated several times. Since instrument leveling is very critical in this procedure, the bubble must be checked before each reading and centered carefully. If it is off center a slight amount, the procedure must be repeated. The difference in elevation between N and F is computed from the readings

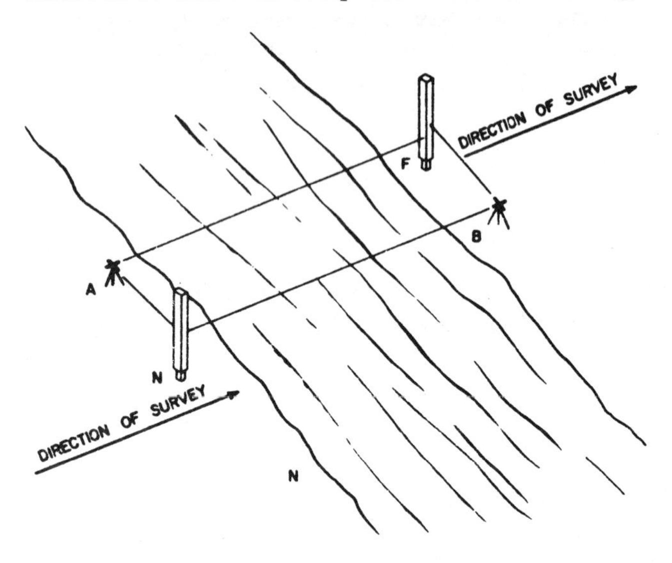

Figure 8-7. Reciprocal leveling.

at A setup and from the readings at B setup separately. Due to the errors in the long sight, each result will have a slightly different value. Note, however, that the long sight is a foresight from A and a backsight from B. The true difference in elevation is the average of both values since the errors have opposite signs and will cancel each other.

b. For more accuracy, several long sight readings are made for each short sight and averaged. A target should be used on the rod, and reset for each reading. Each series of long sights is averaged and combined with corresponding short sights for the computations.

c. Changes in atmospheric density and temperature affect the refraction of a line of sight. The longer the time interval between reciprocal long sights, the greater the chance of an atmospheric change and a variation in the refraction value. For this reason, the time lapse between the long sights should be kept as short as possible.

d. An excellent method of avoiding the time lapse problem is simultaneous-reciprocal observation. The object is to read both long sight values at the same time. This requires two instruments and observers and two rods and rodmen. Some method of communication or sequence of operations must be agreed upon. For example, the sequence can be to take one short sight and five long sights for two cycles. Both observers read their short sights first, then signal the rodmen to turn the rod to face the long sight observer. As soon as the observer sees the rod on the opposite shore facing him, he starts his long sightings. As soon as he takes his five readings, he should make certain that the other observer is also finished. If not, he should continue taking readings for the same length of time even though he may have six or seven to the other observer's five. His seven will bracket the time of the five and the average of each will refer to the same time interval and to the same atmospheric conditions. This procedure can be repeated as many times as agreed upon. The instruments then change stations and the procedure is repeated. The computations will result in a mean value for the difference in elevation that will minimize the long sight errors.

e. The notekeeping for reciprocal leveling is identical to differential leveling. A series of either backsight or foresight readings is taken on the far rod from one setup, and only one sighting on the rear rod. The series of readings are averaged, and a single value is used to make the elevation computation. Each reading of the series is entered on a separate line in the proper column. The computation is made directly below the last reading of the series and the average value is marked. From this point on, the recording follows the regular procedure.

8–11. Water Level Transfer Method

This method has a definite application in areas near large bodies of water, such as lakes or ponds, and in tidal waters. Elevations may be transferred using the water level for distances up to 16 kilometers.

a. Two points are selected—one from which the elevation will be carried forward (near point), and the other to which the elevation will be transferred (far point). A temporary bench mark (para 8–12) is established *below the water surface* at each of the points.

b. Rods are held on each of the bench marks and the elevation of the water surface is read.

c. Observations are made simultaneously at both points at 5-minute intervals for a period of 30 minutes on 2 separate days.

d. The difference in readings is the difference in

elevation between the points, and when added to the elevation of the near point will establish the elevation of the far point. The section of level line which is run to establish the near elevation must be double run to insure the accuracy of the transfer.

e. Observations must follow certain precautions as shown below:

(1) Water level observations must be made on calm days since the wind can build up water on the leeward side of the lake and can also create rough water making reading difficult.

(2) Observations must not be made on opposite sides of causeways or narrow straits, or in lakes connected only at high water stages. These conditions result in a lag in water movement and in different levels of water at the two points.

(3) When observing in tidal waters, tide tables should be consulted to determine if any difference in tide exists between the two points. Using the two nearest tide stations, on either side of the required. TBMs, the difference in time and height of the tides should be computed for both points. The distance between tide stations and the distances to each of the points from the stations should be used to proportion the difference. If the time or the height is not the same at both points, water level transfer should not be used. In this case, tidal observations should be used to compute the elevations.

(4) In all cases, the use of running water or tidal backwater in rivers should be avoided.

8-12. Level Bench Marks

A bench mark is a relatively permanent object, natural or artificial, bearing a marked point whose elevation is known. A bench mark may be further qualified as permanent, temporary, or supplementary. The purpose of a survey normally will govern whether its stations will be permanently or temporarily marked. When it is known that a station may be reused over a period of several years, the station marker should be of a permanent type. Other stations that may never be reused, or reused a few times within a period of 1 or 2 months, usually are marked in a temporary manner. A permanent bench mark normally is abbreviated BM, whereas a temporary or supplementary bench mark is called a TBM.

a. Bench marks set to third order or higher accuracy are intended to form a framework of basic control to which detail surveys are adjusted. The elevations are marked in a permanent manner where they will not be disturbed either by normal native activity or by frost action. The marks are metal caps or disks set in concrete or stone posts, in rock outcrops, or in masonry structures. The disks carry an inscribed name of the organization which set the mark, the name or number of the mark for future identification, and the date the mark was set (fig. 8–8).

b. Bench marks of lower-order accuracies may also be set as permanent markers, and referred to as BMs, but the degree of permanence may only be for a few years. Monumenting, as described in *a* above, is not required. A concrete filled iron pipe, or a chiseled square on a permanent structure (bridge abutment, culvert, or building steps) or rock outcrop serves the purpose. This type of mark is not readily visible and should be carefully described for future recovery. Location of these marks from scanty or inaccurate descriptions can waste a tremendous amount of time. An ideal description should lead a person unfamiliar with the original marks to the spot where it was set. If by chance the mark was destroyed, this should make it obvious that a long search will be a waste of time. Marking this type of bench mark also requires some ingenuity. Since no identification disk is set, the name, number, and elevation of the point cannot appear on the mark. Bench marks can be marked with paint on rock or masonry if it will not create a nuisance or instill a curiosity in some native who will destroy or remove it for some inherent value. A blaze on a tree, a mark on a nearby fencepost, or a stake lettered with a lumber crayon (keel) can be used as a guide which will lead a surveyor with the description to the bench mark.

c. Temporary bench marks are set for comparatively short time use, possibly up to several months. A spike or a 30- to 40-penny nail, driven into a tree or telephone pole and extending about 1 centimeter from the surface, will serve the purpose. A stake driven into the ground can also be used especially if no frost action is expected until it is needed. The detailed description of these points is just as important as the monumented stations.

d. Bench mark systems or *level nets* consist of a series of bench marks which are established within a prescribed accuracy along closed circuits and which are tied to a datum. These nets are adjusted by computations which minimize the effects of accidental errors, and are identified as being of a specific order of accuracy.

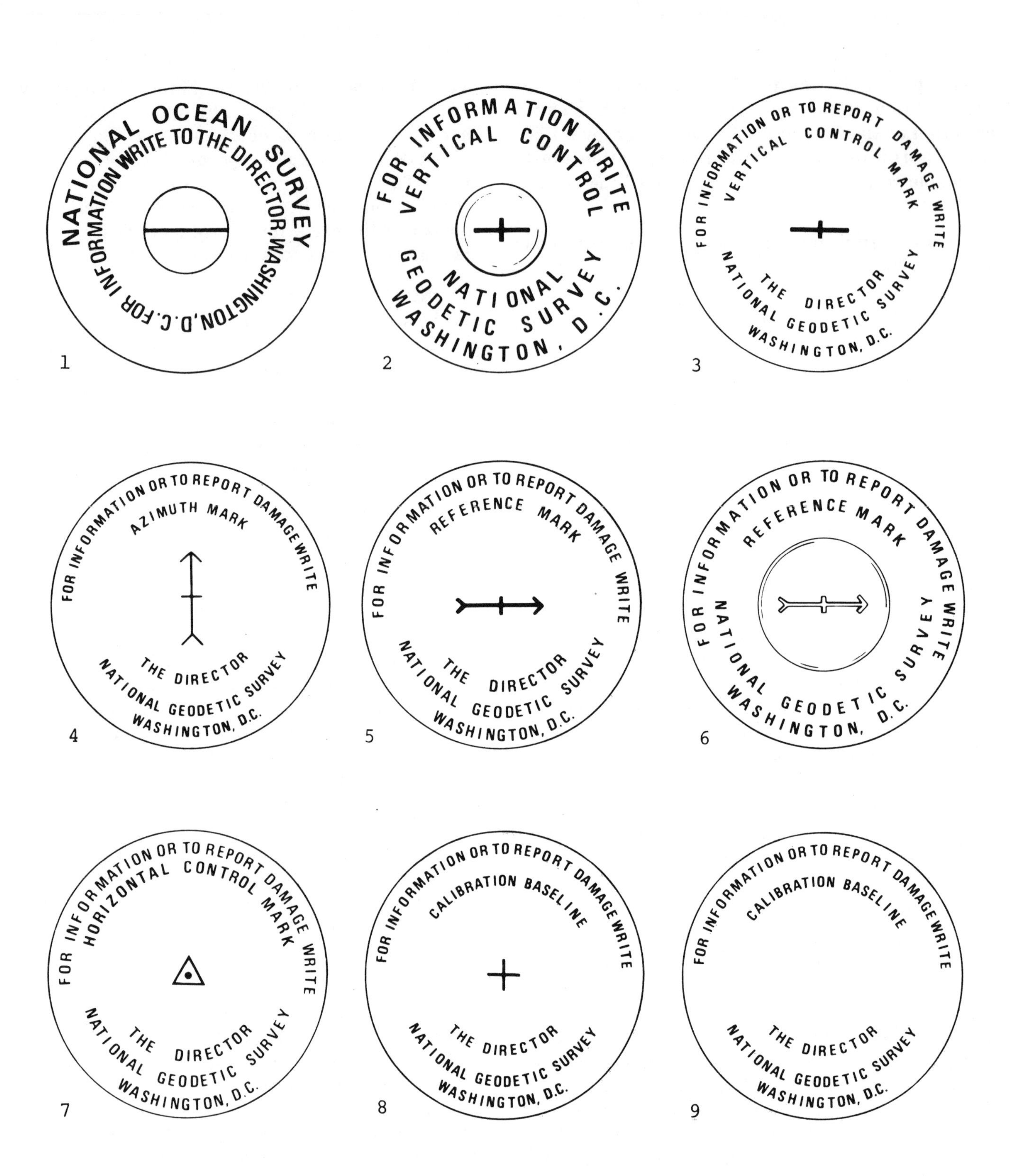

1. Traverse station mark
2. Reference mark
3. Azimuth mark
4. Triangulation station mark
5. Gravity station mark (old type)
6. Magnetic station mark
7. Topographic station mark
8. Geodetic bench mark (new type)
9. Gravity station mark (new type)

Figure 8-8. Standard bronze marks of the National Ocean Survey/National Geodetic Survey.

e. In certain areas, tidal bench marks must be established to obtain a starting datum plane or to check previously established elevations. *Tidal bench marks* are permanent bench marks (*a* above) set on high ground and tied to a tide station near the water surface by accurate spirit levels.

(1) Tide stations are classified as primary and secondary. Primary stations require observations for periods of 19 or more years to derive basic tidal data for a locality. Secondary stations are operated over a limited period (usually less than 1 year) and for a specific purpose (such as checking elevations). The secondary station observations are always compared to and computed from data obtained by primary stations.

(2) A tide station is set up and observations are made for a period determined by the desired accuracy. These observations are compared against a primary tide station in the area and furnish a mean value of sea level in the area.

(3) A closed loop of spirit levels is run from the tide station over the tidal bench marks and tied back to the tide station. This level line must be of the same or higher accuracy than the required bench marks.

(4) For permanency, tidal bench marks usually are set in sets of three away from the shore line where they are not all likely to be disturbed or destroyed by natural acitivity or by future construction.

8–13. Levels With a Direction Instrument or Alidade

The direction (angle) observing instruments can also serve as levels in many surveys. A level bubble is either permanently mounted on the telescope (ch. 4) or is an auxiliary item (para 7–2*b*(5)) issued with the telescopic alidade. The alidade striding level is snapped into place when observing level sights. The sensitivity of these bubbles is not as great as those used on levels, but for lower orders of accuracies they are sufficiently precise. Those bubbles to be used for leveling must be adjusted with their axes parallel to the telescope line of sight. Once this adjustment is made and the bubble leveled, locking the vertical motion converts the instrument into a level. Its operation as a level is the same as described in paragraph 8-8. The only difference is that the vertical slow motion screw may be used to recenter the bubble, if necessary, before reading the rod.

8–14. Profile Leveling

In surveying, a profile is a vertical section of the earth measured along a predetermined or fixed line. The profiles may be used to determine the final grade or alinement in the construction of roads, railroads, and sewers. They may also be used to compute volumes of earthwork.

a. Profiles are a series of ground elevations determined by some method of leveling. In practice, some line such as the centerline of a road, is selected. Differences in elevation are measured and elevations are computed for each station and for each definite change in the slope of the ground. The elevations when plotted to scale on a sheet of cross section (or profile) paper become the profile along the centerline.

b. The leveling method used in profiling depends upon the required accuracy. For rough estimates, an Abney hand level and range pole may be sufficient. Finished construction requires differential leveling with telescopic levels and Philadelphia rods.

c. The profile when drawn can be used to determine the final grade of the road. A series of parallel profiles taken at a fixed distance apart can be used to compute volumes of earth for areas to be cut or filled.

d. Profiles may also be drawn for finished structures to check the final grades against the design specifications. These profiles may or may not be plotted on paper. In some instances, the mathematical computation of grade between points will be sufficient.

8–15. Adjustment of Leveling

The error of closure is the amount by which the quantity obtained from surveying operations fails to agree with the equivalent value accepted from previous surveys. Survey computing is not complete until the closure error is distributed. This process is called adjusting. In precise surveys, the error is adjusted using the theory of least squares, and in some cases, weighted means. The operations discussed in this book normally do not require these precise mathematical adjustments. Instead, they usually take the form of straight line adjusting or prorating. This method assumes that the total error is the result of accumulating a small error at each setup, and that the amount of small error was the same each time. The total error of closure is divided by the number of setups and applied progressively from the beginning to the end of the survey. For example, if 10 setups are made, the total error is divided by 10. The first station or setup gets one-tenth. The error between the first and the second setup would also be one-tenth. Since the first setup is corrected by one-tenth also, the total for the second setup is two-tenths. The third setup would get three-tenths and so on until the correction at the tenth or closing point is ten-tenths or the full amount of error. In leveling, the adjustment is distributed over the turning points which carried the elevation through the survey. All elevations determined as sideshots from an instrument setup must also be corrected. Their correction value is based on the turning point from which the height of instrument was established.

Section III. TRIGONOMETRIC LEVELING

8–16. General

This method applies the fundamentals of trigonometry to determine differences in elevation (fig. 8–9). There are two applications of this method normally used by the surveyor; on long lines of sight for triangulation and electronic traverses and on short lines of sight for conventional traverses and level lines. The procedures and techniques in this chapter pertain only to the short line application. Trigonometric leveling is used only for lower order accuracies where the terrain is prohibitive to differential leveling or when leveling is needed in connection with triangulation and traverses.

8–17. Description

Trigonometric leveling requires a transit, theodolite, or alidade to observe the vertical angles needed in this method. This method is particularly adaptable to uneven terrain, where level sights would be short due to the ground slopes and distance balancing, and for low order surveys where time is a consideration. Distances should be kept below 300 meters when a stadia or standard leveling rod is used, and the curvature and refraction correction (para 8–5) is applied only if the survey accuracy requires it. Trigonometric level surveys should be tied in with sideshots to higher order elevations whenever possible.

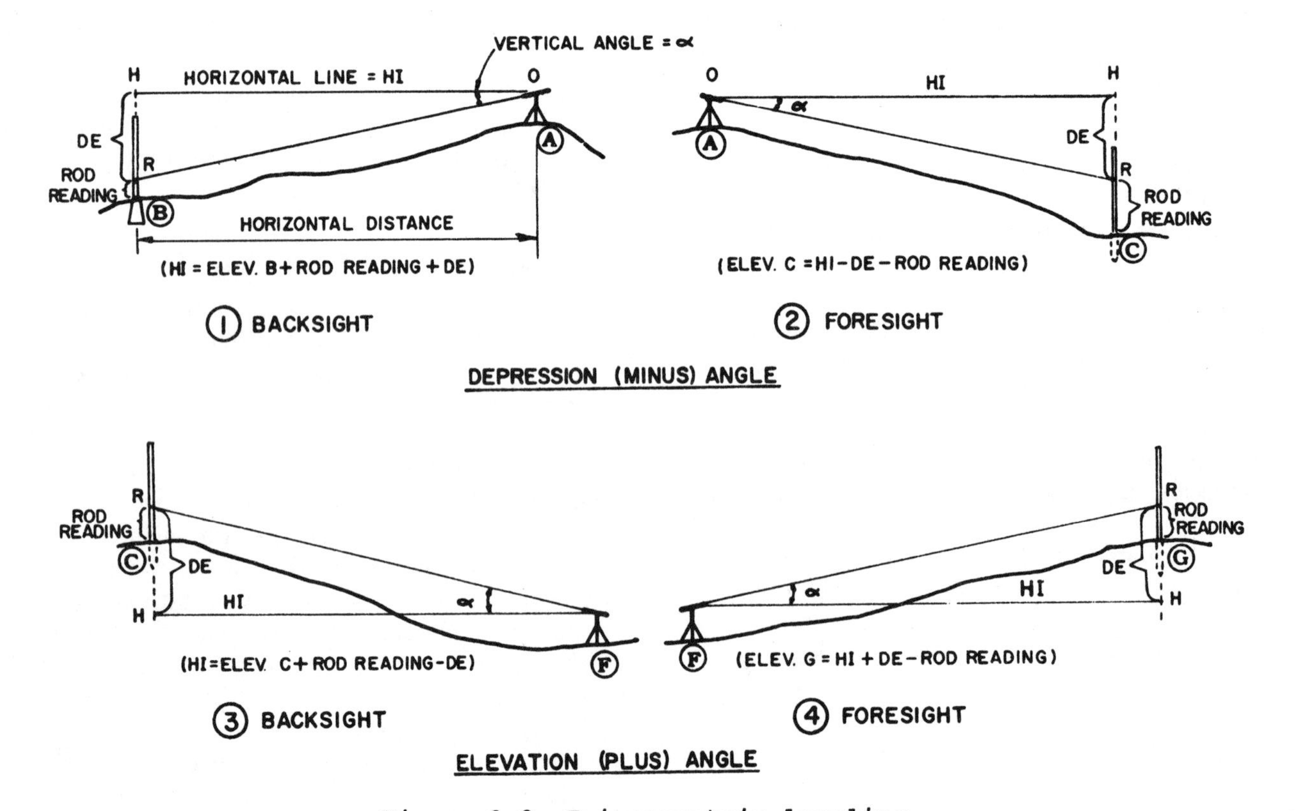

Figure 8-9. Trigonometric leveling.

a. The instrument is set up and leveled at a convenient location to see the starting point and the first turning point. The rod is held on the starting point. The telescope is pointed at some easily read value (a full meter) on the rod, and the vertical angle is read. The distance between the instrument and rod must be determined either by taping, by a stadia reading, or in some instances by triangulation. Now, one side and one angle of a right triangle are known (fig. 8–9). The other sides and angle can be computed. For trigonometric leveling, only the side opposite the measured angle, the difference in elevation or DE, is computed. The computation consists of multiplying the measured distance by the proper trigonometric function of the measured angle (sine; if slope distance, OR, is measured; tangent; if horizontal distance, OH, is measured). The result is the difference in elevation, DE, between the HI and the point on the rod, R. The rod reading (like in differential leveling) is added on backsights and subtracted on foresights. The computed DE is applied in the proper direction to obtain the HI or the elevation as required.

(1) *Depression (minus) angle backsight (①, fig. 8–9)*. The rod is on a point Ⓑ below the instrument. The measured vertical angle (a) is minus or a depression angle. The measured distance is either slope (OR) or horizontal (OH). The required DE (HR) equals the distance multiplied by the sine or tangent of the angle (a). To compute the HI, the rod reading, RB, and the DE are added to the elevation of B, or HI = RB + DE + Elev. B.

(2) *Depression (minus) angle foreseight (②, fig. 8–9)*. The rod is below the instrument, and the vertical angle is minus. The DE is computed as in (1) above. The elevation at C equals the HI minus the DE and minus the rod reading RC, or Elev. C = HI — DE — RC.

(3) *Elevation (plus) angle backsight (③, fig. 8–9)*. The rod is above the instrument, and the vertical angle is plus. The DE is computed as in (1) above. The HI at F equals the elevation at C plus the rod reading, RC, and minus the DE,

or Hi = Elev. C + RC — DE.

(4) *Elevation (plus) angle foresight (④, fig. 8–9)*. The rod is above the instrument and the angle is plus. The DE is again computed as in (1) above. The elevation of G equals the HI plus the DE and minus the rod reading, RG,

or Elev. G = HI + DE — RG.

b. The distance between the instrument and the stations must be known in trigonometric leveling to compute the difference in elevation. This distance may be taped, measured electronically, or read by stadia. It may be a part of another survey (such as traverse) or it may have to be measured during the leveling.

(1) Horizontal distances are simply multiplied by the tangent of the angle to get the difference in elevation. No reduction is required once the proper corrections to the measured distance are applied to get the true horizontal distance.

(2) Slope taping distances must be converted to a horizontal distance before being used in this procedure. See appendix D, for tables which list the following:

(*a*) Inclination corrections for a 50-meter tape (table D–2).

(*b*) Differences in elevation for given horizontal distances and gradients from 0° to 45° (table D–3).

(*c*) Differences in elevation for given slope distances and gradients from 0° to 45° (table D–4).

(*d*) Horizontal distances for given slope distances and gradients from 0° to 45° (table D–5).

(*e*) Differences in elevations and horizontal distances from stadia readings (table D–6).

(3) Electronic distance-measuring devices measure the slope distance between instruments. This is the straight line distance from unit to unit. If the same setup is used, and the electronic equipment is replaced with a theodolite and target or rod, the measured vertical angle can be used to convert the measured distance to a difference in elevation by multiplying by the sine of the angle.

(4) Stadia distances fall into two categories. If the instrument is level when the distance is read, the value is converted directly to a horizontal distance. When the line of sight is moved up or down from the horizontal, the rod reading cannot be converted directly, but requires an additional reduction. Stadia is more fully described in paragraph 6–26. Stadia reduction tables are provided in appendix D. The rod intercept (interval) and vertical angle are used with the tables and yield a horizontal distance and a vertical difference in elevation.

c. The vertical angle used in trigonometric leveling is the angle above or below a horizontal plane, and is designated by a plus or a minus, respectively.

(1) The transit's vertical circle is graduated from 0° to 90° on each of four quadrants. A horizontal line reads zero in either the direct or the reverse position. Vertical angle values will increase whether the telescope is elevated or depressed. The vertical angle is read on the vernier and the sign depends on the telescope position; plus if elevated, and minus if depressed.

(2) The 1-minute and 1-second theodolites use zenith distances; that is, a level sight will read 90° on the circle. As the line of sight is elevated, the value on the circle decreases, and must be subtracted from 90° to give a plus vertical angle. The depressed line of sight reading will be greater than 90° and the amount greater is the minus vertical angle. In the reversed position, the 1-minute and 1-second theodolites read 270° for a horizontal line. The amount above or below 270° is the vertical angle, plus or minus, which is used in trigonometric leveling.

(3) The alidade has a stadia arc with three scales. The center scale is the angle scale and is read by vernier to the nearest minute. Its value for a level line is 30°. Vertical angles are determined by subtracting 30° from the reading. Elevation angles will give a plus result and depression angles, a minus. The other two scales marked H and V can be used in leveling. Their theory and use are described more fully in paragraph 7–3.

Line# 30/FBV /4LL/T DATE: 2 MAY
Obs - J. Doehn Rod C. King Inst
Rec - W. Roe Rod L. Wills T-2 # 14694
From - BM T-6 to BM T-7

SIGHTING	SLOPE DISTANCE	TELESCOPE	ZENITH ° '	DISTANCE "	VERTICAL ° '	ANGLE "		ROD INDEX	HI	ELEV	STATION
											BM T-6
BS	150	D	92 12	13	-02 12	13 ✓		2.0			
		R	267 47	45	-02 12	15 ✓		2.0			
Sum ¢	Mn		359 59	58 ✓	-02 12	14 ✓					
FS	160	D	87 26	42	+02 33	18 ✓		2.0			
		R	272 33	20	+02 33	20 ✓		2.0			
			360 00	02 ✓	+02 33	19 ✓					
											TP#1
BS	135	D	94 32	21	-04 32	21 ✓		2.0			
		R	265 27	36	-04 32	24 ✓		2.0			
			359 59	57 ✓	-04 32	22.5 ✓					
FS	126	D	85 41	33	+04 18	27 ✓		2.0			
		R	274 18	25	+04 18	25 ✓		2.0			
			359 59	58 ✓	+04 18	26 ✓					
											TP#2
Observation to continue in this manner											
until reaching the tie station											

Figure 8-10. Recording notes, stadia method.

8-18. Limiting Factors

The chief source of error in this method is in determining the true distances between the turning points (TP's). This error can be kept to a minimum by accurately determining the stadia contant (para 3-8*c*), limiting the length of the line of sight to 300 meters and the vertical angle to 4 degrees or less. Vertical angles from 6 to 9 degrees are acceptable providing the length of the sight is reduced proportionally. In no case should the vertical angle exceed 9 degrees. Other precautions to be taken are keeping the lower stadia wire at least 0.6 meter above the ground when reading stadia and using a rod level to plumb the rods.

8-19. Party Organization

The organization of the trig-level party is generally that of an instrumentperson (chief of party), a recorder, and two rodpeople, but this will be governed by the terrain and type of vegetation in the area. Where considerable brush cutting is necessary, additional personnel are required.

8-20. Procedures

The allowable error of closure will generally dictate the method and procedures to be used for lower order trigonometric level lines, and this will be specified in the project instructions.

a. Stadia Method.

Line # 30/FBV/ALLIT Date 1 May — Obs & Rec - J. Doehn Instr. #14672; Rod - W. Roe; From BM T-1 ( ) to BM T-3 ( )

DIST FEET	BEAMAN ARC	PRODUCT	ROD CORR	DIFF ELEV	HI	ELEV	Remarks
						413.60	BM T-1
990	50	—	+10.15	+10.15	423.75		
785	50	—	-2.45	-2.45		421.30	TP#1 — TP is Center of bush
450	51 (1)	-4.50	+7.10	+2.60	423.90		
765	52 (2)	+15.30	-3.00	+12.30		436.20	TP#2
695	49 (1)	+6.95	+3.15	+10.10	446.30		
540	48 (2)	-10.80	-9.60	-20.40		425.90	TP#3
995	50	—	+6.18	+6.18	432.08		
880	50	—	-5.00	-5.00		427.08	TP#4
1010	50	—	D 4.70 R 4.80	+4.75	431.83		
660	50	—	+1.90	+1.90		433.73	PPM14-26-1432 A(V)
705	50	—	+4.15	+4.15	437.88		
840	51 (1)	+8.40	-10.00	-1.60		436.28	BM T-3
							Prev. Ad. Elev BM T-3 435.14
							This Line Elev BM T-3 436.28
					JED		Closing error +1.14

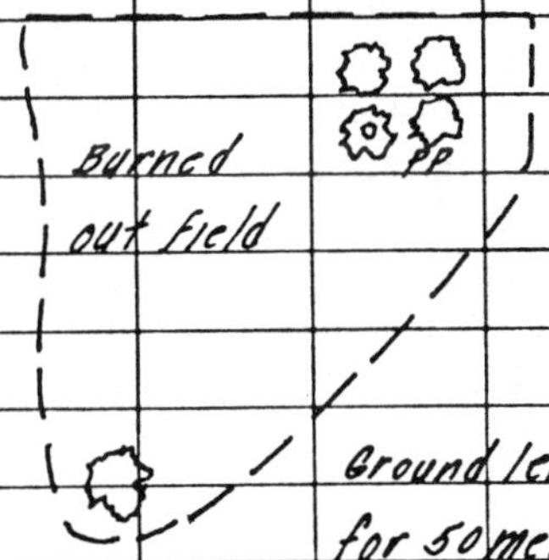

Burned out Field

PP

Ground level for 50 meter radius

Figure 8-11. Recording notes, alidade method.

(1) To run a stadia-trigonometric line of sight, set up the instrument at a convenient point along the proposed line, measure and record the height of instrument (para 7–3), and take a backsight on the rod held on the bench mark. Set the middle wire on an arbitrarily chosen graduation of the rod to be used as an index mark (usually 2.00 meters) and record the value. Then set the bottom wire on the nearest whole meter and read and record the stadia distance. Reset the middle wire on the index mark and observe and record the vertical angle(fig. 8–10). The telescope is reversed and vertical angle read again. The stadia distance should also be checked in the reverse position. The instrument is then turned in the direction of the forward rod and a foresight is taken on the selected turning point using the same instrument procedure. The instrument is moved to a new location ahead of the front turning point as in differential leveling, and the process repeated. When using a theodolite the zenith distance (ZD) is observed and the vertical angle (VA) computed

(2) If the line of sight from the instrument to the rod is inclined, the stadia distance is not the true horizontal distance and must be corrected. This correction depends upon the observed stadia distance and the vertical angle. Distances should not be determined by half-stadia intervals. If it becomes necessary to do so, the separate half intervals should be observed and their sum taken as the total distance.

(3) The recorder must check the stadia and vertical readings before the instrument and rod is moved. The elevation of the forward point (TP, PBM, TBM or RM) is determined by carrying the elevation of the back point to the instrument (HI) and then to the forward point shown in figure 8–9 and described in paragraph 8–17*a*.

(4) Whenever the line of sight with the telescope horizontal will intersect the rod, it is preferable to read the rod as in differential leveling (para 8–13).

b. Short Base Method. When it becomes necessary to use lines of sight over 300 meters and the EDME's are not available, the short base method should be used. This method is performed in a similar manner as the short base traverse. The short base lines must be measured to at least third order traverse accuracy with a ratio between the measured and computed lengths not to exceed one part in 50 parts.

c. Electronic Measuring Methods. Electronic distance measuring may be used in conjunction with the theodolite for trigonometric leveling in the same manner as stadia distances are used in *a* above. Due to the increase in accuracy of the measurement of the slope distance, the length of the line of sight may be extended to meet the situation. At any time the length of the line of sight exceeds 300 meters, simultaneous reciprocal angles must be observed.

d. Alidade Method. The alidade and planetable may be used for lower order elevations. When using the alidade the vertical angle is usually measured in terms of Beamans, instead of degrees. The procedure is the same as in *a* above. The difference in elevation and distance determination using the scales of the stadia arc is described in paragraph 7–3.

Section IV. BAROMETRIC LEVELING

8–21. General

a. On certain survey projects in remote and difficult terrain, the accuracy requirements for vertical control (elevations) may be lowered to such an extent that the barometric method of establishing elevations becomes the most practical and economical, particularly if modern transportation methods such as the helicopter are available. The procedures and techniques discussed in this section are based on the assumption that helicopters will be used to transport the field parties for establishing elevations by the barometric method.

b. Barometric (or altimeter) surveys are run by one of three methods—the leapfrog, the single-base, and the two-base. The single-base method requires a minimum of observers and equipment. However, it needs a series of corrections and is neither as practical nor as accurate as the other two. The two-base method is generally accepted as the standard method for accuracy and is most widely used in engineer surveys. It requires fewer corrections than the single-base method. The leapfrog method uses the same type of corrections as the single-base, but the altimeters are always in close relationship to each other and are operating under reasonably similar atmospheric conditions. The results are more accurate than the single-base method but less accurate than the two-base method.

8-22. Preparation

Before the starting of any barometric survey and regardless of the method used, certain precautions, considerations, and corrections must be observed and/or computed.

a. Weather. Accuracy of barometric surveys is primarily related to the prevailing weather conditions (para 3-21*b*). The weather must be favorable or the results will be inaccurate. Wind velocity is an excellent indicator of the degree of atmospheric stability and should be used as a guide. Wind velocities of 8 knots and below (Beaufort No. 3 or less, table 3-1) are considered favorable during the leveling. In wind velocities between 8 and 13 knots, extreme caution must be used. Barometric leveling should never be attempted in winds exceeding 13 knots. The early morning hours (0600 to 0900), and the early evening hours (1600 to 2000) present the most stable pressure conditions while midday presents the poorest. The surveys should be planned to make the greatest use of the better hours. Overcast days are more suitable than sunny days, especially when the cloud cover is great enough to shut out the sun's rays and create uniform temperatures and stable atmosphere. Avoid observations during thunderstorms and whirlwind or squall conditions.

b. Terrain. Terrain alone does not present serious problems in altimetry. However, its effect on the local atmospheric conditions cause air pressure instabilities and misleading results. As the sun rises, it heats one side of a hill from the top down. The heat from different objects warms the air surrounding them. This warm air rises and the cooler air from the higher altitudes settles, creating air currents and local winds. This air movement, in turn, causes local variations in the atmospheric pressures which are not reflected at any base station altimeters. In mountainous terrain, best results are obtained on or near the tops and ridges and the poorest in the valleys and low spots. Temperature and air pressure differences can also occur between areas in the shade and those in the sun. The most consistant pattern is obtained in flat or gently rolling country with a similar type of vegetation where air disturbances are gentle and temperatures change more or less evenly throughout the day. Altimeters must be shaded during the time of observations.

c. Watch Synchronization. In altimetry, the readings are made on points which normally are not intervisible. To make sure that the field altimeters and those at the base stations are read at the same time and during the same atmospheric conditions, a specific procedure has been developed. This procedure is based on time and requires accurate watches. The watches must be synchronized before the start of the field survey and checked again after the completion of the survey. If time is available, the watches should be checked for at least 1 day prior to the survey, and a rate and correction factor determined for each watch. Knowing the rate, each observer can determine exactly what time the even 5-minute mark is reached for his reading.

d. Time of Readings. All readings at the base stations are taken on the even 5-minute mark. In single- and two-base altimetry, the roving altimeters take 5 readings 2 minutes apart on the even minute. Before the actual reading, the altimeters are set down and the pressure chamber is allowed to stabilize for at least 2 minutes. Thus, when the observer reaches a field point, he sets the altimeters on the selected point, waits at least 2 minutes, and reads the scales when the next even minute mark is reached. Since the base altimeter observers (in the single-base and two-base methods) do not know when the field altimeters will be read, they must take a reading every 5 minutes while the survey is in progress from the time the field party leaves until it returns and a final closing series of readings are taken. In leapfrog surveys, the parties must arrange for the exact time that all altimeters will be read. They must allow travel time to the forward station and altimeter stabilizing time. Observations can prove worthless if a party cannot reach the station in time or if they must hurry their reading before the pressure chamber has stabilized. With radio communication between parties, the time arrangements are simple since the rear observer waits until the forward man notifies him that the altimeters are on the point and he is ready. Both observers can then select the exact time to read.

e. Reading the Scale. The pointer and scales of the altimeter are backed up with a mirror surface. When taking a reading, the observer should lean over the altimeter until he alines the reflected image of the pointer and scale exactly under the direct image. He will also see the reflected image of his eye in the background. This eliminates the parallax or misalinement of the image due to reading the scale at an angle. Some observers find a small hand magnifying glass very useful. The observer should practice reading the altimeter to get the "feel" of alining the needle quickly. Just before making the reading, the observer should

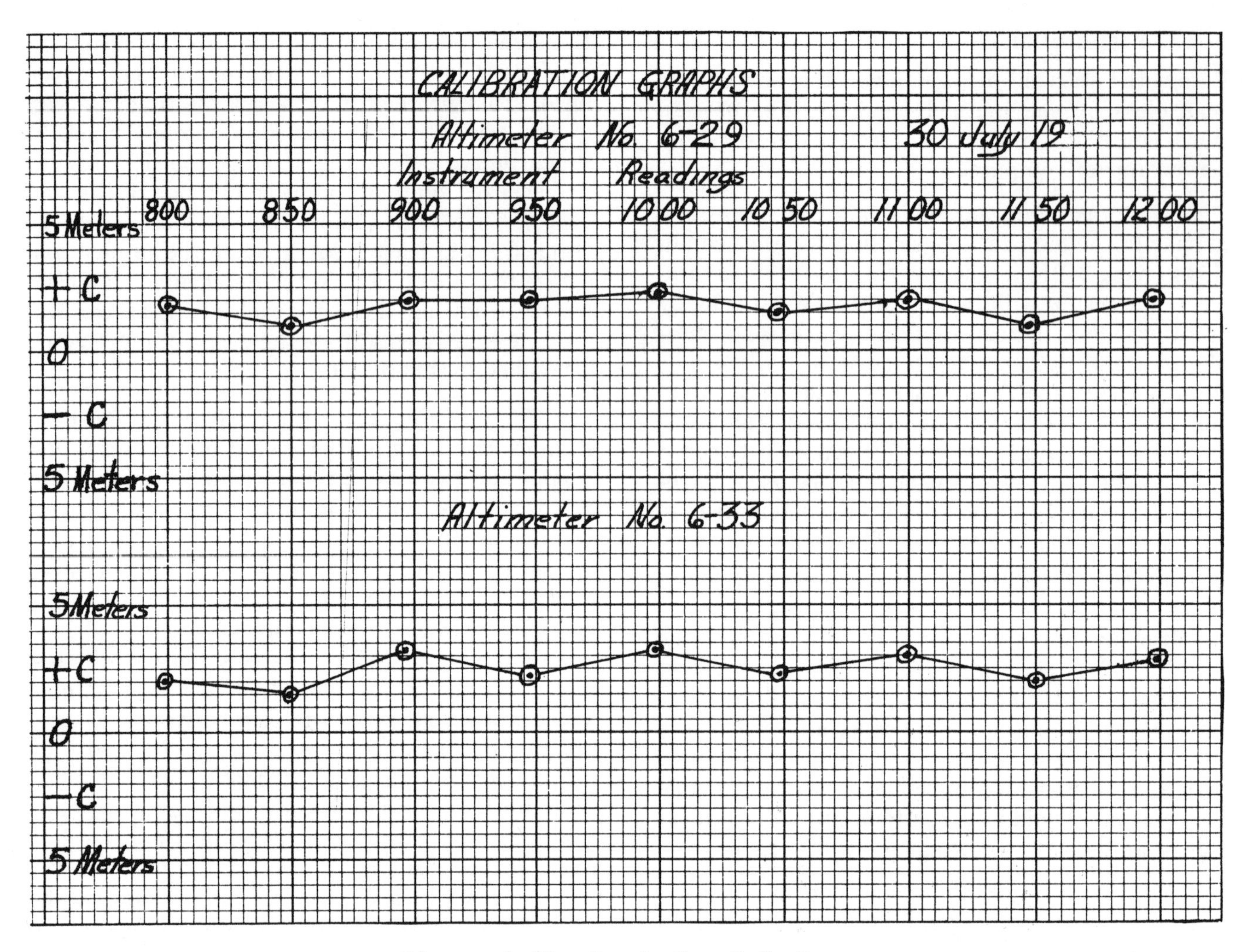

Figure 8-12. Graph for C-factor.

tap the glass gently with the rubber tip of his pencil or with his finger. This takes up the slack in the mechanical movement. The procedure of tap, aline, and read should become a smooth quick operation, since some surveys require one observer to read two altimeters in quick succession when the exact time is reached.

f. Eccentric Points. Occasionally, the required elevation point is not a practical location for an altimeter. It is permissible to move the altimeter to a safer or better closeby location to make the reading. The difference in elevation between the required point and the occupied point should be determined by whatever means are available. A hand level and a tape held vertically over the lower point are very effective. This DE should be recorded with a note explaining the eccentric reading. The DE must *not* be converted into an elevation of the point, but must be left as a correction using a minus sign if the required point is below, and a plus sign if above, the occupied point. For example, the required point is a crossroads, and the occupied point is 18 meters NW of the intersection and 0.2 meter higher. The records will show the readings at the NW corner, and the crossroads will be shown as—

Crossroads, 18 meters SE of station; DE = —0.2 meter.

g. Adjustment. The altimeters must be adjusted as described in paragraph 3–21*e*.

h. Calibration. The altimeters must be calibrated before use in a field survey by one of the two methods described in paragraph 3–21*f*. Upon completion of the calibration, graphs must be constructed for each instrument in order to determine the correction factor, commonly referred to as the C-factor. The difference between the calibrated mean and the actual reading at any stop for each instrument is the C-factor. For each altimeter, the differences at each stop are plotted on a graph using the altitude value as the abscissa and

ALTIMETER CALIBRATION RECORD

PROJECT: 537/215/LAL/B/82 DATE: 30 July

Range of calibration: From 800 Meters To 1200 Meters.

Direction	1		2		3		4		5		6			
	6-25		6-28		6-29		6-35		6-36		6-42		Instr. No.	
	Reading	C	Reading	C	Reading	C	Reading	C	Reading	C	Reading	C	Sum	Mean
U	801	0	803	-2	799	+2	798	+3	800	+1	804	-3	4806	801
U	799	0	800	-1	798	+1	797	+2	798	+1	801	-2	4793	799
D	803	+1	806	-2	801	+3	802	+2	803	+1	807	-3	4822	804
D	800	+1	808	-2	800	+1	800	+1	801	0	804	-3	4808	801
M	801	+.5	803	-1.8	800	+1.8	799	+2	800	+.8	804	-2.8		
U	850	0	853	-3	849	+1	848	+2	850	0	853	-3	5103	850
U	849	0	850	-1	848	+1	847	+2	847	+2	852	-3	5093	849
D	852	0	855	-3	850	+2	851	+1	852	0	855	-3	5115	852
D	850	+1	852	-1	851	0	850	+1	851	0	854	-3	5108	851
M	850	+.2	852	-2	850	+1	849	+1.5	850	+.5	854	-3		
U	904	-1	907	-4	900	+3	899	+4	901	+2	905	-2	5416	903
U	902	-2	903	-3	899	+1	898	+2	897	+3	903	-3	5402	900
D	902	-3	903	-4	897	+2	895	+4	899	0	901	-2	5397	899
D	900	-3	900	-3	895	+2	894	+3	895	+2	900	-3	5384	897
M	902	-2.2	903	-3.5	898	+2	896	+3.2	898	+1.8	902	-2.5		
U	952	-1	954	-3	950	+1	949	+2	951	0	952	-1	5708	951
U	951	-1	953	-3	949	+1	947	+3	947	+3	954	-4	5701	950
D	952	-2	953	-3	947	+3	946	+4	950	0	953	-3	5701	950
D	950	-1	950	-1	946	+3	949	0	947	+2	951	-2	5693	949
M	951	-1.2	952	-2.5	948	+2	948	+2.2	949	+1.2	952	-2.5		
U	1000	-1	1001	-2	997	+2	995	+4	1002	-3	1001	-2	5996	999
U	1001	0	1003	-2	1000	+1	998	+3	1003	-2	1001	0	6006	1001
D	1003	0	1005	-2	999	+4	1000	+3	1006	-3	1004	-1	6017	1003
D	1005	-1	1007	-3	1002	+2	1001	+3	1007	-3	1004	0	6026	1004
M	1002	-.5	1004	-2.2	1000	+2.2	998	+3.2	1004	-2.8	1002	-.8		

NOTE: Recording is continued in this manner until uppermost range is reached.

U= Upward movement D= Downward movement

M =Column mean

COMPUTED BY:	DATE	CHECKED BY:	DATE
E. Gehly	31 July	K. Cita	31 July

Figure 8-13. Altimeter calibration record.

the C-factor as the ordinate. The plotted points are connected with a straight line. During the survey, the reading at each point on the ground is corrected by the C-factor, as taken from the graph, for the altitude read on the altimeter. A sample, recommended format for calibration recording is shown in figure 8–13 and the accompanying C-factor graph is shown in figure 8–12.

8–23. Two-Base Method

At the present time, the two-base method is the only method of altimetry surveying that is acceptable for engineer and mapping survey projects, and then only when one of the other methods of determining differences of elevation is not feasible or impractical (para 8–2*b*(2)).

a. Field Procedure. Bases are established at two points of known elevation. The lower base is selected below the lowest expected elevation point; the upper base above the highest point. The two-base survey consists of reading the altimeters at the unknown elevation at the same time that both the upper and lower base altimeters are read. Since the base altimeters are read on points of known elevation, the atmospheric changes are reflected in their readings. By subtracting the amount due to atmospheric change, the roving altimeter reading is the air pressure difference caused by the difference in elevation from the base stations.

(1) If the difference in elevation between bases is kept within 60 to 90 meters, the average expected error in elevation will be less than 1 meter. Up to 600 meters difference, the expected error will increase to almost 2 meters. The horizontal distance between bases is held below 15 kilometers to minimize errors due to atmospheric variations which may not be reflected in the base readings.

(2) Every effort should be made to select base stations that "overlook" the area to be surveyed. By keeping the base stations in the same basin, atmospheric changes at the bases and at the survey point are more likely to be the same. Selecting a base station on the "other" side of a hill may be more convenient, but can introduce different atmospheric conditions and produce inaccurate results.

(3) The two-base method does not require corrections for temperature, relative humidity, or latitude. The base altimeters read the elevation plus or minus any atmospheric changes. Since the roving altimeter is between the bases, it is assumed that its changes will be similar to the changes at the bases, ((2) above), and in proportion to the vertical distance from each base. Ideally, the altimeter survey should follow a straight line between the bases, and the elevations should be between the base values. This is not always possible, and experience has shown that some de-

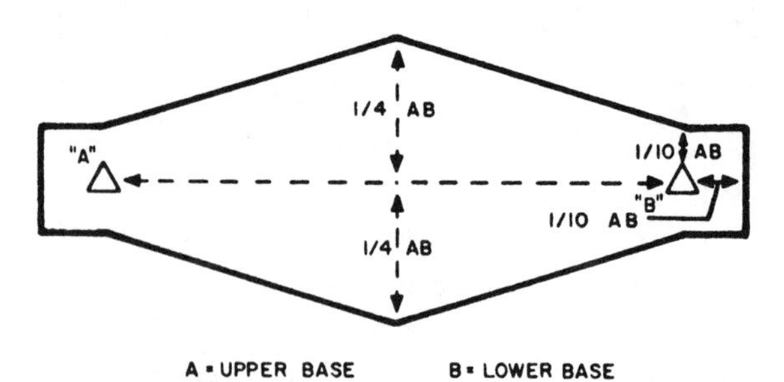

Figure 8-14. Horizontal limits, two-base method.

viation can be tolerated. An elevation above or below the high or low base values by as much as 10 percent of the difference between base elevations can be computed with reasonable accuracy. The limits in horizontal deviation are illustrated in figure 8–12.

(4) Each altimeter line should be run both ways over the course to establish a mean value for each point. If both runs cannot be completed in one continuous survey, the second run should be completed as soon as practicable thereafter, and preferably, in the opposite direction. An example of the procedure is as follows:

(*a*) The location of the two-base stations is selected and the route of the survey planned within the limitations in (2) and (3) above.

(*b*) The entire group of altimeters (two for each base and two for each route) are transported to one of the base locations, for example, the lower base.

(*c*) The altimeters are adjusted and all watches synchronized at point.

(*d*) The survey is started by reading all of the survey altimeters simultaneously while they are at the lower base.

(*e*) Two altimeters remain at the lower base and are read every 5 minutes.

(*f*) The remaining altimeters are transported to the upper base and all are read simultaneously on the next even 5 minutes.

(*g*) Two altimeters remain at the upper base and continue to be read every even 5 minutes.

(*h*) The roving altimeters, in pairs are transported to each of the required points in turn,

Altimeter No. 6-29 (Rover) — Altimeter No. 6-33 (Rover)

DESIGNATION Altimeter line B DATE 1 Aug 19__

Project: 537/US/KAL/81 Obs. K. Cita Party Chief: E. Gehly

Station	Time	Weather	Instr. Temp.	Altimeter Reading	"C" * Factor	Corrected Reading	Altimeter Reading	"C" * Factor	Corrected Reading	Remarks
BM 6A (USGS)60	0900	Lt Wind Clear	78°F	815.5	+ 1.5	817.0	815.0	+1.9	816.9	Lower base Comparison
HALL (USC&GS)30	0930	Lt. Wind Clear	72	1185.0	+ 1.8	1186.8	1184.5	+2.5	1187.0	Upper base Comparison
PPM 14-26 1214A(V)	1000	Lt Wind Clear	73	1034.5	+1.8	1036.3	1034.0	+2.5	1036.5	For sketch and description of the pricked picture points refer to pages 6 through 12 of this book.
PPM 14-26-1214B(V)	1010	Lt Wind Clear	73	1025.5	+1.8	1027.3	1025.0	+2.6	1027.6	
PPM 14-26-1218A(V)	1025	Lt Wind Clear	74	990.5	+2.2	992.7	990.0	+3.0	993.0	
PPM 14-26-1218B(V)	1035	Lt Wind Clear	75	952.5	+2.0	954.5	952.0	+2.3	954.3	
PPM 14-26-1218C(V)	1050	Lt Wind Clear	75	939.0	+2.0	941.0	938.5	+2.4	940.9	
PPM-14-25 1386A(V)	1110	No Wind Clear	76	899.5	+2.0	901.5	898.5	+3.2	901.7	
PPM 14-25 1386B(V)	1125	No Wind Clear	76	875.0	+1.5	876.5	874.5	+2.5	877.0	
HALL (USC&GS)30	1200	No Wind Clear	73	1183.5	+1.8	1185.3	1183.0	+2.5	1185.5	Upper base Comparison
BM 6A (USGS)60	1230	No Wind Clear	80°F	814.5	+1.5	816.0	814.0	+1.9	815.9	Lower base Comparison

* From calibration graph "C-Factor." ✓ JBG

Figure 8-15. Recording notes, two-base method.

are set down for at least 2 minutes to stabilize, tapped to take up the slack, and readings made starting on the next even minute. (Any number of routes may be run simultaneously if instruments and observers are available, and if the routes fall within the horizontal and vertical limits.)

(*i*) After all the points have been occupied, the roving altimeters revisit each point on the way back, and make their second reading.

(*j*) The roving altimeters return to the upper base, and make another set of readings simultaneously with the upper base altimeters for a tie.

(*k*) When all roving altimeters have been returned and tied to the upper base readings, the entire group, including the upper base instruments, are returned to the lower base.

(*l*) The entire set of altimeters, both upper and lower base pairs and all roving pairs are read simultaneously for a closing tie. This completes the survey run.

(*m*) If time does not permit a continuous run out and back to the upper base, the roving altimeters should tie in to the lower base after a one-way trip. The return run should be made preferably from the lower base to the upper base when time permits. The return and closing tie described in (*l*) above, must be made to complete the survey.

(5) Any known elevations which fall within or near to the survey area must be occupied and readings taken. When working near the permissible distance limits, it is advisable to go outside the area to tie in known elevations to the survey. Tie-ins to adjoining altimeter surveys should include at least two points to provide a check between runs.

b. Recording and Computation. Figure 8-15 shows an example of the field recording and is self-explanatory. The columns entitled "C-factor" and "corr reading" are completed in the office by the computer.

8-24. Leapfrog Method

The leapfrog method is conducted in the manner implied by its name (fig. 8-16). Two altimeters, designated A and B, are read simultaneously at the starting base (known) station. Then altimeter A is left at the base station, and altimeter B is moved to the first field station. The two altimeters are again read simultaneously. Then, altimeter A is moved from the base station, by passing the first station, to the second station. The procedure of leapfrogging is done as shown in figure 8-16.

TIME	STATIONS				
	BASE	1	2	3	4
0930	(A)(B)				
0950	(A)	(B)			
1015		(B)	(A)		
1045			(A)(B)		
1110			(A)	(B)	
1140				(B)	(A)
1150					(A)(B)

Figure 8-16. Leapfrog method.

8-25. Single-Base Method

This method of altimetry is the least accurate of the three methods and is very seldom used. The field procedure for the single base method is similar to that of the leapfrog method. After the initial comparison readings are made at the base station, the A altimeter remains in position and readings are taken at even 5-minute intervals throughout the observing period. After the comparison, the B altimeter visits all field stations and makes readings only on even 5-minute marks. Upon completing the field stations the B altimeter returns to the base station for a final comparison with the A altimeter.

CHAPTER 9

ANGLE AND DIRECTION OBSERVING

Section I. DESCRIPTION

9-1. Introduction

Surveying is the science of determining relative positions of points or objects on or near the earth's surface.

a. From the study of geometry, it is evident that the position of a point may be found by measuring only the distances from two known points. The surveyor will find occasions where this two distance method is very practical and even highly desirable. However, in many procedures which require locating a point, the surveyor will use a distance and a direction from one known point, or sometimes, only the direction from two known points without any distances.

b. The definitions in this chapter are simplified in most instances to stay within the scope of plane surveying.

9-2. Direction

In surveying and mapping, by definition, *direction* is the angular relationship of one line to another. This direction must be referred to some other definite line to act as the zero value, and some dimension must be assigned to show the amount of change from the zero line. A horizontal direction can be observed in two ways—clockwise or counterclockwise. Common practice in surveying considers that directions or angles are measured clockwise (to the right) from the reference line, unless otherwise designated.

9-3. Angles

An angle may be defined as the difference in the values of two directions observed from the same initial direction. At times, the directions to several features must be observed from one instrument setup, or the directions as they change over a series of continuous lines must be determined. The series of lines might represent an area boundary which starts from one point, extends around the area, and returns (ties in) to the starting point. On the other hand, the lines might be a road centerline which starts at one point and extends (traverses) to another location or point some distance away. The direction and angle equipment (ch. 4) will measure the changes of directions (angles) between the series of lines. In some types of surveys, only selected angles of a group are observed. To distinguish the different ways the angles are observed, they are given special names, such as station angles, explement angles, or deflection angles in traverse and construction surveys (fig. 9-1). In triangulation, only the differences in directions are observed and used as angles in the computations.

a. Station Angles. When two points of known position are intervisible (Z and A, fig. 9-1), and the survey must start from one of these points (A), the instrument is set up at A, and backsighted on Z as the zero direction. The angle is turned to point B, the first point of the new line. As discussed in paragraph 9-2, the angle normally is turned and read clockwise (to the right). This angle is referred to as the *station angle.* Generally, surveys require at least one direct and one reverse (D/R) pointing on each station to observe the angle. A *direct pointing* is made with the telescope in the direct position. A *reverse pointing* is made with the telescope in the reverse position (plunged) from the direct position. This D/R measurement is made to eliminate the collimation error in the instrument (ch. 4). For a complete D/R reading, the station angle is read, the telescope is reversed and pointed on the forward point (B, fig. 9-1) and read. The angle to the starting point (Z) is turned to the right, read, and is called the *explement angle.* The sum of the station angle and the explement angle should equal 360°.

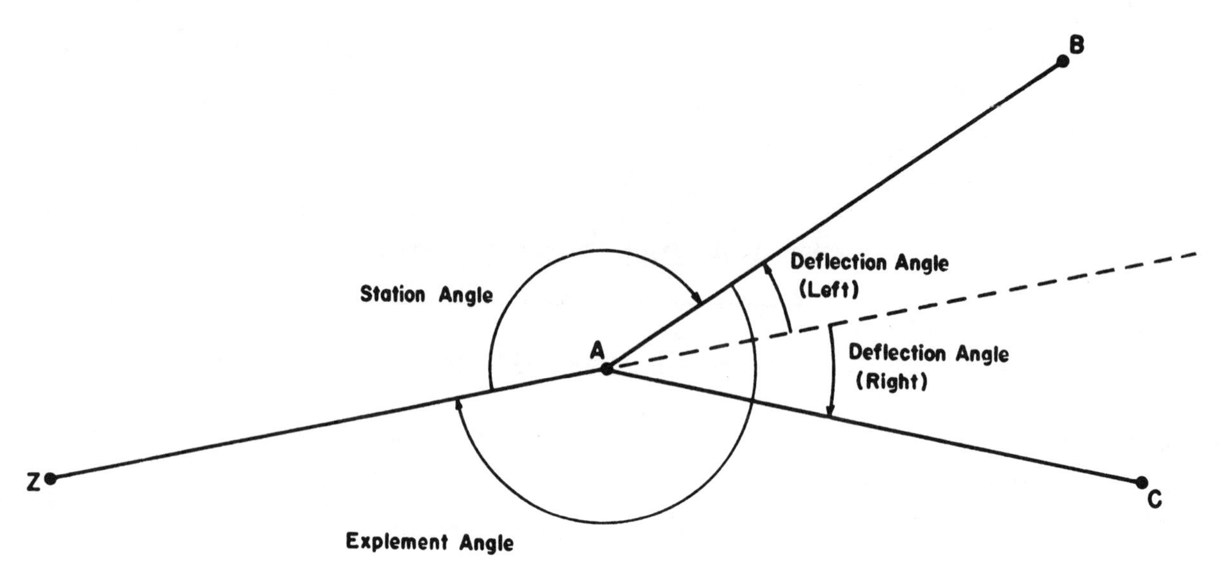

Figure 9-1. Angles.

b. Deflection Angles. Some surveys use deflection angles in their computations. These may be computed from the angles obtained in *a* above, or they may be turned directly using a transit or 1-minute theodolite. The setup at A (fig. 9–1) and backsight on Z are made as in station angles. Instead of the telescope being rotated horizontally, it is plunged first. The angle to B is turned *to the left.* This is called a left deflection angle, and is recorded with a "L" or minus sign in front of the angular value. Figure 9–1 shows a left deflection angle to B, and a right deflection angle to C, if C were the required point. Right deflection angles are recorded with a "R" or plus sign in front of the angular value. When turning and reading deflection angles, special care must be taken to distinguish the left or right values and to record them properly.

c. Interior and Exterior Angles. Some surveys follow around a figure, and close or tie in to the starting point. The angles which are inside the figure may be referred to as the interior angles, while their explements are called exterior angles. Depending upon which direction the survey is run, either the interior or the exterior angles may be read as the direct or station angles. The other angles (exterior or interior) would then be the explement angles. Interior and exterior angles are not recorded in the surveys in this manner; they are read and recorded as direct and reverse directions.

9–4. Azimuth

One of the methods used to describe direction in surveying is called azimuth. *Azimuth* is the horizontal direction observed clockwise from a zero line which runs north or south. Every line has two azimuths, depending on the observer's position. For example, in figure 9–2, a survey is prog-

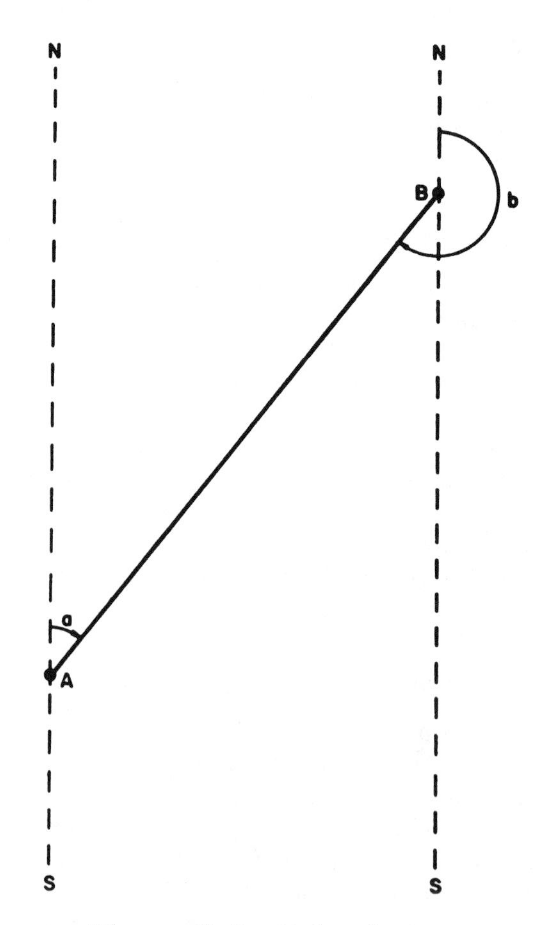

Figure 9-2. Azimuth.

ressing from A toward B. Then, the angle *a* is the forward azimuth for this line. To designate the direction from B to A, the angle *b*, is used. This is known as the back azimuth of the line. For purposes of plane surveying, the forward and back azimuths of a line always differ by exactly 180°. The zero azimuth line can be based on true, grid or magnetic north.

a. True Azimuth. The horizontal direction of a line measured clockwise from a reference plane, usually the meridian. A true azimuth may be established by sighting over a line whose true azimuth is known and subtracting the known value, or by sighting on the stars. A true azimuth is usually referred to as true north azimuth or true south azimuth and is generally determined by observing the stars. The stars observed are usually circumpolar stars, except in the low latitudes (near the equator), and the resulting azimuth is referred to as an astronomic azimuth. By applying a correction (Laplace equation) the astronomic azimuth can be converted to a geodetic azimuth. Both geodetic and astronomic azimuths are considered as true azimuths.

(1) *Astronomic.* The angle between the astronomic meridian plane of the observer and the plane containing the observed point and the true normal (vertical) of the observer, measured in the plane of the horizon, preferably clockwise from north.

(2) *Geodetic.* The angle between the geodetic meridian and the tangent to the geodesic line at the observer, measured in the plane perpendicular to the ellipsoidal normal of the observer, preferably clockwise from south.

b. Grid Azimuth. There are some surveys which are computed on a system of plane coordinates. Such a system is based on the assumption that a small portion of the earth's surface can coincide with the surface of a geometric figure which can be developed into a plane. A system of squares, called a *rectangular grid,* is superimposed over the developed surface. One line (the central meridian) of this grid coincides with a true north-south line. Other north-south grid lines are parallel to this one line, and all east-west grid lines are perpendicular to the north-south grid line. When survey computations are made with reference to a grid system, azimuths use the northerly direction of the north-south grid lines as zero grid azimuth. Grid azimuth is defined as: The angle in the plane of the projection measured clockwise between a straight line and the central meridian of the plane-rectangular coordinate system.

c. Magnetic Azimuth. A method used in surveying to obtain approximate values for azimuths, based on a zero line established with a magnetic compass. Magnetic north as shown by a compass needle is selected as the zero value and the angles are measured from this point, and these are referred to as magnetic azimuths. Although forward and back azimuths for a line should differ by 180°, the compass-measured difference may vary several degrees because of local magnetic attraction. A detailed discussion of the factors that affect measurements with a magnetic compass are found in paragraph 9–6. Magnetic azimuth is defined as: At the point of observation, the angle between the vertical plane through the observed object and the vertical plane in which a freely suspended, symmetrically magnetized needle, influenced by no transient artificial magnetic disturbance, will come to rest. Magnetic azimuth is generally reckoned from magnetic north (0°) clockwise through 360°.

9–5. Bearing

Azimuths (fig. 9–3) are measured from north or south and the angles vary from 0° to 360°, see inner circle of figure. Computations using the trigonometric functions of azimuths may require that the angles be converted to equivalent angles of less than 90°. The use of bearings in the measurements automatically keeps the angles below 90°, outer circle of figure.

a. The *bearing* of a line is its direction within a quadrant, with reference to a meridian (a north-south line). Bearings are measured clockwise or counterclockwise, depending on the quadrant, from either the north or south line. A bearing is identified by first naming the end of the meridian from which it is reckoned (north or south), then the angle value, and the direction (east or west) from the meridian. Thus, a line in the southwest quadrant making an angle of 37° 43′ with the (south) reference meridian has a bearing of S37° 43W. An angle in the northwest quadrant, 47° 25′ from the north meridian has a bearing of N47° 25′W.

b. Bearings, like azimuths, can be true, grid, or magnetic, depending on the reference meridian. If the meridian is a true north-south line, the bear-

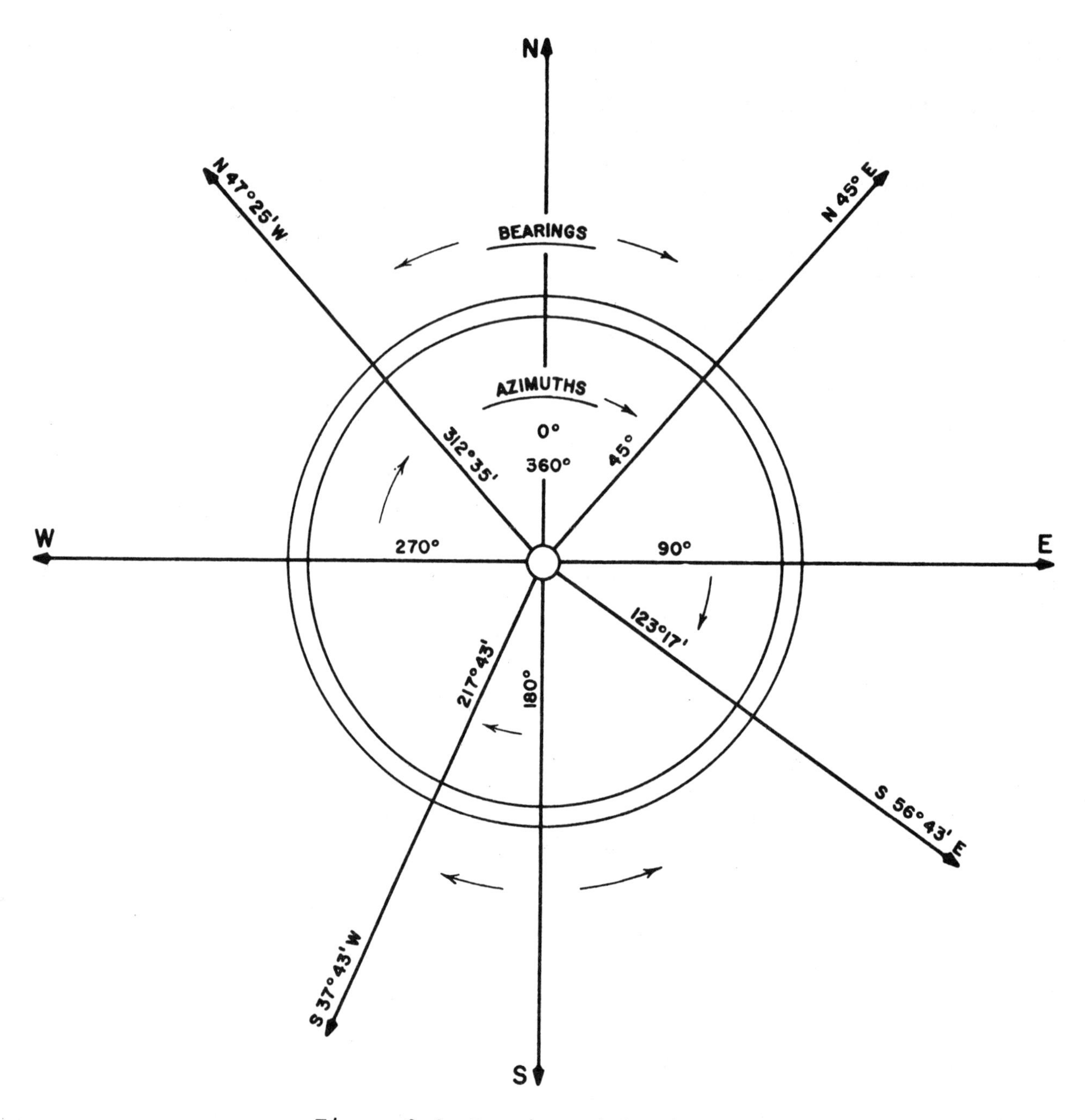

Figure 9-3. Bearings and azimuths.

ing will be true. Grid north-south lines as reference will result in grid bearings and using magnetic north or south will give magnetic bearings.

c. The compasses (sec. VII, ch. 4) included with the transit and the 1-minute theodolite can be used to read magnetic bearings.

9-6. Magnetic Declination

The magnetic needle alines itself with the earth's magnetic field and points toward the earth's magnetic pole. At a few places on the earth's surface, the needle will point exactly north along the true meridian. Everywhere else the magnetic north line indicated by the needle will form a horizontal angle with the true meridian at that point. This angle is called the *magnetic declination.*

a. The magnetic declination is labeled east when the needle points east from true north; and west when the needle points westward. An east declination is added to and a west declination subtracted from the magnetic azimuth to get a true azimuth. Magnetic bearings may also be changed to true bearings by adding east declinations in the NE and SW quadrants, and subtracting them in the SE and NW quadrants. West declinations are added in SE and NW, and subtracted in SW and

NE quadrants.

b. Magnetic declination is not a constant value at any point, but is continually changing. This constant variation affects the reliability of magnetic readings. Even at best, the surveyor must not expect results closer than one-fourth of a degree. Declination is shown on a specially constructed chart, called an isogonic chart, which shows lines connecting points of equal declination. These lines are referred to as *isogonic lines* and are drawn for each degree of declination east and west of zero. The zero line is called the *agonic line.* Lines west of the agonic line show areas where the needle will point east or have an east declination. Lines east of the agonic line show west declination areas, or where the needle will point west. Such charts are published in the United States once every 5 years by the Coast and Geodetic Survey. Due to this 5-year period between printings, a series of annual change lines are added. These lines indicate the amount and direction of the annual change expected in the different locations.

c. Local attraction is a term denoting local influence that causes the magnetic needle to be deflected away from the magnetic meridian for that locality. Some sources of local attraction are: magnetite (loadstone) in the ground, electric power lines, steel structures, railroad rails, surface and underground iron pipes, and other fixed objects of iron and steel. Contrary to common belief, ordinary iron ores do not attract the magnetic needle unless they contain magnetite or pyrrhotite. According to its true definition, local attraction does not include avoidable iron or steel articles which can cause large deflection errors. Some of these avoidable sources are: steel keys, knives, steel rimmed spectacles, spectacle cases, metal buttons, wire stiffeners in hats, and steel helmets; surveying equipment such as axes, steel tapes, and ranging poles; and other items such as motor vehicles, radio receivers and transmitters, and power generators. The smaller of these items should be moved from 3 to 10 meters away from the instrument, and the larger items may require being moved several hundred meters. The surveyor using the magnetic needle should be constantly aware of the influence of surrounding features on the needle.

9-7. Vertical Angles

A *vertical angle* (①, fig. 9-4) is the angle measured vertically up or down from a horizontal plane of reference. When the telescope is pointed in the horizontal plane (level), the value of the vertical angle is zero. When the telescope is pointed up at a higher feature (elevated), the vertical angle increases from zero and is called a plus vertical angle. These values increased from 0

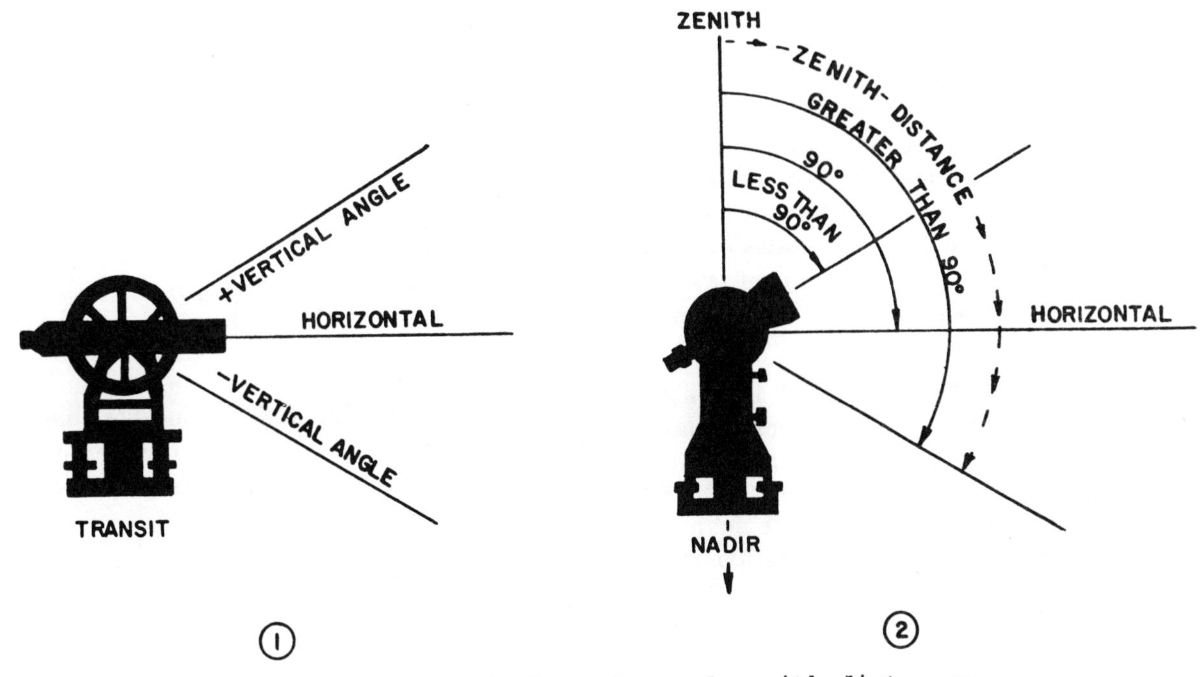

Figure 9-4. Vertical angles and zenith distances.

to +90° when the telescope is pointed straight up. As the telescope is depressed (pointed down), the angle also increases in numerical value. A depressed telescope reading showing that it is below the horizontal plane is called a minus vertical angle. These numerical values increase from 0 to —90° when the telescope is pointed down.

9-8. Zenith Distances

The *zenith* is an imaginary point directly overhead where an extension of the plumb line will intersect an assumed sphere on which the stars appear projected. The equivalent point which is directly below the zenith is called the *nadir.* Use of the zenith permits reading angles in a vertical plane without using a plus or minus. Theodolites (ch. 4), with the exception of the 0.2-second theodolite, have a vertical scale reading zero when the telescope is pointed at the zenith, instead of in a horizontal plane. With the telescope in a direct position and pointed straight up, the reading is 0°; on a horizontal line, the reading is 90°; and straight down, 180°. Vertical angle measurements with the theodolites (②, fig. 9–4), read elevation (plus) angles as values less than 90°, and depression (minus) angles as greater than 90°. These angle measurements with the zenith as the zero value are called the *zenith distances. Double zenith distances* are observations made with the telescope direct and reversed to eliminate errors due to the inclination of the vertical axis and the collimation of the vertical circle. For a discussion of the 0.2-second theodolite refer to paragraph 4–19*c*.

Section II. FIELD PROCEDURE

9-9. Angle Observing Party

The angle (observing) party is very flexible in size, and varies from 1 to 12 or more people. This party may be a subparty of a triangulation or traverse party, or it may work independently. An instrumentperson is required in every angle party, and a recorder is desirable. A rodperson or light-tender is often required at each point to which an angle is being measured. Sometimes, semipermanent targets are erected at some points to avoid stationing someone at each of these points. The chief of the party decides on the dispersal of available personnel.

a. Instrumentperson. The instrumentperson sets his or her transit or theodolite exactly over the selected station or point and observes the angles, horizontal, vertical, or both, as demanded by the survey. If the survey requires elevations by leveling but does not warrant an additional leveling party, the instrumentperson also performs leveling duties.

b. Recorder. The recorder records all field notes as they are read by the instrumentperson, makes sketches and descriptions of all stations, and makes all necessary on station computations. On lower order surveys when personnel are not available, the instrumentperson may act as his or her own recorder. However, this is not recommended because of the lack of an on station check of the field note computations.

c. Rodperson. The rodperson sets or holds a range pole or plumb bob on the rear or the forward stations as directed. Where targets are to be erected, he or she performs this duty making sure that the target is exactly over the point. When leveling is required, he or she will hold the leveling rod on the stations. This person is sometimes a light-tender or radioperson on area triangulation nets.

9-10. Observing Angles or Directions

The value of a horizontal angle between two lines can be determined by setting an index value (possibly zero) on the circle, pointing at a target placed over the terminal on one of the lines, holding the index and moving the telescope to point at a target over the terminal of the other line, reading the circle for this second pointing, and taking the difference between the last reading and the index value. For vertical angles, the index is set with the telescope in level position, the target pointed upon, and the value read on the vertical circle. The one angle value, either horizontal or vertical, determined in this manner is only as accurate as the minimum reading possible on the instrument. It is also subject to numerous errors and blunders. Survey techniques have been developed which minimize these effects.

a. All angles should be measured at least twice, once with the instrument telescope direct and once with the telescope reversed. A greater number of measurements is often made to obtain a mean value with a minimum of residual error.

⊼ at Sta. Knoll

DESIGNATION Project 537/45/LTri/1 DATE 18 Aug 19__

Obs: J. Doehn Instr: K&E 20" Transit # 21345

Rec: M. Cox Hazy, Lt. Wind & Hot

Station	Observed	Tel	Rep	Circle	Vernier A	Vernier B	Mean	Mean of three	Set Mean	Hor Adj	Adj ∡ & Remarks
△ Shoulder		D	0	00° 00'	00"	00"	00"				No eccentricity of instrument or targets.
			1	30 15	40						Time = 1050 hrs
			3	90 46	20	40	30	30° 15' 30.0"			
	△ Knob	R	6	181 32	40	00	50	30 15 26.7	28.3	-01.0	30° 15' 27.3"
△ Knob		R	0	181 33	00	00	00				
			1	264 14	20						
			3	69 37	00	20	10	82° 41' 23.3"			
	△ Top	D	6	317 41	00	20	10	82° 41 20.0	21.6	-01.1	82° 41' 20.5
△ Top		D	0	317 41	00	00	00				
			1	204 44	20						
			3	338 50	40	00	50	247 03 16.7			
	△ Shoulder	R	6	000 00	20	20	20	247 03 10.0	13.2	-01.0	247 03 12.2
							Total =	360 00	03.1	-03.1	360° 00' 00.0"
											Time = 1115 hrs

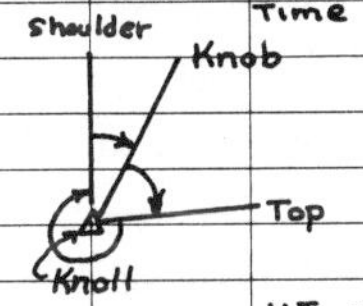

Shoulder
Knob
Top
Knoll

HI = 4.6 Ft.

✓ JEB.

(1)

Project DESIGNATION 537/45/L Trav/3A DATE 20 Aug 19__

Obs: J. Doehn Instr: Dietzgen 20" Transit # 5631

Rec: M. Cox Hazy, Lt. Wind, & Hot

Station	Tel	Rep	Circle	Vernier A	Vernier B	Mean	Def. L or R	Double Def ∡	Def ∡	Remarks
A	D	0	00° 00'	00"	00"	00"				Az MK
	D	1	43 51	20						
B	D/R	1	87 42	20	00	10"	L	87° 42' 10.0"	43° 51' 05.0"	A, B, L
B	D	0	00 00	00	00	00				
	D	1	34 29	20						
C	D/R	1	68 58	40	00	50	R	68° 58' 50.0"	34° 29' 25.0"	A, B, C, R
C	D	0	00 00	00	00	00				
	D	1	42 17	00						
D	D/R	1	84 33	00	20	10	L	84° 33' 10.0"	42° 16' 35.0"	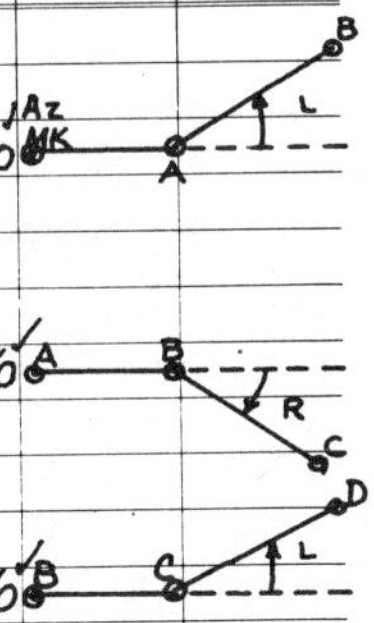 B, C, D, L
D										

✓ JEB.

NOTE: Recording continued in this manner until completed.

(2)

Figure 9-5. Field notes, horizontal angles.

b. There are two basic methods for the multiple measurements of a horizontal angle—the repetition method (para 9–11) and the direction method (para 9–12). The repetition method normally is used with a repeating instrument such as an engineer transit. The direction method may be used with transit, but is always used with a direction-type instrument such as the theodolite. Deflection angles require a different technique, but are usually measured using the repetition method (para 9–11*a*).

9–11. Repetition Method

a. The 1-minute theodolites and the 1-minute and 20-second transits are repeating instruments on which successive observations of a horizontal angle are added by *repetition*, and then a mean taken to obtain an accurate measure of the angle. The transit may be pointed at the target by using either the upper clamp and tangent screws (upper motion), or the lower clamp and tangent screws (lower motion). When a horizontal angle is being measured, the lower motion is always used for pointing at the initial station, and the upper motion is used for pointing at a second target normally to the right of, or clockwise from the first target. The vertical clamp should be tight enough to require a slight pressure to revolve the telescope. Loosen the appropriate upper or lower clamp and turn the transit toward the target while looking along the top of the telescope first, and then through the telescope. When the crosswires are near the target, tighten the clamp that was loosened. With the corresponding tangent screw, perfect the pointing by bisecting the target with the vertical crosswire. The objective lens should be focused before perfecting the pointing.

b. The repetition method results in angles better than the least reading of the instrument. A set of observations may consist of two or more repetitions of the angle with the telescope in one (direct or reversed) position, followed immediately by two or more repetitions of the explement with the telescope in the opposite (reversed or direct) position. ①, figure 9–5 illustrates a set of field notes for six repetitions, three direct, three reverse. Notes for less than six repetitions are kept in the same form. The field procedure is as follows:

(1) Set the A vernier at or near zero.

(2) Point on the initial station and read both verniers (Shoulder, ①, fig. 9–5).

(3) Unclamp the upper motion and point on the right-hand station. (Knob). Read and record the A vernier. Clamp the upper motion and release the lower.

(4) Make two more repetitions, pointing on Shoulder with the lower motion and tangent screw, and on Knob with the upper motion.

(5) Read both verniers.

(6) Without changing the circle settings, reverse the telescope and take three repetitions of the explement, measuring the angle from Knob clockwise to the initial station.

(7) If further angles must be taken from this station, Knob is used as the initial, and the next station clockwise (Top) is used for the second station. This is followed by using Top as the initial and the original station (Shoulder) as the second station. This *closes the horizon* and will permit distributing (adjusting) any small error in each of the three angles equally.

(8) To obtain a check, the six repetitions are meaned and compared against the value of the first angle reading. The A and B vernier are meaned at each reading to get the result (Mean Column). The mean of the initial reading is subtracted from the mean of the three repetitions and the result divided by three. The other mean of three value is obtained by subtracting the total of the three direct from the total of three reverse and dividing by three. The final mean (Set Mean Column) is computed by meaning the two "mean of three" values. Since the horizon was closed, the "set means" are added and compared against 360° 00′ 00″. Any difference is divided by three and distributed among each mean (Hor. Adj. Column). The adjusted value of each angle appears in the Remarks column.

(9) To obtain the required accuracy in the measured value for an angle, it is sometimes necessary to measure the angle by several sets of six repetitions. In this case, each set of repetitions should start with a different circle setting. Recommended initial settings for two, four, and six sets of repetitions with either a 1-minute transit (theodolite) or 20-second transit are given in table 9-1.

c. A deflection angle (para 9–3*b*) is the clockwise or counterclockwise angle from the prolongation of a line. Consequently, deflection angles are always less than 180°. Deflection angles normally are measured with an engineer transit, because most theodolites are not readily adaptable to this use. These angles are often measured in the performance of a route survey, but for other survey-

Table 9-1. Initial Settings of Circle for Repetitions.

Number of sets	Two	Four	Six
1-minute transit—circle graduated to 30 minutes			
1	0°10′00″	0°05′00″	0°05′00″
2	90°20′00″	45°10′00″	30°10′00″
3		90°20′00″	60°15′00″
4		135°25′00″	90°15′00″
5			120°20′00″
6			150°25′00″
20-second transit—circle graduated to 15 minutes			
1	0°05′00″	0°05′00″	0°05′00″
2	90°10′00″	45°05′00″	30°05′00″
3		90°10′00″	60°05′00″
4		135°10′00″	90°10′00″
5			120°10′00″
6			150°10′00″

ing operations they are computed from clockwise angles if required. The field notes and computations are shown in ②, figure 9–5. A mean measured value for the deflection angle is computed in the same way as if the angle were measured by two repetitions. Direct and reverse observations systematically eliminate or neutralize any error. When the value of a deflection angle must be determined with greater accuracy than can be obtained by this procedure, the angle should be measured by the repetition method.

9–12. Direction Method

The direction method measures the direction to each station from one selected as the initial station. The directions are the angles measured clockwise from the initial station to each of the other stations. The angle between any two observed stations is the difference of their directions. A directional theodolite is read by means of micrometer microscopes. In observing, a pointing is made on the initial station and then upon each station around the horizon in a clockwise direction, closing on the initial station. The telescope is then reversed and readings repeated in a counterclockwise direction. The double set of readings constitute one *position.* The D/R readings for one position may be sufficient for angles turned with the 1-minute theodolite. The other theodolites, the 1-second and the two-tenth second, are used in higher accuracy surveys where a specified number of positions must be read. These use progressively different values for the initial reading to distribute slight errors of the horizontal circle.

9–13. Factors Affecting Angle Observations

There are three types of errors that affect angle observations. They may be due to the instrument, to the observer, or to the environment. The survey should try to reduce these errors or to avoid them completely.

a. Instrumental Errors. Instrumental errors that may affect the measured value for a horizontal angle are: the adjustable errors of the transit or theodolite; eccentricity of the horizontal circle; small errors in the graduation of the circle, the verniers, and the micrometer scale; and an error in the apparent length of the micrometer scale. Except for the last error, the effect of all instrumental errors can be eliminated or minimized by proper adjustment of the instrument and by systematic observing procedure.

b. Personal Errors. There are several personal errors that affect horizontal angle measurements. The effects of these errors can all be reduced to a minimum by using correct procedures, and by training.

(1) Errors in centering the instrument and the targets over their stations.

(2) Errors in leveling the instrument.

(3) Errors in pointing the instrument.

(4) Errors in reading the circle and verniers.

(5) Errors in making coincidence for reading.

c. Environmental Errors. The errors that affect horizontal angle measurements are due to differential temperatures within the instrument, horizontal refraction of the line of sight, and phase. Errors due to vertical refraction will not affect horizontal angle measurements within the scope of this manual. Vertical angle measurements which can be influenced by vertical refraction are discussed under leveling (para 8–5).

(1) *Differential temperatures.* Differential temperatures in the instrument are usually caused by direct sunlight. Heat causes variable expansions within the instrument. When using a transit or theodolite, the effect can be minimized by shading the instrument with a surveyor's umbrella.

(2) *Horizontal refraction.* This is the horizontal bending of light rays between a target and an observing instrument. It is caused by differences in the density of the air along the path of the light rays, resulting from temperature variances. Under good observing conditions, lines of sight close to the ground and over sloping terrain may be deflected several seconds horizontally.

Under extreme conditions, the effect of horizontal refraction may amount to over 1 minute of arc. One situation in which horizontal refraction can cause a large error is when the sun is shining on an open field between stands of timber. In this case, the rising column of warmer air over the open field causes a prismatic effect in the air and deflects the line of sight. The force and direction of the wind has considerable influence. A brisk wind will usually reduce the temperature differential and the difference in air density. However, when long lines are to be observed, night observations on lights will usually result in greater accuracy than daytime observation.

(3) *Phase.* Phase, due to the unequal illumination of a target, can often cause an error of considerable magnitude in the measured value of a horizontal angle. The shape of the target and the angle at which the sun's rays intersect the target are the primary causes of phase. When a round target, such as a chimney, is illuminated from one side by the sun, the observer will tend to point the instrument toward the illuminated side rather than toward the center of the chimney. Phase can be minimized by using targets having a flat surface facing the observer.

9–14. Angle Adjustment

The procedures described in paragraphs 9–11 and 9–12 are designed to minimize errors which may accumulate in angle readings. However, some error will still remain. If the survey accuracy requirements will tolerate the amount, the surveyor can accept the readings. Surveys which require six repetitions usually require a horizon closure or station adjustment. When a series of angles around a point are measured, the sum of these angles should equal 360°. The adjustment is known as the *station adjustment* of the angles. This is done by dividing the difference between 360° and the sum of the angles by the number of angles measured. This proportional amount is applied to each angle value. When the adjusted angles are added, the sum will be 360°. One or more odd values should be divided among the angles closest to 90° so that the sum will be exactly 360°. The reason for using angles closest to 90° is that usually the sine of the angles are used in the computations and angles near 90° have the smallest change per second of the sine function. For example, assume 5 angles are measured around a point and the total correction equals 4.″7. The amount to be applied to each angle equals $\frac{4.7}{5}$ or 0.″9 with a 0.″2 remainder. Thus, the two angles nearest to 90° would take a 1.″0 correction and the remaining three a 0.″9 correction. The total correction would add up to 4.″7.

CHAPTER 10

TRAVERSE

Section I. DESCRIPTION AND PREPARATION

10–1. General

A traverse is defined as the measurement of the lengths and directions of a series of straight lines connecting a series of points on the earth. The points connected by the lines of a traverse are known as traverse stations, or as traverse-angle stations. The measurements of the lengths and directions are used to compute the relative horizontal positions of these stations on some system of coordinates. From these computed relative positions, additional data can be measured and computed for layout of new features, such as buildings and roads. Traverse operations are conducted for basic area control; mapping; for large construction projects, such as an airport; for road, railroad, and pipeline alinement; for the control of hydrographic surveys; and for many other projects.

10–2. Classification of Traverses

A traverse is always classified as either a *closed traverse* or an *open traverse* (fig. 10–1). Traverses are also described by the instruments used, such as a transit-tape traverse or electronic traverse; by the purpose served, such as a preliminary route traverse; and by the quality of the results obtained, such as a third order traverse. Previous descriptions, such as precise, primary, and secondary traverse, have been superseded and are replaced by accuracy designations, such as first, second, third and lower order traverse. The distinction between orders is defined by such criteria as the closing error in position, the probable error in the angle measurements, the probable error in the distance measurements, the number of angle stations between astronomic azimuth checks, the azimuth closing error, and the probable error of the astronomic azimuths.

a. Closed Traverse. A closed traverse starts and ends at the same point, or at points whose relative horizontal positions are known. The measurements can be adjusted by computations to minimize the effect of accidental errors made in the measurements. Mistakes (blunders) can be detected by rechecking the computations.

(1) *Loop traverse.* A loop traverse (①, fig. 10–1), as the name implies, forms a continuous loop enclosing an area. The computed circuit closure, or *error of closure* for a loop traverse will normally indicate whether a large mistake was made in the measurements. When mistakes are eliminated, the error of closure will indicate the size of the accidental errors. Systematic errors, however, will seldom show in the error of closure. For example, when the tape used for the distance measurements is longer than its nominal length, all of the recorded lengths will be apportionally too small, and will cause little or no change in the computed error of closure of the traverse.

(2) *Connecting traverse.* A connecting traverse (②, fig. 10–1) starts and ends at separate points whose relative positions have been determined by a survey of an equal or higher order accuracy. A connecting traverse of third order accuracy, for example, may be run and adjusted between two stations whose relative positions were determined by a first, second, or third order traverse or triangulation.

b. Open Traverse. An open traverse (③, fig. 10–1) ends at a station whose relative position is not previously known, and unlike a closed traverse provides no check against mistakes and large errors. An open traverse usually starts at a station determined by a closed traverse or by triangulation, but may start at an assumed position. Open traverses are often used for the preliminary survey for a road or railroad. When the centerline location traverse is run, ties to the preliminary traverse form a series of closed, loop traverses. A random traverse is a special adaptation of an open traverse. It is run between two

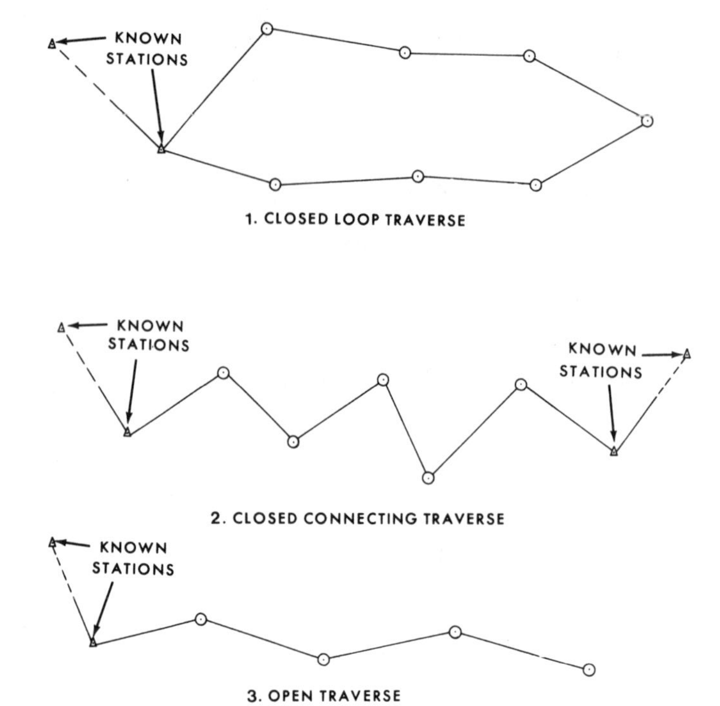

Figure 10-1. Types of traverses.

stations to determine their relative positions, after which the traverse is usually closed by running in the computed line between the two points. One adaptation of the random traverse is used to determine the distance between points on opposite sides of a lake, by running a traverse line around the shore. In this example, the angles of the loop may be closed if a sight can be made across the lake from the final to the initial point. This is *not* considered a closed loop traverse, since the distance between the last and first point is not measured directly. If the distance can be measured by an electronic distance-measuring device, the traverse can become a closed traverse.

10-3. Accuracy

Traverses for engineer and mapping projects are performed under four general orders of accuracy. The highest degree of accuracy, requiring instruments and methods of the highest precision for traverses is first order. Traverses of progressively lower degrees of accuracy for the engineer and mapping projects are second, third, and lower order. The order of accuracy for any traverse is determined by the equipment and methods used in the traverse measurements, by the accuracy attained, and by the accuracy of the starting and terminating stations. The order of accuracy must be specified before the measurements are started. Suitable instruments and methods are then selected to attain this accuracy. A disadvantage of the traverse is the lack of field or computation checks on the accuracy until the traverse is connected to previously established traverse or triangulation networks of the appropriate order of accuracy.

a. Engineer and Mapping Projects. The distance measurement accuracy for both electronic and taped traverses for first, second and third order are 1 part in 35,000, 1 part in 15,000, and 1 part in 7,500 respectively. The order referred to as "lower order" is applied to all traverses of less than third order. The lower order surveys are designed to give sufficient accuracy for the job at hand, usually construction and mapping surveys. In the case of mapping surveys, the linear closure for lower order work is dependent upon the scale of the map for which the control is to be used.

10-4. Reconnaissance

Prior to the actual traverse operations, a reconnaissance must be made to determine the starting point, the route to be followed, the points to be controlled, and the closing station. The best available maps and aerial photographs should be used during the office and field reconnaissance.

a. Selecting the Starting and Closing Points.

Whenever possible, the starting point should be an existing control station that was determined by a survey whose order of accuracy was equal to or greater than the traverse to be run. A second existing control station, visible from the first, should be selected if available for orienting the new traverse. Sometimes no nearby station exists, and the position of the starting point must be determined by some other method. The starting point may be assigned an assumed position, or it is necessary to determine the position from observation of stars. In higher order surveys, star observations for azimuth control at specified intervals are a part of the requirements.

b. Route Selection. Traverses normally are run to establish new stations (control) in an area. The location of the required stations is specified, at least in what vicinity or at what distance between stations. Normally, the clearing of traverse lines should be kept to a minimum. Furthermore, many traverse lines are run through private property, and the indiscriminate cutting of trees will not be permitted.

(1) When the directions of the traverse lines are not fixed prior to the start of the survey, the traverse is run along lines requiring the least clearing, such as railroads, roads, trails, and other open areas. By the proper selection of the route, the traverse can be laid out to pass relatively close to points that must be located or staked out. Aerial photographs are often used as a supplement to maps of the route area. When a loop traverse is to form a long narrow loop, the reconnaissance party should consider the possibility of making one or more crossties to cut the main loop into two or more smaller loops.

(2) On other surveys such as a road centerline layout, the directions of the traverse lines are predetermined, and all obstructions, including large trees, must be cleared from the line. This often requires the assistance of the construction crew.

(3) If the above restrictions are not present and only the station locations are specified, the traverse route is established for ease in field operations and office computations. The number of instrument setups is kept to a minimum, reducing the possibilities of instrument error and the amount of computing required. The traverse line may not necessarily run over every specified station, but should pass relatively close to the ones it misses. These stations can be located during the traverse by using sideshots (para 10–9*c*).

(4) The electronic distance-measuring devices have made traverse reconnaissance even more important. After the general alinement in direction and the planned positioning of stations, the possibility of using electronic equipment should be considered. A platform or tower to clear the surface obstructions will permit comparatively long optical sights for angle reading and at the same time permit determining the distance electronically to avoid taping it by short increments.

(5) For the lower order surveys and where taping must be used, the exact route and station locations normally are selected as the traverse progresses. These stations must be selected so that at any one station, both the rear and forward stations are visible, and only a minimum amount of brush cutting is required to clear the line for taping.

10-5. Station Marks

The purpose of a traverse normally will govern whether its stations will be permanently or temporarily marked. When it is known that a traverse station may be reused over a period of several years, the station marker should be of a permanent type. Other traverse stations that may never be reused, or reused a few times within a period of 1 or 2 months, are usually marked in a temporary manner.

a. Temporary traverse station markers are usually 2-inch by 2-inch wooden stakes, 12 inches or more in length. They are driven flush with the ground, and have a tack or small nail on top to mark the exact point of reference for angular and linear measurements. Such a stake is called a ***hub***. To assist in recovering the hub, a *guard stake* is driven at an angle so that its top is about 1 foot over the hub. The guard stake is usually about ¾ inch thick by 2 or 2½ inches wide and 16 inches or more in length. Lumber crayon (keel) is used to mark letters or numbers on the guard stake to identify the hub. The marked face of the guard stake is toward the hub. Permanent traverse station markers are of various forms such as an iron pipe filled with concrete; a cross cut in concrete or rock; or a hole drilled in concrete or rock and filled with lead, with a tack to mark the exact reference point.

b. For many traverses, such as for a road centerline layout, traverse line distances are referenced to the beginning of the survey. Stakes marked with the distance from the initial point are set at 100-foot intervals, with additional stakes set between the 100-foot stakes when necessary. The stakes at the 100-foot intervals are called *taping stations*, or simply, *stations*; and stakes set between the taping stations are called *plus stations*. This use of the word station should not be confused with the term *traverse station*, which refers to a distance and angle station used in computing positions. The stakes used for marking the taping and plus stations are usually ½ to 1 inch thick, 2 to 3 inches wide, and 16 or more inches long. They are driven vertically into the ground until they are solidly set, with about 12 inches extending above the ground. The wide sides of the stake should be crosswise to the traverse line. The side facing the initial point is marked with lumber crayon from the top down with the station number that indicates the distance from the initial point. The following are examples of stake markings:

(1) At the starting point 0 + 00
(2) At distance of 85 feet 0 + 85
(3) At distance of 700 feet 7 + 00
(4) At distance of 1345.9 feet 13 + 45.9

The guard stake for a hub is marked on its lower inclined face with the taping station number for the hub. For example, the guard stake for a traverse station 1345.92 feet from the initial point would be marked ⊙ 13 + 45.92, the circled dot indicating that this is the taping station number for the hub. Proceeding along the traverse line, the next taping station would be set 54.08 feet from this hub, and the stake would be marked 14 + 00.

c. For other traverses, such as in topographic surveys, the stations are numbered consecutively starting at number one and continuing throughout the traverse. In addition to the number of the station; an abbreviation indicating the type of traverse should be included. For example, ET for electronic traverse or TT for tape traverse.

d. Many of the hubs marking the location of highways, railroads, landing strips, and other projects will require replacement during the construction. *Reference marks* are used to reestablish a station if its marker has been disturbed or destroyed. All permanent survey station markers should be referenced so that they can be replaced in case they are disturbed. Reference marks are placed where they are not likely to be disturbed, and at least two marks should be set for each station. Whenever possible, the marks should be within one tape length of the station, although on large projects, placing the reference marks several hundred feet from the station may be necessary. The marks should be located to allow a good intersection of lines when a station is replaced. At the time of survey, angles are read to each reference mark and the distance is measured to the same precision as the survey requirements.

Section II. MARKING LINES

10-6. Lines Between Corners

The survey is marked upon the ground in the following ways:

(a) The regular corners of the public-land surveys are marked by fixed official monuments.

(b) The relation to natural topographic features is recorded in detail in the field notes.

(c) The locus of the lines is marked upon forest trees by blazing and by hack marks (fig. 10–2 and 3). However, in conformity with the National Environmental Policy Act of 1969, the operational need for the marking of lines should be weighed against possible esthetic damage. If special precautions are to be taken in this regard, they should be set out in the special instructions. Also, in the case of resurveys in areas of mixed public and private lands, it may be necessary to restrict the blazing to trees on public land. The surveyor on the ground should apply good judgment in particular cases not covered by his instructions. Where it has been determined that lines will be marked, the methods discussed here are intended to fix the lines permanently with the minimum environmental effect.

A *blaze* is a smoothed surface cut upon a tree trunk at about breast height. The bark and a small amount of the live wood tissue are removed with an axe or other cutting tool, leaving a flat surface which forever brands the tree. The size of the blaze depends somewhat upon the size of the tree, but should not be made larger than the surface of an axe blade. A blaze five or six inches in height and from two to four inches in width is usually ample.

A *hack* is a horizontal notch cut well into the wood, also made at about breast height. Two hacks are cut to distinguish them from other, accidental marks. A vertical section of the

Figure 10-2. Line blaze.

Figure 10-3. Hack marks on a line tree.

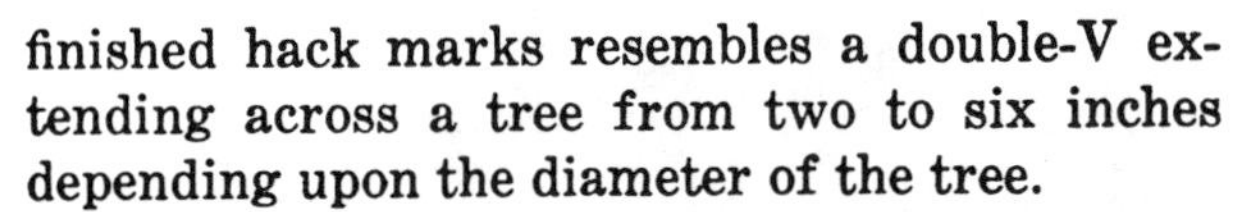

finished hack marks resembles a double-V extending across a tree from two to six inches depending upon the diameter of the tree.

The blaze and hack mark are equally permanent, but so different in character that one mark should never be mistaken for the other. The difference becomes important when the line is retraced in later years.

Trees intersected by the line have two hacks or notches cut on each of the sides facing the line, without any other marks whatever. These are called sight trees or line trees. A sufficient number of other trees standing within 50 links of the line, on either side of it, are blazed on two sides quartering toward the line, in order to render the line conspicuous and readily to be traced in either direction. The blazes are made opposite each other coinciding in direction with the line where the trees stand very near it and approaching nearer each other toward the line the farther the line passes from the blazed trees. Figure 10–4.

The lines should be so well marked as to be readily followed and the blazes plain enough to leave recognizable scars as long as the trees stand. This can be accomplished by blazing just through the bark into the live wood tissue. The blazes should be narrow so that they will heal before decay begins, and special care should be taken not to loosen the cambium layer around the blaze, since this will prevent overgrowth. Where trees have branches growing to the ground, the blazes may be omitted unless it is necessary to remove the branches to permit sighting.

Lines are also marked by cutting away enough of the undergrowth to facilitate correct sighting of instruments. Where lines cross deep wooded valleys, by sighting over the tops, the usual blazing of trees in the low ground when accessible will be performed. The undergrowth will be especially well cut along all lines within distances of 5 chains of corner monuments and within 2 chains of arteries of travel, but the cutting of the undergrowth may be omitted in deep untraveled ravines unless necessary for accurate sighting or measurement.

Line trees and blazing are marked only with reference to the established true line. Where lines are run by the "random and true" line method, the marking of line trees and the blazing is accomplished by returning over the line after all corrections or adjustments to the final line are definitely known. A sufficient number of temporary stakes should be set along a random line to render it generally unnecessary to rerun the true line instrumentally merely for the purpose of blazing the line through timber. This can usually be accomplished by properly estimating the distance from the temporary stakes, but intersections with line trees will be made with precision, and distances thereto accurately measured.

Section III. DISTANCE AND ANGLE MEASURING

10-7. Party Organization

The traverse party may vary from 2 to 15 people. It usually consists of a distance-measuring party, an angle party, and sometimes a level party and other support personnel, all under the supervision of a traverse party chief. The above breakdown of personnel is the ideal, but on many surveys the same personnel will perform functions in one or more of these parties. Therefore, each party member should spend a certain amount of time cross training in each phase of the survey. The organization of the individual par-

Figure 10-4. Marking a line through timber.

ties is described under distance-measuring (para 6–2), and under direction and angle observation (para 9–9). If leveling is also required the leveling party (para 8–6 and 8–19) may be added, or members of the other parties may have to assume the duties and perform these functions.

10-8. Ties to Existing Control

Traverse determines relative horizontal positions of the scheduled stations. This means that the starting point must be a known position or control point. Except for the open or random traverses which may be required occasionally, the traverse must close or tie to another known position or control point. In loop traverses, this may be the starting point. Connecting traverses start at one control point and tie into another control point.

a. As discussed in paragraph 10–4*a*, the starting point should be an existing station, with another station visible for orienting the new traverse. If this starting point is established by traverse, the adjacent stations are intervisible with the instrument at normal tripod height and ties are usually quite easily made. The same is true for the closing point.

b. Adjacent stations, established by triangulation, are not always intervisible from ground level, since they could have been observed from towers erected over one or both stations. Many triangulation stations have an azimuth mark which is visible from the ground at the station and which has been established for orienting new control without the need of erecting towers. When no azimuth mark exists, a supplementary connecting traverse can be run between the two triangulation stations, or supplementary triangulation can be used to tie the traverse to two triangulation stations. The higher orders of traverse use astronomic observations (star shots) to establish the basic azimuth or direction from which the traverse begins.

c. The search for an adjacent station to orient a new traverse should not be too extensive. If the adjacent station is not easily found, an astronomic azimuth should be observed to orient the starting line, and the traverse continued.

d. Any existing control near the traverse line should be tied in to the new work. If possible, at least one of the existing control stations should be occupied and an angle observed between a second station of the existing control and a station of the new traverse. In this way, the new work will have a direction tie to the existing control.

10-9. Linear Measurements

The required traverse accuracy will determine the equipment and the method of measuring the distance. The corrections to be included and the number of measurements that constitute a complete set are determined by the required accuracy.

a. For the lower orders, a single taped distance is sufficient. As the order of accuracy gets higher, double taping (once each way) is required. Steel tapes must be compared to a master tape at specified intervals. For the highest accuracy, electronic equipment or invar tapes are used.

b. Surveys associated with this manual normally will require taping to determine distances for traverse. The taping will follow a direct line between stations. Thus, while the angles are being measured or immediately before or after, the taping party starts its measuring from the occupied station keeping the front end of the tape alined with the forward station. If double taping is required, they start back from the forward station and measure to the occupied station, using the same alinement, but not the same taping points. The second measurement must be independent from the first. The two distances are compared, and if within accuracy requirements, the distance is accepted. If the two measurements disagree by more than the allowable amount, the distance must be retaped. Upon completion, the taping party proceeds to the forward station and uses it as the occupied station for the next line meaurement. This taping is continued until the tie-in control point is reached. The method for keeping taping notes is illustrated in figure 6–1.

c. Required locations off the main traverse line, but comparatively near it, are usually tied-in to the main line by *sideshots.*

(1) Sideshots are single course traverses off the main line to locate one or more points for control. The instrument is set up on the main traverse line and the direction from the main line to the point is observed. The distance from the instrument station to the point is then measured. The position of the point can be computed from this information. The distance can be measured by taping, electronic means, or stadia reading.

(2) One special case of the sideshot is the offset measurement (para 13–4). When the required station falls a short distance (less than a tape length) off the main traverse line, a perpendicular (offset) distance is measured from the station to a point on the line directly opposite the station. The taping party also measures the distance to this point along the traverse line. The resulting distance and 90° offset to the main traverse line can be used to compute the position of the station. Offset measurements usually are restricted to one tape length or less from the main line. Beyond this, the instrument is set up and a sideshot ((1) above) is taken.

d. When the distances are determined by stadia readings, using an angle measuring instrument, the distance reading becomes part of the instrumentman's function. He must also read the vertical angle to convert slope distances to horizontal distances.

10-10. Angular Measurements

The relative directions of the traverse lines are determined from measurements of the horizontal angle formed by the lines at each station. Horizontal angles are measured with a transit or a theodolite, or determined graphically with a planetable and alidade.

a. Traverse Stations. In a traverse, three traverse stations (fig. 10–2) are considered of significance. These are referred to as the *rear station,* the *occupied station,* and the *forward station.* The rear station is that station from which the persons performing the traverse have just moved, or is a point, the azimuth to which is known. The occupied station is the station at which the party is located and over which the surveying instrument is set. The forward station is the next station in succession and constitutes the immediate destination of the party.

b. Horizontal Angles. Horizontal angles are always measured at the occupied station by pointing the instrument toward the rear station and turning the angle clockwise to the forward station for the direct angle, and clockwise from the forward to the rear station for the explement. If a deflection angle (fig. 9–1) is to be used, the instrument, after sighting on the rear station, is plunged (revolved around its horizontal axis) and the angle left or right to the forward station is read. When sighting on the stations, it is good practice to point on that portion of the station target that can cause the least amount of error. If a range pole is used, the lowest visible point should be sighted upon. Thus, any misplumbing of the pole will least affect the angle reading. However, heat waves rising from the ground can also introduce distortions in the line of sight and errors in the measurements. The instrument-person must be aware of these conditions and make every effort to obtain the truest pointing.

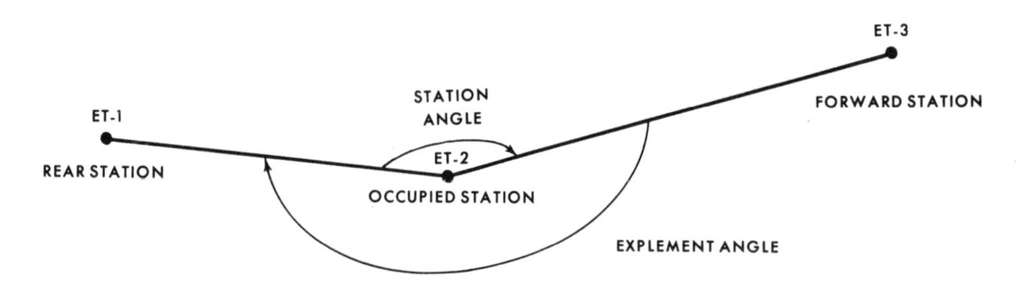

Figure 10-5. Traverse stations and angles.

c. Vertical Angles. Vertical angles are measured at the occupied station either to convert slope distances to horizontal, or to establish elevations of the traverse stations. The measuring of the vertical angles or zenith distances, and the number to be read are a function of the traverse accuracy. Normally, a single direct and reverse reading will minimize instrument errors and is sufficient for surveys associated with this manual. For trigonometric leveling and reciprocal readings, refer to chapter 8.

d. Notekeeping. Keeping notes is dependent upon the method of performing the traverse. The distances for electronic measurements and surveys requiring double taping and slope taping with added tape correction are kept on a separate record. Angle measuring associated with these surveys requires several circle settings or angle repetitions and the notes are kept in their respective notebooks. Sample notes for each of these functions are illustrated in the chapters which discuss these methods. A system of cross reference must be established to guarantee that the angles and distances of the same traverse survey can be located easily even by someone unfamiliar with the survey. For lower order traverses, the angular and distance measurements were usually recorded on the same page in a common notebook.

10–11. Extending Azimuth

Sometimes fences, hedges, pole lines, trees, or other obstructions that cannot be removed, will prevent direct measurements between points when the lengths and azimuths of the lines between these points are required. In such cases, the traverse is run close to the required lines, and ties are made from the traverse to the existing points. When the traverse follows many curves and short tangents, or where the topography necessitates taping through many short lines, an attempt should be made to extend the main traverse angles and azimuths through auxiliary lines extending past several of these shorter lines and bypassing some of the intermediate points. The angles at the short lines must still be measured. However, the azimuth of the main traverse will be the result of a fewer number of sight lines. The short lines are considered short random traverses within the main traverse. The lengths and azimuths of the required lines are then computed.

Section IV. COMPLETION

10–12. Computations

Plane surveying traverses involve the determination of the rectangular grid coordinates of each traverse station and of other points tied to the traverse. First, the grid azimuth of each line of the traverse must be determined progressively along the traverse. This azimuth and the measured distance are converted into values known as latitudes (north-south values) and departures (east-west values) for each station. Using the north and east values for the known starting station, the latitude and departure of each traverse station are progressively added algebraically. The results are the northing and easting of each station from a known or established zero point, and are called the grid coordinates.

a. Azimuths. Azimuths are most easily determined from the forward azimuth and the deflection angle from line to line.

(1) Figure 10–6 shows the computations and adjustments of a connecting traverse. Starting with the fixed or known azimuth of the line from station M15 to station M16, the right (R) deflection angles and the left (L) deflection angles (column 2) are added or subtracted respectively to the previous azimuth value (column 3) to determine the following azimuth.

(2) This traverse has a check azimuth near its midpoint, permitting the adjustment of the azimuths in two sections. The adjustment is determined (para 10–13) and applied to these azimuths to obtain the adjusted azimuth for each line. Note that the first section ties into the azimuth of the azimuth mark of station 9 and starts again with the fixed azimuth of the line from the *azimuth mark* of station 9 for the second station.

b. Latitudes and Departures. The adjusted azimuth (column 5) and the measured or computed horizontal distance (column 6) are combined to determine the departure (ΔE, column 9), and latitude (ΔN, column 10). Natural trigonometric function tables are used in the computations.

(1) The adjusted azimuth can first be converted into an equivalent angle of less than 90°, and the sine and cosine can be taken from a prepared set of trigonometric function tables. The mathematical sign of ΔF and ΔN is determined during this conversion. If bearings are used, this conversion is not necessary.

(2) The conversion is performed in the following manner. Azimuth less than 90° are used directly to enter the tables. The values of both the sine and the cosine in this quadrant will be positive (+). Azimuth values (a) greater than 90° are reduced to an angle of less than 90° (x) by subtracting 90, 180, or 270 as needed. The signs of the functions change in each of the three quadrants. The following list illustrates the sign changes and the equivalent trigonometric value for sin a and cos a:

azimuth (a)	sine a	cosine a
x	+ sin x	+ cos x
90 + x	+ cos x	− sin x
180 + x	− sin x	− cos x
270 + x	− cos x	+ sin x

(3) The appropriate values and signs are entered in columns 7 and 8. To determine the departure (ΔE, column 9) the distance (column 6) is multiplied by the sin a (column 7). The latitude (ΔN, column 10) equals the distance (column 6) times the cos a (column 8).

(4) The original distance and azimuth field readings between any two stations are converted into coordinate distances east and north between the stations. If the sign of the increment is positive (+), the second point is further north or further east; if negative (—), further south or west (less north or less east).

(5) Note that the algebraic sum of the latitudes (ΔN) or the departures (ΔE) between any two stations will give the total difference in coordinates north and east between the stations. This method is used to check the computations through the line. The algebraic total of latitudes (ΣΔN) is added to the starting station value and should equal the final station value. The same applies to the departures (ΣΔE).

(6) After the computation has been checked and the latitudes and departures adjusted (para 10–13), the survey is now tied to a coordinate system.

c. Grid Coordinates. The computed latitudes and departures are a measure of the relative positions between stations of a traverse survey. To make the positions universal, the survey is tied into a grid coordinate system. This system is based on a Universal Transverse Mercator (UTM) Grid for most of the world, with a Universal Polar Stereographic (UPS) Grid

(1)	(2)	(3)	(4)	(5)	(6)	(7)	(8)	(9)	(10)	(11)	(12)	(13)	(14)	(15)
STATION	DEFLECTION ANGLE	UNADJUSTED AZIMUTH	ADJ	ADJUSTED AZIMUTH	DISTANCE	SIN α	COS α	UNADJUSTED ΔE	UNADJUSTED ΔN	ADJUSTED ΔE	ADJUSTED ΔN	GRID COORDINATES E	GRID COORDINATES N	STATION
	° ′ ″	° ′ ″	″	° ′ ″	Feet							Feet	Feet	
M15		192 54 30												
M16	R 38 49 10	231 43 40	+05	231 43 45	547.11	-0.78509	-0.61938	-429.53	-338.87	-429.64	-338.85	73496.28	4876.39	M16
1	R 69 21 00	301 04 40	+10	301 04 50	558.09	-0.85644	+0.51624	-477.97	+288.11	-478.08	+288.13	73066.64	4537.54	1
2	L 22 24 30	278 40 10	+15	278 40 25	556.88	-0.98856	+0.15081	-550.51	+83.98	-550.62	+84.00	72588.56	4825.67	2
3	R 28 03 05	306 43 15	+20	306 43 35	1066.48	-0.80150	+0.59799	-854.78	+637.74	-854.99	+637.79	72037.94	4909.67	3
4	R 28 53 40	335 36 55	+30	335 37 25	805.35	-0.41273	+0.91085	-332.39	+733.55	-332.55	+733.59	71182.95	5547.46	4
5	L 30 08 25	305 28 30	+35	305 29 05	1540.75	-0.81427	+0.58049	-1264.59	+894.39	-1254.89	+894.46	70850.40	6281.05	5
6	L 35 30 40	269 57 50	+40	269 58 30	1263.47	-1.00000	-0.00044	-1263.47	- 0.56	-1263.72	- 0.51	69595.51	7175.51	6
7	L 10 46 10	259 11 40	+45	259 12 25	1105.01	-0.98231	-0.18726	-1085.46	-206.92	-1085.68	-206.87	68331.79	7175.00	7
8	R 2 56 45	262 08 25	+50	262 09 15	1699.69	-0.99064	-0.13651	-1683.78	-232.02	-1684.12	-231.95	67246.11	6968.13	8
9	R 21 17 15	283 25 40	+55	283 26 35								65561.99	6736.18	9
Az.MK		283 26 35	Fixed Az.											
		0 55												
Az.MK		103 26 35												
9	R 166 49 55	270 16 30	+05	270 16 35	853.31	-0.99999	+0.00482	-853.30	+ 4.11	-853.47	+ 4.15	65561.99	6736.18	9
10	L 51 22 30	218 54 00	+10	218 54 10	1067.55	-0.62801	-0.77821	-670.43	-830.78	-670.64	-830.73	64708.52	6740.33	10
11	R 28 53 35	247 47 35	+10	247 47 45	1237.19	-0.92584	-0.37791	-1145.44	-467.55	-1145.68	-467.50	64037.88	5909.60	11
12	L 29 37 30	218 10 05	+15	218 10 20	962.89	-0.61703	-0.78616	-594.13	-756.99	-594.32	-756.95	62892.20	5442.10	12
13	R 73 31 00	291 41 05	+20	291 41 25	1343.28	-0.92920	+0.36959	-1248.18	+496.46	-1248.45	+496.52	62297.88	4685.15	13
14	R 27 30 15	319 11 20	+20	319 11 40	773.18	-0.65349	+0.75693	-505.27	+585.24	-505.42	+585.27	61049.43	5181.67	14
15	L 19 26 00	299 45 20	+25	299 45 45	1149.82	-0.86809	+0.49641	-998.15	+570.78	-998.38	+570.83	60544.01	5766.94	15
16	L 34 49 10	264 56 10	+30	264 56 40	1137.94	-0.99611	-0.08812	-1133.51	-100.28	-1133.73	-100.23	59545.63	6337.77	16
17	L 23 45 55	241 10 15	+35	241 10 50	961.10	-0.87614	-0.48205	-842.06	-463.30	-842.25	-463.26	58411.90	6237.54	17
18	L 81 03 50	160 06 25	+35	160 07 00	1241.57	+0.34011	-0.94039	+422.27	-1167.56	+422.02	-1167.50	57569.65	5774.28	18
N9	R 14 13 00	174 19 25	+40	174 20 05								57991.67	4606.78	N9
N8		174 20 05	Fixed Az					-15500.68	-270.47	-15504.61	-269.61	73496.28	4876.39	M16
		0 40						-15504.61	-269.61			-15504.61	-269.61	
					19870.66			+ 3.93	- 0.86					

Figure 10-6. Computation and adjustment, plane traverse.

in the polar regions.

(1) Basically, a grid is a series of reference lines established at right angles to each other. By accepting one intersection on this grid as an origin and assigning a value to this point, the grid becomes a useful tool in measuring relative positions. For example, let us assume a grid with an origin value of zero north and zero east. The location of a point, six units or blocks north and eight units or blocks east of the origin, is easily identifiable. The point is 10 units slightly east-northeast of the origin.

(2) In the grid systems, the origin values are established. When a control station is computed and a grid value determined for it, the latitude (ΔN) from the established origin to this point is called a *northing*, and the departure (ΔE), an *easting*. The northing and easting of any point are called the *grid coordinates*. Latitudes and departures algebraically added to the grid coordinates establish a new set of grid coordinates for the new station.

(3) In figure 10–6, the adjusted latitudes and departures (ΔN and ΔE) are added algebraically to the known coordinates from the first station (M16) progressively. When the final station (N9) is reached, the coordinates computed through the traverse result in a surveyed value for the position of the station. The difference between the surveyed coordinate value and the known value (of N9) constitutes the closure error of the survey.

(4) The degree of accuracy for the traverse is computed from the linear error of closure and the total length of the traverse. The *linear error of closure* is the straight line distance by which the traverse failed to close. The northing and easting closure is converted to the straight line distance by taking the square root of the sum of the squares of the easting and northing errors. The traverse accuracy is then expressed as the fraction having a numerator equal to 1, and a denominator equal to the total length of the traverse divided by the linear error of closure. For example, let us assume that the easting was off by +6 meters and the northing by +8 meters. The linear error of closure equals $\sqrt{6^2 + 8^2}$ or 10 meters. If the traverse survey extended 50 kilometers, the accuracy of the survey is 1 divided by the result of 50 kilometers divided by 10 meters or:

$$\text{survey accuracy} = \frac{1}{\frac{50\text{ km}}{10\text{ m}}} = \frac{1}{\frac{50{,}000\text{ m}}{10\text{ m}}} = \frac{1}{5{,}000}$$

or as usualy expressed, 1 part in 5,000.

(5) It is often necessary to compute the distance, or the distance and azimuth between two traverse points from the grid coordinates of the two points. The method of squaring the ΔE and ΔN, adding, and taking the square root will only give the distance. There is another method which does not require this arithmetic and results in the distance and the azimuth. The values of ΔE and ΔN for the line are determined from the grid coordinates for the two ends of the line. The tangent of the azimuth of the line is then equal to ΔE divided by ΔN, and the cotangent of the azimuth is equal to ΔN divided by ΔE. The distance (d) is equal to ΔE divided by the sine of the azimuth or ΔN divided by the cosine of the azimuth.

d. Area. Traverses are sometimes run in order to determine the area of a tract of land. This is done by making the traverse legs coincide with the tract boundary where possible, or by using perpendicular offset from the traverse legs to the boundary. Determination and computation of area is covered in section II, chapter 13.

10-13. Adjustments

To eliminate or minimize as many individual measurement errors as possible, adjustments are made at different times during the traverse. When the final adjustment is made, the errors have been distributed and the positions of all points should be within the allowable accuracy. This allowable error is determined by the type of survey as stated in the specifications. The adjustments are made at each station for the angles, after each distance measurement, after the azimuths are checked, and finally after the latitude and departures are computed and tied to the closing station value.

a. Angles. The angles are checked at each traverse station by closing the horizon. Any error is distributed and the angle values are adjusted to reduce the small instrumental errors. This procedure is described in paragraph 9–14.

b. Distances. The distances are computed by applying the necessary corrections after each measurement. If the survey accuracy requires it, several measurements are made over the same course. Adjustment consists of taking a mean of the individual measurements, and in effect, distributing the individual measurement errors equally among all the determinations.

c. Azimuths. The azimuth adjustment distributes the small angle errors accumulated at each traverse station. In most instances, these errors are the small increments of the angle which the instrument cannot read. For example, an angle of 87° 32′ 13″ will be read as 87° 32′ 10″ by a 10″ instrument, thus leaving a 3″ error in the angle. In addition, slight errors in pointing the instrument on the targets result in a small increment of angle error. The combined effect after several stations can add up to a readable amount. When the azimuth is checked against a previously determined "correct" value (column 3, fig. 10–6), the difference in readings is the total amount of this accumulated error. This error can also be referred to as the angular closing error. The angular closing or azimuth error for each traverse line is found by dividing the total error by the number (n) of angles affecting the closure, and multiplying this value by 1, 2, 3, . . . n, in turn. The adjustments (column 4) are applied to the azimuths of column 3 to obtain the adjusted azimuths in column 5, figure 10–6.

d. Latitudes and Departures. Even with the above adjustments made, small residual errors will still remain. The final adjustment is based on the grid coordinates (columns 13 and 14) of the starting station (M16) and the closing station (N9), which have fixed values determined by other surveys.

(1) The algebraic sum of the ΔE values (column 9) should equal the difference between the two given eastings (E, column 13), and the algebraic sum of the ΔN values (column 10) should equal the difference between the two given northings (N, column 14). In this example, the sum of the ΔE's is —15,500.68, while the difference in the eastings for stations M16 and N9 is —15,504.61, or the easting error of closure is +3.93. The sum of ΔN's is —270.47, while the difference in northings for stations M16 and N9 is —269.61, giving a northing error of closure of —0.86.

(2) The error of closure in both easting and northing is distributed proportionately throughout the traverse by a method referred to as the *compass rule method* and computed as follows: divide both the easting error of closure and the northing error of closure by the total length of the traverse (19870.66-summation of column 6), to obtain two factors, E_e for the eastings and E_n for the northings. The correction to the ΔE for any line of the traverse is then equal to E_e multiplied by the length of that line. Similarly, the correction to the ΔN for any line of the traverse is equal to E_n multiplied by the length of that line. The adjusted values for the ΔE's and the ΔN's are shown in columns 11 and 12 (fig. 10–6) and the algebraic sum of each of these columns should equal the correct value, as shown.

(3) In the case of a loop traverse, the algebraic sum of both the unadjusted ΔE's and the unadjusted ΔN's should equal zero. Any difference between the values and zero is the error of closure. The adjustment of a loop traverse is the same as for a connecting traverse, and the adjusted ΔE's and ΔN's should each add to zero algebraically.

CHAPTER 11

TRIANGULATION

Section I. DESCRIPTION

11–1. Introduction

Triangulation is an efficient and accurate method of establishing control points over extensive areas of the earth's surface. In the past, as well as in many surveys today, triangulation has formed and does form the basic horizontal control networks throughout most of the world. Triangulation is the method which employs trigonometric principles to solve the triangles and to determine the horizontal positions and distances. Whereas, in traverse, the distance and the angle are measured at each station, in triangulation only the directions (angles) are measured at each station with the distances measured only at specified intervals. This chapter will discuss only the more general aspects of triangulation as it pertains to lower order.

a. Conventional Triangulation. In the past triangulation had but one basic design and that was the measurement of all directions of geometric figures; the measurement of a base line at specified intervals; and when necessary the observation of astronomic azimuth and position. This type is usually referred to as conventional triangulation and used for lower order surveys.

b. Braced Triangulation. With the advent of electronic distance measuring equipment a modern type of triangulation has evolved which incorporates the use of both direction and distance measurements intermingled within the geometric figure. This type is referred to as braced triangulation and is generally used only for the higher order surveys. Braced triangulation will not be discussed in this chapter.

11–2. General

The discussion for lower order triangulation in this chapter is normally limited to triangles having sides less than 3 kilometers and to triangulation nets that do not extend more than 15 to 25 kilometers. In such nets, a base line is normally measured near each end of the net or tied-in to previously established control and all computations are done on a plane grid system.

a. The simplest form of triangulation consists of one triangle. One side and at least two (usually three) angles are measured, and from this data, the lengths of the other two sides are computed. Then, knowing the position of one point and the azimuth of one line, the relative positions of the other two points (vertices) can be determined in relationship to the known point. The use of one triangle is very effective in crossing obstruction (rivers, ravines, etc) where the distance across cannot be measured easily. A base line, 1–2, (①, fig. 11–1) is established parallel to the obstruction and the distance measured carefully. Then a third point, 3, is established on the other side of the obstruction. Using a direction or angle measuring instrument the three angles of the triangle are observed. From this data, the base line and three angles, the other two sides are computed. Knowing the position of point 1, azimuth of the base line, and all parts of the triangle, then the position of points 2 and 3 can be computed in relationship to point 1.

b. To extend this basic triangle, a fourth point, 4, (②, fig. 11–1) is selected. It must be visible from points 3, and one of the base points, 1. The angles at 1 (from 3 to 4), at 3 (from 1 to 4), and at 4 (from 3 to 1) are observed. Note that one side (1–3) is known from the computations of the first triangle (1–2–3). The new triangle 1–3–4) can be computed and the lengths of the unknown sides (1–4 and 3–4) can be determined. This same extension procedure is carried on by selecting point 5, then 6 through 10 in sequence and observing the appropriate angles. This series of figures is called a triangulation net or chain of triangles. When distance 9–10 (the closing base line) is reached, it is carefully measured. The measurement is compared against the distance as computed through the net and the difference is used to determine the closure error of the net.

c. The basic triangle and the described net using one set of triangles have the disadvantage of insufficient checks. An error in one of the angle readings will distort the triangle and result in a progressive accuracy loss as the net is extended. To permit stronger checks and to reduce the possibility of errors, a different type of net can be used (③, fig. 11–1). For a net of the same length as in ②, figure 11–1, three additional stations with the extra angle readings are required and result in a double line of triangles. The closing base line (9–10) can now be computed in the same manner as in ②, figure 11–1. In addition, it can also be computed independently using the right-hand tier of triangles. The two independent computations offer a much stronger check than the single line of triangles.

d. A still stronger system is illustrated in ④, figure 11–1. This requires more angle observations than the other system, but no more stations than in ②, figure 11–1. The basic figures of this system are quadrilaterals (for example, 1–2–3–4) made up of interlocking triangles (1–2–3, 1–3–4), 1–2–4, and 2–3–4). The stronger check is obtained by observing the angles created by the additional diagonals through the quadrilaterals. The base line distance (1–2) is used to compute the length of 3–4, the 5–6, 7–8, and finally 9–10, the closing base line. Note that in a quadrilateral (1–2–3–4, for instance) to go from 1–2 to 3–4, triangles 1–2–4 and 2–3–4 can be used for one route. For a completely different route, triangles 1–2–3 and 1–3–4 can be used. Each route uses a different set of observed angles. Other routes exist, if needed, although they are not normally used. Triangle 1–2–4 is used to determine side 1–4, then triangle 1–4–3 to get side 2–3, then triangle 2–3–4 to get side 3–4. There are still more possibilities, but these become impractical due to computation problems. The method for determining the best routes through a triangulation net is based on a technique called the *strength of figure* (para 11–4*a*(3)).

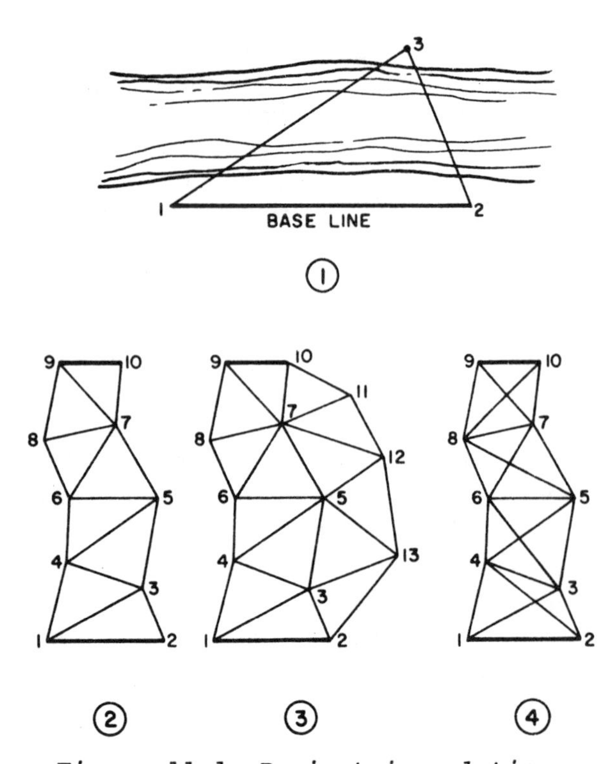

Figure 11-1. Basic triangulation.

11-3. Accuracy

Triangulation for engineer and mapping projects is performed under three primary orders of accuracy and one order for all surveys lower than the primary orders.

a. Engineer and Mapping Projects. The orders of accuracy for these projects are first, second, third and lower order. The three primary orders are subdivided to give a total of six different degrees of precision. First order is divided into class I, II, and III, with length and position accuracies of 1:100,000, 1:50,000, and 1:25,000 respectively. Second order consists of class I and II, with length and position accuracies of 1:20,000 and 1:10,000 respectively. Third order requires a length and position accuracy of 1:5,000. All linear closures are a comparison of the length of a base line as computed through the net and the measured length of the base or the adjusted length of a previously established line, after side and angle equations have been satisfied. The term, lower order is applied to all triangulation using methods and closures of less than third order accuracy. As in traversing, lower order surveys are designed to give sufficient accuracy for the job at hand.

11-4. Reconnaissance

Reconnaissance is essentially the basic design of the survey to be conducted. Station location for a triangulation net is a very important phase of the survey. All station and base line locations should be made prior to the execution of any triangulation. For a lower order triangulation net, the reconnaissance may vary in time from an hour or two for crossing a river, to several days for a network.

a. The reconnaissance party should pay particular attention to the following factors:

(1) Location and length of the base line with respect to the general triangulation system, and to the ease in measuring.

(2) Intervisibility of all stations of the figure being used, and preferably without the use of high towers or stands.

(3) The determination of the strength of figure. Strength of figure is a numerical rating that is used to compare the various routes of computation through a net and to select the strongest. When using a chain of single triangles, there is only one route of computation. There is a limiting factor, however, on the summation of the strength of figure value between base lines or tie-ins for lower order surveys. This limiting factor is dependent upon the accuracy requirement of the new points being established and should seldom exceed 200 for any lower order survey. The strength of figure is based on the size of the distance angles and the number of conditions to be satisfied within a figure. As a general rule for triangulation no triangle should have an angle smaller than 30° or larger than 150°.

(4) Selection of auxiliary stations to be located with respect to the triangulation system. The coordinates of landmarks such as chimneys, church spires, and radio or water towers, even though inaccessible, are often very useful to a subsequent surveying project. The method used to control these auxiliary stations is called intersection (para 13–5*a*).

b. The usefulness of a good reconnaissance can be wasted, especially if the triangulation party does not make full use of the selected points. The reconnaissance party must locate and describe the exact point selected and mark it. A description and sketch should be prepared for all newly established permanent and temporary stations. The description and sketch should be recorded on appropriate current field recording booklets. The description is a narrative report compiled at the station site containing all information necessary to expeditiously locate the station. The sketch should be clear and simple, and contain only the detail required for positive identification. The description and sketch should complement each other and should contain such information as general location, access routes, exact location, whether consent of the landowner has been obtained, type of marker left at station, and whether the setup will be at ground level or require a tower or stand.

11–5. Station Markers

All triangulation stations of third or higher order must be identified on the ground with a permanent station marker. The station mark generally used is a bronze disk embedded in a standard monument, but under some conditions, improvised markers must be used. At all higher order triangulation stations at least two reference marks and if necessary an azimuth mark must also be established. For lower order surveys, unless otherwise required by project instructions, the stations are marked with a 2-inch by 2-inch by 15-inch wooden hub. Triangulation stations are usually given names rather than numbers. Probably the best name is that of a geographic feature or political subdivision at or near the station. Duplicate names must never be used within any one triangulation system. Names of people in the party, nicknames, names without meaning, or names arising from incidents should not be used. Always be sure the station names are spelled correctly on all documents.

Section II. FIELD PROCEDURES

11–6. Party Organization

The triangulation party must be versatile because it is responsible for measuring all base line and directions of the triangulation system. Besides the reconnaissance party, it is composed of one or more direction observation parties, plus additional personnel for the construction of special targets and/or towers and for other required duties. The entire triangulation party is under the direction of the chief of party. The direction observing party consists of an instrumentperson and a recorder, plus the personnel required to display targets, heliotropes, or lights if work is done at night. A distance measuring party is also required for base line measurements. This party may be a taping party or an electronic distance measuring party depending on the survey requirements. In addition there are usually two computers assigned to the triangulation party to complete all field computations. These computers are usually utilized as recorders when not engaged in the computations.

11-7. Targets, Signal, Towers, and Stands

After the station locations have been established, the instrument positioning and the target placement must be considered. The final step before leaving is the monumenting of the station.

a. The reconnaissance party will have selected the location from which the required number of other stations can be seen, and which will be seen from the other stations. An unwooded area on a hilltop would be ideal, but is seldom available. Frequently, woods, brush, and even terrain will obstruct the view in some directions. The reconnaissance party, by moving around the vicinity of the desired location, can select a spot from which one or more points will be completely or partially visible. They will decide whether it is more practical to clear the obstruction (if timber, and permission to cut is available) or to raise the instrument onto a stand or tower for unobstructed viewing. If the obstruction is terrain (another hilltop), it may be necessary to use available maps or other information to determine at which height clearance will be obtained. This information is included in the reconnaissance description. The triangulation party will perform the necessary cutting and construction for the instrument, target setups and the monumenting of the station.

b. A target is any object at which the instrument's telescope is pointed when measuring an angle. An instrumentman, on short sights, may sight on plumb bob strings (up to 75 meters), range poles (up to 300 meters), target sets (up to 1 km) and rods held over a point. As the sight distances increase, these targets become too small and too indistinct to use. Larger targets are required. The possible use of a heliotrope should also be considered if the observations are to be made during daylight hours when the sun is shining.

(1) The size of a target is dependent upon the distance between it and the instrument. It must be large enough to be seen, yet narrow enough to afford a good bisection. Normally, a target which subtends an angle of 4″ to 6″ of arc will fulfill this purpose. This angular dimension converts to a target size as follows. An arc of 1″ equals a width of 0.5 centimeter at a distance of 1 kilometer. For good bisection, a target about 3 centimeters wide is needed at a distance of 1 kilometer, and about 9 centimeters wide at 3 kilometers. With poor lighting or visibility, the size may have to be increased.

(2) The shape of a target easiest to bisect is a triangle or a diamond with a vertex pointing up or down. This vertex must be plumbed exactly over the station and should be high enough for good visibility. If the triangulation requires vertical angle readings, the base of the triangle or the horizontal diagonal of the diamond will afford good bisection. The vertical distance from the ground station to this point is measured and recorded after the target is set up.

(3) The color of the target must be selected for good visibility against the background where it will be viewed. Normally, a target with alternate bands of red and yellow or red and white will be visible against both light and dark backgrounds. If target is to be visible from several directions, it must be made to fit this condition or extra targets may be mounted on the same holder.

(4) One other consideration for targets is possible occupation of the station by one instrument party while the target is being observed by another instrument party from another station. This requires special types of target construction and guying to maintain the target's position, or placing the target at an eccentric position.

(5) When observations are made during daylight hours with the sun shining, a heliotrope (para 4–31) is a very effective target. It requires an attendant to rotate the mirror and keep the sun's rays pointed toward the instrument.

c. During triangulation at night, lights must be used for targets. Up to about a kilometer, the target sets (para 4–30*c*) with built-in illumination are very effective. For longer sight distances, special signal lights (para 4–32) are used. They are available in two sizes and can be stacked in a vertical column for sighting from several instrument stations.

d. During triangulation, auxiliary points may be visible from the different stations, but are impractical to occupy. Such objects as prominent flagpoles, chimneys, and church steeples can be used as intersection points. The angles to them are read from several stations and their positions calculated from these readings. One precaution should be taken. It is very easy to mistake flagpoles and chimneys when they are viewed from several different directions. Even church steeples can lose their unique identity when viewed from a different angle. The instrumentman must be sure that he is using the same point for all sightings.

11-8. Base Lines

In plane surveying (lower order), the triangulation system is small and the base lines are relatively close together. They are measured to the lowest accuracy that will keep the discrepancy from exceeding a specified amount. This accuracy depends on several factors.

a. For simple triangulation, such as the triangulation across a river with one triangle or one quadrilateral, the measured base line is usually one side of the figure. The base line should be of such length and direction that as a general rule, no angle in the triangle, or in the quadrilateral, will be less than 30° or larger than 150°. In any case, the length of the base line should be greater than one-sixth the average length of line in the main scheme.

b. If measured by taping, this length normally is run from one end of the base line to the other, and remeasured in the opposite direction. The mean of the two measured values is used as the base line length. Corrections described in chapter 6 may have to be applied.

c. Electronic distance measuring depends upon the number of fine readings or the number of different frequencies to attain the required accuracy. These variables are discussed in chapter 6. Some instruments have an interchangeable master remote capability. Thus, the distance is measured in "one" direction using A as a master and B as a remote. The functions are changed and the distance is measured in the "opposite" direction with B as a master and A as the remote.

d. The required accuracy of the base line measurement depends on the order of accuracy of the survey. Lower order accuracy requirements are based on the needed accuracy for the survey to be done and are usually specified in project instructions.

11-9. Direction Observations

The directions for triangulation are observed with a theodolite; a transit can be used if accuracy requirements permit its use. The type and/or model of instrument used will depend upon the order of accuracy required by the survey. Chapter 4 describes the many models and types of instruments used for direction observations. The methods used for direction observations are described in chapter 9. On most triangulation nets or single figures there is the additional requirement for the observation of vertical angles (zenith distances), and these are observed with the same instrument that is used for the horizontal directions.

11-10. Recording

The recording procedure for horizontal direction is done in the same manner for all orders of accuracy. The type and model of instrument being used will dictate the method of recording. Base line notes are recorded in the appropriate taping book, leveling book, or in the case of electronic distance measuring on the appropriate single sheet recording form. Paragraph 1-7 outlines the procedure and format to be used for survey recording. For the specific notekeeping refer to the chapters dealing with the particular operation involved.

11-11. Computations

Triangulation parties are normally required to complete unadjusted field computations through the determination of triangle closures and trigonometric leveling computations. There are usually two computers assigned to each triangulation party, whose primary duty is to make these computations. The principal purpose of the field computations are: to insure the completeness of the data; to make a thorough field check of the observations; to inform the chief of party as to the accuracy that is being attained; to reduce the observations to a form readily available for office computation and adjustment; and frequently, to furnish field values for immediate needs. The triangle computations involve, first, the adjustment of measured value for the three angles of each triangle so that their sum will equal 180° and then, the computation of lengths of the unknown sides of each triangle. Computations of triangle side lengths are carried forward from the base line, through the triangles, and tied-in at the closing base line.

a. The sum of the three angles of a plane triangle equals 180°. The difference between the sum of the measured values for the three angles of each triangle and 180° is determined. *One-third* of this difference is then added to, or subtracted from each of the three measured angular values, to make their sums equal 180°.

b. The lengths of the two unknown sides of each triangle are computed by the *law of sines of plane angles*, which states:

$$\frac{a}{\sin A} = \frac{b}{\sin B} = \frac{c}{\sin C}$$

in which A, B, and C are the adjusted plane angles, and a, b, and c are the lengths of the sides opposite the angles A, B, and C respectively. The base line or known side is always letter a and the unknown sides b and c. The length of these unknown sides can be computed using natural functions or logarithms. For these computations, the law of sines is transposed as follows:

$$b = a \sin B / \sin A, \text{ and}$$
$$c = a \sin C / \sin A.$$

c. After finding values for sides b and c of a triangle, each of these values can be used as the known length of a side of another triangle.

CHAPTER 12

SPECIAL SURVEYS

Section I. INTRODUCTION

12–1. Scope

This chapter only familiarizes the surveyor with the different systems and methods, without attempting to discuss procedures, techniques, or specifications for performing these surveys.

Section II. DESCRIPTION

12–2. Trilateration Surveys

Trilateration is the surveying technique which measures the length of sides of triangles rather than angles as in triangulation. A scheme of polygons is designed to extend control over the desired area in the same manner as triangulation. After the stations are selected on the ground, the distances between them are measured with an appropriate electronic measuring device. To date, large-scale trilateration has been used in areas where triangulation was not feasible. This includes long, over-water measurements and networks in regions where the working season is short or much of the country is unexplored. The planning and execution of trilateration are substantially different from triangulation, and at no time can one be substituted directly for the other. The measured distances are normally slope distances and must be reduced to the horizontal, and if accuracy requirements are of high order, they must further be reduced to a common surface of reference, usually mean sea level. These reductions require that the elevation of the stations must be known or obtainable by observation. Altimetry (ch. 8) is used in lower order trilateration to obtain the needed elevations. The following are some of the limitations that affect trilateration measurements and computations.

a. Atmospheric instability influences light and radio measuring waves (para 6–13 and 6–20) and may cause errors in the measured distances.

b. The distance errors become angular distortions in the computations. Normal angle adjustments of triangulation cannot be used for trilateration figures.

c. The measured lengths are slope distances. Vertical control must be carried along with trilateration to reduce the distances to the horizontal, and if necessary to mean sea level. Trilateration during inclement weather prevents trigonometric leveling and requires other methods that are either less accurate or slower.

d. Figures for trilateration are more complicated than those used for triangulation. The reconnaissance party must select more station sites and check more lines of sight for clearance when laying out the net.

e. Because of the limiting factors, trilateration is the least acceptable method of control extension. Every effort should be made to find other solutions for the survey problem before resorting to trilateration.

12–3. Gravity Surveys

a. Gravity is the resultant of two opposing forces; gravitation and centrifugal force. Gravitation is that force of nature which manifests itself as a mutual attraction between mass. Centrifugal force is the component due to the rotation of earth and acts in the opposite direction from gravitational force. The combination of these two forces represents the total gravity at the equator and closely approximates the total at other latitudes. Gravity is measured in units of either force or acceleration. In the centimeter-gram-second (c.g.s.) system of units, the force of gravity is expressed in dyne/gram and acceleration of gravity in units of a gal. The gal (named in honor of Galileo) is an acceleration of one centimeter per second. A milligal (mgal) is 0.001 gal. Gravity data are of paramount importance in geodesy. Gravity determinations are used in measuring the undulations of the geoid and the deflection of the vertical.

b. Gravity observations are divided, generally, into absolute and relative measurements. Absolute gravity determinations are those which measure the entire magnitude of gravity at a point while relative measurements detect the variations in the intensity of the gravity field from point to point.

(1) The absolute acceleration of gravity is usually measured with a pendulum apparatus or mechanism employing the falling body principle. In the former the swinging period of the pendulum is accurately timed. From the time interval and the length of the pendulum, the value of gravity is computed. This type of measurement requires lengthy observations and the average period must be determined from several thousand actual swings of the pendulum. This often requires months of observation and the equipment involved is complex, delicate, and sometimes cumbersome. As a result of the length and complexity of absolute gravity measurements, the absolute gravity is usually determined at a limited number of reference stations which are called base stations. Pendulum measurements obtain absolute gravity to within 3 to 5 parts per 1 million.

(2) The relative gravity measurements are usually made with a gravimeter. The term gravimeter (gravity meter) is used to denote any instrument in which the acceleration of gravity is determined by measuring the force necessary to support a mass in the earth's gravity field. The mass is usually supported by an extremely sensitive spring or by gas pressure and the force measured by the extension of the spring or compression of the gas. In operation, a dial reading which represents the force necessary to support the test mass is made at a base station and again at the desired location. The difference in dial readings between the two stations represents the change in gravity. Gravimeters are highly sensitive allowing reading accuracies of ±0.01 mgal and in exceptional cases ±0.001 mgal.

12–4. Satellite Surveys

Satellite surveys employ the use of artificial earth satellites as a means of extending geodetic control systems. Two systems have been used extensively These systems are used for long line surveys where the distance between stations is from 100 to 1500 miles. They are used for worldwide surveys for the intercontinental, interdatum, and interisland geodetic ties.

a. SECOR. SECOR is an all-weather, mobile geodetic tool designed to provide ranging data to determine the geodetic position of ground points. This system utilizes a satellite-borne transponder and four ground stations measuring the distance between the ground stations and the satellite by phase comparison similar to that of the microwave system used for ground surveys. By knowing the position of three of the ground stations, the position of the fourth station can be determined by what is basically known as a space resection computation.

b. BC–4 Camera. The BC–4 camera is a large camera mounted on a base similar to that of the base section of the Wild T–4 theodolite. The camera is used to photograph an ECHO type satellite while moving through a field of identifiable stars. Using the identifiable stars, the position of the satellite can be determined and hence the position of the camera station.

12–5. Airborne Surveys

HIRAN, SHIRAN, and SHORAN are all airborne survey systems, which employ the trilateration survey principles. Simultaneous measurements from ground stations are taken to an aircraft in flight. The exact coordinates of the aircraft are determined by use of these simultaneous measurements from three known ground stations. The position of the aircraft being known, then a point directly below the aircraft on the ground can be determined, and used in the control of aerial photography for mapping operations. The HIRAN and SHIRAN systems are also used in the same manner as SECOR. The distance measurements from the known stations are used to fix the aircraft by intersection. Then using the aircraft's position and distance measurements from an unknown station, the position of the unknown station is determined by space resection.

12–6. Sodano Azimuth Surveys

The Sodano azimuth surveys are used in conjunction with long line HIRAN surveys. It is a method of establishing an azimuth on a line between two stations that are not intervisible, because of the long distance between the stations. An aircraft is used that is equipped with a light affixed to the fuselage. This aircraft flies between the stations at a distance, where the light is visible simultaneously from both stations. The instrument used for these observations is a camera equipped Wild T–3.

12–7. Astronomic Surveys

Geodetic astronomy is the application of the astronomic science to the determination of astronomic positions and directions (azimuths) between points on earth's surface. These surveys are made to determine astronomic positions, orient survey networks, provide azimuths, and provide deflection components when compared with geodetic and gravity surveys.

a. Celestial Sphere. **The celestial sphere is a mathematical device, or concept, of a sphere of infinite radius whose center is at the center of earth. The earth at the center is considered to be a point rather than as having volume. All celestial bodies are considered as being projected upon this sphere. The radius of earth, being so small in comparison to the distance to the celestial bodies, it is not considered unless observations are made on objects within our own solar system. The points where an extension of the earth's axis pierce the celestial sphere are known as the celestial poles, north and south. The plane of the earth's equator, extended to the celestial sphere, coincides with the celestial equator. Even though the earth rotates and the stars appear stationary among themselves, it is easier to think of the earth being stationary while the celestial sphere with the celestial bodies attached rotates from east to west. This is actually its *apparent motion*. When reference is made to a star's path or motion, it is this apparent motion that is intended(fig. 12–1).**

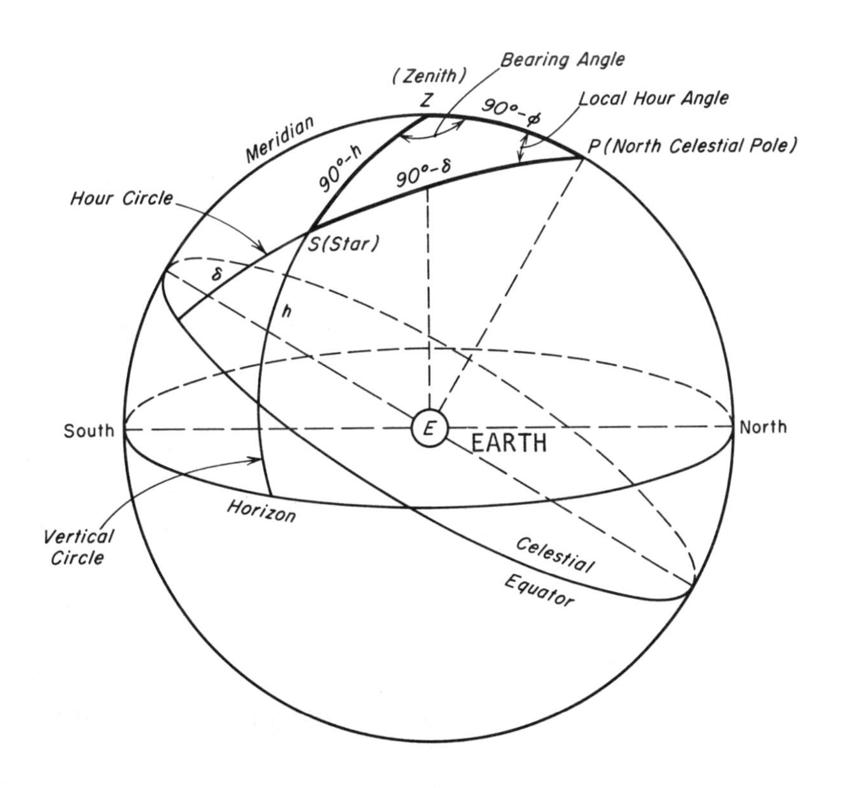

Figure 12-1. Celestial sphere.

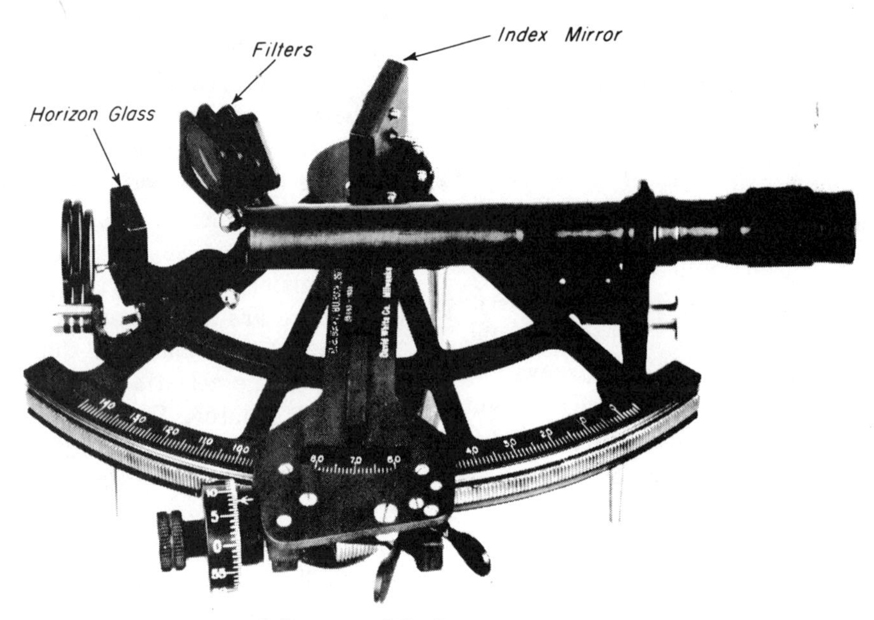

Figure 12-2. Sextant.

The sextant is a portable instrument that is used for measuring horizontal and vertical angles from a ship. In celestial navigation the sextant is used to measure the altitude of various prominent stars above the horizon. In hydrographic surveying the sextant is used to measure the horizontal angles between designated objects on land. This makes possible the subsequent solution of the three-point problem for determining the position of the sounding vessel at various selected times. (see Fig. 12–2)

Section III. PHOTOGRAMMETRY

12–8. Introduction *Photogrammetry* is the science or art of obtaining reliable measurements by means of photography. Photogrammetry is subdivided into various types, but the two basic categories are *terrestrial photogrammetry* and *aerial photogrammetry*. In terrestrial or ground photogrammetry the photographs are taken with the camera mounted on a tripod, and the optical axis of the lens is usually horizontal. Aerial photogrammetry makes use of photographs that have been taken from any airborne vehicle. Such photographs may be either vertical or oblique. *Vertical photographs* are taken with the optical axis pointing vertically downward at the moment of exposure. *Oblique photographs* are obtained when the optical axis is intentionally inclined from the plumb line.

Vertical photography (see Fig. 12–3) is most commonly obtained by cameras mounted in airplanes flying a straight course but with sufficient overlap between adjacent exposures to permit subsequent stereoscopic examination of the pictures. The photographs made on a given course constitute a flight strip, and a sufficient number of strips are taken to cover a given area. The overlap in the direction of flight is called *forward lap,* and the overlap between pictures in adjacent flights is called *side lap*. The amount of forward lap and side lap is commonly specified as 60% and 30%, respectively.

Since aerial photographs are generally taken from moving aircraft, the horizontal position, elevation, and orientation of the camera are not known. Although an attempt is made to keep the camera axis vertical, a small amount of *tilt,* whose direction and magnitude are unknown, is usually present when an exposure is made. Because of these unknown factors, photogrammetric measurements have been beset with many difficulties, but these have been overcome to such an extent that aerial methods have either displaced or considerably modified the ground methods formerly used by all governmental

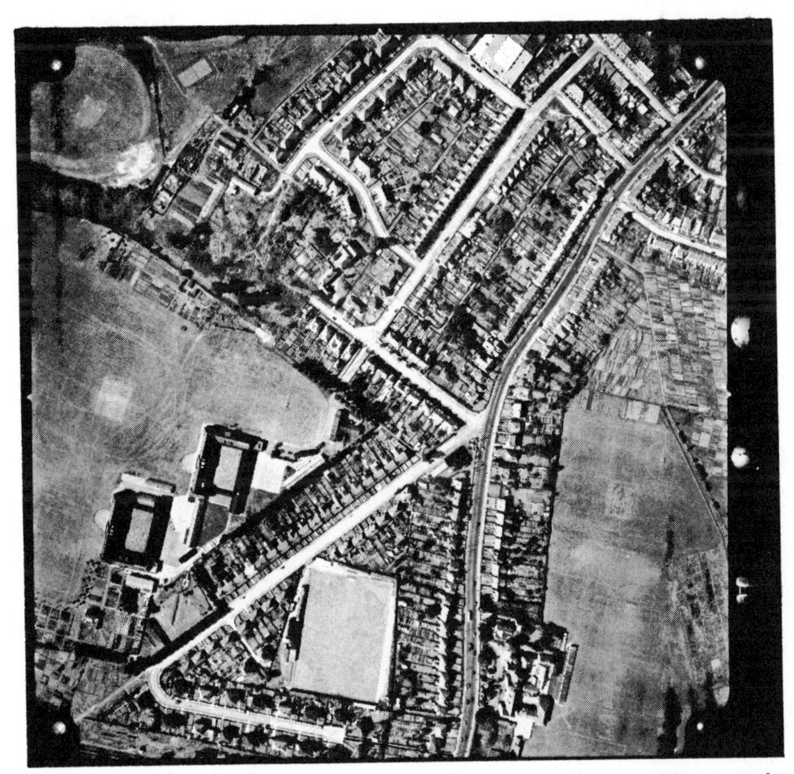

Figure 12-3. Vertical aerial photograph. Wild Heerbrugg Instruments, Inc.

mapping organizations. On practically every important survey or engineering project, by either public or private agencies, aerial photographs are used in some way. For example, wide use is now being made of photogrammetry in the planning, location, and construction of highways. From aerial photographs accurate estimates are made of the costs of the right-of-way, earthwork, and structures along alternate routes, in order to obtain the most feasible and economical location.

Closely associated with photogrammetry, or *metrical photography* as it is sometimes called, is *photographic interpretation.* This subject is concerned with the determination of the nature and description of objects that are imaged on a photograph. Hence, it can be seen that photo interpretation is qualitative in character, whereas photogrammetry is essentially quantitative.

The development of photogrammetry has been intimately connected with that of the camera, photographic materials and processes, and aviation. The first known photograph was the daguerreotype, which was produced in 1839 by Daguerre. Shortly thereafter the science of photogrammetry had its genesis.

12–9. Definitions A few fundamental definitions will be given. Some of the terms are as applicable to the horizontal photograph (Fig. 12–4) as they are to the vertical photograph.

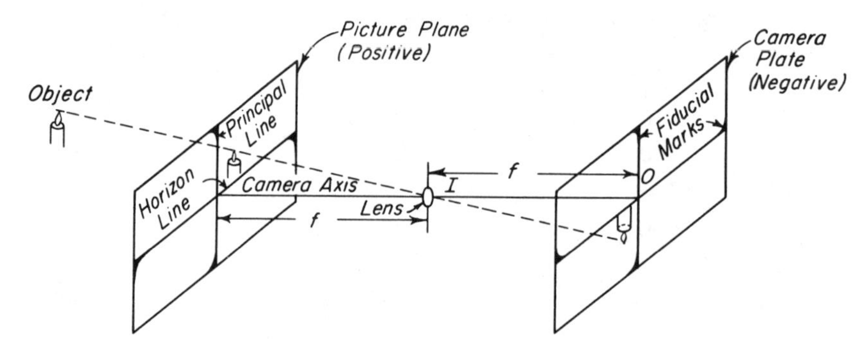

Figure 12-4. Elements of a horizontal photograph.

Point of View, I, is the center of the camera lens.

Camera Axis is the line through the center of the camera lens perpendicular both to the camera plane and the picture plane.

Picture Plane is the plane perpendicular to the camera axis at the focal length distance in front of the lens. It is represented by the positive contact print or photograph taken from a plane or film.

Principal Point is the point of intersection, *O,* of the camera axis with either the picture plane (photograph) or the camera plane (negative).

Focal Length, f, is the perpendicular distance from the center of the camera lens to either the picture plane or the camera plate.

Fiducial Marks are index marks within the camera frame which form images on the edges of the negative. The intersection of straight lines on the photograph connecting these images fixes the principal point of the photograph.

Photograph Nadir is the point of intersection of a vertical (plumb) line through the center of the lens (at the instant of exposure) and the photograph. If there is no tilt of the camera axis when the exposure is made, the photograph nadir and the principal point will be identical.

Ground Nadir is the point of intersection of a vertical line through the center of the lens and the ground surface.

Topographic Map is a map that represents the horizontal and vertical positions of features on a portion of the surface of the earth.

Planimetric Map is a map that presents only the horizontal positions of the features portrayed. It is distinguished from a topographic map by the absence of relief representation.

12–10. Perspective Principles of Vertical Photographs Since any photograph is a perspective view, it is subject to the principles of such views whether it is a terrestrial (horizontal) or an aerial (vertical) photograph. The following principles apply to vertical photographs:

1. The photographic images of all vertical lines of objects on the ground will be radial lines which, if extended, will pass through the principal point O.
2. All parallel level lines on the ground, such as the parallel sides of a square tract of level land, will appear as parallel lines on the photograph.

The first principle stated above has the greatest significance in all photogrammetric uses of vertical photographs. Three examples will be mentioned. (1) In Fig. **12–5** the flagpole represents a vertical line perpendicular to the picture plane, and therefore the image is a straight line which, if extended, passes through the principal point

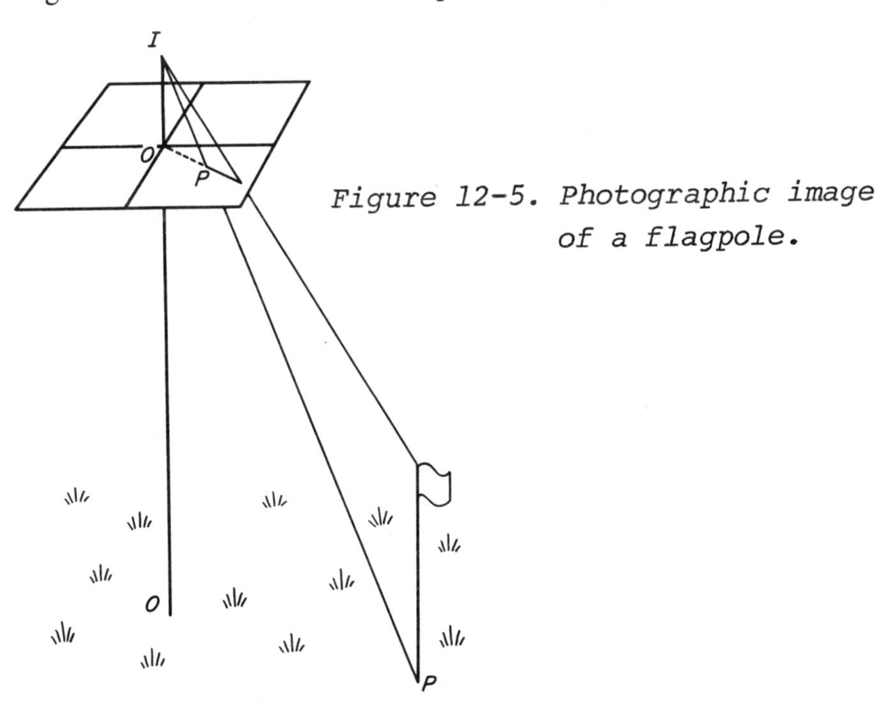

Figure 12-5. Photographic image of a flagpole.

O. (2) In Fig. **12–8** are shown a vertical aerial view and the image p_h of the top of a hill P_h. The vertical projection of P_h down to sea-level datum is at P_o, and the image of this point, if it could be seen in the photograph, would appear at the point p_o. Accordingly, the image of the vertical line P_hP_o is p_hp_o, and this line, if extended, passes through the principal point O. The length of the image d is called the *displacement* of the point p_h because of the elevation of the ground point above the datum plane. (3) In Fig. 12–6 are shown the images

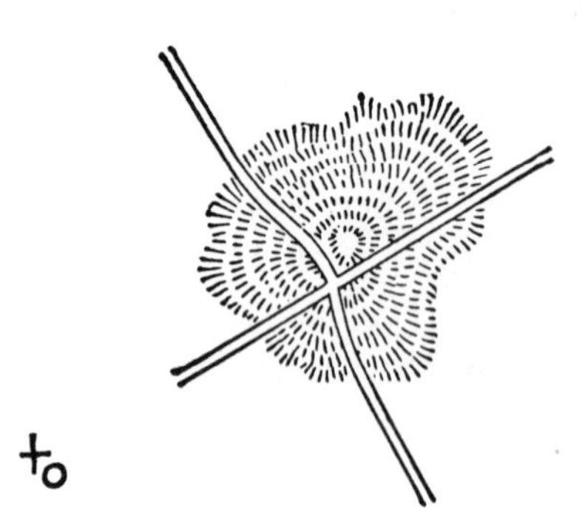

Figure 12-6. Effect of relief displacement.

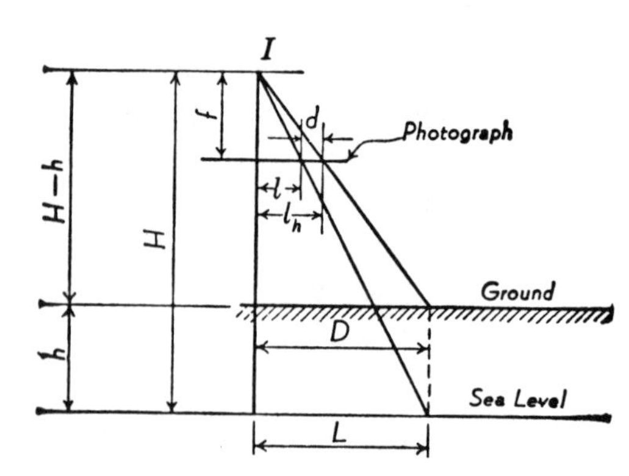

Figure 12-7. Scale relations in a vertical photograph.

of two roads whose alinements on the ground are straight and which intersect at right angles as they pass over a hill. The principal point of the photograph is at O. The displacement of the points on the roadway, whose direction is radial from the principal point O, is along this same direction, and therefore there is no change in the alinement of the photographic image, which remains a straight line. The points in the image of the other roadway are displaced radially from the principal point, and therefore the image of this roadway shows a convex curvature away from the point O.

12–11. Scale of a Photograph It has been shown that the images of ground points are displaced where there are variations in the ground elevation. Hence, there is no uniform scale between the many points on such a photograph; therefore, in discussing the "scale of a photograph" it is assumed that the ground is perfectly horizontal and the camera axis is truly vertical.

The scale of a photograph is the ratio of a given distance on the photograph to the corresponding distance on the ground. Where English units of measure are in use, this ratio is expressed in either of two forms, which may be designated as R, the representative fraction, or S, the map scale. Thus, in Fig. 12–7 for the sea-level elevation,

$$R = \frac{l}{L}$$

in which both l and L are expressed in the same unit, or the same ratio may be expressed as

$$S = \frac{L}{l}$$

in which L is expressed in feet and l is in inches. For example, if $l = 3$ in., and $L = 4500$ ft, then

$$R = \frac{0.25\ \text{(ft)}}{4500\ \text{(ft)}} = \frac{1}{18{,}000}, \text{ and } S = \frac{4500\ \text{(ft)}}{3\ \text{(in.)}} = 1500 \text{ ft per in.}$$

The relationships between the scale of a photograph, the focal length f, and the height of lens H are also shown in Fig. 12–7 It is evident from the similar triangles that

$$\frac{l}{L} = \frac{f}{H}$$

and hence

$$R = \frac{f\,(\text{ft})}{H\,(\text{ft})}$$

also

$$S = \frac{L}{l} = \frac{H\,(\text{ft})}{f\,(\text{in.})}$$

Moreover, for the ground level at an elevation h above sea level, it is evident that

$$R_h = \frac{l_h}{L} = \frac{f}{H - h}$$

and

$$S_h = \frac{L}{l_h} = \frac{H - h}{f}$$

For example, if $f = 6$ in., $H = 9000$ ft, and $h = 500$ ft, then

$$R_h = \frac{0.5}{8500} = \frac{1}{17{,}000}; \text{ and } S_h = \frac{8500}{6} = 1417 \text{ ft per in.}$$

The latter scales are spoken of as the *representative fraction*, or the *scale*, of the photograph at the elevation h, respectively.

It is evident in Fig. 12–7 that the ground distance D is represented by two different distances on the photograph, l and l_h corresponding to the two different elevations, sea level and h above sea level. Since the ground surface is usually characterized by slopes, hills, and valleys, it is obvious that the distance on a photograph that represents a given distance on the ground varies with the different elevations, and there can be no single scale that will apply to all points appearing in the photograph. Thus, to obtain the true ground distance between points of different elevations, it is necessary to refer them to a single plane of reference called the *datum plane*. For special conditions, any assumed elevation may be used as a datum plane; but since sea level is the universal datum for ground elevations it is also commonly used as the datum for aerial photographs. Therefore, unless otherwise specified in the following pages, H will represent the height of the lens above sea level, and h will represent the elevation of a ground point above sea level in all relations dealing with scale factors.

On any given photograph, if the images of two points appear whose ground elevations are equal and if the distance between them is known, then the scale of the photograph, for the known elevation h, is readily determined by the simple relation $S = L/l$, in which L is the known distance, in feet, on the ground, and l is the distance, in inches, on the photograph. The distance L may be measured on the ground, or if a published map is available it may be possible with sufficient accuracy to scale the distance from the map.

EXAMPLE: Two road intersections, which are known to be one mile apart in rather flat terrain, are imaged on a vertical photograph. If the photographic distance is 3.22 in., what is the indicated scale of the photograph?

$$S = \frac{L}{l} = \frac{5280}{3.22} = 1640 \text{ ft per in.}$$

12–12. Number of Photographs Required Because of the overlap required in aerial photographs for mapping purposes, the net area covered by a single photograph will be that included within its full dimensions diminished by the overlap of adjacent prints.

The amount of overlap for two adjacent prints in the direction of the line of flight, called *forward lap,* is usually 60%. The distance between two principal points in a flight series is equal to the size of a print less the amount of forward lap. Thus, if the size of the print is 7 × 7 in., and the forward lap is 60%, then the distance between two adjacent principal points will be 7 in. − (7 × 0.60) = 2.8 in. Likewise, the distance between the principal points of photographs in adjacent flights is given by the size of the print perpendicular to the line of flight, less the amount of the side lap. Thus, if the print is 7 in. wide and the side lap is 25%, then the distance between two adjacent principal points will be 7 in. − (7 × 0.25) = 5.25 in. The number of photographs required, therefore, will be the number required for one strip times the number of strips.

EXAMPLE: A flight mission is to be flown under the following conditions: The area is rectangular, 15 miles by 10 miles in size; the camera negatives are 7 in. square, and the focal length is 6 in. The scale of the photographs will be approximately 1500 ft per in. The forward lap is 60%, and the side lap is 25%. How many photographs will be required?

Solution: The distance between the principal points of two photographs in the line of flight will be 7 in. − (7 × 0.60) = 2.8 in. × 1500 = 4200 ft. The total length of one flight is 15 × 5280 = 79,200 ft. Hence the number of photographs required for one flight is 79,200/4200 = 18.8 or 19. The distance between two flights will be 7 in. − (7 × 0.25) = 5.25 × 1500 = 7875 ft. The number of flights will then be (10 × 5280)/7875 = 6.7 or 7 flights. The number of photographs then is 7 × 19 = 133.

12–13. Image Displacement Caused by Ground Relief If conditions are as shown in Fig. 12–7, i.e., the photograph is truly horizontal and the ground is level, and if other sources of error are disregarded, then the photograph represents, at its proper scale, a true orthographic projection; hence it may be said to be a true map of the ground surface. Also, the photograph will have the same scale throughout the area contained within it. However, these conditions are never fully met in practice, and, since the photograph is a perspective view, any relief of the ground surface will be shown in perspective. Because of this condition, points in the photograph are said to be *displaced* from their true orthographic positions.

The displacement of an image caused by ground relief is shown in Fig. 12–8 where the image of a point P_h on a summit is shown at p_h, and the image of the vertical projection of this point to the datum plane P_o, is shown at p_o. The photographic distance $p_h p_o = d$ is the

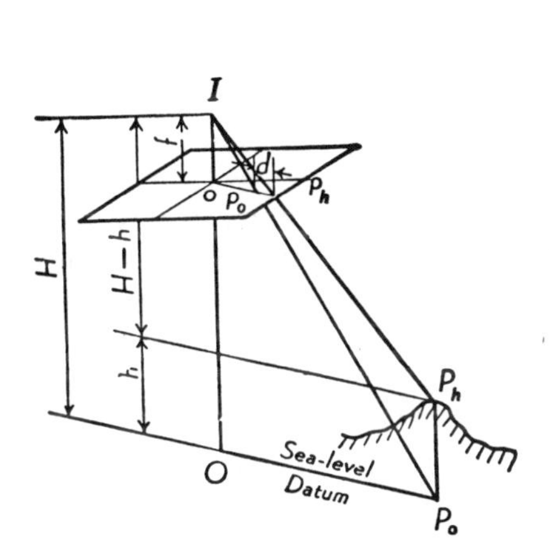

Figure 12-8. Relief displacement.

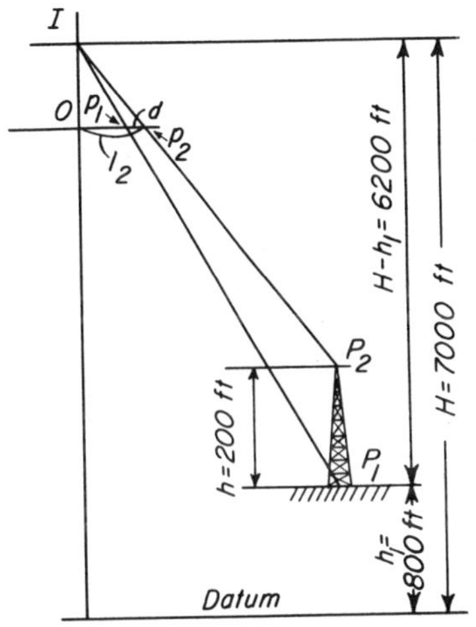

Figure 12-9. Image displacement.

displacement of this image due to its elevation h above the datum.

If we let $op_h = l_h$, $op_o = l$, and $OP_o = OP_h = L$, the following equations may be written:

$$\frac{l}{L} = \frac{f}{H} \quad \text{and} \quad \frac{l_h}{L} = \frac{f}{H-h}, \quad \text{also} \quad d = l_h - l = \frac{Lf}{H-h} - \frac{Lf}{H};$$

from which

$$d = \frac{l_h h}{H}$$

EXAMPLE: The distance from the principal point to an image on a photograph is 2.143 in., and the elevation of the object above the datum (sea level) is 850 ft. The height of lens above the datum is 7200 ft. Then $d = (2.143 \times 850)/7200 = 0.253$ in.

Another example of the use of this relation is shown in Fig. 12–9 where the images of the top and the bottom of a tower, P_2 and P_1, are shown on a photograph at p_1 and p_2. The distance to the point p_2 from the principal point O is indicated as l_2. The other values given are $H = 7000$ ft, $H - h_1 = 6200$ ft, $h = 200$ ft, and $l_2 = 2.120$ in. It is desired to find the displacement of the image of the top of the tower with respect to the image of the bottom, p_1. Following is the computation:

$$d = \frac{l_2 h}{H - h_1} = \frac{2.120 \times 200}{6200} = 0.068 \text{ in.}$$

CHAPTER 13

LOCATION OF DETAIL AND AREA DETERMINATION

Section I. LOCATION OF DETAIL

13–1. General

One of the final steps of field surveying involves the location of detail points relative to the survey lines and/or stations. If, for example, a traverse is run for the location of a new road, buildings and other existing features as well as the points on the road must often be tied into the control to determine their relative positions and/or elevation. In mapping surveys, topographic features that are used as picture points in photogrametric mapping must be tied into the newly established and existing control systems to determine the positions and/or elevation. In general, detail points are located by measuring angles, distances, or a combination of these methods from the lines and stations which have been established. It is sometimes necessary to run a short open traverse or sideshot (para 10–9*c*) from a control station to a detail point, but most points can be located by one of the methods described in the following paragraphs. Usually the best method to use in any particular case is the one that will give the required data and accuracy with the least effort. However, with the advent of many new types of equipment in the surveying field, it is sometimes as easy to use third order methods, as the lower order methods with only a nominal amount of effort, time, and expense. The best method is usually obvious to the surveyor who is familiar with all methods and the project requirements.

13–2. Control Systems

Detail must be located relative to some type of control system, either local, national, or worldwide (para 1–4). In this way, the position of the detail will be established with some degree of permanency. Relative positions of detail points can be determined within a local control system (state or municipal) if the control is tied into geodetic control, the positions of the points can be computed with respect to the national or worldwide systems. The main control scheme is either a network of triangulation or traverses, and located close to the points to be tied in, to reduce the supplementary control requirement. Supplementary control consists of short traverse or triangulation which is run close to or across a project area. These stations must be established to the degree of accuracy required by the purpose of the survey. If a main control scheme must be run first, it should be of an accuracy to permit the supplementary survey to furnish the proper accuracy for the detail. For example, if third order accuracy is required for the detail, and control must be extended for a long distance, the starting control must be of sufficient accuracy to allow this extension without falling below the third order requirements upon reaching the detail point. If the starting control was of third order accuracy, any length of extension beyond one or two figures (triangulation) or one or two kilometers (traverse or leveling), the accuracy will drop below the requirement limits and make the survey useless for the intended purpose. The same reasoning holds true for both horizontal and vertical control.

13–3. Angle and Distance Methods

These methods require the use of both angle and distance measuring to locate detail. The angles may be measured by a transit, theodolite or a properly oriented alidade and the distance by tape, stadia, or electronic devices.

a. Radiation.

(1) *Transit or theodolite.* This method locates points by measuring an angle and distance from the same station. It is an excellent method for locating a number of fairly close points from one instrument setup. In this method, the instrument is set up at one station and backsighted on

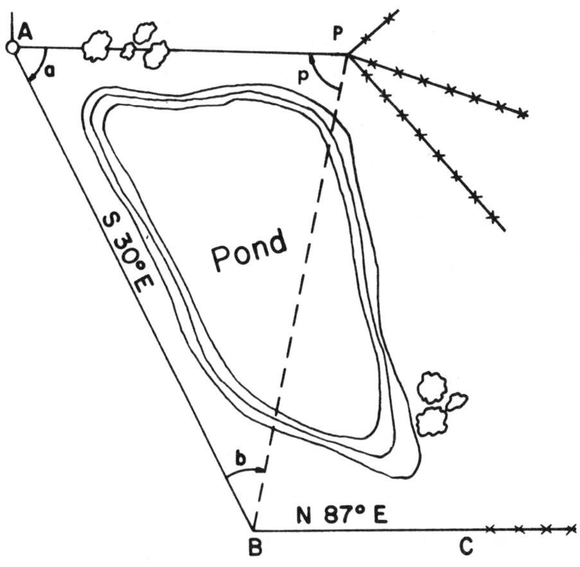

Figure 13-1. Location: angle at one station and distance from another.

another station, the azimuth between the stations being known or determinable. The horizontal angle is observed between each detail point and the backsight station. The horizontal distance is measured between each detail point and the instrument station. The azimuth of the backsight line and the horizontal angle to each detail point will give the azimuth to each new point. Then, with this azimuth and the measured distance, the coordinates of these points can be determined relative to the instrument station's coordinates. When locating points such as building corners, the lengths and widths of the buildings must be measured to locate any corners not visible from the instrument station.

(2) *Alidade.* This method locates the detail points by measured distances (usually stadia) and relationship between lines drawn on the planetable sheet. The planetable is oriented by use of two known stations, the instrument (planetable and alidade) plumbed, leveled and tightened in azimuth at one station; the stadia rod located at the other. By sighting through the alidade at the stadia rod and plotting the rod station on the planetable sheet, then drawing a line between these two points (instrument and rod stations) an orienting line is thus determined. Then by keeping the straight edge of the alidade tightly against a pin through the point on the planetable sheet representing the instrument station and turning the alidade (planetable staying immobile) and sighting on a rod located at the detail point. By reading the stadia rod and beaman arc the relative position of the detail point can be plotted on the planetable sheet by use of the alidade straight edge and determined distance. Refer to chapter 7 for the detailed procedure used with this method.

b. Angle at One Station and Distance From Another. Sometimes it is necessary to locate a point by measuring an angle at one station and a distance from another one. The instrument is setup at station A (fig. 13–1) during angle observation. An obstruction prevents observing the direction to station P. However, the distance A to P can be measured. When the instrument is moved to station B, the angle b can be measured but the distance B to P cannot. In triangle ABP, we now have three known factors, distance AB (previous survey), distance AP (measured) and angle b (observed). To determine the position (coordinates) of P, either the distance BP or angle a must be computed. With either of these values, an angle and distance (a and AP or b and BP) can be used to determine the relative position of P. The values of the unknown are computed as follows:

$$\text{Sin } p = \frac{AB \sin b}{AP}$$

$$a = 180° - (b+p)$$

$$BP = \frac{AP \sin a}{\sin b}$$

c. Short Base. This method is used to locate distant or inaccessible points by observing horizontal angles to the point from each end of a measured line. In figure 13–2, the length AN is measured and the angles n and a_1, are observed Angle w_1 can be computed by $w_1 = 180° - (n + a_1)$. Using the law of sines, either A–WT or N–WT can be computed. Then setting at A and backsighting on a known station and observing the angle to WT the azimuth of line A–WT can be determined. Knowing the azimuth and distance of line A–WT the relative position of the water tower can be determined. A second determination of the position should be made, measuring the distance A to E and observing angles a_2 and e, computing as above. This second determination of position will show if any error was made in the angle observations or distance measuring.

d. Range Ties. A range tie can be used to determine a point's location by observing an angle and measuring a distance. This method requires extra instrument manipulation and should be used only when none of the previous methods can be used satisfactorily. This method will not only establish a corner of a structure, but also the alinement of one of the sides. In figure 13–3, assume that the building is not visible from either station A or B, or that either or both of the distances from A or B to a corner of the building cannot be easily measured. Set the instrument up at either A or B and establish line AB, establish a point R at the intersection of line AB and line 1–2. The instrument is then moved to R and the distance along line (A–B) to R is measured. An angle observation to the building is made using either A or B as the backsight. The range distance (R–2) is measured as well as the building dimensions. The location of point 2 can be computed, and from this the rest of the points located.

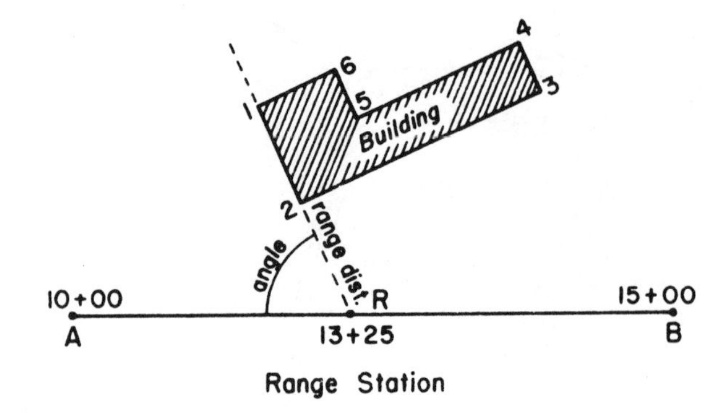

Figure 13-3. Range ties.

13–4. Distance Methods

These methods may or may not need the angle-observing instruments. Measurements can be made using tapes only. If the transit or theodolite is used, it is strictly for alinement and establishing perpendiculars.

a. Distance From Two Stations. Points are located by measuring horizontal distances from two stations when a transit or theodolite are not available. The distance is measured from two stations of known position to the unknown point. Then knowing three sides of a triangle, the angles and the location of the point can be computed by the law of cosines. The measurements must be made carefully since there is no check on them, and errors or mistakes will affect the point's location.

b. Swing Offsets. The swing offset method (fig. 13–4) is used to locate points close to the control lines. Measurement of a swing offset distance provides an accurate determination of the perpendicular distance from the control line to the point being located. It is somewhat similar to the range tie but requires no angle observation. In determining the offset distance, one tapeperson holds the zero mark of the tape at a corner of the building while the other tapeperson swings an arc with the tape to the control line (AB). When the shortest reading on the graduated end of the tape is observed, the swing offset or perpendicular distance to the control line is obtained (a or b). The alinement of AB can be first established at the intersection of the offset points and the tapeperson reads the value. The more common practice is to have the instrumentperson read the shortest distance through the telescope. The distance from the instrument station to the swing offset points (a or b) is measured. A tie or check distance can be measured from some known point (c) along the line or an angle (*a*) can be observed for a tie from either instrument station.

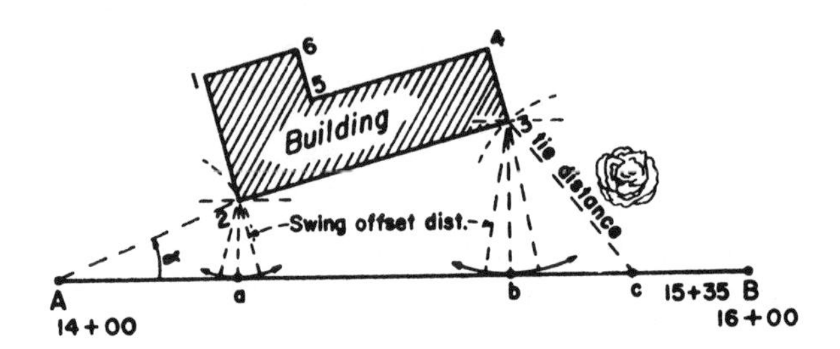

Figure 13-4. Swing offsets.

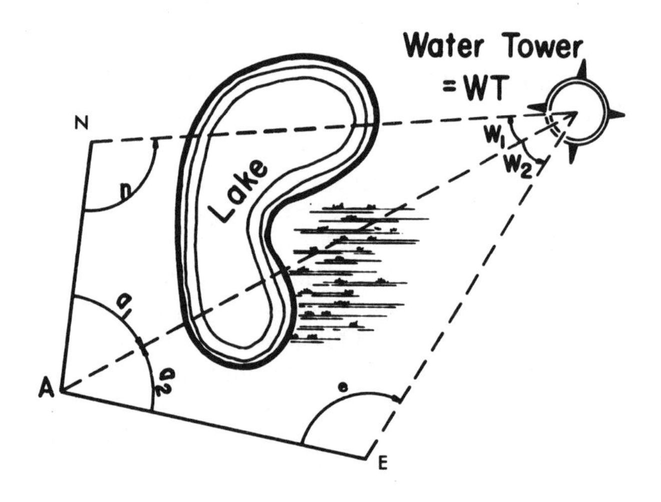

Figure 13-2. Location by short base.

c. Perpendicular Offsets. The method of perpendicular offsets from a control line (fig. 13–5 is similar to swing offsets, but is more suitable for locating detail of irregular objects, such as stream banks and winding roads. The control line is established close to the irregular line to be located and perpendicular offsets (aa', bb', cc', etc) are measured to define the irregular shape. When the offset distances are short, the 90° angles are usually estimated, but when distances are several hundred feet long, the angles should be laid off with an instrument. The distances to the offset

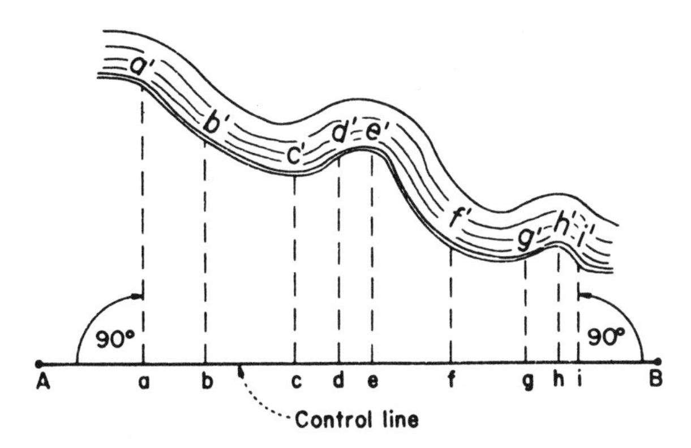

Figure 13-5. Perpendicular offsets.

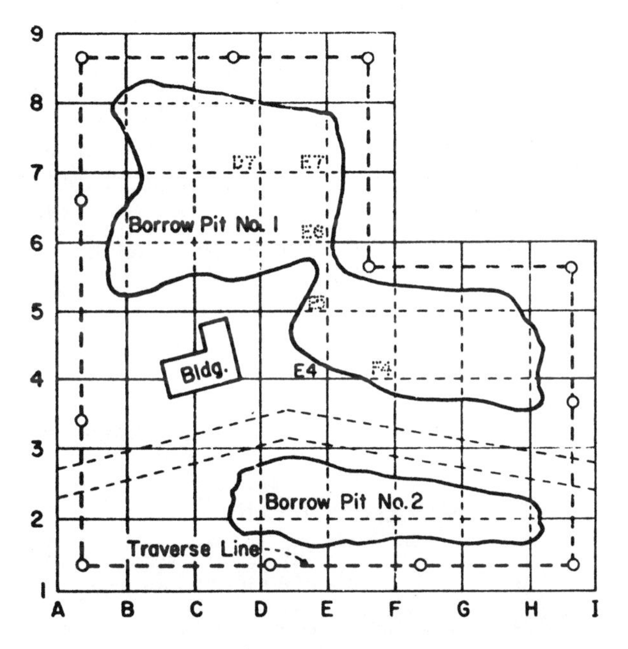

Figure 13-6. Coordinate method.

points (a to i) are measured along the control line.

d. Coordinates. In the coordinate method of locating detail, a system of squares of convenient size is laid out on the ground. The required detail is then located horizontally with respect to this coordinate grid. This method is often used when elevations, as well as horizontal positions are required for detail within a small area such as a borrow pit. The size of the squares is selected to locate the most detail with the least effort. In figure 13–6, assume that the building dimensions are about 25 meters. The size and shape of the borrow pit makes a 25-meter grid practical. A 50-meter grid would require too many extra measurements, whereas a 10-meter grid would require too much time and effort to lay out. Note that the location of the outline of the pit can be started by using only one dimension along a grid line from a marked grid corner. The more or less straight lines between these crossings are then simply sketched in, and only the turns and corners have to be located by the dimensions from two grid lines.

13–5. Angle Methods

The angle methods require the use of a transit or theodolite for the observation of angles. Distances are not measured in these methods.

a. Intersection. In this method of locating detail, at least two stations of known position are required. The instrument is set up at each of the known stations, using the other known station as a backsight and turning the horizontal angle to the detail to be located (fig. 13–7). The distance between the known stations being known or computed and the three angles of the triangle, (two observed and one computed) being known, the law of sines is used to compute the other distances. Knowing all parts of the triangle, the location of the detail is then determined by position computation. In order to have a check on the angle observations a third known point should be used in this method.

b. Resection. In this method of locating detail, at least three stations of known positions are required. However, it is highly recommended that resection be done from four or more points as this not only improves the accuracy, but provides a check against blunders which is impossible on only three. The instrument is set up at the detail being located and angle observations are made to the known stations. Figure 13–8 shows the angles to be observed in this method. Care must be used to select stations from

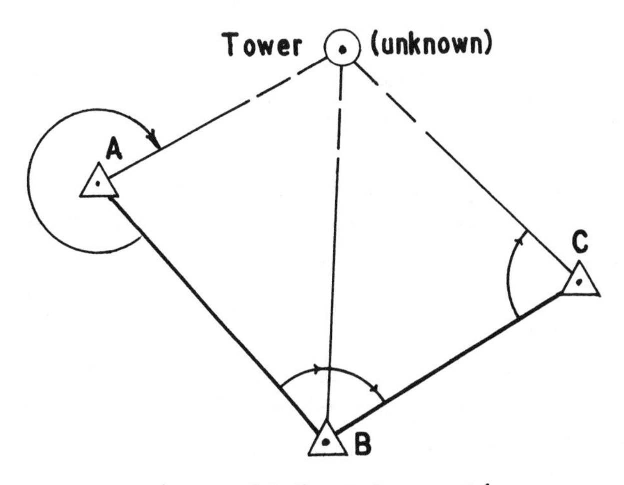

Figure 13-7. Intersection.

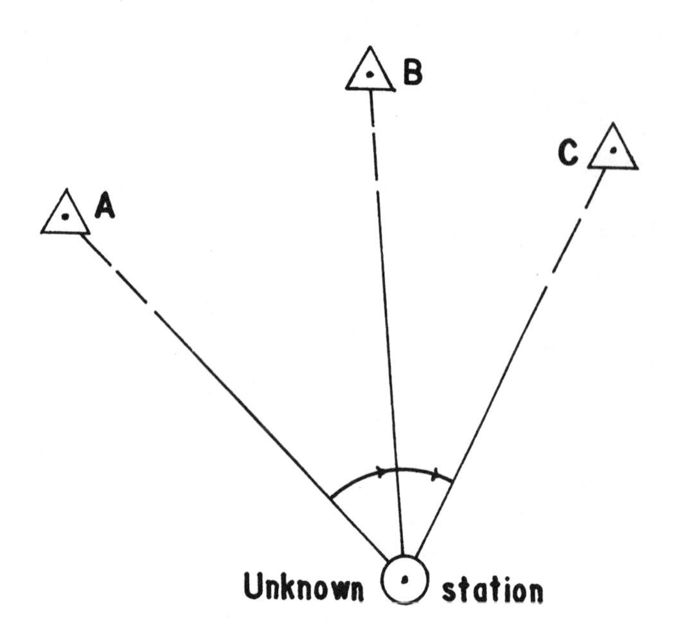

Figure 13-8. Resection.

which the resection is to be made, for if the point to be located falls on the circumference of the circle through the three known stations being observed, the problem is indeterminate. Resection should be avoided in favor of one of the other methods described, as no check can be made on the work until the field data has been completely reduced by computers.

13-6. Field Notes

Location of detail can be recorded as traverse or triangulation with a tabulation of the angles and/or distances measured. However, field notes for detail location are usually in the form of sketches, with all angular and distance measurements recorded directly on the sketch. The measurements from one station are usually shown on a single sheet but two or more sheets are used if additional space is required. Measurements from more than one station may be shown on one sketch, but crowding should be avoided. Tabular field notes that supplement sketches for added clarity, must be referenced to the sketches to avoid confusion.

Section II. AREA DETERMINATION

13-7. General

One of the main objectives of a land or construction survey is to determine the area of a tract of land. Generally, this is done by running a closed traverse. Where the boundaries of the tract are straight lines, the traverse courses coincide with the boundary lines. Where the boundaries are irregular or curved, the area can be determined by perpendicular offsets from the traverse or by arcs of circular curves. The following paragraphs in this section, cover some of the methods and computations that can be used to determine the area of a tract of land.

13-8. Regular Area

Where the area to be determined is bound by a series of straight lines the area is said to be regular. Usually the area of a regular figure is determined by running a closed traverse whose lines coincide with the boundary lines. For area determination of a regular figure one of the following methods can be used.

a. Triangle. From plane geometry and trigonometry the area of a triangle is related to its angles and lengths of sides.

(1) *Two sides and include angle.* When the lengths of two sides and the included angle are known, then—

$$\text{Area} = \frac{1}{2}\, ab \sin C$$

(2) *Three sides.* When the lengths of the three sides of any triangle are known, then—

$$\text{Area} = \sqrt{s(s-a)\,(s-b)\,(s-c)}$$

where: $s = \frac{1}{2}(a+b+c)$

b. Rectangle. The area of a rectangle can be determined by dividing it into triangles and using one of the formulas in *a* above, or knowing its length and width, then, Area = length x width.

c. Coordinates. When the coordinates of the corners of a tract of land are known the area can be determined by the coordinate method. The coordinates of these corners can be determined by triangulation or traverse based on any rectangular coordinate system, assumed or true. The calculation involves the determination of areas of trapezoids formed by projecting lines upon two coordinate axis. The coordinate axis are a true meridian and a parallel of latitude. In figure 13-9, ABCDEF is a tract whose area is to be determined. The coordinates of all points being known; the perpendicular distance from each point to the chosen meridian is the departure distance, and the perpendicular distance from each point to the chosen parallel is the latitude distance. The departure distance for point A is Aa or d_1 and the latitude distance is Aa′ or l_1. Departure distances are positive east of chosen meridian and negative west; latitude distances are positive north of chosen parallel and negative south. In the figure all departure and latitude distances are positive, since all points are located in the northeast quadrant. Twice the area of the trapezoids formed by the projecting lines is determined by the following formula:

$$2\ \text{Area} = [l_1\ (d_6 - d_2) + l_2\ (d_1 - d_3) + l_3\ (d_2 - d_4) + l_4\ (d_3 - d_5) + l_5\ (d_4 - d_6) + l_6\ (d_5 - d_1)]$$

Then based on figure 13–9, we have

$$2\ \text{Area} = -\ [600(300-400)+800(200-600) + (700(400-500)+500(600-700) + 200(500-300)+400(700-200)]$$

$$2\ \text{Area} = +60{,}000 + 320{,}000 + 70{,}000 + 50{,}000 - 40{,}000 - 200{,}000$$

$$2\ \text{Area} = 260{,}000$$

$$\text{Area} = 130{,}000\ \text{sq. ft.}$$

d. Double-Meridian-Distance. The double-meridian-distance (DMD) method is similar to the coordinate method but the computation does not involve the direct use of coordinates. The traverse is run and the adjusted latitudes and departures are determined. This method is based on the theory that the area of a right triangle equals one-half of the product of the two sides. Since latitudes and departures are at right angles to each other, the area bounded by the distance, the latitude and the departure is a right triangle. This area can be determined by taking one-half of the product of the latitude and the departure. However, the triangle may add or subtract from the total area of the figure depending on its location. To avoid determining a plus or minus area for each triangle, a slight refinement is made. The departure is added twice; first in determining the DMD of the course and then when the next course's DMD is determined. Multiplying the DMD of each course by its latitude results in twice the area, but the sign of this product determines whether the area adds to or subtracts from the figure area. A step-by-step procedure to work out a DMD area is given below, and shown in figure 13–10.

(1) All latitudes and departures are computed and adjusted.

(2) The most westerly station (A) is selected as the first point, and line AB as the first course to avoid negative DMD's.

(3) The DMD of the first course equals the departure of the course itself, 200.

(4) The DMD of any other course (BC, for example) equals the DMD of the preceding course (AB), plus the departure of the preceding course (AB), plus the departure of the course (BC) itself or 200 + 200 + 200 = 600. For the next course (CD), the same procedure: 600 + 200 — 100 = 700.

(5) The DMD of the last course is numerically equal to its departure, but with the opposite sign (+100).

(6) Each DMD value is multiplied by its latitude, positive products being entered under north double areas and negative products under south double areas.

(7) The sum of all the north double areas minus the sum of all the south double areas, disregarding the sign, equal twice the area of the tract. If meters are used for the traverse, square meters will be the result; if feet are used, square feet will result.

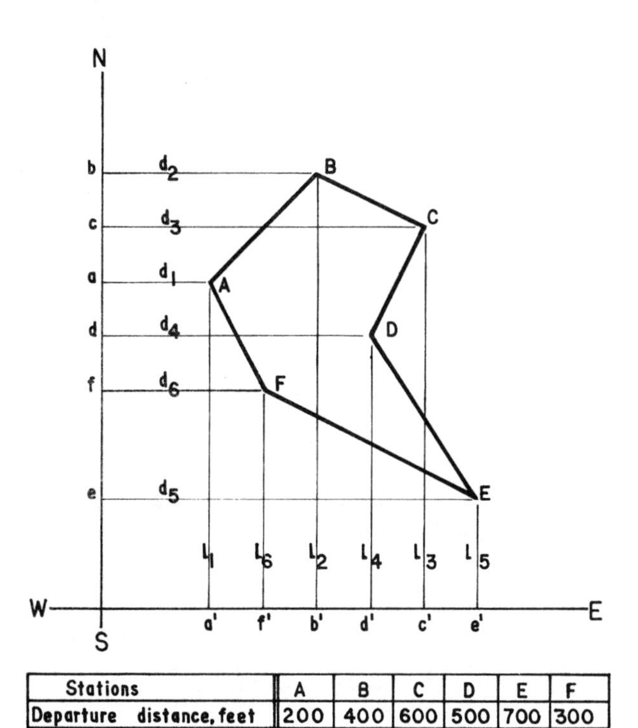

Stations	A	B	C	D	E	F
Departure distance, feet	200	400	600	500	700	300
Latitude distance, feet	600	800	700	500	200	400

Figure 13-9. Area by coordinate method.

13–9. Irregular Area

If the tract of land has a boundary that is irregular, such as a stream bank or ocean shore the area is said to be irregular. The usual procedure is to run a closed traverse on all the straight boundary lines and also run a line at a convenient distance from the irregular boundary and measure perpendicular offsets to boundary. The area of the tract is then equal to the area within the closed traverse (computed as in para 13–8 above) plus the irregular area.

a. Regular Intervals. When the irregular boundary is a gradual curve or evenly spaced sharp changes in direction, the offsets are taken at regular intervals along the traverse to the boundary. The offsets should be taken sufficiently close together, so that the boundary can be assumed as straight between offsets and very little error will accumulate because of this assumption. The area thus enclosed will then be of trapezoidal shape and can be determined by the trapezoidal rule. If the boundaries are of a curved nature as to make up segments of parabolas, then Simpson's one-third rule should be used. Simpson's rule will in general yield more accurate area determinations than the trapezoidal rule, but it is a harder rule to apply.

(1) *Trapezoidal rule.* The average of the end offsets plus the sum of the intermediate offsets multiplied by the common interval between offsets is equal to the area. Figure 13–11 illustrates this rule based on the formula—

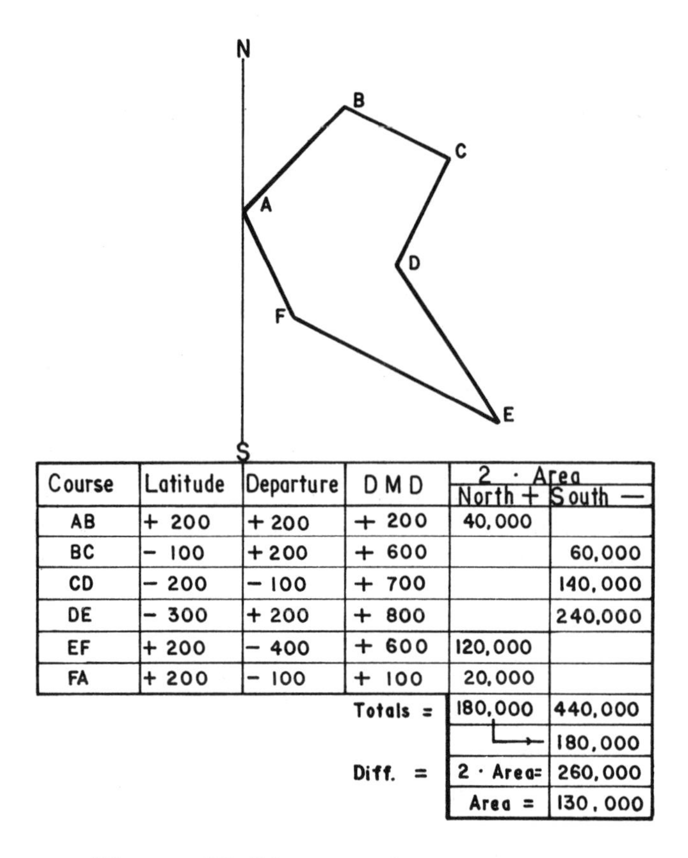

Course	Latitude	Departure	D M D	2 · Area North +	South —
AB	+ 200	+ 200	+ 200	40,000	
BC	– 100	+ 200	+ 600		60,000
CD	– 200	– 100	+ 700		140,000
DE	– 300	+ 200	+ 800		240,000
EF	+ 200	– 400	+ 600	120,000	
FA	+ 200	– 100	+ 100	20,000	
			Totals =	180,000	440,000
					180,000
			Diff. =	2 · Area=	260,000
				Area =	130,000

Figure 13-10. Area by DMD method.

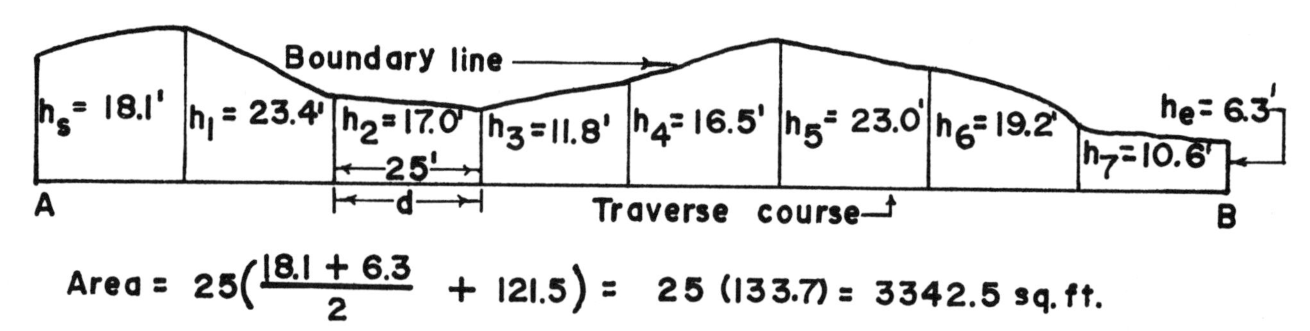

$$\text{Area} = 25\left(\frac{18.1 + 6.3}{2} + 121.5\right) = 25\ (133.7) = 3342.5 \text{ sq. ft.}$$

Figure 13-11. Area by trapezoidal rule.

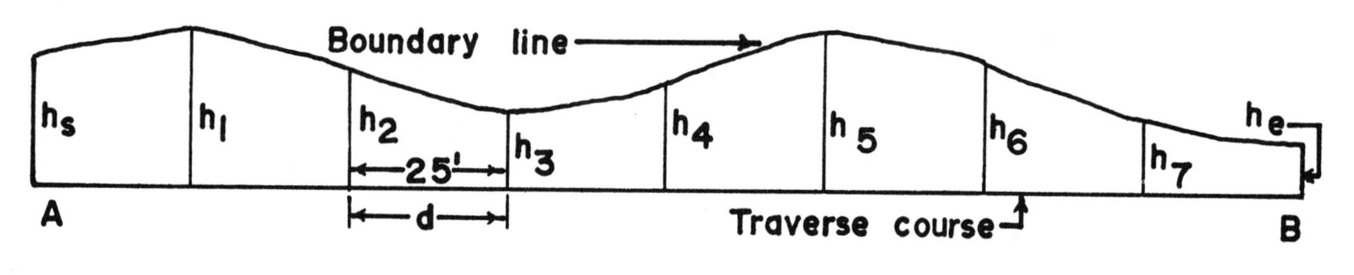

$$\text{Area} = 25/3\ [18.1 + 6.3 + 2(23.4 + 11.8 + 23.0 + 10.6) + 4(17.0 + 16.5 + 19.2)] = 25/3\ [24.4 + 137.6 + 210.8] = 25/3\,(372.8) = 3107 \text{ sq. ft.}$$

Figure 13-12. Area by Simpson's one-third rule.

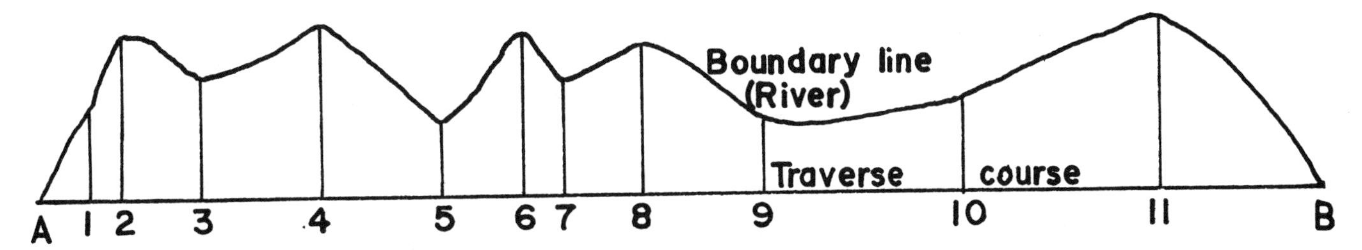

TRAVERSE COURSE A to B					
Station	Distance to A	Length offset	Difference	Products +	Products −
A	0	0			
1	60	11.5	−20.1		1,206.0
2	100	20.1	− 3.8		380.0
3	200	15.3	− 1.4		280.0
4	350	21.5	+ 5.8	2030.0	
5	500	9.5	+ 1.0	500.0	
6	600	20.5	− 4.8		2,880.0
7	650	14.3	+ 2.4	1,560.0	
8	750	18.1	+ 5.9	4,425.0	
9	900	8.4	+ 5.9	5,310.0	
10	1150	12.2	− 13.2		15,180.0
11	1400	21.6	+ 12.2	17,080.0	
B	1600	0	+ 21.6	34560.0	
			Totals =	65,465.0	19,926.0
				19,926.0	←
			2 · Area =	45,539.0	
			Area =	22,769.5	

Figure 13-13. Area by irregular offsets.

$$\text{Area} = d\left(\frac{hs + he}{2} + \Sigma h\right)$$

where:
d = Common interval
hs + he = end offsets
Σ h = Summation of intermediate offsets

(2) *Simpson's one-third rule.* This rule assumes that a curve through three successive points is a part of a parabola. Using this assumption the rule states the following: The sum of the end offsets, plus two times the sum of the odd intermediate offsets, plus four times the sum of the even intermediate offsets multiplied by ⅓ the common interval is equal to the area. When using this rule the number of offsets must be odd. If the number of offsets is even, then one of the partial areas at either end of the area must be dropped and computed separately by the trapezoidal rule. Figure 13–12 is a graphic display and sample computation of this rule based on the following formula: (Offset distances are the same as figure 13–11.)

$$\text{Area} = d/3\,[h_s + h_e + 2(\Sigma h \text{ odd}) + 4(\Sigma h \text{ even})]$$

Where:
d = common interval
hs + he = end offsets
Σh odd = sum of the odd offsets (1st, 3rd, 5th, 7th, etc)
Σh even = sum of the even offsets (2nd, 4th, 6th, etc)

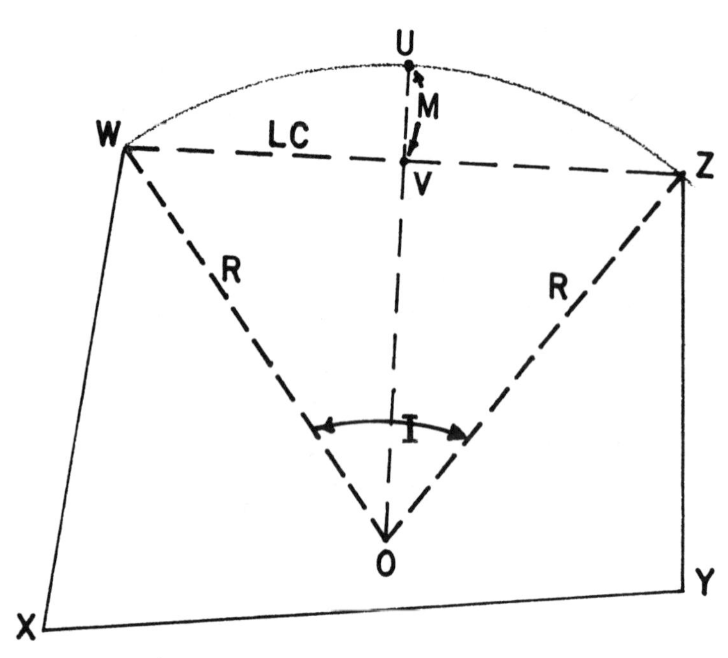

Figure 13-14. Segments of circles method.

b. Irregular Interval. When the nature of the irregular boundary is such that it cannot be assumed that it is a gradual curve or straight line between the offsets, then the offsets must be measured at irregular intervals rather than at a common interval. The rule used for irregular intervals is—The distance along the traverse of each intermediate offset from the first multiplied by the difference between two adjacent offsets (always preceding minus following), plus the distance of the last offset from the first multiplied by the sum of the last two offsets equals twice the area. Figure 3–13 illustrates this rule.

13-10. Area of Segment of Circles

When the needed area is located along a highway or railroad curve and one of the boundary lines is the arc of a circle, then the segment of a circle method is usually used. Generally a closed traverse is run on all straight lines of the boundary with a chord being measured on the circular portion of the boundary. Figure 13–14 illustrates this method and the computation is shown below.

a. Circular Sector. The circular sector of the area shown in the figure is that area enclosed by OWUZ. The area of a circular sector is determined by the formulas—

$$\text{Area (OWUZ)} = \frac{\pi R^2 I^\circ}{360^\circ} \text{ also,}$$

$$\text{Area (triangle OWZ)} = \frac{R^2}{2} \text{Sin I}$$

b. Circular Segment. The circular segment of the area shown in the figure is that area enclosed by WUZ. The area of the circular segment is then determined, exactly by the formula—

$$\text{Area (segment)} = \text{A (sector)} - \text{A (triangle)} = R^2 \left(\frac{\pi I^\circ}{360^\circ} - \frac{\text{Sin I}}{2} \right)$$

c. Field Procedure. Many different combinations of measurements may be made in the field for determining the area of a circular segment. The measurements most often are those of the chord WZ, referred to usually as the long chord (LC), and the middle ordinate UV, referred to as M. The middle ordinate is the distance from the center of the curve to the center of the long chord, measured along a line that bisects the I angle and is perpendicular to the long chord.

d. Computation. In the field, the long chord (LC) and middle ordinate (M) were measured to be 425.0 feet and 20.25 feet respectively. Based on the formulas for horizontal curves and those listed in *a* and *b* above, the computation is done as follows:

$$\tan \tfrac{1}{4} I = \frac{2M}{LC} = \frac{40.50}{425.0} = 0.09529$$

$\tfrac{1}{4}$ I = 5°26′6 $\quad$ $\tfrac{1}{2}$ I = 10°53′2 $\quad$ I = 21°46′4

Sin $\tfrac{1}{2}$ I = 0.18887 $\quad$ Sin I = 0.37094

$$R = \frac{LC}{2 \text{ Sin } \tfrac{1}{2} I} = \frac{425.0}{0.37774} = 1125.113 \text{ ft}$$

$$\text{Area} = R^2 \left(\frac{\pi I^\circ}{360^\circ} - \frac{\text{Sin I}}{2} \right) = 1125.113^2$$

$$\left(\frac{3.142\text{–}21.773}{360} - \frac{0.37094}{2} \right) =$$

$$(1125.113)^2 \, (0.19003 - 0.^\circ 18547) =$$

$$(1125.113)^2 \, (0.00456) =$$

$$\text{Area} = 5772.41 \text{ sq. ft.}$$

e. Parabolic Segment. If the required area of a circular segment is only needed approximately, the formulas for a parabolic segment can be used. The area by parabolic segment is only an approximation and the accuracy of the approximation decreases as the size of the I angle increases. The sample computation used with the following formula uses the same measurements as *d* above.

Area = 2/3 LC M
Area = 2/3 (425.0) (20.25)
Area = 5737.5 sq. ft.

CHAPTER 14

LAND SURVEYS

Section I. INTRODUCTION

14–1. General

Land surveying embraces those surveying operations involved in original surveys to locate and monument the boundaries of a property; the preparation of a legal description of the limits of a property and of the area included; the preparation of a property map; resurveys to recover and remonument property corners; and surveys in connection with the subdivision of a property into two or more parts. Land surveying is one of the oldest branches of engineering known to man, the principles of which have been passed down through the centuries. For this reason, the essentials of land surveying as practiced in various countries are similar in principle. Although the principles pertaining to the survey of public and private lands within the United States are not necessarily directly applicable to foreign areas, a knowledge of these principles will indirectly assist the surveyor to analyze the survey practices of a foreign nation and thereby assist him to conduct the survey in a manner required by the property laws of the nation concerned.

14–2. Description of Land Parcels

Parcels of land may be described in many forms, most of which are legal and binding depending upon the laws that were in effect at the time the survey was accomplished. These lands may be described by metes and bounds; by stating its location and size in a rectangular system; by giving the coordinates of the property corners with reference to a plane coordinate system; by a deed reference to a description in a previously recorded deed; or by reference to block and individual property numbers appearing on a recorded map.

a. Metes and Bounds. When a tract of land is defined by giving the courses and lengths of all boundaries it is said to be described by *metes and bounds.* This is an age-old method of describing land and still forms the basis for the majority of deed descriptions in the eastern United States and in many foreign nations. A good metes and bounds description starts at a point of origin which is usually monumented and referenced by ties from well established monuments or other reference points. The bearing and length of each side is given in turn around the tract to close back on the point of beginning. Bearing may be true, grid or magnetic, the former is usually preferred. When magnetic bearings are read, the magnetic declination and the date of the survey must be stated.

b. Rectangular System. In the early days of the United States, provisions were made to subdivide territorial land into townships and section in the form of rectangles with lines running along the cardinal directions of north-south and east-west.

c. Coordinate System. For many years the triangulation and traverse monuments of various domestic and foreign survey agencies have been defined by their geographic positions, that is, by their latitudes and longitudes. The necessary computations when using this system are long and involved. Too few land surveyors in the early days were sufficiently versed in the theory of geodetic surveying for this system to attain widespread use.

d. Plane Coordinate Systems. In recent years, plane coordinate systems have been developed and used in many states of the United States. The State Plane Coordinate System solution was devised by Dr. Oscar Adams of the United States Geological Survey to the problem of supplying control for locating and surveying property lines and other civil projects within the various states. Dr. Adams organized this rectangular coordinate system to eliminate the confusion that arose from the many and various individual local systems that could not be correlated. Sufficient control points are provided in this system so that local projects can be tied into these points by plane survey methods with a geodetic adjustment. The choice of the projection upon which the State Coordinate system is based depends upon the geographic shape of the state. Generally, states whose greater dimension is east to west are developed on the Lambert Conformal Conic Projection. Those with a greater north-south extent are developed on the Secant Transverse Mercator Projection. Tables are published for both projections for the conversion of coordinates to latitude and longitude or the reverse, to facilitate adaptation of any state to the system.

14-3. Relationship Between Geodetic and Land Surveying

Triangulation, precise traverse, and other geodetic or astronomic control surveys are required to fix the base line and principal meridians used in the rectangular system of land surveys and to provide control points for extending the coordinates of the plane coordinate systems to local land surveys. In many areas, the stations of these basic control networks are used to connect major streets, highways and railroads of the local systems to the national networks. Pairs of control monuments are set at frequent intervals so that it is usually only a short distance to the site of a local property survey. The geographic position of the monuments and their elevation are available in the public records.

14-4. Resurveys of Old Boundaries

The relocation of property boundaries frequently presents difficulties arising from indefinite points of origin; indefinite meridians; failure of distances to check; obliterated corners; and inability to traverse along boundaries.

a. Points of Origin Indefinite or Difficult to Locate. The following are phrases extracted from old deed descriptions: "Beginning at drill hole in a ledge on the easterly shore of Lake Champlain . . ."; "beginning at the point where you and I stood talking yesterday. . ."; "beginning at the northwest corner of Dana's upper pasture." Lake Champlain is over one-hundred miles long and it might prove time consuming to locate the ledge referred to, let alone trying to find a small drill hole. The second phrase is meaningless except to the two parties to the original transaction, both may be long since dead or uncertain of the exact location of the point in question. The upper pasture may have grown over by brush and forest years ago and all traces of its corners obliterated. To resolve such difficulties, the surveyor must obtain information from old residents whose opinions may be biased, whose memories may be faulty or he must scan aerial photographs for evidence of old lines of cultivation or logging, and must make a thorough search of the locality for monuments or evidence of their existence.

b. Indefinite Meridian. Some deed descriptions give bearings of the property lines but fail to indicate whether these are true, grid, or magnetic or to state the magnetic declination or date of survey. It may not be possible to resolve this difficulty until two adjacent monuments have been recovered. Then the true bearing of the line between them can be determined, compared with the bearing given in the deed, and a correction applied to all other bearings in the description as an aid in searching for other corner monuments.

c. Failure of Distances to Check. The lines in many old surveys were measured by inaccurate or make shift methods that are difficult to check. Distances were sometimes measured by counting the revolutions of a wagon wheel, the size of the wheel was usually not mentioned and the tally was often incorrect. Distances measured with a chain were frequently in error because of wear on the links, or because of poor alinement or plumbing. *It is not the function of the surveyor to correct errors in the original survey, but rather to locate the corners as original established.* When two adjacent corners have been relocated, the taped distance between them can be compared with the distance as given in the deed. The other deed distances are then adjusted in proportion to give lengths which will aid in locating the remaining corner monuments.

d. Obliterated Corners. When the original property corner consisted of a tree, a wooden stake, or an iron pipe, the rotting of the wood or rusting of the pipe may have destroyed the corner. Measurements of the bearing and distance from an adjacent located corner can give an approximate location. Careful removal of the surface litter in this area may then disclose evidence of the rotted stake, stump, or pipe. The humus from the wood and the rust from the pipe will generally show a distinct difference in coloration and physical characteristics in contrast to the surrounding soils. Four stakes can then be driven nearby in such positions that cords stretched between diagonally opposite stakes will intersect over the original point. A new permanent concrete or stone monument can be installed and centered under the intersection of the cords.

e. Inability to Traverse Along Boundaries. The actual boundaries of a property may be occupied by fences, stone walls, hedgerows, or a line of shade trees. Such conditions necessitate the use of offset lines or random lines and add to the difficulty of conducting the survey.

14-5. Laws Relating to Boundaries

The body of laws and precedents relating to boundaries is voluminous. A complete coverage of these various laws would fill several volumes. The land surveyor has no judicial function but, if he is to perform his work skillfully and well, he should have a knowledge of the property laws and ruling of the locality in which he is functioning. In this book it is only possible to set forth a few guiding principles and to indicate a few important differences in existing laws.

a. Function. It is the function of the surveyor to recover, if possible, the original boundaries and corners of a property. It is not his function to correct error in the original survey.

b. Errors. When an original survey is incomplete or contains errors, it is assumed that it was the intention of the grantor to transfer a definite tract of land to the grantee and the deed will be so interpreted as to make it effectual rather than void.

c. Calls. Calls in deed descriptions refer to monuments, distances, directions, adjoining owners, and the area of the tract. When all corners or boundaries cannot be relocated or retraced with certainty, the greatest weight must be given in the resurvey to those calls which are most likely to be correct and which therefore take precedence over other calls. The usual order of calls in order of their decreasing precedence is as follows:

(1) Natural monuments and boundaries.

(2) Artificial monuments.

(3) Calls for adjoiners.

(4) Courses and distances.

(5) Area.

An exception to the general rule occurs when a will states that equal or other specified divisions of land to several heirs are intended, in which case the call for area (5) above will take precedence over the call for course and distance, (4) above.

d. Usage Rulings. Local usage and rulings vary with respect to where the actual boundary lies within a fence, party wall, hedge, ditch, or walk. The surveyor should consult local authorities in this matter if the actual corner monuments cannot be recovered.

e. Abutting.

(1) Where property abuts on a highway, ownership in some localities extends to the center line of the road, the public enjoying the use of the roadway as an easement and land reverting to the abutting owners if the roadway is abandoned. In other localities, title to the roadway resides in the state, county, parish, or municipality.

(2) The right of owners of land abutting on the ocean, lakes, or tidal or upland streams or rivers are subject to statutes which vary in different localities. Ownership may extend to the highwater mark, to low water, or to the thread or middle of the channel. The owner of such property may have riparian rights in a waterway or may acquire a riparian grant. Local laws must be consulted.

(3) Owners of waterfront property hold title to the *ripa* or bank, such ownership extending, in general to the mean high water line. When by action of water, the bed of the body of water slowly changes its position, the high water mark also shifts and ownership of adjacent land progresses with it. Riparian owners have numerous other rights, including that of access to navigable water which involves the right to build a wharf or pier out from shore. Such a wharf must not extend beyond the established pierhead line or it would interfere with navigation.

f. Adverse Possession. Land belonging to another may be acquired after a definite period of time by adverse possession. If a property owner *A* occupies not only his own land but also a part of the land of *B*, and if such occupancy has been open, hostile, notorious, and continuous for the statutory period, then *A* gains title to that portion of *B's* land so occupied. The statutory period is generally 20 years. In some localities ownership of part of a public highway can be acquired after 40 years of adverse occupancy; in other localities, adverse possession does not apply against public land.

14-6. Responsibilities Related to Land Surveys

Responsibilities for supervision of surveys of all types are covered in the appropriate survey manuals. Additional responsibilities particularly applicable to lands survey include, but are not limited to—

a. Obtain permission for survey personnel to trespass on private lands.

b. Secure deed descriptions and copies of filed maps covering the property to be surveyed from the local registry of deeds, hall of records, or other repository of public records pertaining to property.

c. Arrange for reasonable compensation to property owners for any damage resulting from the survey operations.

d. Initiate title searches to establish validity of title and any easements or clouds on the title which may exist.

e. Obtain descriptions of the location and coordinates of control monuments in the vicinity.

f. Secure the cooperation of local surveyors or of other agencies who have conducted surveys in the area and who may be able to supply information and assistance.

Section II. SURVEYS OF PRIVATE LANDS

14-7. Functions

In resurveying property boundaries and in carrying out surveys for the subdivision of land, the surveyor has the following functions, resposibilities, and liabilities:

a. Locate in the public records all deed descriptions and maps pertaining to the property and must properly interpret the requirements contained therein.

b. Set and properly reference new monuments and replace obliterated monuments.

c. Liable for damage caused by errors resulting from incompetent professional work.

d. Attempt to follow in the tracks of the original surveyor, relocating the old boundaries and not attempting to correct the original survey.

e. Prepare proper descriptions and maps of the property.

f. May be required to connect a property survey with control monuments so that the grid coordinates of the property corners can be determined.

g. Report all easements, encroachments, or discrepancies discovered during the course of the survey.

h. When original monuments cannot be recovered with certainty from data contained in the deed description, seek additional evidence. Such evidence must be substantial in character and must not be merely personal opinion.

i. In the absence of conclusive evidence as to the location of a boundary, seek agreement between owners as to a mutally acceptable location. The surveyor has no judicial functions; he may serve as an arbiter in relocating the boundary according to prevailing circumstances and procedures set forth by local authority.

j. When a boundary dispute is carried to the courts, he may be called upon to appear as an expert witness.

k. Must respect the laws of trespass. The right to enter upon property in conducting public surveys is provided by law in most localities. In a few political subdivisions, recent laws make similar provision with respect to private surveys. Generally, the surveyor must gain permission from the owner before entry on private property. Lacking permission from an adjoiner, it is sometimes possible to make the survey without trespassing on the adjoiner's land, but such a condition normally adds to the difficulty of the task.

l. Liable for actual damage to private property resulting from his operations.

14-8. Use of Deeds and Records

a. Use of Recorded Deeds. When a surveyor is required to survey a property, the deed may be furnished to him. Frequently, though, he must look in the public records for the recorded copy of the deed, for filed plans or maps, or for the deeds to adjacent property.

b. Registry of Deeds. Land records are listed in a registry of deeds. The statutes of various localities differ in their designation of the recording official. He may be a town or city clerk or registrar, a county registrar, or other legally constituted recording officer. The recording of deeds is usually a function of the county government and each county will maintain one or more registries of deeds or halls of records serving the whole or a designated part of the area of the country.

c. Grantor and Grantee Indexes. The index of deeds at the registry is in two parts, one listing each deed under the name of the grantor or *seller* of the land and the other containing the listing under the name of the grantee or purchaser. The index is frequently further divided by years, so that the surveyor should ascertain the name of the buyer or seller of the property and the approximate date of transfer. He then can enter the index for the proper year and find the number of the deed book, and the page of that book on which the deed is recorded.

d. Deed Book. The appropriate page of the deed book contains a copy of the deed in question. It may contain a reference to a filed map, particularly when the property was once part of a larger tract. Reference numbers in the text or on the margin refer to the next preceding transfer or assignments or attachment of the property. When the deed description shows the name of adjoiners, the deed of these adjoiners may also be looked up. Information from such deeds will often supplement the description of the property in question when such description shows conflicting data or is not complete. The surveyor should be able to carry out the survey using the listed data.

14–9. Preparing Plan and Description

a. Plan. From the field notes, a scale plan is prepared showing all pertinent information. The plan is drawn with a dark reproducible ink on a good grade of tracing cloth. The scale should be of such size as to clearly show all necessary details (fig. 14–1). All pertinent dimensions and bearings must be shown and the coordinates of the corners on the local grid if these have been determined. The plan should also show the names of adjoiners; area of the property; and all easements or encroachments. Additional information to be shown if possible: direction of true meridians; any town, city, county or other political subdivision lines; street or road lines; boundary monuments and lines; fences, walls, buildings, passageways, etc. The title block must contain—

(1) Location.
(2) Scale.
(3) Date of survey.
(4) Magnetic declination, if applicable.
(5) Signature of surveyor.
(6) Surveyors address and seal, when required by law.

b. Description. In preparing the description of a property, the surveyor must bear in mind that the description must clearly identify the location of the property and give all necessary data from which the boundaries can be reestablished at any future date. The written description contains most of the information shown on the plan. Usually both a description and plan are prepared and, when the property is transferred, are recorded according to the laws of the country (or other political division concerned. A description of the property shown in figure 14–1 is given below.

"All that certain tract or parcel of land and premises, hereinafter particularly described, situated, lying and being in the city of Springfield in the county of Fairfax and Commonwealth of Virginia and constituting lot 2 shown on the revised map of the Foster property in said city as filed in the Fairfax County Hall of Records.

Beginning at an iron pipe in the northwesterly line of Jackson Street therein distant along same four hundred point seventy-one one-hundredths (400.71) feet northeasterly from a stone monument at the northerly corner of Wilson Place and Jackson Street; thence running (1) north forty-four degrees thirty-one and one-half minutes west along land of J. M. Jones one hundred and fifty-six point thirty-two one-hundredths (156.32) feet to an iron bar; then turning and running (2) north forty-five degrees twenty-eight and one-half minutes east along land of A. F. Foster one hundred eighteen point zero (118.00) feet to an iron bar; thence turning and running (3) south forty-four degrees thirty-one and one-half minutes east along and of E. G. Rogers one hundred and fifty-six point thirty-two one-hundredth (156.32) feet to an iron bar in the northwesterly line of Jackson Street; thence turning and running (4) south forty-five degrees twenty-eight and one-half minute west along said line of Jackson Street one hundred eighteen point zero-zero (118.00) feet to the point and place of beginning; all bearing being true and the lot containing a calculated area of eighteen thousand four hundred and forty-five (18,445) square feet. This description has been prepared from a survey made by W. C. Keys, Licensed Land Surveyor, Virginia No. 12345."

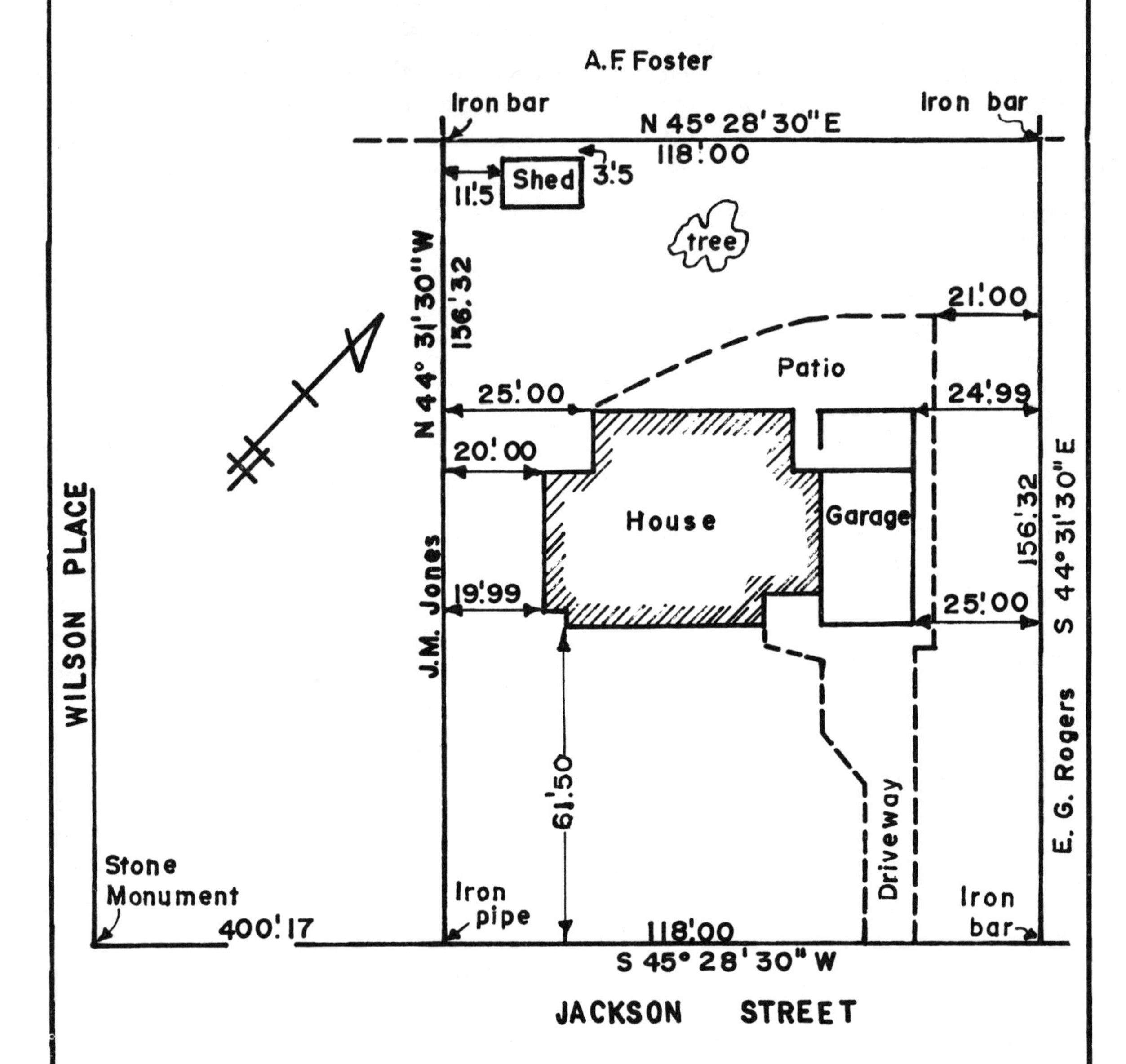

FIELD BOOK M-13 PAGE 12 & 13

MAP BOOK- 9

SCALE 1 inch = 30 feet

Lic. Land Surveyor, Va. 12345

Figure 14-1. Lot plan.

Section III. SURVEYS OF PUBLIC LANDS

14–10. Public Domain

At the time of the establishment of the United States, the thirteen original States retained title to all unappropriated lands within their respective boundaries and retained control of surveys of these lands. Later, such title was also retained by the States of Maine, Vermont, West Virginia, Kentucky, Tennessee and Texas. As successive territories on the North American continent were purchased by or ceded to the Federal Union, title to vacant lands in such territories was vested in the United States Government. These regions (fig. 14–2) became the public domain. To provide for the division of these public lands and their subsequent settlement, the Congress passed legislation to establish a rectangular system of surveys in these areas and to setup procedures for the transfer of land to private ownership. The original legislation, together with revisions enacted by Congress in later years, forms the basis for the system of surveys used to divide the public lands.

14–11. United States System

The legislation provides that the public lands shall be divided by true meridional lines and by parallels of latitude to form townships approximately six miles square. The corners of the townships are marked with progressive numbers from the point of beginning. Each township is further subdivided into sections of approximately 640 acres each by lines running generally parallel to the township boundaries. The sections are numbered, beginning with number 1 in the northeast section of the township and proceeding west and east alternately through the township until all thirty-six sections have been numbered. Lands within an Indian reservation and lands previously transferred to private ownership are not included in the subdivision survey.

a. Responsibility. Surveys of the public domain comes under the jurisdiction of the U.S. Bureau of Land Management formerly the General Land

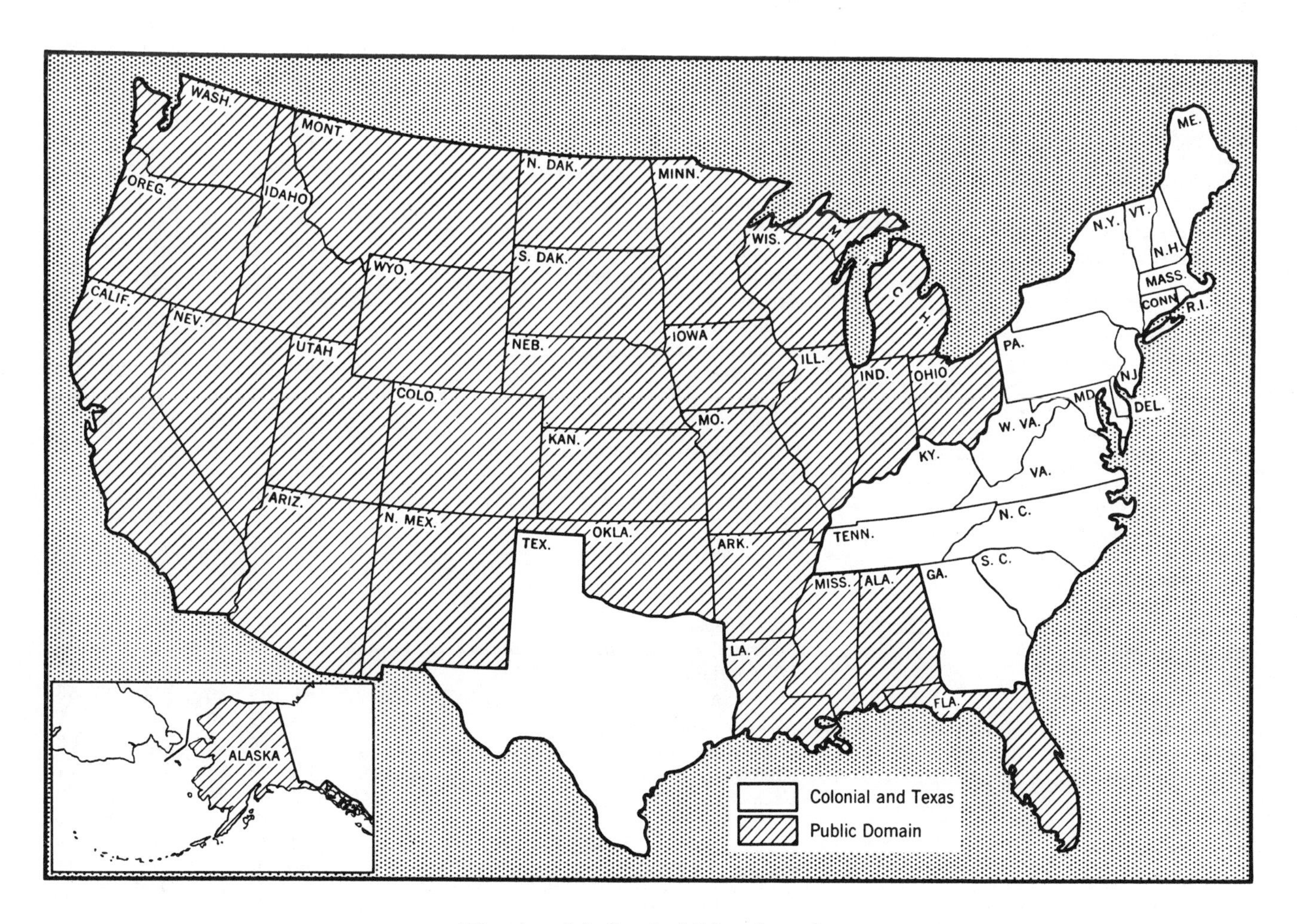

Figure 14-2. Public domain.

Office. The director of this bureau exercises general supervision of surveys of the public domain and of the disposal of such lands. Several field regions have been set up, each under the supervision of a regional cadastral engineer. All remaining original surveys of the lands in the public domain and resurveys of certain types are made by cadastral engineers in the service of the bureau.

b. System of Surveying. The system of surveys used is described in detail in the manual of instructions for the Survey of the Public Lands of the United States, published by the Bureau of Land Management, and is available from the Superintendent of Documents, Washington, D.C. The major features of the system and current field practices are described in subsequent paragraphs.

c. Initial Points. The area embraced by the public domain has been divided into some thirty-five regions for survey purposes. In each region, an *initial point* has been established and monumented, this initial point serving as the origin for all public-land surveys to be extended throughout that region. The geographic positions of these initial points have been determined by ties to geodetic stations. The various initial points, base lines and principal meridians of the United States are designated by a name assigned to the principal meridian, refer to table 14-1 for the thirty-five regions.

d. Principal lines. The surveys are run on cardinal lines, that is, lines running north-south and east-west. All principal lines in the sytem are therefore meridians of longitude (north-south) and parallels of latitude (east-west). The meridi-

Table 14-1. Meridians and Base Lines of U.S. Rectangular Surveys

Meridians	Governing surveys (wholly or in part) in States of—	Longitude of initial points west from Greenwich			Latitude of initial points		
		°			°		
Black Hills	South Dakota	104	03	16	43	59	44
Boise	Idaho	116	23	35	43	22	21
Chicksaw	Mississippi	89	14	47	35	01	58
Choctaw	do	90	14	41	31	52	32
Cimarron	Oklahoma	103	00	07	36	30	05
Copper River	Alaska	145	18	13	61	49	21
Fairbanks	do	147	38	26	64	51	50
Fifth Principal	Arkansas, Iowa, Minnesota, Missouri, North Dakota, and South Dakota.	91	03	07	34	38	45
First Principal	Ohio and Indiana	84	48	11	40	59	22
Fourth Principal	Illinois[1]	90	27	11	40	00	50
Do	Minnesota and Wisconsin	90	25	37	42	30	27
Gila and Salt River	Arizona	112	18	19	33	22	38
Humboldt	California	124	07	10	40	25	02
Huntsville	Alabama and Mississippi	86	34	16	34	59	27
Indian	Oklahoma	97	14	49	34	29	32
Louisiana	Louisiana	92	24	55	31	00	31
Michigan	Michigan and Ohio	84	21	53	42	25	28
Mount Diablo	California and Nevada	121	54	47	37	52	54
Navajo	Arizona	108	31	59	35	44	56
New Mexico Principal	Colorado and New Mexico	106	53	12	34	15	35
Principal	Montana	111	39	33	45	47	13
Salt Lake	Utah	111	53	27	40	46	11
San Bernardino	California	116	55	17	34	07	20
Second Principal	Illinois and Indiana	86	27	21	38	28	14
Seward	Alaska	149	21	24	60	07	36
Sixth Principal	Colorado, Kansas, Nebraska, South Dakota, and Wyoming.	97	22	08	40	00	07
St. Helena	Louisiana	91	09	36	30	59	56
St. Stephens	Alabama and Mississippi	88	01	20	30	59	51
Tallahassee	Florida and Alabama	84	16	38	30	26	03
Third Principal	Illinois	89	08	54	38	28	27
Uintah	Utah	109	56	06	40	25	59
Ute	Colorado	108	31	59	39	06	23
Washington	Mississippi	91	09	36	30	59	56
Willamette	Oregon and Washington	122	44	34	45	31	11
Wind River	Wyoming	108	48	49	43	00	41

[1] The numbers are carried to fractional township 29 north in Illinois, and are repeated in Wisconsin, beginning with the south boundary of the State; the range numbers are given in regular order.

ans are straight lines but they are not parallel since they converge to meet at the poles. The parallels of latitude are at right angles to the meridian at any point. They are, therefore, curved lines.

(1) *Base line.* The base line of a region is the parallel of latitude that is extended from the initial point true east and west (table 14–1).

(2) *Principal meridian.* The principal meridian of a region is the meridian of longitude that is extended from the initial point either true north or south, or in both directions, as required by the position of the initial point within the region concerned (table 14–1).

e. Distance Along Cardinal Lines. The unit of distance measurement in the public domain surveys is the Surveyor's (or Gunter's) Chain of 66 feet. Steel tapes, graduated in links and chains, are used today rather than the old type chains. Distances are measured along both the base line and the principal meridian, and monuments are placed at intervals of 40 chains (½ mile) to mark quarter-section or section corners. Township corners are placed every 6 miles along these lines. Distance shall be measured twice and remeasurements made if the discrepancy of the two measurements exceeds the specified limit.

14-12. Division of Lands

The regions are divided into smaller areas as discussed below.

a. Tracts. The regions are divided first into tracts approximately 24 miles square. The division is done by placing monuments along the principal meridian at intervals of 24 miles from the initial point, *standard parallels* (also called *correction lines*) are run east and west from the principal meridian. These are monumented at intervals of 40 chains. The standard parallels are designated with respect to their relationship to the base line and become the First Standard Parallel North, Second Standard Parallel North, First Standard Parallel South, and so on until the limits of the region have been reached. *Guide meridians* are extended *north* from the base line, and from each standard parallel, at intervals of 24 miles east and west from the principal meridian. Each guide meridian is terminated at its intersection with the next standard parallel to the north of its point of origin. The guide meridians become known as the First Guide Meridian East, First Guide Meridian West and so on, depending upon their locations with respect to the principal meridian. As with the other principal lines, the guide meridians are monumented at intervals of 40 chains. The tract of land bounded by successive pairs of standard parallels and guide meridians will have theoretical boundary lengths of 24 miles on the 'west, south, and east sides. The north boundary, because of the convergence of the meridians, will be less than 24 miles. Because of errors of field measurements, the boundaries will seldom equal their theoretical lengths.

b. Townships. The 24 miles square tracts are divided into *townships. Range lines* are run north along true meridians from the monuments (known as *standard township corners*) placed at 6 mile intervals, east and west of the principal meridian, along the base line and each standard parallel. As is the case with the guide meridians, each range line terminates at its intersection with the first standard parallel to the north of its point of origin. These intersection points, at the termini of the range lines and guide meridians, are known as *closing township corners.* Township corners are established at 6 mile intervals on these range lines, and quarter-section and section monuments are placed at intervals of 40 chains. *Township lines,* or latitudinal lines, are then run along parallels of latitude to join the township corners marked by monuments previously set at 6 mile intervals along the principal meridian, guide meridians, and range lines. These range lines and township lines divide the tract into townships. The lengths of the east and west boundaries of a township are theoretically equal to 6 miles. The north and south boundaries of a township vary from a maximum length of 6 miles for its southern boundary at the base line or standard parallel forming the south line of a 24 mile quadrilateral, to a minimum length at its northern boundary. Figure 14–3 shows the division of four such tracts into 16 townships by a base line, principal meridian, standard parallels, guide meridians, range lines, and township lines.

c. Sections. Each township is then divided into sections, each having an area of approximately one square mile or 640 acres. *Meridional* section lines are initiated at the section corners which have been set at intervals of 80 chains (1 mile) along the south boundary of the township and are run from the south to the north boundary parallel to the east boundary of the township. Quartersection and section corners are established alternately at intervals of 40 chains along the meridional section lines. From the section corners so established, *latitudinal* section lines are run from west to east as random lines parallel to the south boundaries of the respective sections to the corresponding section corners on the next meridional section line (or township line) to the east. The falling on the objective corner is noted, and the true line established from east to west. These section lines will divide the township into 36 sections, rhomboidal in shape but nearly square. All sections, with the exception of the six in the most westerly range of the township, will have theoretical dimensions of 80 chains on all four sides. In practical application of this plan, the accumulated error of measurement is placed in the last interval closing on the north boundary of the township. Those sections in the westerly range will be less than 80 chains in width from east to west, the variation depending upon the distance of the section boundary from the southerly boundary of the tract.

14-13. Numbering of Townships and Sections

a. Townships. A series of adjacent townships running east and west is known as a *tier.* An adjacent series running north and south is known as a *range.* The tiers of townships are numbered consecutively, both to the north and south of the base line. The ranges of townships are likewise numbered, both to the east and west of the principal meridian. A township is designated by: the serial number of its tier and the letter N or S to indicate the position of the tier north or south of the base line; the serial number of its range and the letter E or W to indicate the position of the range east or west of the principal meridian. Assuming that figure 14–3 represents the start of the division along the Fourth Principal Meridian, the cross-hatched township shown would be designated "Township 3 north, Range 2 west, of the Fourth Principal Meridian." This would be abbreviated "T3N, R2W, 4th P.M."

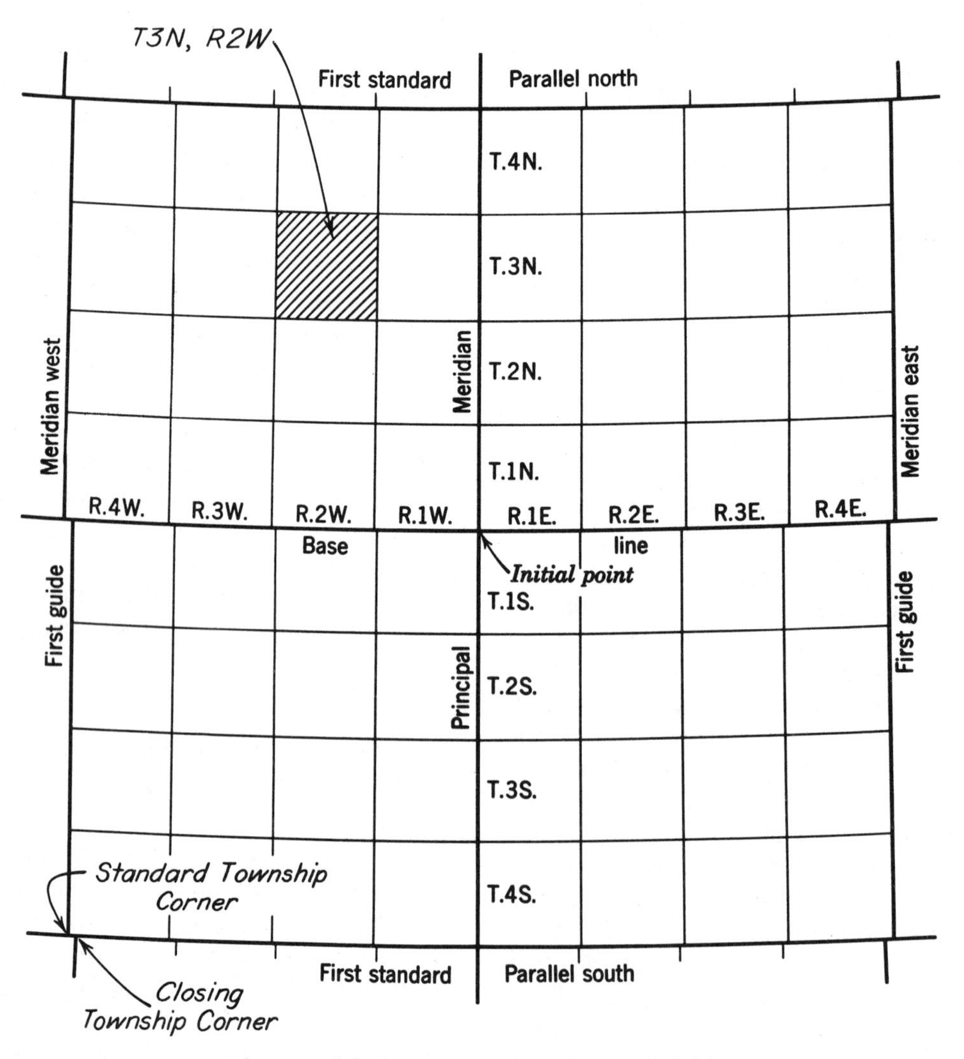

Figure 14-3. Tracts and townships.

b. Sections. The sections are numbered beginning with 1 in the northeast section of the township and progressing west and east alternately along the tiers of sections until all have been numbered. Figure 14–4 shows the numbering of one township into sections.

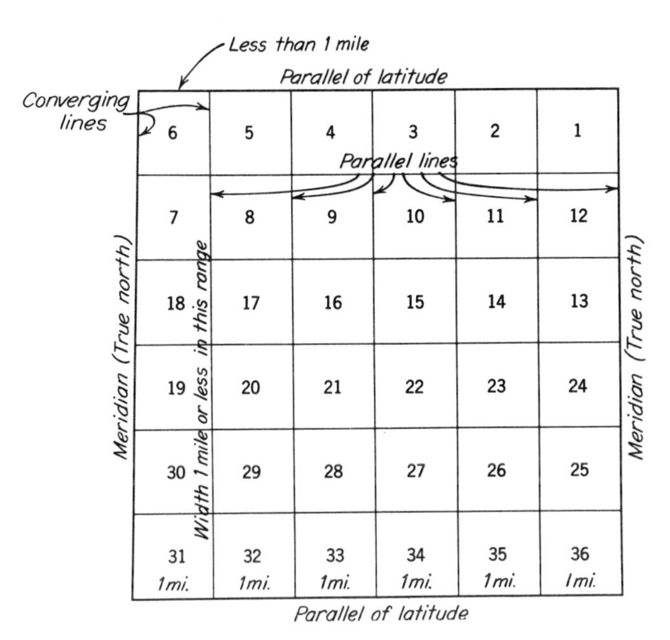

Figure 14-4. Numbering of sections.

14-14. Subdividing of Sections

Sections are usually subdivided into quarter-sections, except for those sections on the westerly range and on the northerly tier of the township. These sections will, in general, contain either more or less than 640 acres because of accumulated survey errors in the northerly 20 chains of each range or meridional section line or because of the convergence of the meridians. Such sections (fig. 14–6) are commonly subidivided into quarter-sections, half-sections, quarter-quarter-sections, and numbered lots. Similarly special subdivisions are required in the southern tier or in the eastern range of sections when the south or the east boundary of the township is defective in alinement (fig. 14–5). Fractional subdivisions are also required when there is a lake or other navigable body of water within the section or when there are patented mineral claims, private land claims, reservations, state lines and other special surveys in the area. These subsidisions of section lines are protracted on the official plat of the township by the cadastral engineers of the Bureau of Land Management. The subdivision of sections and fractional sections is not a normal function of the Bureau of Land Management. These surveys are customarily executed by local surveyors. Before a section is subdivided it is necessary to identify the section and quarter-section corners or to relocate them.

a. Subdividing into Quarter-Sections. When the quarter-section corners have been identified, two straight lines are run between opposite quarter-section corners to divide the section into four quarters. The intersection of these lines is monumented and is the legal center of the section. When opposite quarter-section corners have not or cannot be fixed, as when one falls in a lake or river or falls outside of the public domain, subdi-

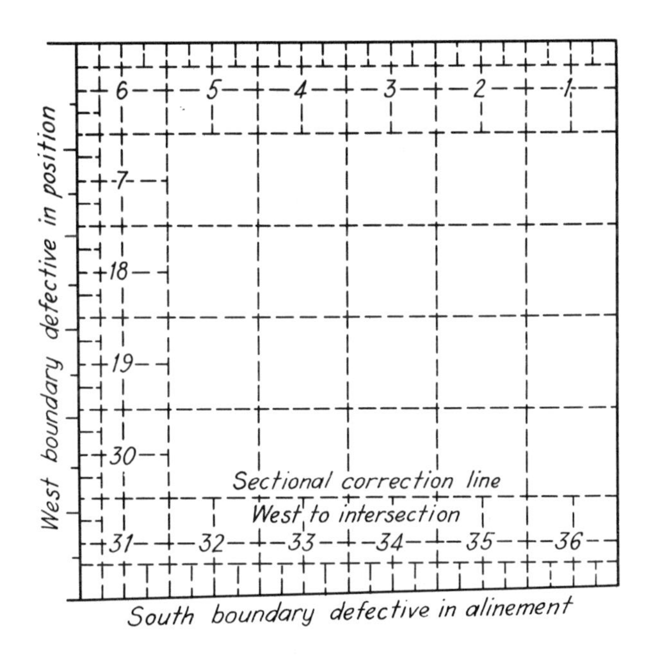

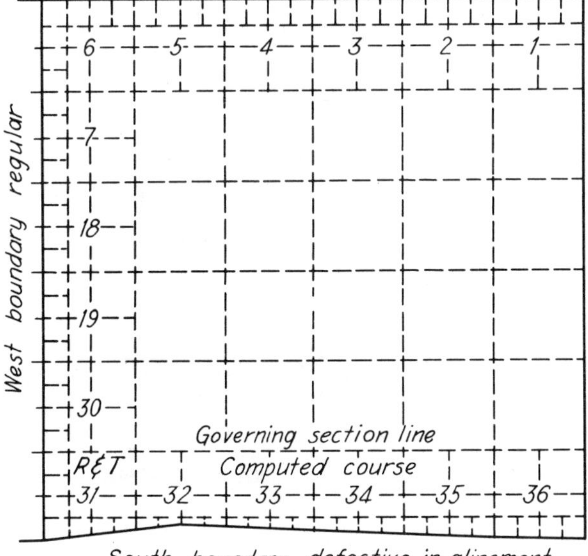

Figure 14-5. Subdividing townships with defective boundaries.

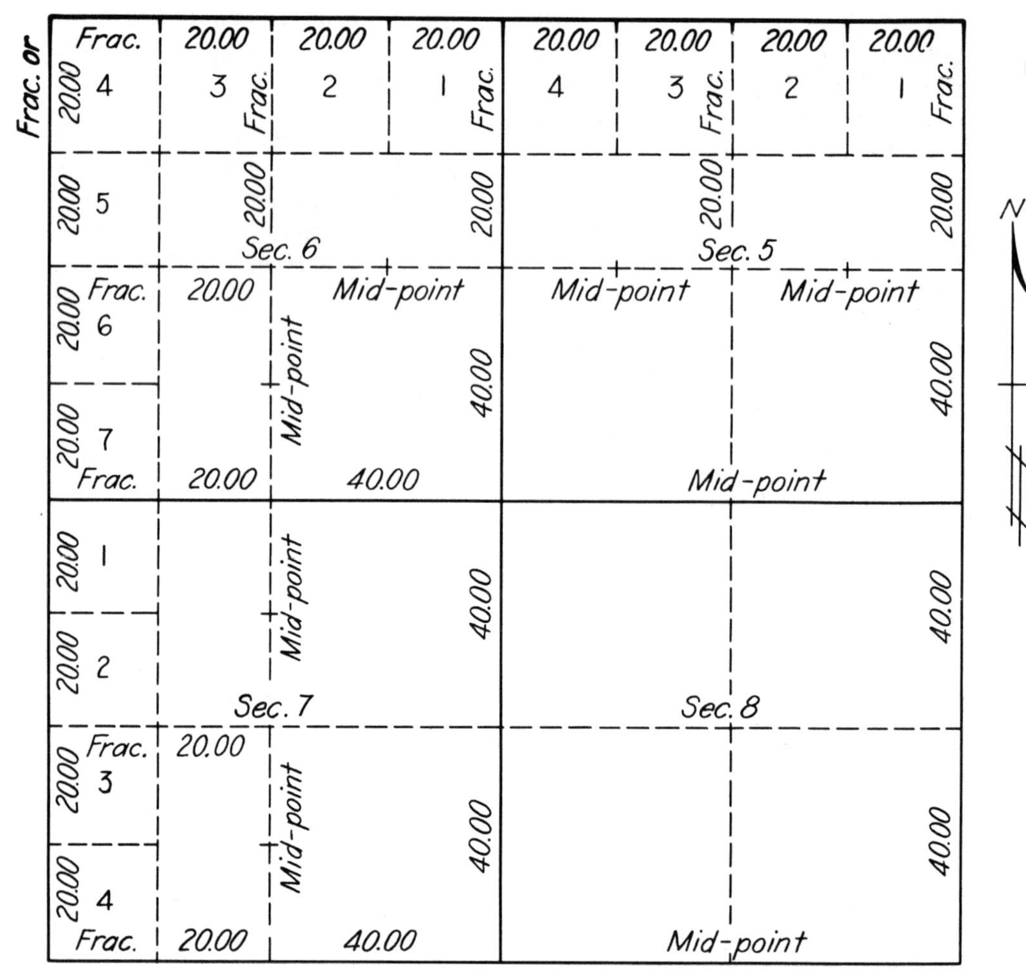

Showing normal subdivision of sections

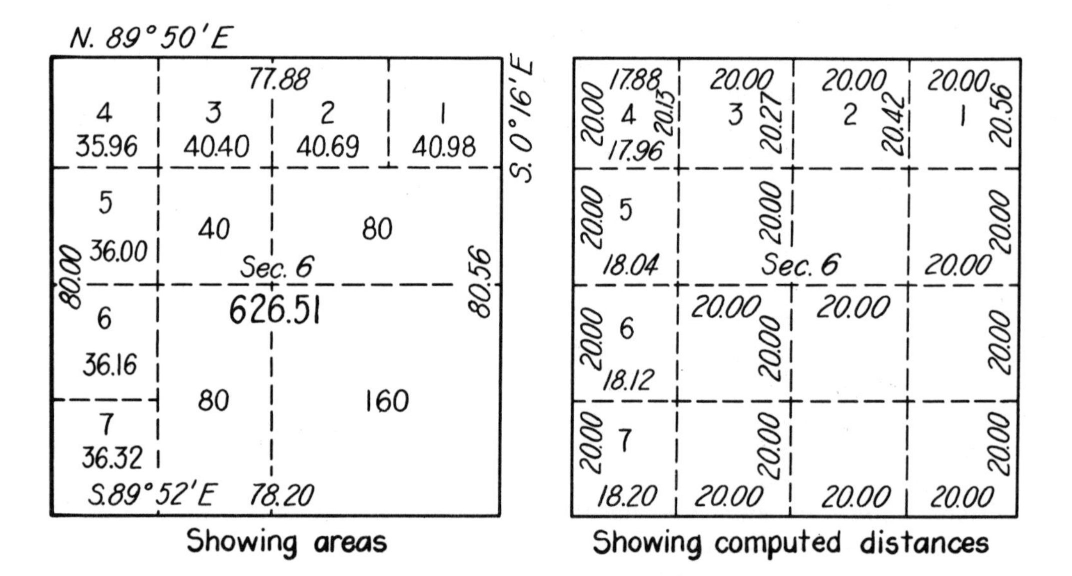

Figure 14-6. Subdividing of sections.

vision-of-section lines are run from those quarter-section corners located, parallel to the north, south, east, or west lines of the section, in accordance with the lines indicated upon the official plat.

b. Subdividing of Quarter-Sections. In subdividing quarter-sections, the quarter-quarter- or sixteenth-section corners are established at points midway between the section and quarter-section corners, and between the quarter-section corners and the center of the section. An exception to this is made on the last half-mile of those lines closing on township boundaries. Here the corners are set at 20 chains, proportionate measurement, from the preceding quarter-section corner, so that survey discrepancies are placed in the last distance. When the sixteenth-section corners have been established, straight lines are run between opposite corners to divide the quarter-section into sixteenths of a section. In a manner similar to that outline above, sections can be subdivided into parts which are, respectively, 1/64, 1/256, and 1/1024 of a section.

c. Fractional Lots. The provisions of *a* and *b* above, apply to sections of normal size and to fractional parts thereof. Those sections bordering on the north and west boundaries of a township, excepting section 6, are subdivided by protractions on the official plate into parts representing two half-quarter sections and four lots (fig. 14–6). The lots are numbered from 1 to 4 in an east to west direction or in a north to south direction except in section 6 which lies in the northwest corner of the township. In section 6 there are 7 fractional lots as indicated in figure 14–6. Similar fractional lots may be formed on the south and east sides of a township when these boundaries are defective (fig. 14–5).

14-15. Townsites

Townsites in the public domain are those areas within one or more townships which are divided into streets, alleys, and blocks of lots. Many townsites, particularly those developed on the relatively flat praires and plains, follow along rectangular lines closely correlated with the regular lines of the survey system. Other irregular townsites came into being, prior to subdivision surveys in the area, around trading posts and military establishments. The townsite boundaries are surveyed, closed, and tied in with the lines of the public land system, monuments being placed at the intersections of all such lines with the townsite boundaries. All street, block, and lot lines are likewise staked out and a plat of the area prepared, showing the relationship of public survey lines to all lines in the townsite.

14-16. Location of Islands

The beds of navigable bodies of water below the high water line do not form a part of the public domain. The sovereignty to such lands lies in the individual states in which the waterways are located. However, in the case of islands which were above high water prior to the date of admission into the Union of the state in which the islands are located, title was vested in the United States. Such islands are part of the public domain and are subject to survey. The islands are therefore to be located by triangulation, direct measurement, or other suitable method, their shore lines located by a meander traverse, and the islands shown on the official plat or plats. Township or section lines which traverse the island are located and regular township, section, and meander corners are set. When the size of the island warrants, subdivision-of-section lines are protracted normally thereon.

14-17. Procedure for Establishing Section Lines

Figure 14–7 shows the normal order of running lines in subdividing a township. Starting at the southwest corner of section 36, line 1 is run northerly parallel to the east boundary of the township. Quarter-section and section corners are set along this line at distances of 40 and 80 chains respectively. From the section corner just established, a random line, 2, is run easterly parallel to the south boundary of the township, a temporary quarter-section corner being set on this line. The "falling" of this line is determined. This falling is the distance from the point where the random line intersects the guide meridian to the standard corner previously set on the guide meridian. If the "falling" of the line is within allowable limits, the direction of the return course which will join the section corners is computed. This return line is then run, a permanent quarter-section corner being set midway along it. This process is continued for section 25 by running line 3 northerly and parallel to the east township boundary and then running random line 4 easterly and closing back. Successive sections in the range are monumented until section 1 is reached. As indicated by the numbered lines in figure 14–7, the process is repeated for each range in the township, the layout of the sections in the second range starting with line 12. On the last range, the random line 46 is run to the east and random line 47 is run to the west.

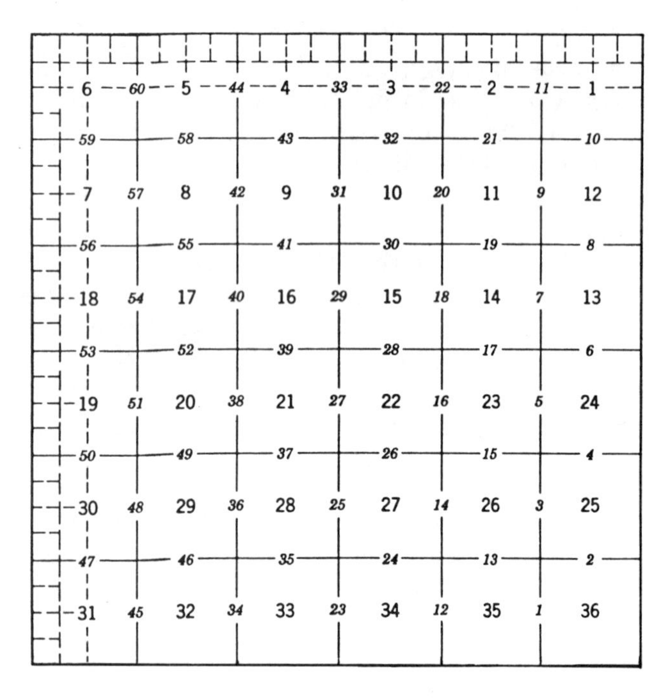

Figure 14-7. Running section lines.

14-18. Monuments

The marking of corners in land surveys used in present day surveys has been standardized by the Bureau of Land Management. All types of corner monuments are systematically marked to furnish identification of the monument.

a. Post and Tree Monuments. The usual wrought-iron post monument carries a brass cap. Identifying capital letters and Arabic numerals are stamped on this cap with a stamping set. Similar letters and numbers are marked on tree monuments, on the trunk just above the root crown, using a timber scribe. If the tree is smooth barked, the markings are cut directly into the bark; if it is rough barked, a long blaze is made on the trunk to receive the markings. Table 14–2 is a list of the commonly used markings that apply to all classes of corners. The marking on the brass cap are made to read from the south side.

b. Stone Monuments. Where stone monuments are used, the letters, numbers, and other markings are chiseled into one or more of the vertical faces of the stone rather than upon its top. Stone monuments are set with either their faces or their edges in the cardinal directions, depending upon the location of the corner in the survey. Stone township corners bear the letters and numbers indicating the townships and ranges to which the monument is common. If the monument is on a base line or standard parallel, it must also show the letters SC or CC to distinguish between a standard or closing township corner. Section corners are likewise distinguished as standard or closing corners. In addition, they bear a number of grooves on the faces or a number of notches on the edges of the monument. These grooves or notches indicate, for exterior section corners, the number of miles from the monument to the adjoining township corners. For subdivisional corners, they show the number of miles from the monument to the east and south township boundary lines, furnishing a means of identifying the sections which meet at the monument. Figure 14–8 shows the typical letters and numerals on monuments.

Table 14–2. Monument Markings

Marks	*Indicate*
AM	Amended monument
AMC	Auxiliary meander corner
AP	Angle point
BO	Bearing object
BT	Bearing tree
C	Center
CC	Closing corner
E	East
LM	Location monument
M	Mile
MC	Meander corner
N	North
NE	Northeast
NW	Northwest
PL	Public land (unsurveyed)
R	Range
RM	Reference monument
S	Section
S	South
SC	Standard corner
SE	Southeast
SMC	Special meander corner
SW	Southwest
T	Township
TR	Tract
W	West
WC	Witness corner
WP	Witness point
¼	Quarter-section
1/16	Sixteenth-section
1916	Date (year)

c. Meander Corners. All navigable bodies of water and other important rivers and lakes are segregated from the public lands at mean high water level. When a survey line intersects such a body of water a meander corner is established. The distance from the nearest section or quarter-section monument on the survey line is measured

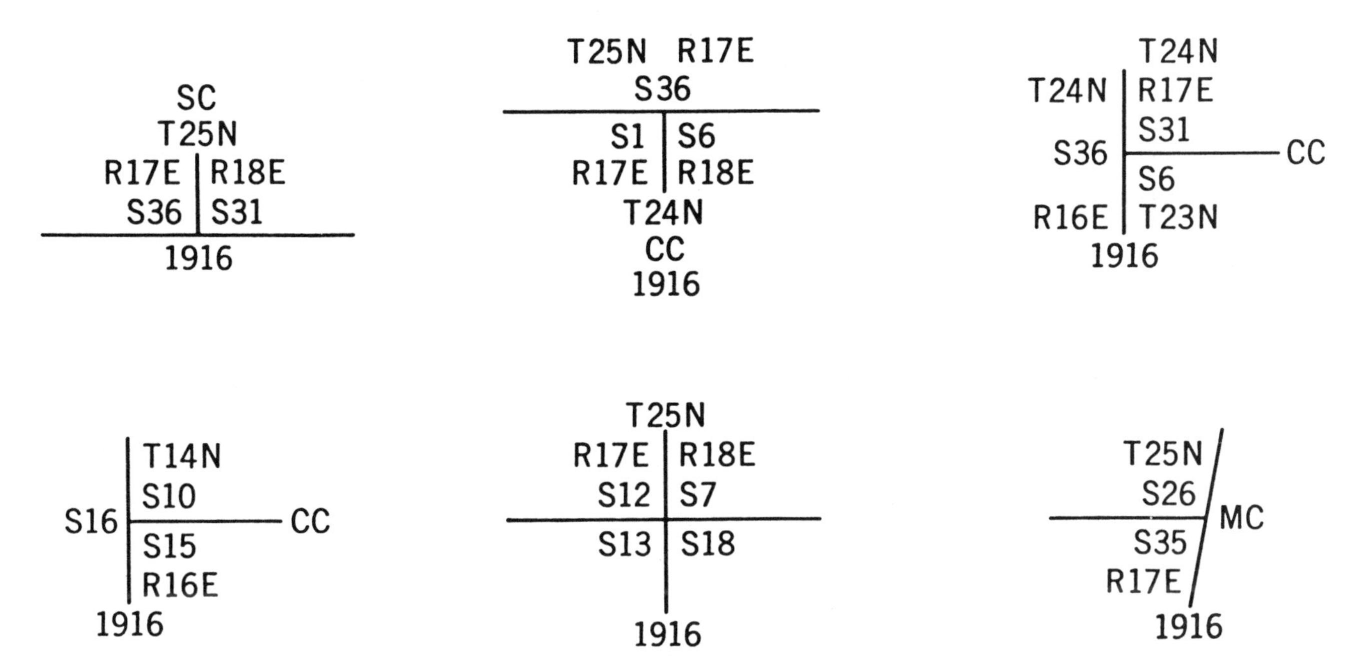

Figure 14-8. Markings on monuments.

and recorded. The monument is marked MC on the side facing the water. The other markings that refer to township, sections, and date of establishment are also stamped or chiseled. Such a monument is referenced by bearings to trees or other convenient witnesses. Where the width of the water course permits, the survey line is carried across to a meander corner on the far shore.

14-19. Restoration of Lost Corners

A lost corner is a point of a survey the position of which cannot be determined, beyond reasonable doubt, either from traces of the original marks or from acceptable evidence or testimony. In the search for a corner, retracements will have been made from located monuments on all lines which meet at the corner. The total resurvey distances between located monuments on either side of the lost corner will be known and can be compared with distances shown in the original notes. This comparison will normally permit the restoration of the corner by application of the principle of proportionate measurements. The methods used in restoring lost corners are given below.

a. Double Proportionate Measurement. The method of double proportionate measurement is usually applicable to the restoration of a corner which is common to four townships or which is common to four sections. A proportionate measurement is one that gives a concordant relationship between all parts of a line, that is, the new values given to the several parts, as the result of a remeasurement, will bear the same relation to the lengths of the same parts of the original survey as the new measurement of the total length bears to the total length of the original survey. The term double proportionate measurement is applied to measurements made between two known corners on a latitudinal line and between two known corners on a meridional line for the purpose of reestablishing the corner at the intersection. The effect of double proportional measurement is that the distances will control the relocation rather than the direction of the lines shown in the original record. Figure 14-9, shows a lost corner X which is common to four townships. Points A and B represent recovered corners on the range line and C and D represent corners on the latitudinal line. A retracement is first made between the nearest known corners (A and B) on the meridional line, north and south of the missing corner, and upon that line a temporary point E, is placed at the proper proportionate distance. This point E determines the latitude of the lost corner. In setting point E, the distance BE is to the original record distance BX as the total retracement distance BA is to the original distance BA. Next the nearest corners (C and D) on the latitudinal line are connected and a temporary point F is placed on this line at the proper proportionate distance. This point F determines the departure or longitude of the lost corner. Then a line is run, east or west from the temporary point E and a line is run,

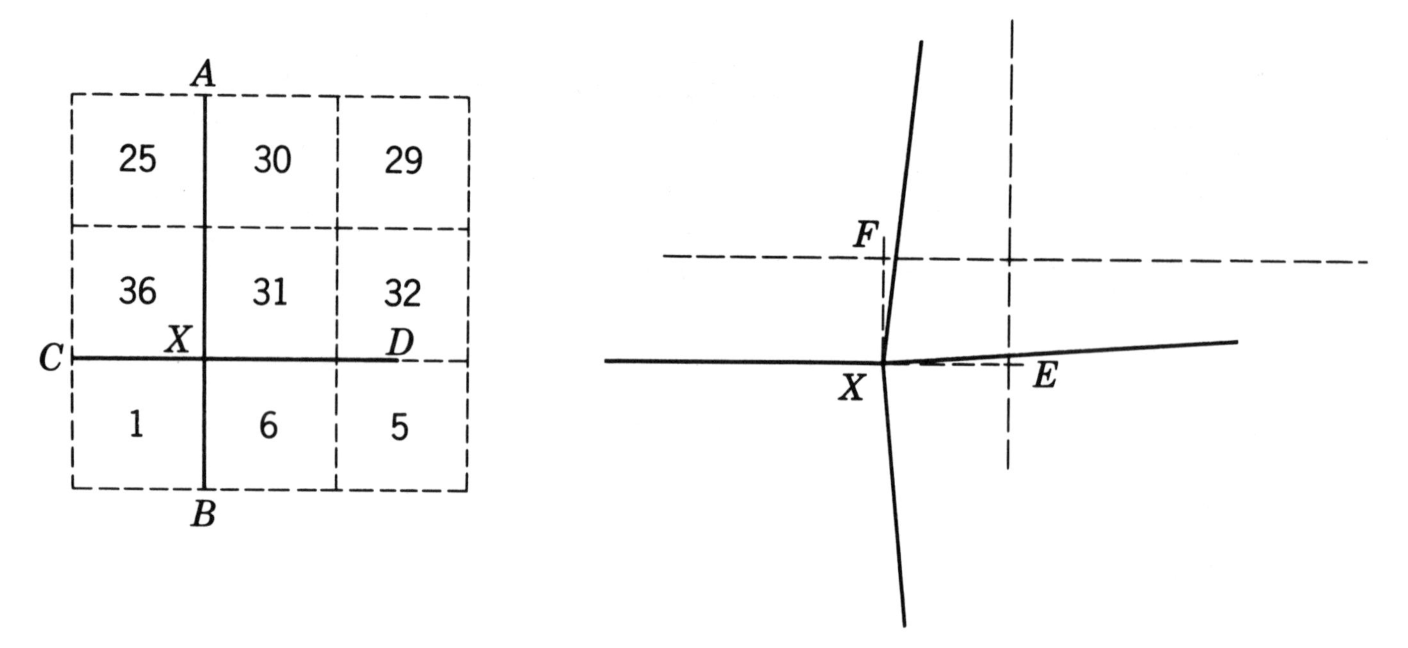

Figure 14-9. Lost corner by double proportional measurement.

north or south from the temporary point F. The intersection of these two lines at X fixes the position at which a monument is to be set to restore the lost corner.

b. Single Proportionate Measurement. The method of single proportionate measurement is generally applicable to the restoration of lost corners on standard parallels and other lines established with reference to definite alinement in one direction only. Intermediate corners on township exteriors and other controlling boundaries are also restored by this method. A retracement is made connecting the nearest identified regular corners on the line and on opposite sides of the missing corner. In running the retracement line, a temporary stake is set on the trial line of the original record distance for the missing corner. The total distance between recovered monuments is measured and the "falling" at the objective corner is determined. On meridional township lines an adjustment is made at the temporary point for the proportional distance along the line, and then the point is set over to the east or west a distance which bears the same relationship to the total "falling" at the objective corner as the distance of the missing corner from the point of beginning bears to the total length of the retracement line. On standard parallels and on latitudinal township lines three adjustments are necessary. The temporary point is first set forward or back the proprotional part of the difference between the record distance and the retracement measurement; it is then set over to place it on the true latitudinal curve, and lastly, a correction is made for the proportional part of the "falling" at the objective corner. The adjusted position is thus placed on the true line which connects the nearest identified corners, and at the same proportionate distance from each as existed in the original survey. Any number of intermediate lost corners may be restored during a single retracement survey by setting temporary points for each and making appropriate adjustment to the position of each such point. Lost meander corners, originally established on a line projected across the meandering body of water and monumented on both shores, are restored by this single proportionate measurement method, after the section or quarter-section corners upon opposite sides of the missing meander corner have been identified. Where these adjacent corners have been obliterated as well, but where there is evidence of the stability of the shore and no indication of erosion or accretion, the shore itself may be taken as an identified natural feature and the lost corner replaced with reference to the shore.

c. Restoring Closing Corners. A lost closing corner is reestablished on the true line that was originally closed upon, and at the correct proportionate distance between the nearest regular corners to the right and left. In restoring a lost closing corner on a standard parallel or other controlling line, this controlling line is retraced, beginning at the corner from which the line was originally run. A temporary point is set at the record distance, and the total distance and "falling" determined at the next recovered regular corner beyond the missing monument. The temporary point is then adjusted in position as in single proportionate method.

d. Restoring Broken Boundaries. In certain instances it is necessary to restore the angle points, within a section, of the meander courses for a stream or other body of water. The positions of the meander corners on the section boundaries are first recovered and restored. The record meander courses and distances are run, setting temporary angle points. If the end of the last course run fails to fall on the objective meander corner (fig. 14–10), the distance and bearing from the end of the last course to the closing meander corner is measured. Each temporary angle point is moved in a direction to reduce the closing error along a line having the same bearing as the closing error. The distance which each point is to be moved bears the same relationship to the total closing error as the distance of that point from the point of beginning (measured along the meander courses) bears to the sum of the lengths of all the courses.

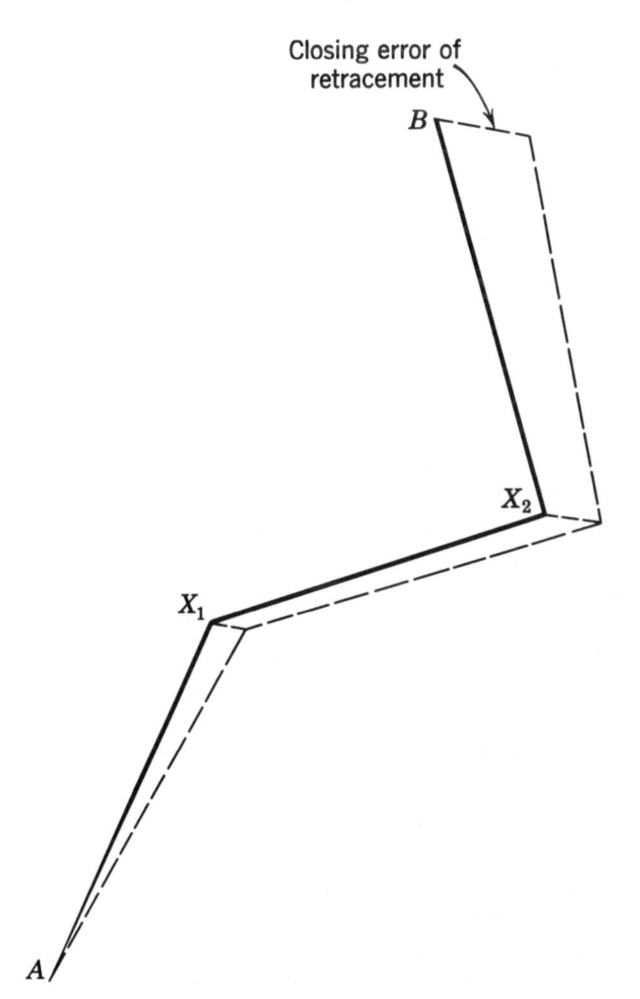

Figure 14-10. Retracement of broken boundary.

e. Special Cases. Much experience, pains-taking attention to detail and all valid corroborative evidence, and sound judgment are essential if the surveyor is to correctly retrace the boundaries of a tract on which many of the corner monuments have become obscured, obliterated, or lost, and when accessories have been destroyed. Through accumulated years of experience with situations of all types, the cadastral engineers of the Bureau of Land Management have developed competence with respect to resurveys of public lands beyond that which the surveyor can reasonably expect to acquire. When faced with such a situation, the regional offices of the Bureau of Land Management should be contacted for assistance.

14-20. Plats of Surveyed Lands

The official plat of a township or other subdivision is the drawing on which is shown the direction and length of each line surveyed, established, retraced, or resurveyed; the relationship to adjoining official surveys; the boundaries, designation, and area of each parcel of land; and insofar as practicable, a delineation of the topography of the area and a representation of the culture and works of man within the survey limits. A subdivision of public lands is not deemed to have been surveyed or identified until the notes of the field survey have been approved, a plat prepared, the survey accepted by the Director of the Bureau of Land Management as evidenced by a certification to that effect on the plat, and the plat has been filed in the district land office. The original drawing shows both a graphic scale and a representative fraction for both the township as a whole and for the enlarged diagram. Plats are drawn on sheets of uniform size, 19 x 24 inches, in trimmed dimensions, for convenience in filing. The usual scale is 1 inch equals 40 chains, equivalent to a representative fraction of 1:31,680. Where detail drawings of a portion of the survey area are required, scales of 1 inch equals 20 chains or 1 inch equals 10 chains may be used. A detail of a small area may be shown as an inset on the main plat. Larger details are drawn on separate sheets. When the drawing is simple, with few topographic or hydrographic features or works of man to be shown, the entire drawing is in black ink. When the features other than survey lines are quite extensive, color printing is used.

Section IV. CONSTRUCTION OF MONUMENTS

14-21. Iron Posts and Stone Monuments

Both iron posts and stone monuments will always be set a depth of three-fourths their length unless it is impossible to complete the excavation, in which case the monument will be planted as deep as conditions will permit, and the necessary support will be secured by a stone mound.

a. The caps of the iron posts are marked with steel dies at the time when used. The posts are set in the ground about three-fourths of their length, and earth and stone, if the latter is at hand, are tamped into the excavation to give the post a solid anchorage (fig. 14-11).

b. A stone monument is marked with a steel chisel or punch with such letters, figures, grooves, or notches as may be required, and is firmly set in the ground about three-fourths of its length.

In loose, wind-blown soil, the monument is much more stable if surrounded with stone, the mound being built with a wide base, and up to the height of the post. This will be even more secure if clay soil can be procured for filling the voids. The location may be of enough importance to justify the construction of a concrete base surrounding the iron post to prevent the blowing of the soil (fig. 14-11).

Underpinning or other special means for supporting the iron post may be required when constructing a stable monument in marshland. Encasement of the post in concrete to prevent corrosion may be necessary in the case of alkaline soils, salts, salt-water marshes, organic-acid water in swamp areas, or similar situations.

14-22. Corners

Where the corner point falls upon surface rock, preventing excavation, a cross (X) is cut at the exact corner point, and, if feasible the monument is erected in the same position, supported by a large mound of stone with broad base, so well constructed that it will possess thorough stability.

a. The tablet is used for marking corners which fall upon rock outcrops on slopes where a stable mound would be impracticable. A drill hole is made to receive the stem, and a recess is made for the top so that the tablet may be securely

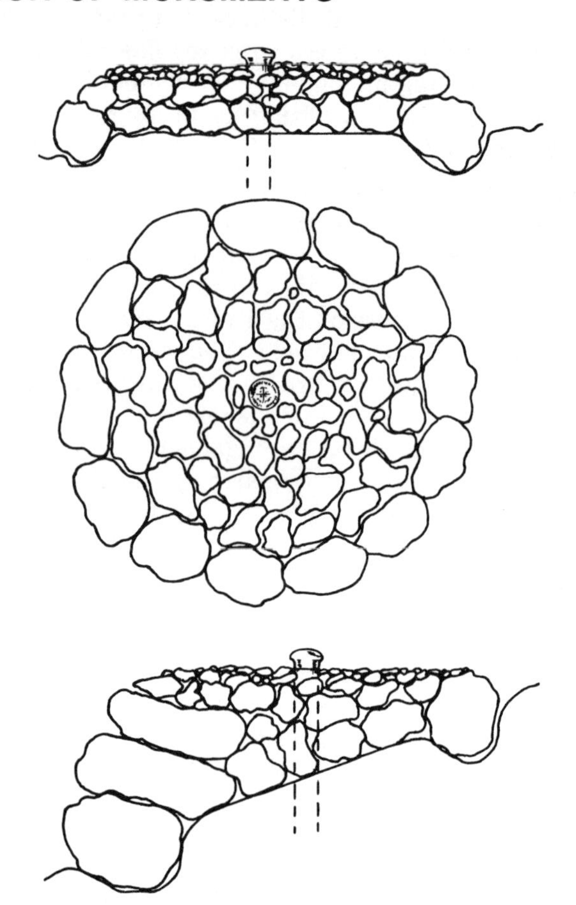

Figure 14-11. A mound of stone. Base should not be less than 3 ft. in diameter.

cemented in place and sealed against moisture. To be permanent the cementing must be done with clean first-class materials, carefully proportioned. The tablet is marked in the same manner as the iron-post monument.

On slopes too steep to allow construction of a stable mound, but covered with a thin layer of soil which would hide the tablet if it were set directly in the bedrock, a short length of zinc-coated iron pipe of 1-inch outside diameter is forced into the drill hole. The stem of the tablet is then wedged securely into the top of the pipe in a position above the layer of soil.

b. Where the corner point falls at the position occupied by a tree, the tree is marked as the corner monument, even if fully matured. The full quota of accessories should be secured, including reference monuments.

A tree too small to receive the usual marks without injury is marked with an "X" only at breast height on the *south* side. The X should be scribed in the bark on smooth-barked trees. On rough-barked trees the X should be made by two axe cuts reaching just into the live wood tissue. It is advantageous to clear out nearby trees of similar size in order that the marked tree will not be smothered by faster growing uninjured trees.

c. Monuments marking corners that fall in cultivated fields or meadows are so constructed as to interfere with farming operations as little as possible and conform to the wishes of the owner insofar as practicable.

Generally an iron post, brass tablet in a concrete block, a marked stone, or some suitable article may be buried at the corner point and witnessed by a substantial guard post. Bearing objects or bearing trees within a reasonable distance are employed, and accurate bearings to one or more available distant objects are recorded.

Section V. CORNER ACCESSORIES

14–23. Locating Corners

a. Accessories. The purpose of an accessory is to evidence the position of the corner monument. A connection is made from the corner monument to fixed natural or artificial objects in its immediate vicinity, whereby the corner may be relocated from the accessory. Thus, if the monument is destroyed or removed, its position may be identified by any remaining evidence of the accessories. One or more kinds of accessory are employed at each corner established in the public-land surveys (except for corners of minor subdivisions and where specifically not required by the Manual, or omitted by the special instructions).

Accessories consist of (1) bearing trees or other natural objects such as notable cliffs and boulders, permanent improvements, reference monuments; (2) mounds of stone; or (3) pits and memorials. Aside from availability, selection is based on their order of permanence.

b. The surveyor cannot perform any more important service than that of establishing permanent and accurate evidence of the location of the corners in his survey. Where the accessories cannot be employed, other means should be adopted that will best serve the purpose.

14–24. Bearing Trees and Bearing Objects

Bearing trees are selected for marking when available, ordinarily within a distance of 3 chains of the corner; a greater distance if important. One tree is marked in each section unless a tree in one or more positions may not be available. A full description of each bearing tree is given in the field notes. This includes the species of each tree, its diameter at breast height, the exact direction from the monument, the horizontal distance counting to the center of the tree at its root crown; and, the exact marks scribed for the identification of the corner.

Almost any nearby natural object that can be readily identified should be recorded by description, course, and distance. Such objects may not be of a character that can be marked, excepting in the case of a rock cliff or boulder. These are supplemental to the marking of bearing trees, or to fill out a quota where trees are not available in some sections. The description of the cliff or boulder should provide ready identification, including the marking of a cross (X) plainly and deeply chiseled at the exact point to which the direction and distance are recorded.

Another desirable accessory, especially where the usual types are not available nor suitable on account of the site conditions such as at a corner that falls in cultivated land, is to record accurate bearings to two or more prominent landmarks.

a. The marks upon a bearing tree are made upon the side facing the monument, scribed in the manner already outlined for marking tree corner monuments. The marks embrace the information suggested in the schedule hereinafter given, with such letters and figures as may be appropriate for a particular corner, and will include the letters "BT". A tree will always be marked to agree with the section in which it stands, and will be marked in a vertical line reading downward, ending in the letters "BT" at the lower end of the blaze approximately 6 inches above the root crown.

Figure 14-12. Marking a bearing tree.

b. There is a great difference in the longevity of trees, and in their rate of decay; trees should therefore be selected, if possible, with a view to the length of their probable life, their soundness, favorable site conditions and size. Sound trees, not matured, of the most hardy species, favorably located, are preferred for marking. Trees 5 inches or less in diameter should not be selected for marking if larger trees are available, and it is generally better to avoid marking fully matured trees, especially those showing signs of decay. Trees 4 inches in diameter, or less, if no better trees are available, are marked with the letter "BT" only at the base, and an "X" at breast height, facing the monument. The species, size and exact position of the bearing trees are of vital importance, as this data will generally serve to identify a bearing tree without uncovering the marks, or even to identify two or more stumps after all evidence of the marks has disappeared (see Fig. 14–12).

c. Generally only one tree is marked in each section at a particular corner, but in certain instances two trees are required in a section. In such cases it is better to select trees of different species or of widely different size, direction or distance. If the trees are of the same species, in order that confusion may be avoided in the future identification of a remaining tree where the companion tree has disappeared, one is marked with an "X" only (and "BT" at the base).

d. A cross (X) and the letters "BO" are chiseled into a bearing object, if it is a rock cliff or boulder; the record should enable another surveyor to determine just where the marks will be found. The rock bearing object is the most permanent of all accessories; it is used wherever practicable, and within a distance of 5 chains.

e. A connection to any permanent artificial object or improvement may be included in this general class of corner accessories. The field notes should be explicit in describing such objects, and should indicate the exact point to which a connection is made, as "southwest corner of foundation of Smith's house," "center of Smith's well," "pipe of Smith's windmill," etc. No marks will be made upon private property without the consent of the owner.

14–25. Memorials

Where there is no tree or other bearing object, as above described, and where a mound of stone or pits are impracticable, a suitable memorial is deposited alongside the monument. A memorial may consist of any durable article which will serve to identify the location in case the monument is destroyed. Such articles as glassware, stoneware, a marked (X) stone, a charred stake, a quart of charcoal, or pieces of metal constitute a suitable memorial. A full description of such articles is embodied in the field notes wherever they are employed as a memorial. When replacing an old monument with a new one, such as substituting an iron post for an old marked stone, the old marker is preserved as a memorial.

14–26. Stone Mounds

a. Where native stone is available and the surface of the ground is favorable, a mound of stone is employed as an accessory to a corner monument, or to surround it, even though a full quota of trees or other bearing objects can be utilized. A mound of stone erected as a corner accessory should be built as stable as possible, should consist of not fewer than five stones, and should be not less than 2 feet base and 1½ feet high. Where the ground is suitable, the stone mound is improved by first digging a circular trench, 4 to 6 inches deep, for an outer ring, then placing the base of the larger stones in the trench. In stony ground the size of the mound is sufficiently increased to make it conspicuous. The position of the accessory mound is shown in the schedule following. The nearest point on its base should be about 6 inches distant from the monument. The field notes show the size and position of the mound.

b. Where it is necessary to support a monument in a stone mound, and if bearing trees or other objects are not available, a marked (X) stone or other memorial is deposited alongside the monument.

A stone mound accessory, in addition to the mound surrounding a monument, is built wherever this will aid materially in making the location conspicuous.

14–27. Pits

a. Where the full quota of trees or other bearing objects are unavailable for marking, the position of the monument is, under favorable conditions, evidenced by pits. No pits should be dug in a roadway, or where the ground is overflowed for any considerable period, or upon steep slopes, or where the earth will wash, or in loose or light soil, or where there is no native sod, or where suitable stone for a mound is at hand.

A firm soil covered with a healthy native sod is most favorable for a permanent pit. Under such conditions the pits will gradually fill with a material slightly different from the original soil, and a new species of vegetation will generally take the place of the native grass; these characteristics, under favorable conditions, make it possible to identify the original location of the pits after the lapse of many years.

b. All pits should be dug 18 inches square and 12 inches deep, with the nearest side 3 feet distant from the corner monument, oriented with a square side (and not a corner) towards the monument. The earth removed is scattered in such a way that it will not again fill the pits. A description of the pits is embodied in the field notes, and should include a statement of their size and position.

APPENDIX A

PRACTICAL PROBLEMS

A-1. Laying off Given Distances

In surveying, it is often necessary to lay off a given distance in the field. The normal procedure is horizontal taping. If corrections described in chapter 6 are not required, the methods described in paragraph 6–4 can be used. However, if corrections are necessary, the method described below is probably the simplest to understand.

a. Laying off a given distance involves the following:

(1) Measurement of a distance that, when corrected, will approximate the required distance.

(2) Correction of the measured distance to its true horizontal distance.

(3) Measurement of an additional distance, forward or backward, to establish the point at the true given distance from the starting point.

b. Assume that a horizontal distance of 137.59 meters is to be laid off along a given line. Measure along the line for 137.59 meters (para 6–4) and mark this point. Determine all the necessary corrections of this measured distance to obtain the true distance. If, for example, this true distance, after correction, is found to be 133.78, an additional distance of 3.81 meters must be laid off. Using the marked point as a starting point, this additional 3.81 meters is measured and a new point is marked. This new point is then 137.59 meters from the original starting point.

c. One of the most useful procedures for laying off a given distance is pacing. Pacing is the measurement of distance using the standard pace or stride; the stride being equal to two paces. The standard pace is 2.5 feet; the stride, 5 feet. For distances over 100 feet, it is not necessary to count the total number of paces for the entire distance. Rather, it is thought of as so many 40 paces plus the remainder as increments. For example, assume a distance of 357 feet (meters and yards are quickly converted into feet). The distance (357 feet) will take 40 paces three times, plus 20 paces, plus 3 more. With practice, distances can be paced to surprising accuracy. Pacing can be used to measure the size of a structure, to locate control monuments from the descriptions, to estimate distances during taping operations, to determine areas, and for many operations.

A-2. Extend a Straight Line

It is often necessary to extend a straight line marked by two points on the ground. One of three methods can be used depending on whether there are obstacles in the line ahead or not, and whether a small or a large obstacle is encountered.

a. No Obstacles. The method used to accomplish the extension is known as *double centering, double reversing*, or *reversing in azimuth and altitude.* For example, the straight line through the marked points A and B (①, fig. A–1) is extended to, and marked at point C by the following steps:

(1) Set up and center the instrument over B.

(2) Backsight on A.

(3) Plunge the telescope and mark the point C′.

(4) Reverse the instrument, and again backsight on A.

(5) Plunge the telescope and mark the point C″.

(6) Mark the point C halfway between C′ and C″. This is the required point on the line AB extended.

(7) If the instrument is in perfect adjustment (which is seldom, if ever, the case) points C′ and C″ will coincide with point C.

(8) For further extension, the instrument is moved to C and the procedure repeated to obtain D.

b. Beyond a Small Obstacle. When a tree or other small obstacle is encountered in the line of sight between two points, the solution below should be used.

(1) The transit or theodolite is set up at

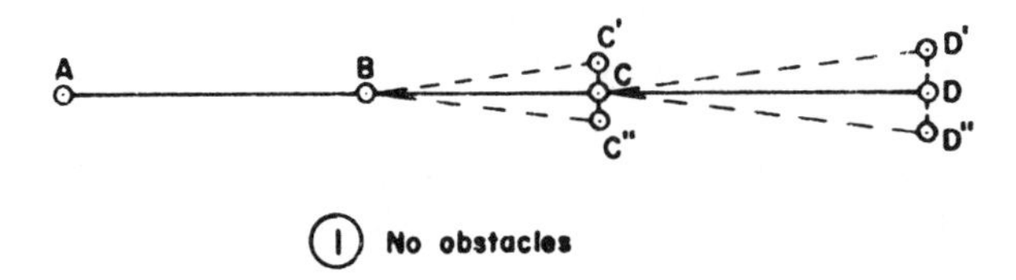

① No obstacles

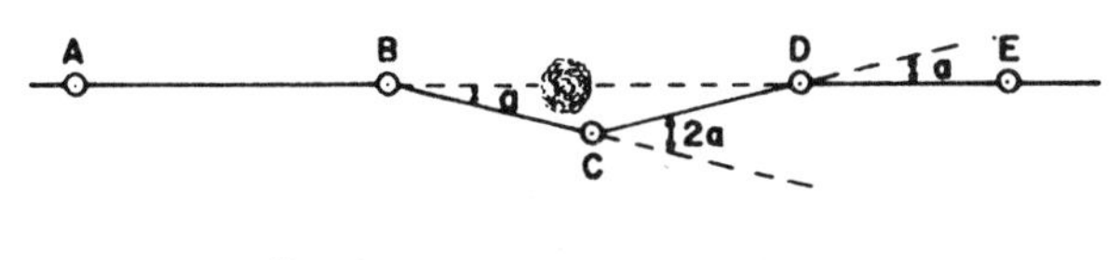

② Beyond a small obstacle

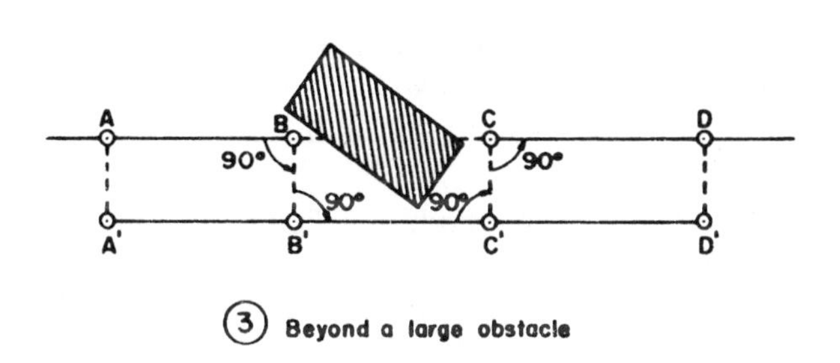

③ Beyond a large obstacle

Figure A-1. Extending a straight line.

point B, ②, figure A–1, as far from the obstacle as practical.

(2) Point C is set *off* the line, near the obstacle, and where the line BC will clear the obstacle.

(3) At B, measure the deflection angle a.

(4) Move the instrument to C, and lay off the deflection angle 2a.

(5) Measure the distance BC, and lay off the distance CD equal to BC.

(6) Move the instrument to D, and lay off the deflection angle a.

(7) Mark the point E. The line DE is the prolongation of the line AB.

c. Beyond a Large Obstacle. When a line is being run between two fixed points and a large obstacle, such as a building is encountered, there are numerous methods that can be used to extend the line beyond the obstacle. The commonly used method to solve this problem is shown in ③, figure A–1. The solution establishes a line parallel to the original line and at a distance to clear the obstruction. Once this parallel line passes the obstruction, another parallel line is established using the same distance value in the other direction. This second parallel line is the extension of the original line. There are two ways to establish the parallel lines. One measures angles and distances; the other measures only distances and uses the instrument to extend the lines.

(1) The instrument is set up at B (③, fig. A–1), and a 90° angle turned from line AB. The distance BB′ is carefully measured and recorded. The instrument is moved to B′ and another 90° angle turned. B′C′ is laid off to clear the obstacle. The instrument is moved to C′ and a third 90° angle turned. Distance CC′ equal to BB′ is measured and marked. This establishes a point C, on the original line. The instrument is moved to C and a fourth 90° angle turned to establish the alinement CD which is the extension of AB beyond the obstacle.

(2) When the obstacle-clearing distance, BB′ or CC′, is less than a tape length, the turning of four 90° angles can be avoided. An easily readable distance is selected on the tape to clear the obstacle. The instrument marks the point B and then is moved approximately to C′. The selected distance is swung from A and from B while the instrumentman moves the instrument until it is exactly on the same line with the perpendicular offset points A′ and B′. The point under the instrument is C′. Point D′ is then marked as an extension of A′ B′ C′. Perpendicular offsets are again swung from C′ and D′, and the instrument is moved to either point. CD is the extension of AB.

A–3. Laying Off a Horizontal Angle

When an angle value is given and the requirement is to lay off this value from a given line, several methods may be used. Some are dependent on whether the angle-measuring instrument is a repeating type, such as a transit, or a direction type, such as a theodolite. The next consideration is the accuracy to which the angle must be laid out. It is also possible to lay out angles using the tape alone.

a. With a transit and the accuracy requirement within the instrument's reading capability, the method is as follows.:

(1) By use of the upper motion, set the A vernier to read zero.

(2) By use of the lower motion, take a backsight on the target identifying the given line.

(3) By use of the upper motion, set the A vernier to the value of the given angle.

(4) Set a marker on the line indicated by the line of sight of the telescope.

b. A horizontal angle can be laid off with a directional theodolite in the same manner as with a transit except for the manipulations of the instrument.

(1) Backsight on the target.

(2) Set the horizontal circle to read 0°00′ 05″ ± 05″.

(3) Check the backsight and determine the circle reading for this pointing.

(4) To this circle reading, add the value of the angle to be laid off, obtaining the circle reading for pointing along the line to be established.

(5) Turn the instrument to this reading and mark the point on the new line.

c. The angles laid off by the methods above, are of sufficient accuracy for many purposes. When greater accuracy is necessary, the method below should be used.

(1) Lay off the given angle as close as possible by either method explained above, thus setting point C′ (①, fig. A–2).

(2) By the method explained in paragraph 9–10, measure the angle laid off (angle ABC′).

(3) Determine the difference, Δa, (in seconds of arc), between the values of the angle as measured and the angle required to be laid off (Δa = ABC — ABC′).

(4) Measure the horizontal distance (BC = 1).

(5) Compute d (= C′C) by the following equation: d= 0.000004851Δa.

(6) From C′, measure the distance d, and set point C.

(7) As a check, measure the angle ABC by the same method as in (2) above.

d. In this method of laying off an angle, it is not necessary to measure the horizontal distance BC′. This permits higher precision in the measurement.

(1) Lay off the method explained in *a*, or *b* above, thus setting point C′ (②, fig. A–2).

(2) By the method explained in paragraph 9–10, measure the angle ABC′.

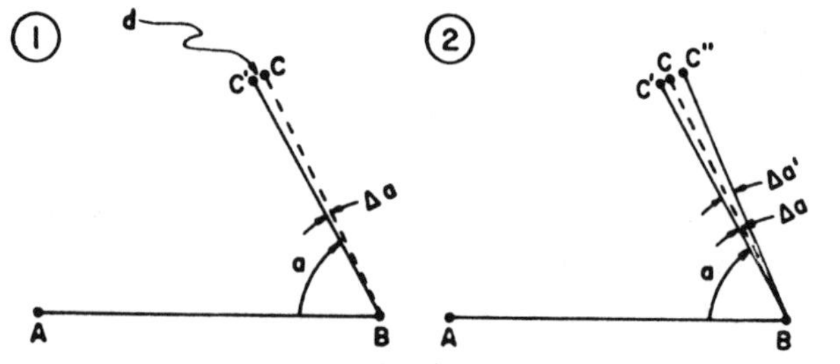

Figure A–2. Laying off an angle with an instrument.

(3) Determine the difference, Δa, between the values of the measured angle and the required angle (Δa = ABC — ABC′).

(4) Mark a second point C″ near the point C′ so that the line BC (forming the required angle with line AB) will pass between C′ and C″.

(5) Measure the angle Δa′ = C′BC″, by method used in (2) above.

(6) Measure the comparatively short horizontal distance C′C″ (=D).

(7) Compute d (= C′C) by the following equation: d = D.

(8) On the line C′C″, mark point C at the distance d from C′.

(9) To check the position of C, measure the horizontal angle ABC.

e. The methods described above depend upon the use of a transit or theodolite to measure the angle. It is possible to lay off or measure an angle using only the tape.

(1) Assume that angle β (①, fig. A–3) is to be laid off. Measure along line AB an easily readable distance (50 or 100 feet or meters) and mark point C. In the trigonometric tables, find the tangent of the required angle. Establish a perpendicular at C in the direction of the required angle. Along this perpendicular, measure the amount resulting by multiplying the measured distance (BC) by the tangent of angle β, and mark point D. The required angle B is equal to ABD.

(2) If the angle (a, ②, fig. A–3) is already established on the ground and its value must be determined, the procedure still uses trigonometry as above. Let the angle BAC, be the given angle, a. From A, measure a distance of 100 units (feet or meters) first along the line AB and then along the line AC and mark points B and C. Measure the distance BC. Determine the size of the angle by using the following equation:

$$\sin \tfrac{1}{2} a = \frac{BC/2}{100} = \frac{BC}{200}$$

f. To establish a perpendicular using the tape, or in effect to lay out a 90° angle, a special trigonometric or geometric principle is used.

(1) In any right triangle whose three sides are in the proportion of three, four, five; the angle opposite the longest side is a 90° angle. Then, with two points spaced, for example; 40 feet apart, a point 30 feet from one of these points and 50 feet from the other point (③, fig. A–3) will form a triangle having a 90° angle opposite the 50-foot side. To establish the perpendicular to line AB at a selected point (A), along AB

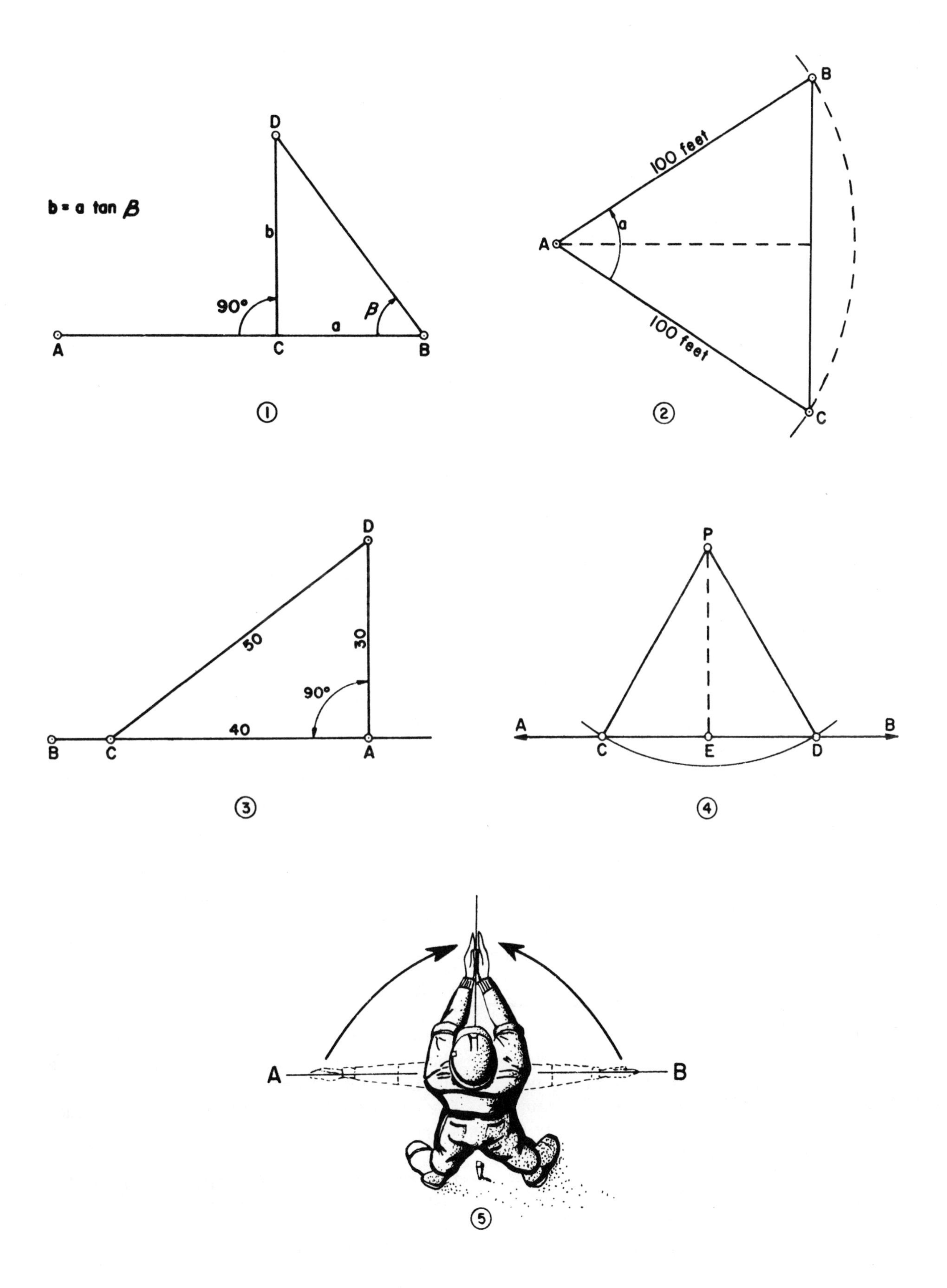

Figure A-3. Laying off angles with a tape.

measure a distance of 40 units (feet or meters) from A and mark point C (AC = 40 units). From A, swing an arc using a distance equal to 30 units, and from C, a distance equal to 50 units. These two arcs will intersect at point D. AD is the perpendicular to line AB at point A.

(2) If the point (P, ④, fig. A–3) through which the perpendicular must pass is not on the line (AB), a geometric construction is used. With a tape length slightly greater than the perpendicular distance from P to the line AB, and holding the tape at P, mark two points C and D on the line AB so that PC and PD are equal distances. Measure the distance between points C and D and divide by two. Establish point E exactly midway between C and D and on line AB. E is the foot of the perpendicular (PE) from P to the line AB.

(3) For cross-section leveling, strip topography, and some other purposes, it is necessary to lay off a 90° angle at numerous points along a line. This 90° angle can often be established by estimation with sufficient accuracy for the particular job. The surveyor straddles the point on the line, arms extended sideward along the marked line (⑤, fig. A–3). By looking alternately right then left, he adjusts the position of his feet until his body is in the line (AB). He then brings his hands together in front of him, thus pointing along a line approximately 90° from the marked line. An experienced operator can lay off a 90° angle by this method so that a point 100 feet away will be within 1 foot of the true perpendicular.

A–4. Setting Up On a Line Between TWO POINTS

A problem often encountered in surveying is to find a point exactly on the line between two other points when neither can be occupied, or when an obstruction such as a hill lies between the two points. The point to be occupied must be located so that both of the other points are visible from it. The process of establishing the intermediate point is known as *wiggling-in* or *ranging-in*.

a. The approximate position of the line between the two points, at the instrument station is first estimated by using two range poles. The range poles are lined in alternately in the following manner. In ①, figure A–4, range pole 1 is set and range pole 2 is moved until it is exactly on line between pole 1 and point A. This is done by sighting along the edge of pole 1 at the station A until pole 2 seems to be on line. Range pole 2 is set and pole 1 is moved until it is on line between pole 2 and point C. Now pole 2 is moved into line again

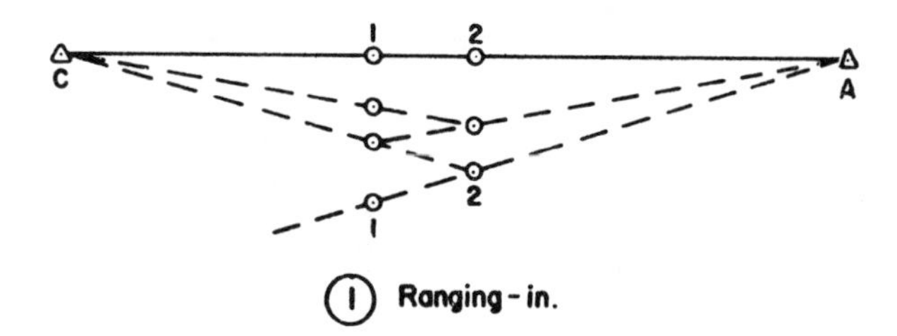

① Ranging-in.

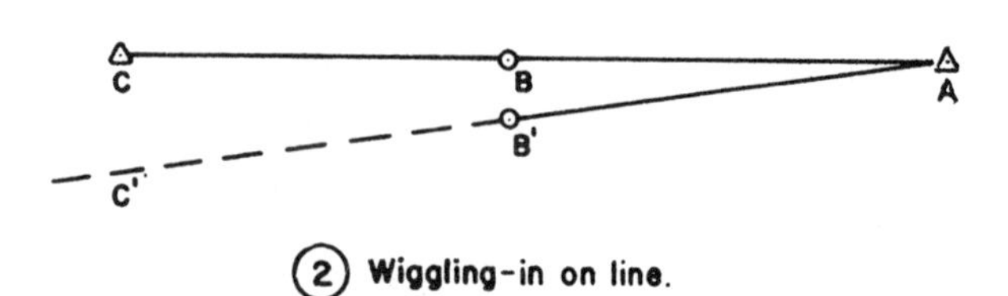

② Wiggling-in on line.

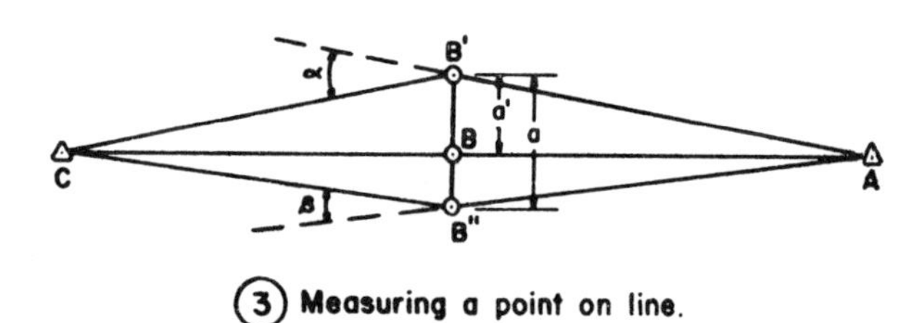

③ Measuring a point on line.

Figure A–4. Setting up on a line between two points.

and then pole 1 alternately until both are on line AC. The line will appear to pass through both poles and both stations from either viewing position.

b. After finding the approximate position of the line between the two points, set up the instrument on this line. The instrument probably will not be exactly on line, but will be over a point such as B′, ②, figure A–4. With the instrument at B′, backsight on A, and plunge the telescope and notice where the line of sight (C′) passes the point C. Estimate this distance (CC′), and also the distance that B′ would be away from C and A. Estimate the amount to move the instrument to place it on the required line. Thus if B′ is midway between A and C, and C′ misses C by about 1 meter to the left, B′ must be moved about one-half meter to the right to reach B. Continue the sequence of backsighting, plunging the telescope, and moving the instrument until the line of sight passes through both A and C. During this procedure, the telescope is reversed but the instrument is not rotated, that is, if the telescope is reversed for backsighting on A, all sightings on A are made with telescope reversed. Mark a point on the ground directly under the instrument. Then, repeat the procedure with the telescope direct for each backsight on A. Mark a second point on the ground. The required point on the line AC is then the midpoint between the two marked points.

c. The procedure outlined above is usually time consuming. Even though the shifting head of the instrument is used in the final instrument movements, the instrument may have to be picked up and moved several times. The following procedure often saves time. After finding the approximate position of the line between the two points, two points (B′ and B″, ③, fig. A–4) are marked 1 or 2 feet apart where they are known to straddle the line AC. Set up over each of these two points in turn, and measure the deflection angles α and β. Also measure the horizontal distance a, between points B′ and B″. Then the position (B) on the line AC can be found by using the following equation:

$$a' = a \frac{\alpha}{\alpha + \beta}$$

in which a′ is the proportionate offset distance from B′ toward B″ for the required point B, and α and β are both expressed in minutes, or in seconds.

A-5. Locating The Point of Intersection of Two Lines

In many surveys it is necessary to find the point of intersection between two lines, for example, the intersection between a traverse line and a property-boundary line. Figure A–5 illustrates the determination and marking of the point of intersection between the line AB and the line CD. Both of these lines are run out and marked close to the point of intersection. The transit is then set up on the line CD at the point D, and two stakes E and F are set about 1 to 2 meters apart, on the line CD, straddling the extension of line AB. A string is stretched between E and F. The transit is moved to a set up at B and a stake (G) is set on the line AB where it intersects the line between E and F. The exact point of intersection can be marked on the top of the stake by running a pencil along the string until the instrumentman signals stop.

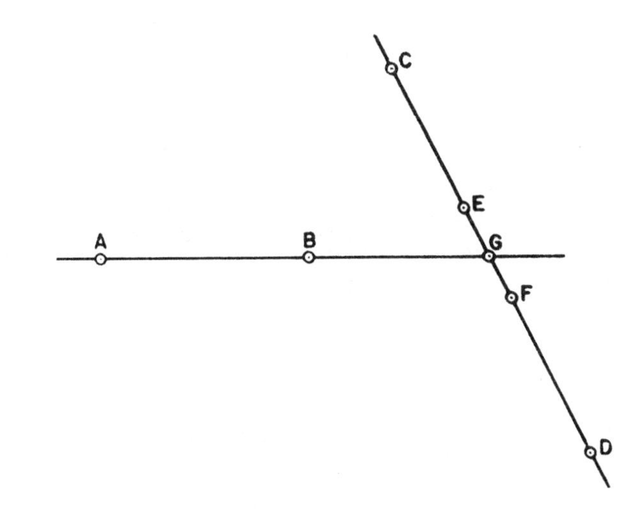

Figure A-5. Intersection of two lines.

A-6. Setting a Point Near a Setup

Sometimes it is necessary to set a point too close to an occupied station to permit focusing the telescope, or even to point the telescope. In this case, a point is set on the required line about 8 to 10 feet away from the instrument. A string is then stretched from the occupied station to the set point, and the required point is marked on this line along the string.

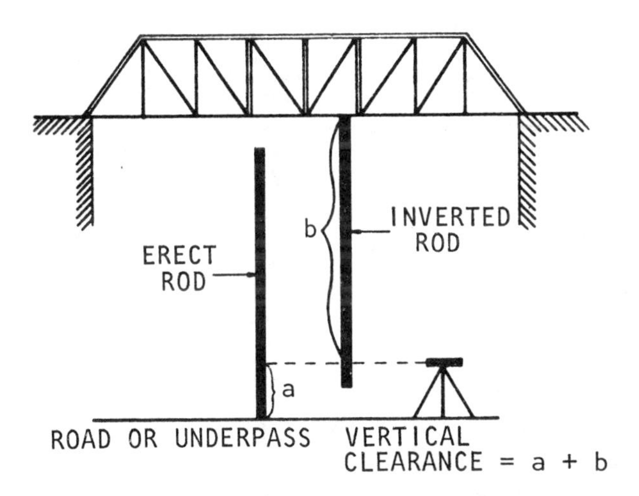

Figure A-6. Measuring vertical clearance.

A-7. Measuring a Vertical Clearance

One practical problem which the surveyor will meet at times, is to measure a vertical clearance of an underpass or a bridge. The ground elevation beneath is determined by normal leveling procedures (ch. 8). The elevation of the bottom of the bridge stringers is determined by holding an inverted rod against the bottom of the overhead structure as shown in figure A–6. The erect rod reading plus the inverted rod reading (a + b) equals the vertical clearance between the two surfaces.

A-8. Typical Field Problems

Figure A–7 illustrates a series of typical field problems in a graphical form using some of the techniques described above.

SITUATION 1: To project line AB beyond obstacle C

PRINCIPLE	DIAGRAM	COMMENT
(a) Two right angles and two swing offsets	(plan view)	Measure DE= GF= Swing A= Swing H. Sights to Swing A and Swing H must be 100 ft minimum
(b) Four right angles	(plan view)	Measure DE = GF
(c) Equilateral triangle	(plan view)	Measure DE = EF
(d) Isosceles triangle	(plan view)	Measure DE = EF and angle GEF = 360° - 2k
(e) One right angle and one convenient angle	(plan view)	Measure EF = DE ÷ cosine angle e. Lay off angle k = 90° + angle e
(f) Pass line over obstacle	(elevation view)	Use available high ground
(g) Pass line over obstacle	(elevation view)	Mount instrument on obstacle

SITUATION 2: To determine line between two nonintervisible points A and B

PRINCIPLE	DIAGRAM	COMMENT
(a) Wiggle in one instrument to a point on line	(elevation view)	Select intermediate point C from which A and B are visible
(b) Line in two intermediate points	(elevation view) (plan view)	Use two instruments and two rodmen Instrument A sets trial line ACD Instrument B sights C and lines in D' Instrument A sights D' and lines in C' Continue process until both rods C and D line in from instruments A and B

SITUATION 3: To replace obstructed line AB with an equivalent usable line

PRINCIPLE	DIAGRAM	COMMENT
(a) Parallel offset	(plan view)	Estimate D opposite B, sight on Swing offset from B, and measure Swing A. If Swing A is large, move C'. If Swing A is small, add to offset measurement
(b) Random line	(plan view)	Set off convenient line AC, measure angle ACB, and project to obstructed line by similar triangles
(c) Random traverse	(plan view)	Run traverse and reduce lengths and directions by trigonometry

Figure A-7. Typical field problems.

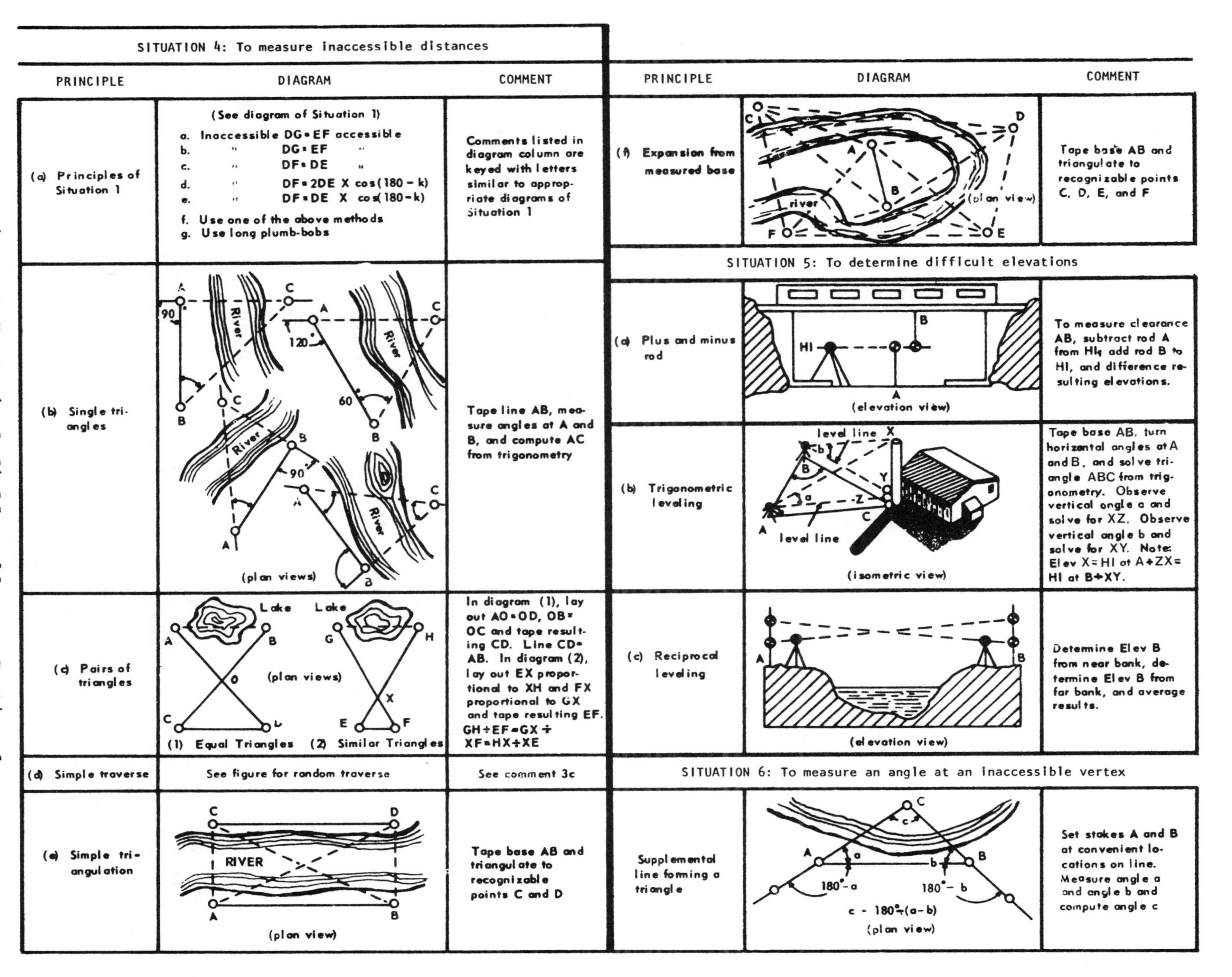

SITUATION 4: To measure inaccessible distances

PRINCIPLE	DIAGRAM	COMMENT
(a) Principles of Situation 1	(See diagram of Situation 1) a. Inaccessible DG=EF accessible b. " DG=EF " c. " DF=DE " d. " DF=2DE X cos(180 − k) e. " DF=DE X cos(180−k) f. Use one of the above methods g. Use long plumb-bobs	Comments listed in diagram column are keyed with letters similar to appropriate diagrams of Situation 1
(b) Single triangles	(plan views)	Tape line AB, measure angles at A and B, and compute AC from trigonometry
(c) Pairs of triangles	(1) Equal Triangles (2) Similar Triangles (plan views)	In diagram (1), lay out AO=OD, OB=OC and tape resulting CD. Line CD=AB. In diagram (2), lay out EX proportional to XH and FX proportional to GX and tape resulting EF. GH÷EF=GX÷XF=HX+XE
(d) Simple traverse	See figure for random traverse	See comment 3c
(e) Simple triangulation	(plan view)	Tape base AB and triangulate to recognizable points C and D

PRINCIPLE	DIAGRAM	COMMENT
(f) Expansion from measured base	(plan view)	Tape base AB and triangulate to recognizable points C, D, E, and F

SITUATION 5: To determine difficult elevations

PRINCIPLE	DIAGRAM	COMMENT
(a) Plus and minus rod	(elevation view)	To measure clearance AB, subtract rod A from HI, add rod B to HI, and difference resulting elevations.
(b) Trigonometric leveling	(isometric view)	Tape base AB, turn horizontal angles at A and B, and solve triangle ABC from trigonometry. Observe vertical angle a and solve for XZ. Observe vertical angle b and solve for XY. Note: Elev X=HI at A+ZX=HI at B+XY.
(c) Reciprocal leveling	(elevation view)	Determine Elev B from near bank, determine Elev B from far bank, and average results.

SITUATION 6: To measure an angle at an inaccessible vertex

PRINCIPLE	DIAGRAM	COMMENT
Supplemental line forming a triangle	c = 180°−(a−b) (plan view)	Set stakes A and B at convenient locations on line. Measure angle a and angle b and compute angle c

Figure A-7. Typical field problems - Continued.

APPENDIX B

PRECISION: ERRORS, ACCURACY, AND PROBABILITY

B–1. Errors in Measurements

Precision is the degree of refinement in the performance of an operation or in the statement of a result. Precision relates to the quality of execution whereas accuracy relates to the quality of the result. *Accuracy* is the degree of conformity with a standard. Every measurement the surveyor makes will be in error to the extent that no measurement is exact. The surveyor's mission is to accomplish the survey by taking measurements, the errors of which are within prescribed limits. To accomplish this mission, it is essential for the surveyor to know what errors will occur, their effect, and their evaluation. The magnitude of errors in any measured value depends on the instruments used, the method used, the number and type of observations made, and the competence of the personnel. Mistakes or blunders are quite distinct from errors. They result from a misunderstanding of the problem, poor judgment, confusion in the mind of the observer, or are indications of incompetence. Mistakes are detected and avoided by the systematic checking of all work. The necessary checks to prevent mistakes are discussed in the manual under the descriptions of the operations. Only errors, not mistakes, are included in this appendix.

B–2. Sources of Errors

Errors may come from one or more of three sources:

a. Instrumental. Instrumental errors arise from imperfections in the instrument. For example, the circle graduations on a theodolite may not be spaced exactly the same amount around the full circle. The effect of most instrumental errors can be minimized by proper field observing procedure or by corrections mathematically computed and applied.

b. Personal. Personal errors arise from the limitations of the human senses, principally sight, touch, and hearing. For example, when the vertical crosshair of the transit's telescope is not perfectly centered on the target, a small error results in the measured value of an angle.

c. Natural or Environmental. Natural errors arise from variations in temperature, humidity, wind, gravity, refraction, and magnetic declination. For example, side wind will cause an apparent shortening of the tape.

B–3. Types of Errors

Each error, regardless of source is classified as to type.

a. Systematic Errors. Systematic errors occur from well-understood causes and can be reduced by taking proper precautions to eliminate them. For example, the length of a steel tape varies between the 0 and 50 meter graduations. Its length was calibrated as 49.99, but it actually is 49.98 meters. Each tape length would be 0.01 meter too long. Thus, a systematic error of 1:5000 is introduced in each measured course. Systematic errors can be greatly reduced by methods of observation, and by eliminating calibration and adjustment errors in the equipment.

b. Accidental Errors. Accidental errors are caused by a combination of factors beyond the control of the observor. For example, when determining a horizontal distance with a steel tape, small uncontrollable errors will result from applying tension to the tape, determining the temperature of the tape, measuring the slope of the tape, or transferring the positions of the end graduations to the ground. The magnitude and algebraic sign of these errors are matters of chance. A compensating effect results when as many accidental errors are positive in algebraic sign as are negative. Accidental errors remain after mistakes have been eliminated, and systematic errors have been reduced. The theory of probability (para B–5 below) is based on a study of the occurrence of accidental errors. Accidental errors are also called *random errors, irregular errors,* and *erratic errors.*

c. Comparison Between Systematic and Accidental Errors. The relative importance of systematic and accidental errors depends on the purpose of the survey. When the purpose tolerates low accuracy measurements, all errors are often treated as accidental errors, and no effort is made to correct for the systematic errors. If the required accuracy is increased, a corresponding increase in effort must be made to determine the values and correct the effects of systematic errors. In a survey requiring high accuracy measurements, after the effects of all systematic errors have been corrected, the probable value of the accidental errors must be determined.

B-4. Discrepancy, Resultant Error, and Residual Error

Besides the adjectives used to describe the source and type of error, the surveyor will hear errors referred to as residual or resultant, or as discrepancies. These terms do not define the error, but are more or less comparative.

a. Discrepancy. Discrepancy is the difference between two values for the same quantity, either between a measured and a known value for a quantity or between two measured values. A small discrepancy indicates that no mistakes were made and that accidental errors were small, but it does not indicate the magnitude of the systematic errors. For example, two measurements of a line 10 kilometers long may show a discrepancy of only 1 meter, but systematic errors of 5 meters, caused by slope and temperature of the tape, may still be common to both measurements.

b. Resultant Error. The resultant error in any measurement is the difference between the measured value and the true value for a quantity. Since the true value of a quantity can never be known, the value of the resultant error can never be known. However, its definition is given to establish a plane of reference for further discussion.

c. Residual Error. A residual error is the difference between the value of a quantity obtained by a single measurement and the mean value of that quantity obtained by a number of measurements. The residual error most nearly approximates the resultant error, since the greater the number of measurements, the closer is the approach to the true value. The residual errors for a series of measured values are used in determining probable error.

B-5. Probability

The most probable value is determined from a series of measurements each of which most nearly approximates the true value. Accidental errors cannot be eliminated because they exist in every observation of a quantity. After the elimination of the effect of mistakes and systematic errors, the evaluation of the accuracy of the measurements is based on the theory of probability.

a. Theory of Probability. The theory of probability deals with the accidental errors in a series of like or related measurements. It assumes that—

(1) Small accidental errors occur more frequently than large ones.

(2) Accidental errors are as likely to be positive as negative.

(3) Very large accidental errors do not occur.

(4) The probable number of uncompensated accidental errors is equal to the square root of the number of opportunities for these errors to occur.

(5) The mean of an infinitely large number of like observations would be a true value.

b. Most Probable Value. The most probable value for a quantity is mathematically determined as described below:

(1) For a series of measurements of a single quantity, made under similar conditions, the most probable value is the arithmetical mean of the measured values.

(2) When there is a discrepancy between the sum of several measured quantities and a known value (that the sum should equal) and each has been measured with the same degree of precision, the most probable value is obtained by distributing the discrepancy equally among the measured values. For example, if the sum of the measured values for the three angles of a triangle is 180° minus 12 seconds, the most probable value is equal to the measured value plus 4 seconds of arc for each of the angles.

(3) When the sum of the measured values fails to equal their measured sum, and each has been measured with the same degree of precision, the most probable value is obtained by distributing the discrepancy equally among all the measred values *including their measured sum.* If the adjustment is subtracted from each of the several measurements that were added together, it must be added to the measurement representing their sum, and vice versa. For example, three angles totaling less than 360° are measured at a point, and the angle representing the sum of the three

angles is also measured. If the sum of the measured values of the three angles is 20 seconds greater than the measured value representing the sum, 5 seconds are subtracted from each of the three component angles, and 5 seconds are added to the measured value representing the sum of the three angles.

(4) When a different degree of precision has been used to measure the values which, combined, equal a measured sum, the most probable value for each is determined by an adjustment of weighted observations.

B–6. Adjustment of Weighted Observations

The *weight* of an observation is the *relative value* of the observation when compared with other observations of the same or related quantities. The value measured by the most reliable method is always assigned the greatest weight. The most probable value is then determined by using the weights.

a. Theory. When distances between points along a line are measured, the full length of the line is also measured. Or, when adjacent angles about a point are measured, the total angle is also measured. It is evident that any adjustment to the measured value for the full length of line or arc should be greater than that applied to any portion. The weights assigned to the measured values are inversely proportional to their magnitude.

b. Assignment of Weights. Factors influencing the surveyor's decision include method of measurement, terrain, and weather conditions. Weights may be assigned to observed values by any of the following methods:

(1) *To horizontal angles.*

(*a*) Directly proportional to the number of measurements in each set of readings for a single angle, and directly proportional to the number of sets of readings for each angle in a series of angles.

(*b*) Inversely proportional to the squares of the probable errors.

(*c*) As designated by the surveyor.

(2) *To horizontal distances.*

(*a*) Directly proportional to the number of times each distance is measured, and inversely proportional to each distance in a series of distances.

(*b*) Inversely proportional to the squares of the probable errors.

(*c*) As designated by the surveyor.

(3) *To elevations.*

(*a*) Directly proportional to the number of runs over a route, and inversely proportional to the length of each route in a series of routes.

(*b*) Inversely proportional to the squares of the probable errors.

(*c*) As designated by the surveyor.

c. Use of Weights. In the determination of most probable values, or the adjustments to be applied to measured values, weights are used as follows:

(1) The most probable value (weighted mean value) of a single quantity is the sum of the measured values, each multiplied by its weight, divided by the sum of the weights. For example, the value of an angle was determined to be 47°24′30″ by a series of four measurements, then it was redetermined to be 47°24′25″ by a series of six measurements. The weighted mean value is—

$$\frac{(47°24'30'' \times 4) + (47°24'25'' \times 6)}{4 + 6} = 47°24'27''$$

(2) Adjustments to the measured values of a series of related quantities are inversely proportional to the corresponding weights. When the adjustments are applied, the results are the most probable values of the several quantities. For example, the three angles, A, B, and C, of a triangle were measured at different times, angle A by 16 measurements, angle B by 12 measurements, and angle C by 8 measurements. The sum of these three angles was found to be 15 seconds of arc less than 180°. The adjustments to these three angles, to obtain their most probable values, are determined as follows: the weight for angle A equals 16, for angle B equals 12, and for angle C equals 8. As the adjustments are inversely proportional to the weights, these values are—for angle A, 1/16; for angle B, 1/12; for angle C, 1/8; and for the sum of the angles, 1/16 + 1/12 + 1/8 = 13/48 = 0.2708. The adjustments are—

$$A' = \frac{15}{0.2708} \times 1/16 = +3.5 \text{ seconds of arc}$$

$$B' = \frac{15}{0.2708} \times 1/12 = +4.6 \text{ seconds of arc}$$

$$C' = \frac{15}{0.2708} \times 1/8 = +6.9 \text{ seconds of arc}$$

The values A′, B′, and C′, when added to angles A, B, and C respectively, give the most probable value for each of these angles.

B–7. Probable Error

The probable error is the most probable value of the resultant error in the measurement. This is a plus-or-minus quantity that may be larger or smaller than the resultant error, and its probability of being larger is equal to its probability of being smaller.

a. In determining the value of the probable error, the residual error is computed for each measured value. The value of the probable error is then computed, by using the sum of the squares of the residuals as shown below.

b. The following equations are based on the assumption that a large number of observations of each quantity have been made. Experiments indicate that good results can be obtained from as few as 10 measurements. In practice, the equations are often applied to a series as small as two measurements, but the results of such computations are not reliable.

c. When each value in a series of measurements has the same weight—

(1) The probable error, E_m, in the value of the quantity is—

$$Em = 0.6745 \sqrt{\frac{\Sigma v^2}{n(n-1)}} = \frac{0.6745}{\sqrt{n(n-1)}}\sqrt{\Sigma v^2}$$

in which Σv^2 is the sum of the squares of the residuals, and n is the number of measurements of the quantity.

(2) The probable error, E, in each value of the series of measurement is—

$$E = 0.6745 \sqrt{\frac{\Sigma v^2}{n-1}} = \frac{0.6745}{\sqrt{n-1}}\sqrt{\Sigma v^2}$$

in which Σv^2 is the sum of the squares of the residuals, and n is the number of measurements of the quantity.

(3) When solving equation (1) or equation (2), the values of

$$\frac{0.6745}{\sqrt{n(n-1)}} \text{ and of } \frac{0.6745}{\sqrt{n-1}}$$

can be taken from table B–1 for any value of n between 2 and 35 inclusive.

Table B–1. Probable-Error Factors

n	$\frac{0.6745}{\sqrt{n-1}}$	$\frac{0.6745}{\sqrt{n(n-1)}}$
2	0.6745	0.4769
3	0.4769	0.2754
4	0.3894	0.1947
5	0.3372	0.1508
6	0.3016	0.1231
7	0.2754	0.1041
8	0.2549	0.0901
9	0.2385	0.0795
10	0.2248	0.0711
11	0.2133	0.0643
12	0.2034	0.0587
13	0.1947	0.0540
14	0.1871	0.0500
15	0.1803	0.0465
16	0.1742	0.0435
17	0.1686	0.0409
18	0.1636	0.0386
19	0.1590	0.0365
20	0.1547	0.0346
21	0.1508	0.0329
22	0.1472	0.0314
23	0.1438	0.0300
24	0.1406	0.0287
25	0.1377	0.0275
26	0.1349	0.0265
27	0.1323	0.0255
28	0.1298	0.0245
29	0.1275	0.0237
30	0.1252	0.0229
31	0.1231	0.0221
32	0.1211	0.0214
33	0.1192	0.0208
34	0.1174	0.0201
35	0.1157	0.0196

d. When the values in a series of measurements of a quantity have different weights, the probable error, Rm, in the weighted mean value is

$$Rm = 0.6745 \sqrt{\frac{\Sigma v^2 W}{(n-1)\Sigma W}}$$

in which Σv^2 W is the sum of the squares of the residuals each multiplied by its weight, n is the number of observations, and ΣW is the sum of the weights.

e. The probable error, Ea, in an area represented by the produce of two lengths, L_1 and L_2 having probable errors E_1 and E_2 respectively, is—

$$Ea = \sqrt{L_1^2\, E_2^2 + L_2^2\, E_1^2}$$

f. The probable error, Ep, in the produce of a constant, or known quantity, C, and a measured quantity for which the probable error is E, is—

$$Ep = CE$$

g. The probable error, Es, of the sum of several independent measurements for which the probable errors ar E_1, E_2, E_3 . . . E_n,

$$Es = \sqrt{E_1^2 + E_2^2 + E_3^2 + \ldots + E_n^2} = \sqrt{\Sigma E^2}$$

h. The probable error, E_d, in the difference between two independent measured values having probable errors E_1 and E_2 is—

$$E_d = \sqrt{E_1^2 + E_2^2}$$

B–8. Degree of Accuracy

The degree of accuracy is a *ratio* that expresses in general terms, the accuracy attained in survey measurements. *For example:* A degree of accuracy of 1 in 5,000 indicates that there are errors in the measurements amounting to one unit in each 5,000 units measured, or in the same proportion for a greater or lesser number of measured units.

a. Horizontal Distance Measurements. The degree of accuracy for a horizontal distance measurement is determined from the measured value and its probable error. For eaxmple: If a line was determined to be 493.68 feet long with a probable error of ±0.08 foot, the degree of accuracy is equal to one part in 493.68 divided by 0.08, or one part in 6171 parts. This is a plus-or-minus quantity, and is usually written: ±1:6171, or 1/6171. Then, if the line was extended to be 6171 feet long, and the measurement of this distance was made by the same personnel using the same equipment and methods as for the 493.68-foot distance, the probable error for the 6171-foot distance would be 1 foot.

b. Angular Measurements. For angular measurements the degree of accuracy is usually expressed as the ratio of the angular closing error to the square root of the number of angles. For example, a loop traverse containing 12 angles has an angular closing error of 0°02′30″. Then the degree of accuracy; $k\sqrt{n}$, attained in measuring the 12 angles is determined as follows:

$$k\sqrt{n} = 2'30' = 150''$$

$$k = \frac{150''}{\sqrt{n}} = \frac{150''}{\sqrt{12}} = 43''$$

$$k\sqrt{n} = 43''\sqrt{n} = \text{degree of accuracy}$$

in which n is the number of angles affecting the closure. Then, if a series of 20 angles were measured by the same personnel, using the same instruments and methods under the same observing conditions as for the 12 angles, the expected closure would be: $\pm 43''\sqrt{n} = \pm 43''\sqrt{20} = \pm 192'' = \pm 3'12''$.

c. Traverse. The degree of accuracy for a traverse is usually expressed as the ratio of the linear error of closure to the total length of the traverse. For example: In a traverse 15,476.8 feet long with a linear error of closure of 2.54 feet, the degree of accuracy is determined as follows:

$$\frac{2.54}{15{,}476.8} = \frac{1}{6093+}$$

The denominator is often rounded off, and the degree of accuracy is expressed as $\pm\frac{1}{6100}$, or ± 1:6100, or 1/6100. This indicates that the errors would cause an error of closure by approximately 1 foot in a traverse 6,100 feet long.

d. Leveling. In leveling surveys, the degree of accuracy is expressend as the ratio of the error of closure in elevation to the square root of the horizontal length of the route over which the leveling was accomplished. For example, in a 12-mile circuit, the error of closure was 0.14 foot. The degree of accuracy, $k\sqrt{L}$, for this circuit is determined as follows:

$$k\sqrt{L} = 0.14$$

$$k = \frac{0.14}{\sqrt{L}} = \frac{0.14}{\sqrt{12}} = 0.0405$$

$$k\sqrt{L} = 0.0405\sqrt{L}$$

in which L is the length of the route in miles.

B–9. Orders of Accuracy

When writing the specifications for a survey, it is very impractical to specify the exact degree of accuracy that is to be attained in each of the measurements. For this reason, the specifications are based on the minimum degree of accuracy allowed for the particular survey, and the range between the allowed degrees is called an order of accuracy. The *orders of accuracy* for surveys are called first order, second order, third order, and lower order. The measurements for first order accuracy surveys are the most accurate, and the measurements for the other orders are progressively less accurate.

a. Orders of accuracy are specified for triangulation, traverse, and leveling. For the measurements made in mapping, orders of accuracy are also specified for the astronomic observations made to establish position and azimuth. As an example of the range between orders of accuracy let us consider the allowed traverse position closure. First order specifies 1:25,000 or better; second order, 1:10,000; third order 1:5,000; fourth order, 1:3,000; fifth order, 1:1,000, and the lowest order, 1:500.

c. The following are specifications for the order of accuracy associated with some surveys:

	$\frac{1}{3000}$ to $\frac{1}{5000}$	$\frac{1}{1500}$ to $\frac{1}{3000}$
(1) *Traverse.*		
Closing error in position not to exceed	1:3000	1:1500
Probable error in main scheme angles	10 seconds	20 seconds
Azimuth angles computed and adjusted to	10 seconds	20 seconds
Slope of tape not to exceed	20%	----------
Marking tape end position	±0.01 foot	±0.03 foot
Angle measurements:		
20-second transit	4 D/R	2 D/R
1-minute transit	6 D/R	3 D/R
(2) *Triangulation.*		
Base line accuracy	1:25,000 (comparative)	
Triangle closure:		
Average	10 seconds	
Maximum	15 seconds	
Log tables for computations	5 place	
(3) *Leveling.*		
Length of sight not to exceed		500 feet
Method of rod reading		One hair
Precision of rod reading		0.02 foot

B–10. Examples

The principles discussed in the preceding paragraphs of this appendix are of primary importance to the surveyor. The following examples illustrate these principles.

a. Example 1. Two measurements of the horizontal distance between two points have been made. Determine the most probable horizontal distance, its probable error, and its degree of accuracy.

Measured value	v	v^2
573.28	− 0.09	0.0081
573.46	+ 0.09	0.0081
Sum = 1146.74		0.0162 = Σv^2

Mean = 573.37 = *Most probable distance*

$$E_m = \frac{0.6745}{\sqrt{n(n-1)}}\sqrt{\Sigma v^2} = 0.4769\sqrt{0.0162}$$

= ± 0.0607 (par. C–7*c* above) = *probable error*

Degree of accuracy = 0.0607/573.37 = 1/9440.

b. Example 2. A line was measured in three sections. Each section was measured twice and the most probable length and the probable error, E, for each section has been determined by the method illustrated in *a* above. Compute the most probable length of the line, its probable error, and its degree of accuracy.

Section	Measured Length	E	E^2
A	1024.3	±0.14	0.0196
B	1219.8	±0.22	0.0484
C	993.6	±0.16	0.0256
	3237.7		0.0936 = ΣE^2

Most probable length = 3237.7

$E_s = \sqrt{\Sigma E^2} = \sqrt{0.0936} = \pm 0.306$ (par. C–7*g above*) = *probable error*

Degree of accuracy = 0.306/3237.72 = 1/10,590

c. Example 3. The difference in elevation between two points has been determined over three separate routes. Determine the most probable difference in elevation between the two points, the probable error in this difference in elevation, and the degree of accuracy for each of the three routes.

(1) *Given data.*

Route	Length = L	Difference in elevation	d
A	4.5	412.46	0.46
B	3.0	412.37	0.37
C	12.5	412.18	0.18

The factor *d* is equal to the difference in elevation minus a selected constant which in this case is 412.00. This allows multiplication of smaller values in the adjustment computations ((2) below) and can be done with the slide rule.

(2) *Adjustment computations.*

Route	W = 1/L	Wd
A	0.2225	0.1024
B	0.3333	0.1232
C	0.0800	0.0144
	ΣW = 0.6358	0.2400 = ΣWd

Most probable difference in elevation = 412.00 +0.2400/0.6385 = 412.00 + 0.377 = 412.377 = 412.38

(3) *Probable error computations.*

Route	v	v^2	v^2W
A	+0.08	0.0064	0.00142
B	−0.01	0.0001	0.00033
C	−0.20	0.0400	0.00320
			Σv^2W = 0.00495

Where v = *measured difference in elevation for each route minus the most probable difference in elevation.*

$$R_m = \pm 0.6745 \sqrt{\frac{\Sigma v^2 W}{(n-1)\Sigma W}} = \pm 0.6745 \sqrt{\frac{0.00495}{2 \times 0.6358}} = + 0.6745 \sqrt{0.00389} = \pm 0.6745 \times 0.0197$$

$$= \pm 0.0133 \text{ (par. C–7}d\text{ above)} = \textit{probable error}$$

(4) *Degree of accuracy computations.* The basic equation for the degree of accuracy in leveling is, $k\sqrt{L}$, (par. C–8*d* above). For route A, $k\sqrt{L} = 0.08$; for route B, $k\sqrt{L} = 0.01$; and for route C, $k\sqrt{L} = 0.20$. Solving each of these equations for k: for route A, equals 0.0377; for route B, 0.0058; and for route C, 0.0566. The degree of acuracy is: for route A, $0.0377\sqrt{L}$; for route B, $0.0058\sqrt{L}$; and for route C, $0.0566\sqrt{L}$.

APPENDIX C

UNITS OF MEASUREMENT

C–1. Introduction

The surveyor deals in measurements. He must be familiar with the numerous units of measure and methods of converting one unit to another. He must also be able to compute area and volume when necessary. In addition to these measurements which are based on length (linear measurements), the surveyor must also understand angular measure, which has three systems: the sexagesimal or degree, minute, and seconds; the centesimal (metric) using grads; and the mil system. Besides the units for distances and angles, there are also units for time, weight, temperature, and pressure.

C–2. General

Currently within the United States there exists two basic systems of weights and measures; the Customary System and the International System (SI).

a. Customary System. This system as its name implies, is the system that is customarily used. This system had as its basis, the British Imperial System. However, through the years changes took place and the customary system in use today in the United States differs from the original system. This system includes those measurement units, such as inches, feet, yards, pounds, ounces, etc.

b. International System. This system is based on the French metric system, originated in the eighteenth century. Delegates of the 1960 Eleventh General Conference on Weights and Measures defined and officially sanctioned an international system of units, the Système International d'Unités (designated SI in all languages). This system, based on the meter, kilogram, second, ampere, kelvin, and candela, was adopted as the preferred system of units by the official representatives of the thirty-six participating countries, including the United States. The common standard for this system for the inch-pound in metric terms are 1 inch equals 2.54 centimeters and 1 pound equals 0.453 592 37 kilogram, both are exacted values. The Coast and Geodetic Survey, which had used a slightly different conversion factor previously, retained their old established relationship of 1 inch equaling 2.540 005 centimeters because of the extensive revisions which would be necessary to their charts and measurement records. The resulting foot based on this retained conversion is known as the U.S. survey foot. Therefore care must be taken when doing survey work to insure that the proper notations are made on all records of conversions from feet to meters and inverse.

C–3. Linear

Linear measure is used to express distances and to indicate the differences in elevation. The standard units of linear measure are the *foot* and the *meter*. In surveying operations, both these standard units are frequently divided into tenths, hundredths, and thousandths for measurements. When dealing with the long distances, the foot is expanded into the statute mile, and the meter into the kilometer. Tables D–7a and b are the conversion of feet to meters and inverse based on the international system. Tables D–8a and b are the conversion of feet to meters and inverse based on the U.S. survey foot. Table C–1 shows the conversion factors for the most commonly used linear measurements based on the international system (SI). Table C–2 shows the conversion factors for the most commonly used linear measurements based on the U.S. survey foot.

C–4. Area

Surveys are often made to obtain measured data from which area can be computed. The names of area units are frequently derived from the linear units, since area is the product of two linear measurements. The following tabulation shows the *exact comparative values* (International System) of some of the units that the surveyor may

Table C–1. Linear Conversion Factors (SI)

	Inches	Feet	Yard	Statute Miles	Centimeter	Meter	Kilometer
Inch	1*	0.083333	0.0277	------------	2.54*	0.0254*	------------
Foot	12*	1*	0.333	------------	30.48*	0.3048*	------------
Yard	36*	3*	1*	0.000568	91.44*	0.9144*	0.000914
Statute Mile	63,360*	5280*	1760*	1*	------------	1609.344*	1.609344*
Centimeter	0.3937	0.032808	0.010936	------------	1*	0.01*	------------
Decimeter	3.937	0.32808	0.10936	------------	10*	0.1*	------------
Meter	39.37008	3.280840	1.093613	0.000621	100*	1*	0.001*
Kilometer	------------	------------	------------	.62137	------------	------------	1*

*These are exact values, all others are close approximations.

Table C–2. Linear Conversion Factors (U.S. Survey Foot)

	Inches	Feet	Yard	Statute Mile	Centimeter	Meter	Kilometer
Inch	1	.083333	.0277	-----------	2.540005	.0254	-----------
Foot	12	1	.333	-----------	30.48006	.304801	-----------
Yard	36	3	1	.000568	91.44018	.914402	.000914
Statute Mile	63,360	5280	1760	1	-----------	1609.347	1.609347
Centimeter	.3937	-----------	-----------	-----------	1	.01	-----------
Decimeter	3.937	.328	-----------	-----------	-----------	.1	-----------
Meter	39.37	3.280833	1.093611	-----------	100	1	.001
Kilometer	-----------	-----------	-----------	0.62137	-----------	1000	1

be required to use:

1 square inch (sq in) = 6.4516 square centimeters (cm^2)

1 square foot (sq ft) = 144 sq in = 0.09290304 square meter (m^2)

1 square yard (sq yd) = 9 sq ft = 0.836 127 36 m^2

1 acre = 43,560 sq ft = 4840 sq yd = 4,046.8564224 m^2

1 cm^2 = 0.1550003 sq in

1 m^2 = 10.76391 sq ft = 1.195990 sq yd

The surveyor should not confuse such terms as 7 square feet and 7-foot square. The expression 7 square feet denotes an area of unspecified shape, containing that number of square feet. The term 7-foot square specifies an area contained in a square having sides 7 feet long. The area is 7 feet times 7 feet, or 49 square feet.

C–5. Volume

The surveyor is interested in units of volume measure as they apply to earthwork volumes, reservoir capacities, rainfall runoff, and other volumes. The following tabulation shows the *exact comparative values* (International System) of some of the basic units of volume measure:

1 cubic inch (cu in) = 16.387 064 milliliters (ml)

1 cubic foot (cu ft) = 1,728 cu in = 0.028 316 846 592 cubic meters (m^3)

1 cubic yard (cu yd) = 27 cu ft = 0.764 554 857 984 m^3

1 ml = 0.061 023 74 cu in = 0.001 liter

1 m^3 = 61.023.47 cu in = 35.31467 cu ft = 1,000 liter

1 gallon = 231 cu in = 0.003 785 411 784 m^3

C–6. Angular

Angular or circular measure is used for designating the value of horizontal and vertical angles. For general use in the measurement of angles, the circumference of the circle is divided into some even number of equal parts. The unit of angular measure is the angle at the center of the circle subtended by one of the small subdivisions of the circumference. The various units of angular measure are known as *units of arc*. There are three systems of angular measure that may be encountered by personnel in the use of surveying instruments.

a. Sexagesimal System. In the sexagesimal system, the circle is divided into 360 equal parts known as *degrees* of arc, each degree into 60 equal parts known as *minutes* of arc, and each minute into 60 equal parts known as seconds of arc. Angles in this system are written, for example, as

263° 47′ 16″ .48, which is read as "two hundred sixty-three degrees, forty-seven minutes, and sixteen point four eight seconds of arc." In the United Stabes, this is the most commonly used system of angular measure.

b. Centesimal, or Metric System. In the centesimal system, the full circle is divided into four quadrants and each quadrant is divided into 100 equal parts known as *grads* or grades. Each grad is further divided into decimal parts. Angles in this system are written, for example: as 376.7289 grads, or as 376^g 72′ 89″ which is read as "three hundred seventy-six point seven two eight nine grads," or as "three hundred seventy-six grads, seventy-two centesimal minutes, and eighty-nine centesimal seconds."

c. Mil System. In the mil system, the circle is divided into 6,400 equal parts known as *mils*. The mil (symbolized by m) is divided into decimal parts. Angles in this system are written, for example: as 1728.49m which is read as "one thousand seven hundred twenty-eight point four nine mils." Refer to table D–1, appendix D, for conversion of degrees to mils and inverse.

d. Comparison of Systems. The relationship among values in the three systems of angular measure are—

1 circle = 360 degrees = 400 grads = 6400 mils
1 degree = 1.1111111 grads = 17.7777778 mils
1 minute = 0.29630 mils
1 grad = 0.9 degree = 0° 54′ 00″ = 16 mils
1 mil = 0.05625 degree = 0° 03′ 22″ .5 or 3.37500 minutes = 0.0625 grad

C–7. Time

For practical purposes in everyday affairs and in surveying, the measurement of time intervals is of great concern. The time used in everyday life is known as *standard time,* and is based on the mean apparent revolution of the sun about the earth caused by the earth's rotation on its axis. Standard time is used in surveying to regulate the normal day's operations, but when it is necessary to observe the sun or the stars to determine the azimuth of a line or the position of a point on the earth's surface, three other kinds of time are used with which the surveyor should be familiar. These are: *Apparent (true solar) time, civil (mean solar) time, and sideral (star) time.*

a. Differences in Duration. In all four kinds of time, the basic units of measure are the year, day, hour, minute, and second of time. The duration of any one of these units is not the same for all kinds of time. For example: The sidereal day is approximately 4 minutes shorter than a standard- or civil-time day.

b. Notation. In the practice of surveying it is customary to say, or write the time of day as the number of hours, minutes, and seconds since midnight. Then the recorded time would appear, for example, as $16^h37^m52^s.71$—which is read as "sixteen hours, thirty-seven minutes, and fifty-two point seven one seconds of time."

c. Angles and Time. Units of time measure are sometimes used to designate the sizes of angles. The longitude of a point on the earth's surface is often expressed in this manner. The relationship between the units of time measure and the units of angular measure in the sexagesimal system are—

1 hour = 15 degrees ($1^h = 15°$)
1 minute of time = 15 minutes of arc ($1^m = 15'$)
1 second of time = 15 seconds of arc ($1^s = 15''$)
1 degree = 4 minutes of time ($1° = 4^m$)
1 minute of arc = 4 seconds of time ($1' = 4^s$)
1 second of arc = 0.0667 second of time ($1'' = 0.0667^s$)

C–8. Weight

The surveyor is interested in the weight of his instruments and the pull to be applied to the tape to give correct linear measurements. The common units of weight in surveying are the avoirdupois ounce, the avoirdupois pound, the gram, and the kilogram. The following tabulation gives the *exact comparative values* (International System) between these units.

1 ounce (oz) = 28.349 523 125 grams (g)
1 pound (lb) = 16 oz = 453.592 27g = 0.453 592 37 kilograms (kg)
1 kg = 1000g = 35.273 96 oz (approximately)

C–9. Temperature

When the existing temperature differs from a standard temperature for certain types of measurement, the measured values will be in error and must be corrected. In each of the several temperature-measurement scales, the unit of measure is called a degree, which varies for the different temperature scales. When the scale extends below zero, values below zero are identified by a minus sign. Temperatures are written, for example, as 23° F. or —5° C., the letter designating the particular temperature scale. Some of the most commonly used temperature scales are described below.

a. Celsius Scale. On the Celsius (centigrade) scale, zero is the temperature of the melting point of ice, and *plus* 100 is the temperature of water at the boiling point. Temperatures on the Celsius scale are written: 18°C., or —21°C.

b. Fahrenheit Scale. On the Fahrenheit scale, the temperature of the melting point of ice is *plus* 32°, and the temperature of the boiling point of water is *plus* 212°.

c. Comparison of Scales. A Fahrenheit degree represents five-ninths of the change in heat intensity indicated by a degree on the celsius scale. Temperatures on either of the two scales can be converted to the other by the following:

Degrees C. = 5/9 (degrees F. — 32)
Degrees F. = (9/5 degrees C.) + 32

When converting Fahrenheit to Celsius, the 32° are subtracted first, then the 5/9 is taken. When converting Celsius to Fahrenheit, the 9/5 is computed first, then the 32° are added. Refer to table D-9, appendix D, for conversion of temperature.

C–10. Pressure

Measurements of atmospheric pressure are used in surveying to determine approximate differences in elevation between points on the earth's surface, and to determine the best approximate correction for the effect of atmospheric refraction. The units of measure for atmospheric pressure, and their relationships are—

1 atmosphere = 29.9212 inches of mercury
= 760 millimeters of mercury
= 14.6960 pounds per square inch
= 1.03323 kilograms per square centimeter
= 33.899 feet of water
= 1.01325 bars, or 1013.25 millibars

APPENDIX D

TABLES

D–1	Conversion of degrees to mils and inverse
D–2	Inclination corrections, 50 meter tape
D–3	Difference in elevation, given horizontal distances and gradients from 0° to 45°
D–4	Difference in elevation, given slope distances and gradients from 0° to 45°
D–5	Horizontal distance, given slope distances and gradients from 0° to 45°
D–6	Stadia reduction
D–7a	Conversion meters to feet (International System)
D–7b	Conversion feet to meters (International System)
D–8a	Conversion meters to feet (US Survey foot)
D–8b	Conversion feet to meters (US Survey foot)
D–9	Conversion degrees Celsius to degrees Fahrenheit and inverse
D–10	Temperature correction, steel tape
D–11	Inclination correction, percent of slope

Table D-1. Conversion of Degrees to Mils and Inverse

Conversion of degrees to mils

(Conversion factor—1 degree = 17.77778 mils; 1 minute = 0.29630 mils)

Degrees	Mils	Degrees	Mils	Degrees	Mils	Minutes	Mils	Minutes	Mils
1	17.8	31	551.1	61	1084.4	1	0.3	31	9.2
2	35.6	32	568.9	62	1102.2	2	0.6	32	9.5
3	53.3	33	586.7	63	1120.0	3	0.9	33	9.8
4	71.1	34	604.4	64	1137.8	4	1.2	34	10.1
5	88.9	35	622.2	65	1155.6	5	1.5	35	10.4
6	106.7	36	640.0	66	1173.3	6	1.8	36	10.7
7	124.4	37	657.8	67	1191.1	7	2.1	37	11.0
8	142.2	38	675.6	68	1208.9	8	2.4	38	11.3
9	160.0	39	693.3	69	1226.7	9	2.7	39	11.6
10	177.8	40	711.1	70	1244.5	10	3.0	40	11.9
11	195.6	41	728.9	71	1262.2	11	3.3	41	12.1
12	213.3	42	746.7	72	1280.0	12	3.6	42	12.4
13	231.1	43	764.4	73	1297.8	13	3.9	43	12.7
14	248.9	44	782.2	74	1315.6	14	4.1	44	13.0
15	266.7	45	800.0	75	1333.3	15	4.4	45	13.3
16	284.4	46	817.8	76	1351.1	16	4.7	46	13.6
17	302.2	47	835.6	77	1368.9	17	5.0	47	13.9
18	320.0	48	853.3	78	1386.7	18	5.3	48	14.2
19	337.8	49	871.1	79	1404.5	19	5.6	49	14.5
20	355.6	50	888.9	80	1422.2	20	5.9	50	14.8
21	373.3	51	906.7	81	1440.0	21	6.2	51	15.1
22	391.1	52	924.4	82	1457.8	22	6.5	52	15.4
23	408.9	53	942.2	83	1475.6	23	6.8	53	15.7
24	426.7	54	960.0	84	1493.3	24	7.1	54	16.0
25	444.5	55	977.8	85	1511.1	25	7.4	55	16.3
26	462.2	56	995.6	86	1528.9	26	7.7	56	16.6
27	480.0	57	1013.3	87	1546.7	27	8.0	57	16.9
28	497.8	58	1031.1	88	1564.5	28	8.3	58	17.2
29	515.6	59	1048.9	89	1582.2	29	8.6	59	17.5
30	533.3	60	1066.7	90	1600.0	30	8.9	60	17.8

Examples: 1. Convert 64° 29′ to mils.

64° = 1137.8 mils
29′ = 8.6

64°29′ = 1146.4 mils.

2. Convert 87.95° to mils.

87° = 1546.7 mils
.90° = 16.0
.05° = 0.889 = 0.9

87.95° = 1563.6 mils.

Table D-1. Conversion of Degrees to Mils and Inverse—Continued

Conversion of mils to degrees and minutes
(Conversion factor—1 mil equals 0.05625 degrees equals 3.37500 minutes)

Mils	00	10	20	30	40
0	00°00′.00	00°33′.75	1°07′.50	1°41′.25	2°15′.00
100	5 37.50	6 11.25	6 45.00	7 18.75	7 52.50
200	11 15.00	11 48.75	12 22.50	12 56.25	13 30.00
300	16 52.38	17 26.25	18 00.00	18 33.75	19 07.50
400	22 30.00	23 03.75	23 37.50	24 11.25	24 45.00
500	28 07.50	28 41.25	29 15.00	29 48.75	30 22.50
600	33 45.00	34 18.75	34 52.50	35 26.25	36 00.00
700	39 22.50	39 56.25	40 30.00	41 03.75	41 37.50
800	45 00.00	45 33.75	46 07.50	46 41.25	47 15.00
900	50 37.50	51 11.25	51 45.00	52 18.75	52 52.50
1000	56 15.00	56 48.75	57 22.50	57 56.25	58 30.00
1100	61 52.50	62 26.25	63 00.00	63 33.75	64 07.50
1200	67 30.00	68 03.75	68 37.50	69 11.25	69 45.00
1300	73 07.50	73 41.25	74 15.00	74 48.75	75 22.50
1400	78 45.00	79.18.75	79 52.50	80 26.25	81 00.00
1500	84 22.50	84 56.25	85 30.00	86 03.75	86 37.50

Mils	50	60	70	80	90
0	2°48′.75	3°22′.50	3°56′.25	4°30′.00	5°03′.75
100	8 26.25	9 00.00	9 33.75	10 07.50	10 41.25
200	14 03.75	14 37.50	15 11.25	15 45.00	16 18.75
300	19 41.25	20 15.00	20 48.75	21 22.50	21 56.25
400	25 18.75	25 52.50	26 26.25	27 00.00	27 33.75
500	30 56.25	31 30.00	32 03.75	32 37.50	33 11.25
600	36 33.75	37 07.50	37 41.25	38 15.00	38 48.75
700	42 11.25	42 45.00	43 18.75	43 52.50	44 26.25
800	47 48.75	48 22.50	48 56.25	49 30.00	50 03.75
900	53 26.25	54 00.00	54 33.75	55 07.50	55 41.25
1000	59 03.75	59 37.50	60 11.25	60 45.00	61 18.75
1100	64 41.25	65 15.00	65 48.75	66 22.50	66 56.25
1200	70 18.75	70 52.50	71 26.25	72 00.00	72 33.75
1300	75 56.25	76 30.00	77 03.75	77 37.50	78 11.25
1400	81 33.75	82 07.50	82 41.25	83 15.00	83 48.75
1500	87 11.25	87 45.00	88 18.75	88 52.50	89 26.25

Mils	Minutes	Mils	Minutes	Mils	Minutes
1	3.38	4	13.50	7	23.63
2	6.75	5	16.88	8	27.00
3	10.13	6	20.25	9	30.38

Example: Convert 1254 mils to degrees and minutes.
1250 mils = 70° 18.75′
4 mils = 13.50′

1254 mils = 70° 32.25′.

Table D-2. Inclination Corrections, 50 meter tape

Difference of elevation		Correction	Difference of elevation		Correction	Difference of elevation		Correction	Difference of elevation		Correction
Meters	Feet	Mm.	Meters	Feet	Mm.	Meters	Feet	Mm.	Meters	Feet	Mm.
0.00	0.000	0.0	0.50	1.640	2.5	1.00	3.281	10.0	1.50	4.921	22.5
.01	.033	.0	.51	1.673	2.6	1.01	3.314	10.2	1.51	4.954	22.8
.02	.066	.0	.52	1.706	2.7	1.02	3.346	10.4	1.52	4.987	23.1
.03	.098	.0	.53	1.739	2.8	1.03	3.379	10.6	1.53	5.020	23.4
.04	.131	.0	.54	1.772	2.9	1.04	3.412	10.8	1.54	5.052	23.7
.05	.164	.0	.55	1.804	3.0	1.05	3.445	11.0	1.55	5.085	24.0
.06	.197	.0	.56	1.837	3.1	1.06	3.478	11.2	1.56	5.118	24.3
.07	.230	.0	.57	1.870	3.2	1.07	3.510	11.4	1.57	5.151	24.6
.08	.262	.1	.58	1.903	3.4	1.08	3.543	11.7	1.58	5.184	25.0
.09	.295	.1	.59	1.936	3.5	1.09	3.576	11.9	1.59	5.217	25.3
.10	.328	.1	.60	1.968	3.6	1.10	3.609	12.1	1.60	5.249	25.6
.11	.361	.1	.61	2.001	3.7	1.11	3.642	12.3	1.61	5.282	25.9
.12	.349	.1	.62	2.034	3.8	1.12	3.675	12.5	1.62	5.315	26.2
.13	.427	.2	.63	2.067	4.0	1.13	3.707	12.8	1.63	5.348	26.6
.14	.459	.2	.64	2.100	4.1	1.14	3.740	13.0	1.64	5.381	26.9
.15	.492	.2	.65	2.133	4.2	1.15	3.773	13.2	1.65	5.413	27.2
.16	.525	.3	.66	2.165	4.4	1.16	3.806	13.5	1.66	5.446	27.6
.17	.558	.3	.67	2.198	4.5	1.17	3.839	13.7	1.67	5.479	27.9
.18	.591	.3	.68	2.231	4.6	1.18	3.871	13.9	1.68	5.512	28.2
.19	.623	.4	.69	2.264	4.8	1.19	3.904	14.2	1.69	5.545	28.6
.20	.656	.4	.70	2.297	4.9	1.20	3.937	14.4	1.70	5.577	28.9
.21	.689	.4	.71	2.329	5.0	1.21	3.970	14.6	1.71	5.610	29.2
.22	.722	.5	.72	2.362	5.2	1.22	4.003	14.9	1.72	5.643	29.6
.23	.755	.5	.73	2.395	5.3	1.23	4.035	15.1	1.73	5.676	29.9
.24	.787	.6	.74	2.428	5.5	1.24	4.068	15.4	1.74	5.709	30.3
.25	.820	.6	.75	2.461	5.6	1.25	4.101	15.6	1.75	5.741	30.6
.26	.853	.7	.76	2.493	5.8	1.26	4.134	15.9	1.76	5.774	31.0
.27	.886	.7	.77	2.526	5.9	1.27	4.167	16.1	1.77	5.807	31.3
.28	.919	.8	.78	2.559	6.1	1.28	4.199	16.4	1.78	5.840	31.7
.29	.951	.8	.79	2.592	6.2	1.29	4.232	16.6	1.79	5.873	32.0
.30	.984	.9	.80	2.625	6.4	1.30	4.265	16.9	1.80	5.906	32.4
.31	1.017	1.0	.81	2.657	6.6	1.31	4.298	17.2	1.81	5.938	32.8
.32	1.050	1.0	.82	2.690	6.7	1.32	4.331	17.4	1.82	5.971	33.1
.33	1.083	1.1	.83	2.723	6.9	1.33	4.364	17.7	1.83	6.004	33.5
.34	1.115	1.2	.84	2.756	7.1	1.34	4.396	18.0	1.84	6.037	33.9
.35	1.148	1.2	.85	2.789	7.2	1.35	4.429	18.2	1.85	6.070	34.2
.36	1.181	1.3	.86	2.822	7.4	1.36	4.462	18.5	1.86	6.102	34.6
.37	1.214	1.4	.87	2.854	7.6	1.37	4.495	18.8	1.87	6.135	35.0
.38	1.247	1.4	.88	2.887	7.7	1.38	4.528	19.0	1.88	6.168	35.3
.39	1.280	1.5	.89	2.920	7.9	1.39	4.560	19.3	1.89	6.201	35.7
.40	1.312	1.6	.90	2.953	8.1	1.40	4.593	19.6	1.90	6.234	36.1
.41	1.345	1.7	.91	2.986	8.3	1.41	4.626	19.9	1.91	6.266	36.5
.42	1.378	1.8	.92	3.018	8.5	1.42	4.659	20.2	1.92	6.299	36.9
.43	1.411	1.8	.93	3.051	8.6	1.43	4.692	20.4	1.93	6.332	37.2
.44	1.444	1.9	.94	3.084	8.8	1.44	4.724	20.7	1.94	6.365	37.6
.45	1.476	2.0	.95	3.117	9.0	1.45	4.757	21.0	1.95	6.398	38.0
.46	1.509	2.1	.96	3.150	9.2	1.46	4.790	21.3	1.96	6.430	38.4
.47	1.542	2.2	.97	3.182	9.4	1.47	4.823	21.6	1.97	6.463	38.8
.48	1.575	2.3	.98	3.215	9.6	1.48	4.856	21.9	1.98	6.496	39.2
.49	1.608	2.4	.99	3.248	9.8	1.49	4.888	22.2	1.99	6.529	39.6

Table D-2. Inclination Corrections, 50 Meter Tape—Continued

Difference of elevation		Correction	Difference of elevation		Correction	Difference of elevation		Correction	Difference of elevation		Correction
Meters	Feet	Mm.	Meters	Feet	Mm.	Meters	Feet	Mm.	Meters	Feet	Mm.
2.00	6.562	40.0	2.50	8.202	62.5	3.00	9.842	90.1	3.50	11.483	122.7
2.01	6.594	40.4	2.51	8.235	63.0	3.01	9.875	90.7	3.51	11.516	123.4
2.02	6.627	40.8	2.52	8.268	63.5	3.02	9.908	91.3	3.52	11.549	124.1
2.03	6.660	41.2	2.53	8.301	64.0	3.03	9.941	91.9	3.53	11.581	124.8
2.04	6.693	41.6	2.54	8.333	64.6	3.04	9.974	92.5	3.54	11.614	125.5
2.05	6.726	42.0	2.55	8.366	65.1	3.05	10.007	93.1	3.55	11.647	126.2
2.06	6.759	42.5	2.56	8.399	65.6	3.06	10.039	93.7	3.56	11.680	126.9
2.07	6.791	42.9	2.57	8.432	66.1	3.07	10.072	94.3	3.57	11.713	127.6
2.08	6.824	43.3	2.58	8.465	66.6	3.08	10.105	95.0	3.58	11.745	128.3
2.09	6.857	43.7	2.59	8.497	67.1	3.09	10.138	95.6	3.59	11.778	129.0
2.10	6.890	44.1	2.60	8.530	67.6	3.10	10.171	96.2	3.60	11.811	129.8
2.11	6.923	44.5	2.61	8.563	68.2	3.11	10.203	96.8	3.61	11.844	130.5
2.12	6.955	45.0	2.62	8.596	68.7	3.12	10.236	97.4	3.62	11.877	131.2
2.13	6.988	45.4	2.63	8.629	69.2	3.13	10.269	98.1	3.63	11.909	131.9
2.14	7.021	45.8	2.64	8.661	69.7	3.14	10.302	98.7	3.64	11.942	132.7
2.15	7.054	46.2	2.65	8.694	70.3	3.15	10.335	99.3	3.65	11.975	133.4
2.16	7.087	46.7	2.66	8.727	70.8	3.16	10.367	100.0	3.66	12.008	134.1
2.17	7.119	47.1	2.67	8.760	71.3	3.17	10.400	100.6	3.67	12.041	134.9
2.18	7.152	47.5	2.68	8.793	71.9	3.18	10.433	101.2	3.68	12.073	135.6
2.19	7.185	48.0	2.69	8.825	72.4	3.19	10.466	101.9	3.69	12.106	136.3
2.20	7.218	48.4	2.70	8.858	73.0	3.20	10.499	102.5	3.70	12.139	137.1
2.21	7.251	48.9	2.71	8.891	73.5	3.21	10.531	103.1	3.71	12.172	137.8
2.22	7.283	49.3	2.72	8.924	74.0	3.22	10.564	103.8	3.72	12.205	138.6
2.23	7.316	49.8	2.73	8.957	74.6	3.23	10.597	104.4	3.73	12.238	139.3
2.24	7.349	50.2	2.74	8.989	75.1	3.24	10.630	105.1	3.74	12.270	140.1
2.25	7.382	50.7	2.75	9.022	75.7	3.25	10.663	105.7	3.75	12.303	140.8
2.26	7.415	51.1	2.76	9.055	76.2	3.26	10.696	106.4	3.76	12.336	141.6
2.27	7.447	51.6	2.77	9.088	76.8	3.27	10.728	107.0	3.77	12.369	142.3
2.28	7.480	52.0	2.78	9.121	77.3	3.28	10.761	107.7	3.78	12.402	143.1
2.29	7.513	52.5	2.79	9.154	77.9	3.29	10.794	108.4	3.79	12.434	143.8
2.30	7.546	52.9	2.80	9.186	78.5	3.30	10.827	109.0	3.80	12.467	144.6
2.31	7.579	53.4	2.81	9.219	79.0	3.31	10.860	109.7	3.81	12.500	145.4
2.32	7.612	53.9	2.82	9.252	79.6	3.32	10.892	110.3	3.82	12.533	146.1
2.33	7.644	54.3	2.83	9.285	80.2	3.33	10.925	111.0	3.83	12.566	146.9
2.34	7.677	54.8	2.84	9.318	80.7	3.34	10.958	111.7	3.84	12.598	147.7
2.35	7.710	55.3	2.85	9.350	81.3	3.35	10.991	112.4	3.85	12.631	148.4
2.36	7.743	55.7	2.86	9.383	81.9	3.36	11.024	113.0	3.86	12.664	149.2
2.37	7.776	56.2	2.87	9.416	82.4	3.37	11.056	113.7	3.87	12.697	150.0
2.38	7.808	56.7	2.88	9.449	83.0	3.38	11.089	114.4	3.88	12.730	150.8
2.39	7.841	57.2	2.89	9.482	83.6	3.39	11.122	115.1	3.89	12.762	151.6
2.40	7.874	57.6	2.90	9.514	84.2	3.40	11.155	115.7	3.90	12.795	152.3
2.41	7.907	58.1	2.91	9.547	84.8	3.41	11.188	116.4	3.91	12.828	153.1
2.42	7.940	58.6	2.92	9.580	85.3	3.42	11.220	117.1	3.92	12.861	153.9
2.43	7.972	59.1	2.93	9.613	85.9	3.43	11.253	117.8	3.93	12.894	154.7
2.44	8.005	59.6	2.94	9.646	86.5	3.44	11.286	118.5	3.94	12.926	155.5
2.45	8.038	60.1	2.95	9.678	87.1	3.45	11.319	119.2	3.95	12.959	156.3
2.46	8.071	60.6	2.96	9.711	87.7	3.46	11.352	119.9	3.96	12.992	157.1
2.47	8.104	61.0	2.97	9.744	88.3	3.47	11.384	120.6	3.97	13.025	157.9
2.48	8.136	61.5	2.98	9.777	88.9	3.48	11.417	121.3	3.98	13.058	158.7
2.49	8.169	62.0	2.99	9.810	89.5	3.49	11.450	122.0	3.99	13.091	159.5

Table D-2. Inclination Corrections, 50 Meter Tape—Continued

Difference of elevation		Correction	Difference of elevation		Correction	Difference of elevation		Correction	Difference of elevation		Correction
Meters	Feet	Mm.	Meters	Feet	Mm.	Meters	Feet	Mm.	Meters	Feet	Mm.
4.00	13.123	160.3	4.50	14.764	202.9	5.00	16.404	250.6	5.50	18.045	303.4
4.01	13.156	161.1	4.51	14.797	203.8	5.01	16.437	251.6	5.51	18.077	304.5
4.02	13.189	161.9	4.52	14.829	204.7	5.02	16.470	252.6	5.52	18.110	305.6
4.03	13.222	162.7	4.53	14.862	205.6	5.03	16.503	253.6	5.53	18.143	306.7
4.04	13.255	163.5	4.54	14.895	206.5	5.04	16.535	254.7	5.54	18.176	307.9
4.05	13.287	164.3	4.55	14.928	207.5	5.05	16.568	255.7	5.55	18.209	309.0
4.06	13.320	165.1	4.56	14.961	208.4	5.06	16.601	256.7	5.56	18.241	310.1
4.07	13.353	165.9	4.57	14.993	209.3	5.07	16.634	257.7	5.57	18.274	311.2
4.08	13.386	166.7	4.58	15.026	210.2	5.08	16.667	258.7	5.58	18.307	312.3
4.09	13.419	167.6	4.59	15.059	211.1	5.09	16.699	259.8	5.59	18.340	313.5
4.10	13.451	168.4	4.60	15.092	212.0	5.10	16.732	260.8	5.60	18.373	314.6
4.11	13.484	169.2	4.61	15.125	213.0	5.11	16.765	261.8	5.61	18.405	315.7
4.12	13.517	170.0	4.62	15.157	213.9	5.12	16.798	262.8	5.62	18.438	316.8
4.13	13.550	170.9	4.63	15.190	214.8	5.13	16.831	263.9	5.63	18.471	318.0
4.14	13.583	171.7	4.64	15.223	215.8	5.14	16.863	264.9	5.64	18.504	319.1
4.15	13.615	172.5	4.65	15.256	216.7	5.15	16.896	265.9	5.65	18.537	320.2
4.16	13.648	173.4	4.66	15.289	217.6	5.16	16.929	267.0	5.66	18.570	321.4
4.17	13.681	174.2	4.67	15.321	218.6	5.17	16.962	268.0	5.67	18.602	322.5
4.18	13.714	175.0	4.68	15.354	219.5	5.18	16.995	269.0	5.68	18.635	323.7
4.19	13.747	175.9	4.69	15.348	220.7	5.19	17.028	270.1	5.69	18.668	324.8
4.20	13.780	176.7	4.70	15.420	221.4	5.20	17.060	271.1	5.70	18.701	326.0
4.21	13.812	177.6	4.71	15.453	222.3	5.21	17.093	272.2	5.71	18.734	327.1
4.22	13.845	178.4	4.72	15.486	223.3	5.22	17.126	273.2	5.72	18.766	328.3
4.23	13.878	179.2	4.73	15.518	224.2	5.23	17.159	274.3	5.73	18.799	329.4
4.24	13.911	180.1	4.74	15.551	225.2	5.24	17.192	275.3	5.74	18.832	330.6
4.25	13.944	181.0	4.75	15.584	226.1	5.25	17.224	276.4	5.75	18.865	331.7
4.26	13.976	181.8	4.76	15.617	227.1	5.26	17.257	277.4	5.76	18.898	332.9
4.27	14.009	182.7	4.77	15.650	228.0	5.27	17.290	278.5	5.77	18.930	334.0
4.28	14.042	183.5	4.78	15.682	229.0	5.28	17.323	279.6	5.78	18.963	335.2
4.29	14.075	184.4	4.79	15.715	230.0	5.29	17.356	280.6	5.79	18.996	336.4
4.30	14.108	185.2	4.80	15.748	230.9	5.30	17.388	281.7	5.80	19.029	337.5
4.31	14.140	186.1	4.81	15.781	231.9	5.31	17.421	282.8	5.81	19.062	338.7
4.32	14.173	187.0	4.82	15.814	232.9	5.32	17.454	283.8	5.82	19.094	339.9
4.33	14.206	187.8	4.83	15.846	233.8	5.33	17.487	284.9	5.83	19.127	341.0
4.34	14.239	188.7	4.84	15.879	234.8	5.34	17.520	286.0	5.84	19.160	342.2
4.35	14.272	189.6	4.85	15.912	235.8	5.35	17.552	287.0	5.85	19.193	343.4
4.36	14.304	190.5	4.86	15.945	236.8	5.36	17.585	288.1	5.86	19.226	344.6
4.37	14.337	191.3	4.87	15.978	237.7	5.37	17.618	289.2	5.87	19.258	345.8
4.38	14.370	192.2	4.88	16.010	238.7	5.38	17.651	290.3	5.88	19.291	346.9
4.39	14.403	193.1	4.89	16.043	239.7	5.39	17.684	291.4	5.89	19.324	348.1
4.40	14.436	194.0	4.90	16.076	240.7	5.40	17.716	292.5	5.90	19.357	349.3
4.41	14.468	194.9	4.91	16.109	241.7	5.41	17.749	293.5	5.91	19.390	350.5
4.42	14.501	195.7	4.92	16.142	242.7	5.42	17.782	294.6	5.92	19.423	351.7
4.43	14.534	196.6	4.93	16.175	243.6	5.43	17.815	295.7	5.93	19.455	352.9
4.44	14.567	197.5	4.94	16.207	244.6	5.44	17.848	296.8	5.94	19.488	354.1
4.45	14.600	198.4	4.95	16.240	245.6	5.45	17.881	297.9	5.95	19.521	355.3
4.46	14.633	199.3	4.96	16.273	246.6	5.46	17.913	299.0	5.96	19.554	356.5
4.47	14.665	200.2	4.97	16.306	247.6	5.47	17.946	300.1	5.97	19.587	357.7
4.48	14.698	201.1	4.98	16.339	248.6	5.48	17.979	301.2	5.98	19.619	358.9
4.49	14.731	202.0	4.99	16.371	249.6	5.49	18.012	302.3	5.99	19.652	360.1

Table D-2. Inclination Corrections, 50 Meter Tape—Continued

Difference of elevation		Correction	Difference of elevation		Correction	Difference of elevation		Correction	Difference of elevation		Correction
Meters	Feet	Mm.	Meters	Feet	Mm.	Meters	Feet	Mm.	Meters	Feet	Mm.
6.00	19.685	361.3	6.40	20.997	411.3	6.80	22.310	464.5	7.20	23.622	521.1
6.01	19.718	362.5	6.41	21.030	412.6	6.81	22.342	465.9	7.21	23.655	522.5
6.02	19.751	363.7	6.42	21.063	413.9	6.82	22.375	467.3	7.22	23.688	524.0
6.03	19.783	364.9	6.43	21.096	415.2	6.83	22.408	468.7	7.23	23.720	525.5
6.04	19.816	366.1	6.44	21.129	416.5	6.84	22.441	470.0	7.24	23.753	526.9
6.05	19.849	367.4	6.45	21.161	417.8	6.85	22.474	471.4	7.25	23.786	528.4
6.06	19.882	368.6	6.46	21.194	419.1	6.86	22.507	472.8	7.26	23.819	529.9
6.07	19.915	369.8	6.47	21.227	420.4	6.87	22.539	474.2	7.27	23.852	531.3
6.08	19.947	371.0	6.48	21.260	421.7	6.88	22.572	475.6	7.28	23.884	532.8
6.09	19.980	372.3	6.49	21.293	423.0	6.89	22.605	477.0	7.29	23.917	534.3
6.10	20.013	373.5	6.50	21.325	424.3	6.90	22.638	478.4	7.30	23.950	535.7
6.11	20.046	374.7	6.51	21.358	425.6	6.91	22.671	479.8	7.31	23.983	537.2
6.12	20.079	375.9	6.52	21.391	426.9	6.92	22.703	481.2	7.32	24.016	538.7
6.13	20.112	377.2	6.53	21.424	428.2	6.93	22.736	482.6	7.33	24.049	540.2
6.14	20.144	378.4	6.54	21.457	429.5	6.94	22.769	484.0	7.34	24.081	541.7
6.15	20.177	379.7	6.55	21.489	430.9	6.95	22.802	485.4	7.35	24.114	543.1
6.16	20.210	380.9	6.56	21.522	432.2	6.96	22.835	486.8	7.36	24.147	544.6
6.17	20.243	382.1	6.57	21.555	433.5	6.97	22.867	488.2	7.37	24.180	546.1
6.18	20.276	383.4	6.58	21.588	434.8	6.98	22.900	489.6	7.38	24.213	547.6
6.19	20.308	384.6	6.59	21.621	436.2	6.99	22.933	491.0	7.39	24.245	549.1
6.20	20.341	385.9	6.60	21.654	437.5	7.00	22.966	492.4	7.40	24.278	550.6
6.21	20.374	387.1	6.61	21.686	438.8	7.01	22.999	493.8	7.41	24.311	552.1
6.22	20.407	388.4	6.62	21.719	440.2	7.02	23.031	495.2	7.42	24.344	553.6
6.23	20.440	389.6	6.63	21.752	441.5	7.03	23.064	496.6	7.43	24.377	555.1
6.24	20.472	390.9	6.64	21.785	442.8	7.04	23.097	498.1	7.44	24.409	556.6
6.25	20.505	392.2	6.65	21.818	444.2	7.05	23.130	499.5	7.45	24.442	558.1
6.26	20.538	393.4	6.66	21.850	445.5	7.06	23.163	500.9	7.46	24.475	559.6
6.27	20.571	394.7	6.67	21.883	446.9	7.07	23.195	502.3	7.47	24.508	561.1
6.28	20.604	395.9	6.68	21.916	448.2	7.08	23.228	503.8	7.48	24.541	562.6
6.29	20.636	397.2	6.69	21.949	449.6	7.09	23.261	505.2	7.59	24.573	564.1
6.30	20.669	398.5	6.70	21.982	450.9	7.10	23.294	506.6	7.50	24.606	565.7
6.31	20.702	399.7	6.71	22.014	452.3	7.11	23.327	508.1			
6.32	20.735	401.0	6.72	22.047	453.6	7.12	23.360	509.5			
6.33	20.768	402.3	6.73	22.080	455.0	7.13	23.392	511.0			
6.34	20.800	403.6	6.74	22.113	456.3	7.14	23.425	512.4			
6.35	20.833	404.9	6.75	22.146	457.7	7.15	23.458	513.8			
6.36	20.866	406.1	6.76	22.178	459.1	7.16	23.491	515.3			
6.37	20.899	407.4	6.77	22.211	460.4	7.17	23.524	516.7			
6.38	20.932	408.7	6.78	22.244	461.8	7.18	23.556	518.2			
6.39	20.965	410.0	6.79	22.277	463.2	7.19	23.589	519.6			

Table D-3. Difference in Elevation, Given Horizontal Distances and Gradients from 0° to 45°

From the line of the given gradient take out the tabular numbers corresponding to each of the numbers of the given distance. Point off the value found in the table, with as many places to the left of the decimal point as the distance has places. Add the values thus found. The result is the difference in elevation in the same unit as the distance. Example: Find the difference in elevation corresponding to a horizontal distance of 6,273 feet and a gradient of 3°.

For 3, opposite 3°, and under 3, point off 1 place	0.1572
For 70, opposite 3°, and under 7, point off 2 places	03.668
For 200, opposite 3°, and under 2, point off 3 places	010.48
For 6,000, opposite 3°, and under 6, point off 4 places	0314.4
For 6,273, the difference in elevation is	328.7052

Gradient in degrees	Difference in elevation for horizontal distance of—								
	1	2	3	4	5	6	7	8	9
¼	00044	00087	00131	00174	00218	00262	00305	00349	00392
½	00087	00174	00261	00348	00435	00522	00609	00696	00783
¾	00131	00262	00393	00524	00654	00785	00916	01047	01178
1	00174	00349	00523	00698	00872	01047	01221	01396	01570
1¼	00218	00436	00655	00873	01091	01309	01527	01746	01964
1½	00262	00524	00786	01048	01310	01572	01834	02096	02358
1¾	00306	00611	00916	01222	01527	01833	02138	02444	02749
2	00349	00698	01047	01396	01745	02094	02443	02792	03141
2¼	00393	00786	01179	01572	01964	02357	02750	03143	03536
2½	00436	00872	01308	01744	02180	02616	03052	03488	03924
2¾	00480	00961	01441	01921	02402	02882	03362	03842	04323
3	00524	01048	01572	02096	02620	03144	03668	04192	04716
3½	00612	01223	01835	02446	03058	03670	04281	04893	05504
4	00699	01398	02097	02797	03496	04195	04894	05594	06293
4½	00787	01574	02361	03148	03935	04722	05509	06296	07083
5	00875	01750	02625	03500	04375	05250	06125	07000	07875
5½	00963	01926	02889	03852	04814	05777	06740	07703	08666
6	01051	02102	03153	04204	05255	06306	07357	08408	09459
7	01228	02456	03684	04912	06140	07368	08596	09824	11052
8	01405	02810	04216	05621	07027	08432	09837	11243	12648
9	01584	03168	04752	06336	07920	09504	11088	12672	14256
10	01763	03526	05289	07053	08816	10579	12343	14106	15869
11	01944	03888	05831	07775	09719	11663	13607	15550	17494
12	02125	04251	06376	08502	10628	12753	14879	17004	19130
13	02309	04617	06926	09235	11543	13852	16161	18470	20778
14	02493	04986	07479	09973	12466	14959	17453	19946	22439
15	02679	05359	08038	10718	13397	16077	18756	21436	24115
16	02867	05734	08602	11469	14337	17204	20071	22939	25806
17	03057	06115	09172	12229	15287	18344	21401	24458	27516
18	03249	06498	09747	12996	16245	19494	22743	25992	29241
19	03443	06887	10330	13773	17216	20660	24103	27546	30990
20	03639	07279	10919	14558	18198	21838	25477	29117	32757
21	03839	07677	11516	15354	19193	23032	26870	30709	34547
22	04040	08080	12120	16161	20201	24241	28282	32322	36362
23	04248	08489	12734	16979	21224	25468	29713	33958	38202
24	04452	08904	13356	17809	22261	26713	31166	35818	40070
25	04663	09326	13989	18652	23315	27979	32642	37305	41970
26	04877	09754	14631	19509	24386	29263	34141	39018	43895
27	05095	10191	15286	20381	25476	30572	35667	40762	45858
28	05317	10634	15951	21268	26585	31902	37219	42536	47853
29	05543	11086	16629	22172	27715	33259	38802	44345	49888
30	05773	11547	17320	23094	28867	34641	40414	46188	51961
31	06009	12017	18026	24034	30043	36052	42060	48069	54077
32	06249	12497	18746	24995	31243	37492	43741	49990	56238
33	06494	12988	19482	25976	32470	38965	45459	51953	58447
34	06745	13490	20235	26980	33725	40471	47216	53961	60706
35	07002	14004	21006	28008	35010	42013	49015	56017	63019
37	07536	15071	22607	30142	37678	45213	52749	60284	67820
39	08098	16196	24293	32391	40489	48587	56685	64782	72880
41	08693	17386	26079	34772	43464	52157	60850	69543	78236
43	09325	18650	27976	37301	46626	55951	65276	74602	83927
45	10000	20000	30000	40000	50000	60000	70000	80000	90000

The difference in elevation for any gradient and any horizontal distance may be obtained by multiplying the distance by the tangent of the angle or gradient.

Table D-4. Difference in Evaluation, Given Slope Distances and Gradients from 0° to 45°

Gradient in degrees	Differences in elevation for horizontal distances of—								
	1	2	3	4	5	6	7	8	9
¼	00044	00087	00131	00174	00218	00262	00305	00349	00392
½	00087	00174	00262	00349	00436	00523	00611	00698	00785
¾	00131	00262	00393	00524	00655	00785	00916	01047	01178
1	00174	00349	00523	00698	00873	01047	01222	01396	01571
1¼	00218	00436	00654	00872	01091	01309	01527	01745	01963
1½	00262	00523	00785	01047	01309	01571	01832	02094	02356
1¾	00305	00611	00916	01222	01527	01832	02138	02443	02749
2	00349	00698	01047	01396	01745	02094	02443	02792	03141
2¼	00393	00785	01178	01570	01963	02356	02748	03141	03533
2½	00436	00872	01308	01745	02181	02617	03059	03489	03926
2¾	00480	00960	01439	01919	02399	02879	03359	03838	04318
3	00523	01047	01570	02093	02617	03140	03663	04187	04710
3½	00611	01221	01832	02442	03053	03663	04274	04884	05495
4	00697	01395	02093	02790	03488	04185	04883	05880	06278
4½	00785	01569	02354	03138	03923	04708	05492	06277	07061
5	00871	01743	02615	03486	04358	05229	06101	06972	07844
5½	00959	01917	02876	03834	04783	05751	06710	07668	08627
6	01045	02090	03136	04181	05226	06272	07317	08362	09407
7	01219	02437	03656	04875	06093	07312	08531	09749	10968
8	01392	02783	04175	05567	06959	08350	09742	11134	12525
9	01564	03129	04693	06257	07822	09386	10950	12515	14079
10	01736	03473	05209	06946	08682	10419	12155	13892	15628
11	01908	03816	05724	07632	09541	11449	13357	15265	17173
12	02079	04158	06237	08316	10395	12475	14554	16633	18712
13	02250	04499	06749	08998	11248	13497	15747	17996	20246
14	02419	04838	07258	09677	12096	14515	16934	19354	21773
15	02588	05176	07765	10353	12941	15529	18117	20706	23294
16	02756	05513	08269	11025	13782	16538	19294	22051	24807
17	02924	05847	08771	11695	14619	17542	20466	23390	26313
18	03090	06180	09270	12361	15451	18541	21631	24721	27811
19	03256	06511	09767	13023	16279	19534	22790	26046	29301
20	03420	06840	10261	13681	17101	20521	23941	27362	30782
21	03584	07167	10751	14335	17919	21502	25086	28670	32253
22	03746	07492	11238	14984	18730	22476	26222	29968	33714
23	03907	07815	11722	15629	19537	23444	27351	31258	35166
24	04067	08135	12202	16269	20337	24404	28471	32539	36606
25	04226	08452	12679	16905	21131	25357	29583	33810	38036
26	04384	08767	13151	17535	21918	26302	30686	35070	39453
27	04540	09080	13620	18160	22700	27239	31779	36319	40859
28	04695	09389	14084	18779	23473	28168	32863	37558	42252
29	04848	09696	14544	19392	24241	29089	33937	38785	43633
30	05000	10000	15000	20000	25000	30000	35000	40000	45000
31	05150	10301	15451	20602	25752	30902	36053	41203	46354
32	05299	10598	15898	21197	26496	31795	37094	42394	47693
33	05446	10893	16339	21786	27232	32678	38125	43571	49018
34	05592	11184	16776	22368	27960	33551	39143	44735	50327
35	05736	11472	17207	22943	28679	34415	40151	45886	51622
37	06018	12036	18055	24073	30091	36109	42127	48146	54164
39	06293	12586	18880	25173	31466	37759	44052	50346	56639
41	06561	13121	19682	26242	32803	39364	45924	52485	59045
43	06820	13640	20460	27280	34100	40920	47740	54560	61380
45	07071	14142	21213	28284	35356	42427	49498	56569	63640

The difference of elevation for any sloping distance and any angle or gradient may be found by multiplying the distance by the sine of the angle.

Table D-5. Horizontal Distance, Given Slope Distances and Gradients from 0° to 45°

Gradient in degrees	Horizontal distances for slope distances of—								
	1	2	3	4	5	6	7	8	9
¼	09999+	19999	29999	39999	49999	59999	69999	79999	89999
½	09999+	19999	29999	39998	49998	59998	69997	79997	89996
¾	09999	19998	29997	39996	49996	59995	69994	79993	89992
1	09998	19997	29995	39994	49992	59991	69989	79988	89986
1¼	09998	19995	29993	39990	49988	59986	69983	79981	89978
1½	09997	19993	29990	39986	49983	59980	69976	79973	89969
1¾	09995	19991	29986	39981	49977	59972	69967	79962	89958
2	09994	19988	29982	39976	49969	59963	69957	79951	89947
2¼	09992	19985	29977	39969	49961	59954	69946	79938	89931
2½	09990	19981	29971	39962	49952	59943	69933	79924	89914
2¾	09988	19977	29965	39954	49942	59931	69920	79908	89896
3	09986	19973	29959	39945	49931	59918	69904	79890	89877
3½	09981	19963	29944	39925	49906	59888	69869	79850	89832
4	09976	19951	29927	39902	49878	59854	69829	79805	89780
4½	09969	19938	29908	39877	49846	59815	69784	79754	89723
5	09962	19924	29886	39848	49810	59772	69733	79695	89657
5½	09954	19908	29862	39816	49770	59724	69678	79632	89586
6	09945	19890	29836	39781	49726	59671	69616	79562	89507
7	09925	19851	29776	39702	49627	59553	69478	79404	89329
8	09903	19805	29708	39611	49513	59416	69319	79221	89124
9	09877	19754	29631	39507	49384	59261	69138	79015	88892
10	09848	19696	29544	39392	49240	59088	68936	78785	88633
11	09816	19633	29449	39265	49082	58898	68714	78530	88347
12	09781	19563	29344	39126	48907	58689	68470	78252	88033
13	09744	19487	29221	38975	48719	58462	68206	77950	87693
14	09703	19406	29108	38812	48515	58218	67921	77624	87326
15	09659	19319	28978	38637	48297	57956	67615	77274	86938
16	09613	19225	28838	38450	48063	57676	67288	76901	86513
17	09563	19126	28689	38252	47815	57378	66941	76504	86067
18	09510	19021	28532	38042	47553	57063	66574	76084	85595
19	09455	18910	28365	37820	47275	56730	66185	75640	85095
20	09397	18794	28191	37588	46985	56381	65778	75175	84572
21	09336	18672	28007	37343	46679	56015	65351	74686	84022
22	09272	18544	27815	37087	46359	55631	64903	74175	83446
23	09205	18410	27615	36820	46025	55230	64435	73640	82845
24	09135	18271	27406	36542	45677	54813	63948	73084	82219
25	09063	18126	27189	36252	45315	54378	63441	72505	81568
26	08988	17976	26964	35952	44940	53928	62915	71903	80891
27	08910	17820	26730	35640	44550	53460	62370	71820	80190
28	08829	17659	26488	35318	44147	52977	61806	70636	79465
29	08746	17492	26238	34985	43731	52477	61223	69969	78716
30	08660	17320	25981	34641	43301	51961	60622	69282	77942
31	08572	17143	25715	34287	42859	51430	60002	68574	77145
32	08481	16961	25442	33922	42403	50883	59364	67844	76325
33	08387	16773	25160	33547	41934	50320	58707	67094	75480
34	08290	16581	24871	33162	41452	49742	58033	66323	74614
35	08192	16383	24575	32766	40958	49149	57341	65532	73724
37	07986	15973	23959	31946	39932	47918	55905	63891	71878
39	07772	15543	23315	31086	38858	46629	54401	62172	69944
41	07547	15094	22641	30188	37736	45283	52830	60377	67924
43	07314	14627	21941	29254	36568	43881	51195	58508	65822
45	07071	14142	21213	28284	35356	42427	49498	56569	63640

The horizontal distance corresponding to any given slope distance and any angle or gradient may be found by multiplying the slope distance by the cosine of the angle.

Table D–6. Stadia Reduction

Stadia work involves observing . . .

a. The angle by which the line of sight departs from the horizontal line. This reading is the argument for entering the table.

b. The rod interval intercepted by the stadia wires, which are usually adjusted so that the distance to the rod is exactly 100 times the reading on the rod when the telescope is level. This distance, however, is measured from the principal focus of the object glass, which is sometimes an appreciable distance to the front. Hence the reading must be increased by two small increments; first, the distance from the principal focus to the object glass, and, second, the distance from the object glass to the instrument's plumb line.

c. The table gives the horizontal distances and differences of elevation for unit readings on the rod and angles of elevation from 0° to 30°. The three lines at the foot of the table gives three values for "C"; the value selected depends upon the instrument used; the quantity "C" includes the two increments described in *b* above.

d. Example: Rod reading is 3.25 and the angle of inclination is 5° 35′; focal distance (C) = 1.00.

Horizontal distance = 99.05 x 3.25 + 0.99 = 322.90 feet

Difference of elevation = 9.68 x 3.25 + 0.09 = 31.55 feet

Table D-6. Stadia Reduction—Continued

(Horizontal distances and elevations from stadia readings)

Minutes	0°		1°		2°		3°	
	Hor. dist.	Diff. elev.	Hor. dist.	Diff. elev.	Hor. dist.	Diff. elev.	Hor. dist.	Diff. elev.
0	100.00	0.00	99.97	1.74	99.88	3.49	99.73	5.23
2	100.00	0.06	99.97	1.80	99.87	3.55	99.72	5.28
4	100.00	0.12	99.97	1.86	99.87	3.60	99.71	5.34
6	100.00	0.17	99.96	1.92	99.87	3.66	99.71	5.40
8	100.00	0.23	99.96	1.98	99.86	3.72	99.70	5.46
10	100.00	0.29	99.96	2.04	99.86	3.78	99.69	5.52
12	100.00	0.35	99.96	2.09	99.85	3.84	99.69	5.57
14	100.00	0.41	99.95	2.15	99.85	3.90	99.68	5.63
16	100.00	0.47	99.95	2.21	99.84	3.95	99.68	5.69
18	100.00	0.52	99.95	2.27	99.84	4.01	99.67	5.75
20	100.00	0.58	99.95	2.33	99.83	4.07	99.66	5.80
22	100.00	0.64	99.94	2.38	99.83	4.13	99.66	5.86
24	100.00	0.70	99.94	2.44	99.82	4.18	99.65	5.92
26	99.99	0.76	99.94	2.50	99.82	4.24	99.64	5.98
28	99.99	0.81	99.93	2.56	99.81	4.30	99.63	6.04
30	99.99	0.87	99.93	2.62	99.81	4.36	99.63	6.09
32	99.99	0.93	99.93	2.67	99.80	4.42	99.62	6.15
34	99.99	0.99	99.93	2.73	99.80	4.48	99.62	6.21
36	99.99	1.05	99.92	2.79	99.79	4.53	99.61	6.27
38	99.99	1.11	99.92	2.85	99.79	4.59	99.60	6.33
40	99.99	1.16	99.92	2.91	99.78	4.65	99.59	6.38
42	99.99	1.22	99.91	2.97	99.78	4.71	99.59	6.44
44	99.98	1.28	99.91	3.02	99.77	4.76	99.58	6.50
46	99.98	1.34	99.90	3.08	99.77	4.82	99.57	6.56
48	99.98	1.40	99.90	3.14	99.76	4.88	99.56	6.61
50	99.98	1.45	99.90	3.20	99.76	4.94	99.56	6.67
52	99.98	1.51	99.89	3.26	99.75	4.99	99.55	6.73
54	99.98	1.57	99.89	3.31	99.74	5.05	99.54	6.78
56	99.97	1.63	99.89	3.37	99.74	5.11	99.53	6.84
58	99.97	1.69	99.88	3.43	99.73	5.17	99.52	6.90
60	99.97	1.74	99.88	3.49	99.73	5.23	99.51	6.96
C=0.75	0.75	0.01	0.75	0.02	0.75	0.03	0.75	0.05
C=1.00	1.00	0.01	1.00	0.03	1.00	0.04	1.00	0.06
C=1.25	1.25	0.02	1.25	0.03	1.25	0.05	1.25	0.08

Table D-6. Stadia Reduction—Continued

Minutes	4°		5°		6°		7°	
	Hor. dist.	Diff. elev.	Hor. dist.	Diff. elev.	Hor. dist.	Diff. elev.	Hor. dist.	Diff. elev.
0	99.51	6.96	99.24	8.68	98.91	10.40	98.51	12.10
2	99.51	7.02	99.23	8.74	98.90	10.45	98.50	12.15
4	99.50	7.07	99.22	8.80	98.88	10.51	98.48	12.21
6	99.49	7.13	99.21	8.85	98.87	10.57	98.47	12.26
8	99.48	7.19	99.20	8.91	98.86	10.62	98.46	12.32
10	99.47	7.25	99.19	8.97	98.85	10.68	98.44	12.38
12	99.46	7.30	99.18	9.03	98.83	10.74	98.43	12.43
14	99.46	7.36	99.17	9.08	98.82	10.79	98.41	12.49
16	99.45	7.42	99.16	9.14	98.81	10.85	98.40	12.55
18	99.44	7.48	99.15	9.20	98.80	10.91	98.39	12.60
20	99.43	7.53	99.14	9.25	98.78	10.96	98.37	12.66
22	99.42	7.59	99.13	9.31	98.77	11.02	98.36	12.72
24	99.41	7.65	99.11	9.37	98.76	11.08	98.34	12.77
26	99.40	7.71	99.10	9.43	98.74	11.13	98.33	12.83
28	99.39	7.76	99.09	9.48	98.73	11.19	98.31	12.88
30	99.38	7.82	99.08	9.54	98.27	11.25	98.29	12.94
32	99.38	7.88	99.07	9.60	98.71	11.30	98.28	13.00
34	99.37	7.94	99.06	9.65	98.69	11.36	98.27	13.05
36	99.36	7.99	99.05	9.71	98.68	11.42	98.25	13.11
38	99.35	8.05	99.04	9.77	98.67	11.47	98.24	13.17
40	99.34	8.11	99.03	9.83	98.65	11.53	98.22	13.22
42	99.33	8.17	99.01	9.88	98.64	11.59	98.20	13.28
44	99.32	8.22	99.00	9.94	98.63	11.64	98.19	13.33
46	99.31	8.28	98.99	10.00	98.61	11.70	98.17	13.39
48	99.30	8.34	98.98	10.05	98.60	11.76	98.16	13.45
50	99.29	8.40	98.97	10.11	98.58	11.81	98.14	13.50
52	99.28	8.45	98.96	10.17	98.57	11.87	98.13	13.56
54	99.27	8.51	98.94	10.22	98.56	11.93	98.11	13.61
56	99.26	8.57	98.93	10.28	98.54	11.98	98.10	13.67
58	99.25	8.63	98.92	10.34	98.53	12.04	98.08	13.73
60	99.24	8.68	98.91	10.40	98.51	12.10	98.06	13.78
C=0.75	0.75	0.06	0.75	0.07	0.75	0.08	0.74	0.10
C=1.00	1.00	0.08	0.99	0.09	0.99	0.11	0.99	0.13
C=1.25	1.25	0.10	1.24	0.11	1.24	0.14	1.24	0.16

Table D–6. Stadia Reduction—Continued

Minutes	8°		9°		10°		11°	
	Hor. dist.	Diff. elev.	Hor. dist.	Diff. elev.	Hor. dist.	Diff. elev.	Hor. dist.	Diff. elev.
0	98.06	13.78	97.55	15.45	96.98	17.10	96.36	18.73
2	98.05	13.84	97.53	15.51	96.96	17.16	96.34	18.78
4	98.03	13.89	97.52	15.56	96.94	17.21	96.32	18.84
6	98.01	13.95	97.50	15.62	96.92	17.26	96.29	18.89
8	98.00	14.01	97.48	15.67	96.90	17.32	96.27	18.95
10	97.98	14.06	97.46	15.73	96.88	17.37	96.25	19.00
12	97.97	14.12	97.44	15.78	96.86	17.43	96.23	19.05
14	97.95	14.17	97.43	15.84	96.84	17.48	96.21	19.11
16	97.93	14.23	97.41	15.89	96.82	17.54	96.18	19.16
18	97.92	14.28	97.39	15.95	96.80	17.59	96.16	19.21
20	97.90	14.34	97.37	16.00	96.78	17.65	96.14	19.27
22	97.88	14.40	97.35	16.06	96.76	17.70	96.12	19.32
24	97.87	14.45	97.33	16.11	96.74	17.76	96.09	19.38
26	97.85	14.51	97.31	16.17	96.72	17.81	96.07	19.43
28	97.83	14.56	97.29	16.22	96.70	17.86	96.05	19.48
30	97.82	14.62	97.28	16.28	96.68	17.92	96.03	19.54
32	97.80	14.67	97.26	16.33	96.66	17.97	96.00	19.59
34	97.78	14.73	97.24	16.39	96.64	18.03	95.98	19.64
36	97.76	14.79	97.22	16.44	96.62	18.08	95.96	19.70
38	97.75	14.84	97.20	16.50	96.60	18.14	95.93	19.75
40	97.73	14.90	97.18	16.55	96.57	18.19	95.91	19.80
42	97.71	14.95	97.16	16.61	96.55	18.24	95.89	19.86
44	97.69	15.01	97.14	16.66	96.53	18.30	95.86	19.91
46	97.68	15.06	97.12	16.72	96.51	18.35	95.84	19.96
48	97.66	15.12	97.10	16.77	96.49	18.41	95.82	20.02
50	97.64	15.17	97.08	16.83	96.47	18.46	95.79	20.07
52	97.62	15.23	97.06	16.88	96.45	18.51	95.77	20.12
54	97.61	15.28	97.04	16.94	96.42	18.57	95.75	20.18
56	97.59	15.34	97.02	16.99	96.40	18.62	95.72	20.23
58	97.57	15.40	97.00	17.05	96.38	18.68	95.70	20.28
60	97.55	15.45	96.98	17.10	96.36	18.73	95.68	20.34
C = 0.75	0.74	0.11	0.74	0.12	0.74	0.14	0.73	0.15
C = 1.00	0.99	0.15	0.99	0.16	0.98	0.18	0.98	0.20
C = 1.25	1.23	0.18	1.23	0.21	1.23	0.23	1.22	0.25

Table D–6. Stadia Reduction—Continued

Minutes	12°		13°		14°		15°	
	Hor. dist.	Diff. elev.	Hor. dist.	Diff. elev.	Hor. dist.	Diff. elev.	Hor. dist.	Diff. elev.
0	95.68	20.34	94.94	21.92	94.15	23.47	93.30	25.00
2	95.65	20.39	94.91	21.97	94.12	23.52	93.27	25.05
4	95.63	20.44	94.89	22.02	94.09	23.58	93.24	25.10
6	95.61	20.50	94.86	22.08	94.07	23.63	93.21	25.15
8	95.58	20.55	94.84	22.13	94.04	23.68	93.18	25.20
10	95.56	20.60	94.81	22.18	94.01	23.73	93.16	25.25
12	95.53	20.66	94.79	22.23	93.98	23.78	93.13	25.30
14	95.51	20.71	94.76	22.28	93.95	23.83	93.10	25.35
16	95.49	20.76	94.73	22.34	93.93	23.88	93.07	25.40
18	95.46	20.81	94.71	22.39	93.90	23.93	93.04	25.45
20	95.44	20.87	94.68	22.44	93.87	23.99	93.01	25.50
22	95.41	20.92	94.66	22.49	93.84	24.04	92.98	25.55
24	95.39	20.97	94.63	22.54	93.81	24.09	92.95	25.60
26	95.36	21.03	94.60	22.60	93.79	24.14	92.92	25.65
28	95.34	21.08	94.58	22.65	93.76	24.19	92.89	25.70
30	95.32	21.13	94.55	22.70	93.73	24.24	92.86	25.75
32	95.29	21.18	94.52	22.75	93.70	24.29	92.83	25.80
34	95.27	21.24	94.50	22.80	93.67	24.34	92.80	25.85
36	95.24	21.29	94.47	22.85	93.65	24.39	92.77	25.90
38	95.22	21.34	94.44	22.91	93.62	24.44	92.74	25.95
40	95.19	21.39	94.42	22.96	93.59	24.49	92.71	26.00
42	95.17	21.45	94.39	23.01	93.56	24.55	92.68	26.05
44	95.14	21.50	94.36	23.06	93.53	24.60	92.65	26.10
46	95.12	21.55	94.34	23.11	93.50	24.65	92.62	26.15
48	95.09	21.60	94.31	23.16	93.47	24.70	92.59	26.20
50	95.07	21.66	94.28	23.22	93.45	24.75	92.56	26.25
52	95.04	21.71	94.26	23.27	93.42	24.80	92.53	26.30
54	95.02	21.76	94.23	23.32	93.39	24.85	92.49	26.35
56	94.99	21.81	94.20	23.37	93.36	24.90	92.46	26.40
58	94.97	21.87	94.17	23.42	93.33	24.95	92.43	26.45
60	94.94	21.92	94.15	23.47	93.90	25.00	92.40	26.50
C=0.75	0.73	0.16	0.73	0.17	0.73	0.19	0.72	0.20
C=1.00	0.98	0.22	0.97	0.23	0.97	0.25	0.96	0.27
C=1.25	1.22	0.27	1.21	0.29	1.21	0.31	1.20	0.34

Table D-6. Stadia Reduction—Continued

Minutes	16°		17°		18°		19°	
	Hor. dist.	Diff. elev.	Hor. dist.	Diff. elev.	Hor. dist.	Diff. elev.	Hor. dist.	Diff. elev.
0	92.40	26.50	91.45	27.96	90.45	29.39	89.40	30.78
2	92.37	26.55	91.42	28.01	90.42	29.44	89.36	30.83
4	92.34	26.59	91.39	28.06	90.38	29.48	89.33	30.87
6	92.31	26.64	91.35	28.10	90.35	29.53	89.29	30.92
8	92.28	26.69	91.32	28.15	90.31	29.58	89.26	30.97
10	92.25	26.74	91.29	28.20	90.28	29.62	89.22	31.01
12	92.22	26.79	91.26	28.25	90.24	29.67	89.18	31.06
14	92.19	26.84	91.22	28.30	90.21	29.72	89.15	31.10
16	92.15	26.89	91.19	28.34	90.18	29.76	89.11	31.15
18	92.12	26.94	91.16	28.39	90.14	29.81	89.08	31.19
20	92.09	26.99	91.12	28.44	90.11	29.86	89.04	31.24
22	92.06	27.04	91.09	28.49	90.07	29.90	89.00	31.28
24	92.03	27.09	91.06	28.54	90.04	29.95	88.96	31.33
26	92.00	27.13	91.02	28.58	90.00	30.00	88.93	31.38
28	91.97	27.18	90.99	28.63	89.97	30.04	88.89	31.42
30	91.93	27.23	90.96	28.68	89.93	30.09	88.86	31.47
32	91.90	27.28	90.92	28.73	89.90	30.14	88.82	31.51
34	91.87	27.33	90.89	28.77	89.86	30.19	88.78	31.56
36	91.84	27.38	90.86	28.82	89.83	30.23	88.75	31.60
38	91.81	27.43	90.82	28.87	89.79	30.28	88.71	31.65
40	91.77	27.48	90.79	28.92	89.76	30.32	88.67	31.69
42	91.74	27.52	90.76	28.96	89.72	30.37	88.64	31.74
44	91.71	27.57	90.72	29.01	89.69	30.41	88.60	31.78
46	91.68	27.62	90.69	29.06	89.65	30.46	88.56	31.83
48	91.65	27.67	90.66	29.11	89.61	30.51	88.53	31.87
50	91.61	27.72	90.62	29.15	89.58	30.55	88.49	31.92
52	91.58	27.77	90.59	29.20	89.54	30.60	88.45	31.96
54	91.55	27.81	90.55	29.25	89.51	30.65	88.41	32.01
56	91.52	27.86	90.52	29.30	89.47	30.69	88.38	32.05
58	91.48	27.91	90.48	29.34	89.44	30.74	88.34	32.09
60	91.45	27.96	90.45	29.39	89.40	30.78	88.30	32.14
C=0.75	0.72	0.21	0.72	0.23	0.71	0.24	0.71	0.25
C=1.00	0.96	0.28	0.95	0.30	0.95	0.32	0.94	0.33
C=1.25	1.20	0.35	1.19	0.38	1.19	0.40	1.18	0.42

Table D-6. Stadia Reduction—Continued

Minutes	20°		21°		22°		23°	
	Hor. dist.	Diff. elev.	Hor. dist.	Diff. elev.	Hor. dist.	Diff. elev.	Hor. dist.	Diff. elev.
0	88.30	32.14	87.16	33.46	85.97	34.73	84.73	35.97
2	88.26	32.18	87.12	33.50	85.93	34.77	84.69	36.01
4	88.23	32.23	87.08	33.54	85.89	34.82	84.65	36.05
6	88.19	32.27	87.04	33.59	85.85	34.86	84.61	36.09
8	88.15	32.32	87.00	33.63	85.80	34.90	84.57	36.13
10	88.11	32.36	86.96	33.67	85.76	34.94	84.52	36.17
12	88.08	32.41	86.92	33.72	85.72	34.98	84.48	36.21
14	88.04	32.45	86.88	33.76	85.68	35.02	84.44	36.25
16	88.00	32.49	86.84	33.80	85.64	35.07	84.40	36.29
18	87.96	32.54	86.80	33.84	85.60	35.11	84.35	36.33
20	87.93	32.58	86.77	33.89	85.56	35.15	84.31	36.37
22	87.89	32.63	86.73	33.93	85.52	35.19	84.27	36.41
24	87.85	32.67	86.69	33.97	85.48	35.23	84.23	36.45
26	87.81	32.72	86.65	34.01	85.44	35.27	84.18	36.49
28	87.77	32.76	86.61	34.06	85.40	35.31	84.14	36.53
30	87.74	32.80	86.57	34.10	85.36	35.36	84.10	36.57
32	87.70	32.85	86.53	34.14	85.31	35.40	84.06	36.61
34	87.66	32.89	86.49	34.18	85.27	35.44	84.01	36.65
36	87.62	32.93	86.45	34.23	85.23	35.48	83.97	36.69
38	87.58	32.98	86.41	34.27	85.19	35.52	83.93	36.73
40	87.54	33.02	86.37	34.31	85.15	35.56	83.89	36.77
42	87.51	33.07	86.33	34.35	85.11	35.60	83.84	36.80
44	87.47	33.11	86.29	34.40	85.07	35.64	83.80	36.84
46	87.43	33.15	86.25	34.44	85.02	35.68	83.76	36.88
48	87.39	33.20	86.21	34.48	84.98	35.72	83.72	36.92
50	87.35	33.24	86.17	34.52	84.94	35.76	83.67	36.96
52	87.31	33.28	86.13	34.57	84.90	35.80	83.63	37.00
54	87.27	33.33	86.09	34.61	84.86	35.85	83.59	37.04
56	87.24	33.37	86.05	34.65	84.82	35.89	83.54	37.08
58	87.20	33.41	86.01	34.69	84.77	35.93	83.50	37.12
60	87.16	33.46	85.97	34.73	84.73	35.97	83.46	37.16
C=0.75	0.70	0.26	0.70	0.27	0.69	0.29	0.69	0.30
C=1.00	0.94	0.35	0.93	0.37	0.92	0.38	0.92	0.40
C=1.25	1.17	0.44	1.16	0.46	1.15	0.48	1.15	0.50

Table D-6. Stadia Reduction—Continued

Minutes	24°		25°		26°		27°	
	Hor. dist.	Diff. elev.	Hor. dist.	Diff. elev.	Hor. dist.	Diff. elev.	Hor. dist.	Diff. elev.
0	83.46	37.16	82.14	38.30	80.78	39.40	79.39	40.45
2	83.41	37.20	82.09	38.34	80.74	39.44	79.34	40.49
4	83.37	37.23	82.05	38.38	80.69	39.47	79.30	40.52
6	83.33	37.27	82.01	38.41	80.65	39.51	79.25	40.55
8	83.28	37.31	81.96	38.45	80.60	39.54	79.20	40.59
10	83.24	37.35	81.92	38.49	80.55	39.58	79.15	40.62
12	83.20	37.39	81.87	38.53	80.51	39.61	79.11	40.66
14	83.15	37.43	81.83	38.56	80.46	39.65	79.06	40.69
16	83.11	37.47	81.78	38.60	80.41	39.69	79.10	40.72
18	83.07	37.51	81.74	38.64	80.37	39.72	78.96	40.76
20	83.02	37.54	81.69	38.67	80.32	39.76	78.92	40.79
22	82.98	37.58	81.65	38.71	80.28	39.79	78.87	40.82
24	82.93	37.62	81.60	38.75	80.23	39.83	78.82	40.86
26	82.89	37.66	81.56	38.78	80.18	39.86	78.77	40.89
28	82.85	37.70	81.51	38.82	80.14	39.90	78.73	40.92
30	82.80	37.74	81.47	38.86	80.09	39.93	78.68	40.96
32	82.76	37.77	81.42	38.89	80.04	39.97	78.63	40.99
34	82.72	37.81	81.38	38.93	80.00	40.00	78.58	41.02
36	82.67	37.85	81.33	38.97	79.95	40.04	78.54	41.06
38	82.63	37.89	81.28	39.00	79.90	40.07	78.49	41.09
40	82.58	37.93	81.24	39.04	79.86	40.11	78.44	41.12
42	82.54	37.96	81.19	39.08	79.81	40.14	78.39	41.16
44	82.49	38.00	81.15	39.11	79.76	40.18	78.34	41.19
46	82.45	38.04	81.10	39.15	79.72	40.21	78.30	41.22
48	82.41	38.08	81.06	39.18	79.67	40.24	78.25	41.26
50	82.36	38.11	81.01	39.22	79.62	40.28	78.20	41.29
52	82.32	38.15	80.97	39.26	79.58	40.31	78.15	41.32
54	82.27	38.19	80.92	39.29	79.53	40.35	78.10	41.35
56	82.23	38.23	80.87	39.33	79.48	40.38	78.06	41.39
58	82.18	38.26	80.83	39.36	79.44	40.42	78.01	41.42
60	82.14	38.30	80.78	39.40	79.39	40.45	77.96	41.45
C=0.75	0.68	0.31	0.68	0.32	0.67	0.33	0.66	0.35
C=1.00	0.91	0.41	0.90	0.43	0.89	0.45	0.89	0.46
C=1.25	1.14	0.52	1.13	0.54	1.12	0.56	1.11	0.58

Table D-6. Stadia Reduction—Continued

Minutes	28°		29°		30°	
	Hor. dist.	Diff. elev.	Hor. dist.	Diff. elev.	Hor. dist.	Diff. elev.
0	77.96	41.45	76.50	42.40	75.00	43.30
2	77.91	41.48	76.45	42.43	74.95	43.33
4	77.86	41.52	76.40	42.46	74.90	43.36
6	77.81	41.55	76.35	42.49	74.85	43.39
8	77.77	41.58	76.30	42.53	74.80	43.42
10	77.72	41.61	76.25	42.56	74.75	43.45
12	77.67	41.65	76.20	42.59	74.70	43.47
14	77.62	41.68	76.15	42.62	74.65	43.50
16	77.57	41.71	76.10	42.65	74.60	43.53
18	77.52	41.74	76.05	42.68	74.55	43.56
20	77.48	41.77	76.00	42.71	74.49	43.59
22	77.42	41.81	75.95	42.74	74.44	43.62
24	77.38	41.84	75.90	42.77	74.39	43.65
26	77.33	41.87	75.85	42.80	74.34	43.67
28	77.28	41.90	75.80	42.83	74.29	43.70
30	77.23	41.93	75.75	42.86	74.24	43.73
32	77.18	41.97	75.70	42.89	74.19	43.76
34	77.13	42.00	75.65	42.92	74.14	43.79
36	77.09	42.03	75.60	42.95	74.09	43.82
38	77.04	42.06	75.55	42.98	74.04	43.84
40	76.99	42.09	75.50	43.01	73.99	43.87
42	76.94	42.12	75.45	43.04	73.93	43.90
44	76.89	42.15	75.40	43.07	73.88	43.93
46	76.84	42.19	75.35	43.10	73.83	43.95
48	76.79	42.22	75.30	43.13	73.78	43.98
50	76.74	42.25	75.25	43.16	73.73	44.01
52	76.69	42.28	75.20	43.18	73.68	44.04
54	76.64	42.31	75.15	43.21	73.63	44.07
56	76.59	42.34	75.10	43.24	73.58	44.09
58	76.55	42.37	75.05	43.27	73.52	44.12
60	76.50	42.40	75.00	43.30	73.47	44.15
C=0.75	0.66	0.36	0.65	0.37	0.65	0.38
C=1.00	0.88	0.48	0.87	0.49	0.86	0.51
C=1.25	1.10	0.60	1.09	0.62	1.08	0.64

Table D–7a. Conversion Meters to Feet (International System)

LENGTH—METERS TO FEET 1 METER = 3.280 839 895 FEET

METERS	FEET	METERS	FEET	METERS	FEET	METERS	FEET	METERS	FEET
0	0.0000	50	164.0420	100	328.0840	150	492.1260	200	656.1680
1	3.2808	51	167.3228	101	331.3648	151	495.4068	201	659.4488
2	6.5617	52	170.6037	102	334.6457	152	498.6877	202	662.7297
3	9.8425	53	173.8845	103	337.9265	153	501.9685	203	666.0105
4	13.1234	54	177.1654	104	341.2073	154	505.2493	204	669.2913
5	16.4042	55	180.4462	105	344.4882	155	508.5302	205	672.5722
6	19.6850	56	183.7270	106	347.7690	156	511.8110	206	675.8530
7	22.9659	57	187.0079	107	351.0499	157	515.0919	207	679.1339
8	26.2467	58	190.2887	108	354.3307	158	518.3727	208	682.4147
9	29.5276	59	193.5696	109	357.6115	159	521.6535	209	685.6955
10	32.8084	60	196.8504	110	360.8924	160	524.9344	210	688.9764
11	36.0892	61	200.1312	111	364.1732	161	528.2152	211	692.2572
12	39.3701	62	203.4121	112	367.4541	162	531.4961	212	695.5381
13	42.6509	63	206.6929	113	370.7349	163	534.7769	213	698.8189
14	45.9318	64	209.9738	114	374.0157	164	538.0577	214	702.0997
15	49.2126	65	213.2546	115	377.2966	165	541.3386	215	705.3806
16	52.4934	66	216.5354	116	380.5774	166	544.6194	216	708.6614
17	55.7743	67	219.8163	117	383.8583	167	547.9003	217	711.9423
18	59.0551	68	223.0971	118	387.1391	168	551.1811	218	715.2231
19	62.3360	69	226.3780	119	390.4199	169	554.4619	219	718.5039
20	65.6168	70	229.6588	120	393.7008	170	557.7428	220	721.7848
21	68.8976	71	232.9396	121	396.9816	171	561.0236	221	725.0656
22	72.1785	72	236.2205	122	400.2625	172	564.3045	222	728.3465
23	75.4593	73	239.5013	123	403.5433	173	567.5853	223	731.6273
24	78.7402	74	242.7822	124	406.8241	174	570.8661	224	734.9081
25	82.0210	75	246.0630	125	410.1050	175	574.1470	225	738.1890
26	85.3018	76	249.3438	126	413.3858	176	577.4278	226	741.4698
27	88.5827	77	252.6247	127	416.6667	177	580.7087	227	744.7506
28	91.8635	78	255.9055	128	419.9475	178	583.9895	228	748.0315
29	95.1444	79	259.1863	129	423.2283	179	587.2703	229	751.3123
30	98.4252	80	262.4672	130	426.5092	180	590.5512	230	754.5932
31	101.7060	81	265.7480	131	429.7900	181	593.8320	231	757.8740
32	104.9869	82	269.0289	132	433.0709	182	597.1129	232	761.1549
33	108.2677	83	272.3097	133	436.3517	183	600.3937	233	764.4357
34	111.5486	84	275.5905	134	439.6325	184	603.6745	234	767.7165
35	114.8294	85	278.8714	135	442.9134	185	606.9554	235	770.9974
36	118.1102	86	282.1522	136	446.1942	186	610.2362	236	774.2782
37	121.3911	87	285.4331	137	449.4751	187	613.5171	237	777.5591
38	124.6719	88	288.7139	138	452.7559	188	616.7979	238	780.8399
39	127.9528	89	291.9947	139	456.0367	189	620.0787	239	784.1207
40	131.2336	90	295.2756	140	459.3176	190	623.3596	240	787.4016
41	134.5144	91	298.5564	141	462.5984	191	626.6404	241	790.6824
42	137.7953	92	301.8373	142	465.8793	192	629.9213	242	793.9632
43	141.0761	93	305.1181	143	469.1601	193	633.2021	243	797.2441
44	144.3570	94	308.3989	144	472.4409	194	636.4829	244	800.5249
45	147.6378	95	311.6798	145	475.7218	195	639.7638	245	803.8058
46	150.9186	96	314.9606	146	479.0026	196	643.0446	246	807.0866
47	154.1995	97	318.2415	147	482.2835	197	646.3255	247	810.3674
48	157.4803	98	321.5223	148	485.5643	198	649.6063	248	813.6483
49	160.7612	99	324.8031	149	488.8451	199	652.8871	249	816.9291

Table D-7a. Conversion Meters to Feet (International System)—Continued

Continued—METERS TO FEET 1 METER = 3.280 839 895 FEET

METERS	FEET	METERS	FEET	METERS	FEET	METERS	FEET	METERS	FEET
250	820.2100	300	984.2520	350	1148.2940	400	1312.3360	450	1476.3779
251	823.4908	301	987.5328	351	1151.5748	401	1315.6168	451	1479.6588
252	826.7717	302	990.8136	352	1154.8556	402	1318.8976	452	1482.9396
253	830.0525	303	994.0945	353	1158.1365	403	1322.1785	453	1486.2205
254	833.3333	304	997.3753	354	1161.4173	404	1325.4593	454	1489.5013
255	836.6142	305	1000.6562	355	1164.6982	405	1328.7401	455	1492.7821
256	839.8950	306	1003.9370	356	1167.9790	406	1332.0210	456	1496.0630
257	843.1758	307	1007.2178	357	1171.2598	407	1335.3018	457	1499.3438
258	846.4567	308	1010.4987	358	1174.5407	408	1338.5827	458	1502.6247
259	849.7375	309	1013.7795	359	1177.8215	409	1341.8635	459	1505.9055
260	853.0184	310	1017.0604	360	1181.1024	410	1345.1443	460	1509.1863
261	856.2992	311	1020.3412	361	1184.3832	411	1348.4252	461	1512.4672
262	859.5800	312	1023.9220	362	1187.6640	412	1351.7060	462	1515.7480
263	862.8609	313	1026.9029	363	1190.9449	413	1354.9869	463	1519.0289
264	866.1417	314	1030.1873	364	1194.2257	414	1358.2677	464	1522.3097
265	869.4226	315	1033.4646	365	1197.5065	415	1361.5486	465	1525.5905
266	872.7034	316	1036.7454	366	1200.7874	416	1364.8294	466	1528.8714
267	875.9842	317	1040.0262	367	1204.0682	417	1368.1102	467	1532.1522
268	879.2651	318	1043.3071	368	1207.3491	418	1371.3911	468	1535.4331
269	882.5459	319	1046.5876	369	1210.6299	419	1374.6719	469	1538.7139
270	885.8268	320	1049.8688	370	1213.9108	420	1377.9527	470	1541.9947
271	889.1076	321	1053.1496	371	1217.1916	421	1381.2336	471	1545.2756
272	892.3884	322	1056.4304	372	1220.4724	422	1384.5144	472	1548.5564
273	895.6693	323	1059.7113	373	1223.7533	423	1387.7953	473	1551.8373
274	898.9501	324	1062.9921	374	1227.0341	424	1391.0761	474	1555.1181
275	902.2310	325	1066.2729	375	1230.3150	425	1394.3569	475	1558.3989
276	905.5118	326	1069.5538	376	1233.5958	426	1397.6378	476	1561.6798
277	908.7926	327	1072.8346	377	1236.8766	427	1400.9186	477	1564.9606
278	912.0735	328	1076.1155	378	1240.1575	428	1404.1995	478	1568.2415
279	915.3543	329	1079.3963	379	1243.4383	429	1407.4803	479	1571.5223
280	918.6352	330	1082.6772	380	1246.7191	430	1410.7611	480	1574.8031
281	921.9160	331	1085.9580	381	1250.0000	431	1414.0420	481	1578.0840
282	925.1968	332	1089.2388	382	1253.2808	432	1417.3228	482	1581.3648
283	928.4777	333	1092.5197	383	1256.5617	433	1420.6037	483	1584.6457
284	931.7585	334	1095.8005	384	1259.8425	434	1423.8845	484	1587.9265
285	935.0394	335	1099.0814	385	1263.1234	435	1427.1653	485	1591.2073
286	938.3202	336	1102.3622	386	1266.4042	436	1430.4462	486	1594.4882
287	941.6010	337	1105.6430	387	1269.6850	437	1433.7270	487	1597.7690
288	944.8819	338	1108.9239	388	1272.9659	438	1437.0079	488	1601.0499
289	948.1627	339	1112.2047	389	1276.2467	439	1440.2887	489	1604.3307
290	951.4436	340	1115.4855	390	1279.5275	440	1443.5695	490	1607.6115
291	954.7244	341	1118.7664	391	1282.8084	441	1446.8504	491	1610.8924
292	958.0052	342	1122.0472	392	1286.0892	442	1450.1312	492	1614.1732
293	961.2861	343	1125.3281	393	1289.3701	443	1453.4121	493	1617.4541
294	964.5669	344	1128.6089	394	1292.6509	444	1456.6929	494	1620.7349
295	967.8478	345	1131.8898	395	1295.9317	445	1459.9737	495	1624.0157
296	971.1286	346	1135.1706	396	1299.2126	446	1463.2546	496	1627.2966
297	974.4094	347	1138.4514	397	1302.4934	447	1466.5354	497	1630.5774
298	977.6903	348	1141.7323	398	1305.7743	448	1469.8163	498	1633.8583
299	980.9711	349	1145.0131	399	1309.0551	449	1473.0971	499	1637.1391

Table D-7a. Conversion meters to feet (International System)—continued

Continued—METERS TO FEET 1 METER = 3.280 839 895 FEET

METERS	FEET	METERS	FEET	METERS	FEET	METERS	FEET	METERS	FEET
500	1640.4199	550	1804.4619	600	1968.5039	650	2132.5459	700	2296.5879
501	1643.7008	551	1807.7428	601	1971.7848	651	2135.8268	701	2299.8687
502	1646.9816	552	1811.0236	602	1975.0656	652	2139.1076	702	2303.1496
503	1650.2625	553	1814.3045	603	1978.3465	653	2142.3884	703	2306.4304
504	1653.5433	554	1817.5853	604	1981.6273	654	2145.6693	704	2309.7113
505	1656.8241	555	1820.8661	605	1984.9081	655	2148.9501	705	2312.9921
506	1660.1050	556	1824.1470	606	1988.1890	656	2152.2310	706	2316.2729
507	1663.3858	557	1827.4278	607	1991.4698	657	2155.5118	707	2319.5538
508	1666.6667	558	1830.7086	608	1994.7506	658	2158.7926	708	2322.8346
509	1669.9475	559	1833.9895	609	1998.0315	659	2162.0735	709	2326.1155
510	1673.2283	560	1837.2703	610	2001.3123	660	2165.3543	710	2329.3963
511	1676.5092	561	1840.5512	611	2004.5932	661	2168.6352	711	2332.6772
512	1679.7900	562	1843.8320	612	2007.8740	662	2171.9160	712	2335.9580
513	1683.0709	563	1847.1129	613	2011.1548	663	2175.1968	713	2339.2388
514	1686.3517	564	1850.3937	614	2014.4357	664	2178.4777	714	2342.5197
515	1689.6325	565	1853.6745	615	2017.7165	665	2181.7585	715	2345.8005
516	1692.9134	566	1856.9554	616	2020.9974	666	2185.0393	716	2349.0814
517	1696.1942	567	1860.2362	617	2024.2782	667	2188.3202	717	2352.3622
518	1699.4751	568	1863.5170	618	2027.5591	668	2191.6010	718	2355.6430
519	1702.7559	569	1866.7979	619	2030.8399	669	2194.8819	719	2358.9239
520	1706.0367	570	1870.0787	620	2034.1207	670	2198.1627	720	2362.2047
521	1709.3176	571	1873.3596	621	2037.4016	671	2201.4435	721	2365.4855
522	1712.5984	572	1876.6404	622	2040.6824	672	2204.7244	722	2368.7664
523	1715.8793	573	1879.9212	623	2043.9632	673	2208.0052	723	2372.0472
524	1719.1601	574	1883.2021	624	2047.2441	674	2211.2861	724	2375.3281
525	1722.4409	575	1886.4829	625	2050.5249	675	2214.5669	725	2378.6089
526	1725.7218	576	1889.7638	626	2053.8058	676	2217.8477	726	2381.8897
527	1729.0026	577	1893.0446	627	2057.0866	677	2221.1286	727	2385.1706
528	1732.2834	578	1896.3255	628	2060.3674	678	2224.4094	728	2388.4514
529	1735.5643	579	1899.6063	629	2063.6483	679	2227.6903	729	2391.7323
530	1738.8451	580	1902.8871	630	2066.9291	680	2230.9711	730	2395.0131
531	1742.1260	581	1906.1680	631	2070.2100	681	2234.2520	731	2398.2939
532	1745.4068	582	1909.4488	632	2073.4908	682	2237.5328	732	2401.5748
533	1748.6877	583	1912.7296	633	2076.7716	683	2240.8136	733	2404.8556
534	1751.9685	584	1916.0105	634	2080.0525	684	2244.0945	734	2408.1365
535	1755.2493	585	1919.2913	635	2083.3333	685	2247.3753	735	2411.4173
536	1758.5302	586	1922.5722	636	2086.6142	686	2250.6562	736	2414.6982
537	1761.8110	587	1925.8530	637	2089.8950	687	2253.9370	737	2417.9790
538	1765.0919	588	1929.1339	638	2093.1758	688	2257.2178	738	2421.2598
539	1768.3727	589	1932.4147	639	2096.4567	689	2260.4987	739	2424.5406
540	1771.6535	590	1935.6955	640	2099.7375	690	2263.7795	740	2427.8215
541	1774.9344	591	1938.9764	641	2103.0183	691	2267.0604	741	2431.1024
542	1778.2152	592	1942.2572	642	2106.2992	692	2270.3412	742	2434.3832
543	1781.4960	593	1945.5380	643	2109.5800	693	2273.6220	743	2437.6640
544	1784.7769	594	1948.8189	644	2112.8609	694	2276.9029	744	2440.9449
545	1788.0577	595	1952.0997	645	2116.1417	695	2280.1837	745	2444.2257
546	1791.3386	596	1955.3806	646	2119.4225	696	2283.4645	746	2447.5065
547	1794.6194	597	1958.6614	647	2122.7034	697	2286.7454	747	2450.7874
548	1797.9003	598	1961.9422	648	2125.9842	698	2290.0262	748	2454.0682
549	1801.1811	599	1965.2231	649	2129.2651	699	2293.3071	749	2457.3491

Table D-7a. Conversion Meters to Feet (International System)—continued

Continued—METERS TO FEET 1 METER = 3.280 839 895 FEET

METERS	FEET	METERS	FEET	METERS	FEET	METERS	FEET	METERS	FEET
750	2460.6299	800	2624.6719	850	2788.7139	900	2952.7559	950	3116.7979
751	2463.9107	801	2627.9527	851	2791.9947	901	2956.0367	951	3120.0787
752	2467.1916	802	2631.2336	852	2795.2756	902	2959.3176	952	3123.3596
753	2470.4724	803	2634.5144	853	2798.5564	903	2962.5984	953	3126.6404
754	2473.7533	804	2637.7953	854	2801.8372	904	2965.8792	954	3129.9212
755	2477.0341	805	2641.0761	855	2805.1181	905	2969.1601	955	3133.2021
756	2480.3149	806	2644.3569	856	2808.3989	906	2972.4409	956	3136.4829
757	2483.5958	807	2647.6378	857	2811.6798	907	2975.7218	957	3139.7638
758	2486.8766	808	2650.9186	858	2814.9606	908	2979.0026	958	3143.0446
759	2490.1575	809	2654.1995	859	2818.2415	909	2982.2834	959	3146.3254
760	2493.4383	810	2657.4803	860	2821.5223	910	2985.5643	960	3149.6063
761	2496.7191	811	2660.7611	861	2824.8031	911	2988.8451	961	3152.8871
762	2500.0000	812	2664.0420	862	2828.0840	912	2992.1260	962	3156.1680
763	2503.2808	813	2667.3228	863	2831.3648	913	2995.4068	963	3159.4488
764	2506.5617	814	2670.6037	864	2834.6457	914	2998.6877	964	3162.7296
765	2509.8425	815	2673.8845	865	2837.9265	915	3001.9685	965	3166.0105
766	2513.1234	816	2677.1653	866	2841.2073	916	3005.2493	966	3169.2913
767	2516.4042	817	2680.4462	867	2844.4882	917	3008.5302	967	3172.5721
768	2519.6850	818	2683.7270	868	2847.7690	918	3011.8110	968	3175.8530
769	2522.9659	819	2687.0078	869	2851.0498	919	3015.0919	969	3179.1339
770	2526.2467	820	2690.2887	870	2854.3307	920	3018.3727	970	3182.4147
771	2529.5275	821	2693.5695	871	2857.6115	921	3021.6535	971	3185.6955
772	2532.8084	822	2696.8504	872	2860.8924	922	3024.9344	972	3188.9763
773	2536.0892	823	2700.1312	873	2864.1732	923	3028.2152	973	3192.2572
774	2539.3701	824	2703.4120	874	2867.4540	924	3031.4960	974	3195.5380
775	2542.6509	825	2706.6929	875	2870.7349	925	3034.7769	975	3198.8189
776	2545.9317	826	2709.9737	876	2874.0157	926	3038.0577	976	3202.0997
777	2549.2126	827	2713.2546	877	2877.2966	927	3041.3386	977	3205.3806
778	2552.4934	828	2716.5354	878	2880.5774	928	3044.6194	978	3208.6614
779	2555.7743	829	2719.8163	879	2883.8582	929	3047.9002	979	3211.9422
780	2559.0551	830	2723.0971	880	2887.1391	930	3051.1811	980	3215.2231
781	2562.3359	831	2726.3779	881	2890.4199	931	3054.4619	981	3218.5039
782	2565.6168	832	2729.6588	882	2893.7008	932	3057.7428	982	3221.7848
783	2568.8976	833	2732.9396	883	2896.9816	933	3061.0236	983	3225.0656
784	2572.1785	834	2736.2205	884	2900.2625	934	3064.3044	984	3228.3464
785	2575.4593	835	2739.5013	885	2903.5433	935	3067.5853	985	3231.6273
786	2578.7401	836	2742.7821	886	2906.8241	936	3070.8661	986	3234.9081
787	2582.0210	837	2746.0630	887	2910.1050	937	3074.1470	987	3238.1890
788	2585.3018	838	2749.3438	888	2913.3858	938	3077.4278	988	3241.4698
789	2588.5827	839	2752.6247	889	2916.6667	939	3080.7086	989	3244.7506
790	2591.8635	840	2755.9055	890	2919.9475	940	3083.9895	990	3248.0315
791	2595.1443	841	2759.1863	891	2923.2283	941	3087.2703	991	3251.3123
792	2598.4252	842	2762.4672	892	2926.5092	942	3090.5511	992	3254.5932
793	2601.7060	843	2765.7480	893	2929.7900	943	3093.8320	993	3257.8740
794	2604.9868	844	2769.0288	894	2933.0709	944	3097.1129	994	3261.1548
795	2608.2677	845	2772.3097	895	2936.3517	945	3100.3937	995	3264.4357
796	2611.5485	846	2775.5905	896	2939.6325	946	3103.6745	996	3267.7165
797	2614.8294	847	2778.8714	897	2942.9134	947	3106.9554	997	3270.9973
798	2618.1102	848	2782.1522	898	2946.1942	948	3110.2362	998	3274.2782
799	2621.3911	849	2785.4330	899	2949.4750	949	3113.5170	999	3277.5590

Table D–7b. Conversion Feet to Meters (International System)

LENGTH—FEET TO METERS 1 FOOT = 0.3048 METER

FEET	METERS	FEET	METERS	FEET	METERS	FEET	METERS	FEET	METERS
0	0.0000	50	15.2400	100	30.4800	150	45.7200	200	60.9600
1	0.3048	51	15.5448	101	30.7848	151	46.0248	201	61.2648
2	0.6096	52	15.8496	102	31.0896	152	46.3296	202	61.5696
3	0.9144	53	16.1544	103	31.3944	153	46.6344	203	61.8744
4	1.2192	54	16.4592	104	31.6992	154	46.9392	204	62.1792
5	1.5240	55	16.7640	105	32.0040	155	47.2440	205	62.4840
6	1.8288	56	17.0688	106	32.3088	156	47.5488	206	62.7888
7	2.1336	57	17.3736	107	32.6136	157	47.8536	207	63.0936
8	2.4384	58	17.6784	108	32.9184	158	48.1584	208	63.3984
9	2.7432	59	17.9832	109	33.2232	159	48.4632	209	63.7032
10	3.0480	60	18.2880	110	33.5280	160	48.7680	210	64.0080
11	3.3528	61	18.5928	111	33.8328	161	49.0728	211	64.3128
12	3.6576	62	18.8976	112	34.1376	162	49.3776	212	64.6176
13	3.9624	63	19.2024	113	34.4424	163	49.6824	213	64.9224
14	4.2672	64	19.5072	114	34.7472	164	49.9872	214	65.2272
15	4.5720	65	19.8120	115	35.0520	165	50.2920	215	65.5320
16	4.8768	66	20.1168	116	35.3568	166	50.5968	216	65.8368
17	5.1816	67	20.4216	117	35.6616	167	50.9016	217	66.1416
18	5.4864	68	20.7264	118	35.9664	168	51.2064	218	66.4464
19	5.7912	69	21.0312	119	36.2712	169	51.5112	219	66.7512
20	6.0960	70	21.3360	120	36.5760	170	15.8160	220	67.0560
21	6.4008	71	21.6408	121	36.8808	171	52.1208	221	67.3608
22	6.7056	72	21.9456	122	37.1856	172	52.4256	222	67.6656
23	7.0104	73	22.2504	123	37.4904	173	52.7304	223	67.9704
24	7.3152	74	22.5552	124	37.7952	174	53.0352	224	68.2752
25	7.6200	75	22.8600	125	38.1000	175	53.3400	225	68.5800
26	7.9248	76	23.1648	126	38.4048	176	53.6448	226	68.8848
27	8.2296	77	23.4696	127	38.7096	177	53.9496	227	69.1896
28	8.5344	78	23.7744	128	39.0144	178	54.2544	228	69.4944
29	8.8392	79	24.0792	129	39.3192	179	54.5592	229	69.7992
30	9.1440	80	24.3840	130	39.6240	180	54.8640	230	70.1040
31	9.4488	81	24.6888	131	39.9288	181	55.1688	231	70.4088
32	9.7536	82	24.9936	132	40.2336	182	55.4736	232	70.7136
33	10.0584	83	25.2984	133	40.5384	183	55.7784	233	71.0184
34	10.3632	84	25.6032	134	40.8432	184	56.0832	234	71.3232
35	10.6680	85	25.9080	135	41.1480	185	56.3880	235	71.6280
36	10.9728	86	26.2128	136	41.4528	186	56.6928	236	71.9328
37	11.2776	87	26.5176	137	41.7576	187	56.9976	237	72.2376
38	11.5824	88	26.8224	138	42.0624	188	57.3024	238	72.5424
39	11.8872	89	27.1272	139	42.3672	189	57.6072	239	72.8472
40	12.1920	90	27.4320	140	42.6720	190	57.9120	240	73.1520
41	12.4968	91	27.7368	141	42.9768	191	58.2168	241	73.4568
42	12.8016	92	28.0416	142	43.2816	192	58.5216	242	73.7616
43	13.1064	93	28.3464	143	43.5864	193	58.8264	243	74.0664
44	13.4112	94	28.6512	144	43.8912	194	59.1312	244	74.3712
45	13.7160	95	28.9560	145	44.1960	195	59.4360	245	74.6760
46	14.0208	96	29.2608	146	44.5008	196	59.7408	246	74.9808
47	14.3256	97	29.5656	147	44.8056	197	60.0456	247	75.2856
48	14.6304	98	29.8704	148	45.1104	198	60.3504	248	75.5904
49	14.9352	99	30.1752	149	45.4152	199	60.6552	249	75.8952

Table D-7b. Conversion Feet to Meters (International System)—Continued

Continued—FEET TO METERS 1 FOOT = 0.3048 METER

FEET	METERS	FEET	METERS	FEET	METERS	FEET	METERS	FEET	METERS
250	76.2000	300	91.4400	350	106.6800	400	121.9200	450	137.1600
251	76.5048	301	91.7448	351	106.9848	401	122.2248	451	137.4648
252	76.8096	302	92.0496	352	107.2896	402	122.5296	452	137.7696
253	77.1144	303	92.3544	353	107.5944	403	122.8344	453	138.0744
254	77.4192	304	92.6592	354	107.8992	404	123.1392	454	138.3792
255	77.7240	305	92.9640	355	108.2040	405	123.4440	455	138.6840
256	78.0288	306	93.2688	356	108.5088	406	123.7488	456	138.9888
257	78.3336	307	93.5736	357	108.8136	407	124.0536	457	139.2963
258	78.6384	308	93.8784	358	109.1184	408	124.3584	458	139.5984
259	78.9432	309	94.1832	359	109.4232	409	124.6632	459	139.9032
260	79.2480	310	94.4880	360	109.7280	410	124.9680	460	140.2080
261	79.5528	311	94.7928	361	110.0328	411	125.2728	461	140.5128
262	79.8576	312	95.0976	362	110.3376	412	125.5776	462	140.8176
263	80.1624	313	95.4024	363	110.6424	413	125.8824	463	141.1224
264	80.4672	314	95.7072	364	110.9472	414	126.1872	464	141.4272
265	80.7720	315	96.0120	365	111.2520	415	126.4920	465	141.7320
266	81.0768	316	96.3168	366	111.5568	416	126.7968	466	142.0368
267	81.3816	317	96.6216	367	111.8616	417	127.1016	467	142.3416
268	81.6864	318	96.9264	368	112.1664	418	127.4064	468	142.6464
269	81.9912	319	97.2312	369	112.4712	419	127.7112	469	142.9512
270	82.2960	320	97.5360	370	112.7760	420	128.0160	470	143.2560
271	82.6008	321	97.8408	371	113.0808	421	128.3208	471	143.5608
272	82.9056	322	98.1456	372	113.3856	422	128.6256	472	143.8656
273	83.2104	323	98.4504	373	113.6904	423	128.9304	473	144.1704
274	83.5152	324	98.7552	374	113.9952	424	129.2352	474	144.4752
275	83.8200	325	99.0600	375	114.3000	425	129.5400	475	144.7800
276	84.1248	326	99.3648	376	114.6048	426	129.8448	476	145.0848
277	84.4296	327	99.6696	377	114.9096	427	130.1496	477	145.3896
278	84.7344	328	99.9744	378	115.2144	428	130.4544	478	145.6944
279	85.0392	329	100.2792	379	115.5192	429	130.7592	479	145.9992
280	85.3440	330	100.5840	380	115.8240	430	131.0640	480	146.3040
281	85.6488	331	100.8888	381	116.1288	431	131.3688	481	146.6088
282	85.9536	332	101.1936	382	116.4336	432	131.6736	482	146.9136
283	86.2584	333	101.4984	383	116.7384	433	131.9784	483	147.2184
284	86.5632	334	101.8032	384	117.0432	434	132.2832	484	147.5232
285	86.8680	335	102.1080	385	117.3480	435	132.5880	485	147.8280
286	87.1728	336	102.4128	386	117.6528	436	132.8928	486	148.1328
287	87.4776	337	102.7176	387	117.9576	437	133.1976	487	148.4376
288	87.7824	338	103.0224	388	118.2624	438	133.5024	488	148.7424
289	88.0872	339	103.3272	389	118.5672	439	133.8072	489	149.0472
290	88.3920	340	103.6320	390	118.8720	440	134.1120	490	149.3520
291	88.6968	341	103.9368	391	119.1768	441	134.4168	491	149.6568
292	89.0016	342	104.2416	392	119.4816	442	134.7216	492	149.9616
293	89.3064	343	104.5464	393	119.7864	443	135.0264	493	150.2664
294	89.6112	344	104.8512	394	120.0912	444	135.3312	494	150.5712
295	89.9160	345	105.1560	395	120.3960	445	135.6360	495	150.8760
296	90.2208	346	105.4608	396	120.7008	446	135.9408	496	151.1808
297	90.5256	347	105.7656	397	121.0056	447	136.2456	497	151.4856
298	90.8304	348	106.0704	398	121.3104	448	136.5504	498	151.7904
299	91.1352	349	106.3752	399	121.6152	449	136.8552	499	152.0952

Table D–7b. Conversion Feet to Meters (International System)—Continued

Continued—FEET TO METERS 1 FOOT=0.3048 METER

FEET	METERS	FEET	METERS	FEET	METERS	FEET	METERS	FEET	METERS
500	152.4000	550	167.6400	600	182.8800	650	198.1200	700	213.3600
501	152.7048	551	167.9448	601	183.1848	651	198.4248	701	213.6648
502	153.0096	552	168.2496	602	183.4896	652	198.7296	702	213.9696
503	153.3144	553	168.5544	603	183.7944	653	199.0344	703	214.2744
504	153.6192	554	168.8592	604	184.0992	654	199.3392	704	214.5792
505	153.9240	555	169.1640	605	184.4040	655	199.6440	705	214.8840
506	154.2288	556	169.4688	606	184.7088	656	199.9488	706	215.1888
507	154.5336	557	169.7736	607	185.0136	657	200.2536	707	125.4936
508	154.8384	558	170.0784	608	185.3184	658	200.5584	708	215.7984
509	155.1432	559	170.3832	609	185.6232	659	200.8632	709	216.1032
510	155.4480	560	170.6880	610	185.9280	660	201.1680	710	216.4080
511	155.7528	561	170.9928	611	186.2328	661	201.4728	711	216.7128
512	156.0576	562	171.2976	612	186.5376	662	201.7776	712	217.0176
513	156.3624	563	171.6024	613	186.8424	663	202.0824	713	217.3224
514	156.6672	564	171.9072	614	187.1472	664	202.3872	714	217.6272
515	156.9720	565	172.2120	615	187.4520	665	202.6920	715	217.9320
516	157.2768	566	172.5168	616	187.7568	666	202.9968	716	218.2368
517	157.5816	567	172.8126	617	188.0616	667	203.3016	717	218.5416
518	157.8864	568	173.1264	618	188.3664	668	203.6064	718	218.8464
519	158.1912	569	173.4312	619	188.6712	669	203.9112	719	219.1512
520	158.4960	570	173.7360	620	188.9760	670	204.2160	720	219.4560
521	158.8008	571	174.0408	621	189.2808	671	204.5208	721	219.7608
522	159.1056	572	174.3456	622	189.5856	672	204.8256	722	220.0656
523	159.4104	573	174.6504	623	189.8904	673	205.1304	723	220.3704
524	159.7152	574	174.9552	624	190.1952	674	205.4352	724	220.6752
525	160.0200	575	175.2600	625	190.5000	675	205.7400	725	220.9800
526	160.3248	576	175.5648	626	190.8048	676	206.0448	726	221.2848
527	160.6296	577	175.8696	627	191.1096	677	206.3496	727	221.5896
528	160.9344	578	176.1744	628	191.4144	678	206.6544	728	221.8944
529	161.2392	579	176.4792	629	191.7192	679	206.9592	729	222.1992
530	161.5440	580	176.7840	630	192.0240	680	207.2640	730	222.5040
531	161.8488	581	177.0888	631	192.3288	681	207.5688	731	222.8088
532	162.1536	582	177.3936	632	192.6336	682	207.8736	732	223.1136
533	162.4584	583	177.6984	633	192.9384	683	208.1784	733	223.4184
534	162.7632	584	178.0032	634	193.2432	684	208.4832	734	223.7232
535	163.0680	585	178.3080	635	193.5480	685	208.7880	735	224.0280
536	163.3728	586	178.6128	636	193.8528	686	209.0928	736	224.3328
537	163.6776	587	178.9176	637	194.1576	687	209.3976	737	224.6376
538	163.9824	588	179.2224	638	194.4624	688	209.7024	738	224.9424
539	164.2872	589	179.5272	639	194.7672	689	210.0072	739	225.2472
540	164.5920	590	179.8320	640	195.0720	690	210.3120	740	225.5520
541	164.8968	591	180.1368	641	195.3768	691	210.6168	741	225.8568
542	165.2016	592	180.4416	642	195.6816	692	210.9216	742	226.1616
543	165.5064	593	180.7464	643	195.9864	693	211.2264	743	226.4664
544	165.8112	594	181.0512	644	196.2912	694	211.5312	744	226.7712
545	166.1160	595	181.3560	645	196.5960	695	211.8360	745	227.0760
546	166.4208	596	181.6608	646	196.9008	696	212.1408	746	227.3808
547	166.7256	597	181.9656	647	197.2056	697	212.4456	747	227.6856
548	167.0304	598	182.2704	648	197.5104	698	212.7504	748	227.9904
549	167.3352	599	182.5752	649	197.8152	699	213.0552	749	228.2952

Table D–7b. Conversion Feet to Meters (International System)—Continued

Continued—FEET TO METERS 1 FOOT = 0.3048 METER

FEET	METERS	FEET	METERS	FEET	METERS	FEET	METERS	FEET	METERS
750	228.6000	800	243.8400	850	259.0800	900	274.3200	950	289.5600
751	228.9048	801	244.1448	851	259.3848	901	274.6248	951	289.8648
752	229.2096	802	244.4496	852	259.6896	902	274.9296	952	290.1696
753	229.5144	803	244.7544	853	259.9944	903	275.2344	953	290.4744
754	229.8192	804	245.0592	854	260.2992	904	275.5392	954	290.7792
755	230.1240	805	245.3640	855	260.6040	905	275.8440	955	291.0840
756	230.4288	806	245.6688	856	260.9088	906	276.1488	956	291.3888
757	230.7336	807	245.9736	857	261.2136	907	276.4536	957	291.6936
758	231.0384	808	246.2784	858	261.5184	908	276.7584	958	291.9984
759	231.3432	809	246.5832	859	261.8232	909	277.0632	959	292.3032
760	231.6480	810	246.8880	860	262.1280	910	277.3680	960	292.6080
761	231.9528	811	247.1928	861	262.4328	911	277.6728	961	292.9128
762	232.2576	812	247.4976	862	262.7376	912	277.9776	962	293.2176
763	232.5624	813	247.8024	863	263.0424	913	278.2824	963	293.5224
764	232.8672	814	248.1072	864	263.3472	914	278.5872	964	293.8272
765	233.1720	815	248.4120	865	263.6520	915	278.8920	965	294.1320
766	233.4768	816	248.7168	866	263.9568	916	279.1968	966	294.4368
767	233.7816	817	249.0216	867	264.2616	917	279.5016	967	294.7416
768	234.0864	818	249.3264	868	264.5664	918	279.8064	968	295.0464
769	234.3912	819	249.6312	869	264.8712	919	280.1112	969	295.3512
770	234.6960	820	249.9360	870	265.1760	920	280.4160	970	295.6560
771	235.0008	821	250.2408	871	265.4808	921	280.7208	971	295.9608
772	235.3056	822	250.5456	872	265 7856	922	281.0256	972	296.2656
773	235.6104	823	250.8504	873	266.0904	923	281.3304	973	296.5704
774	235.9152	824	251.1552	874	266.3952	924	281.6352	974	296.8752
775	236.2200	825	251.4600	875	266.7000	925	281.9400	975	297.1800
776	236.5248	826	251.7648	876	267.0048	926	282.2448	976	297.4848
777	236.8296	827	252.0696	877	267.3096	927	282.5496	977	297.7896
778	237.1344	828	252.3744	878	267.6144	928	282.8544	978	298.0944
779	237.4392	829	252.6792	879	267.9192	929	283.1592	979	298.3992
780	237.7440	830	252.9840	880	268.2240	930	283.4640	980	298.7040
781	238.0488	831	253.2888	881	268.5288	931	283.7688	981	299.0088
782	238.3536	832	253.5936	882	268.8336	932	284.0736	982	299.3136
783	238.6584	833	253.8984	883	269.1384	933	284.3784	983	299.6184
784	238.9632	834	254.2032	884	269.4432	934	284.6832	984	299.9232
785	239.2680	835	254.5080	885	269.7480	935	284.9880	985	300.2280
786	239.5728	836	254.8128	886	270.0528	936	285.2928	986	300.5328
787	239.8776	837	255.1176	887	270.3576	937	285.5976	987	300.8376
788	240.1824	838	255.4224	888	270.6624	938	285.9024	988	301.1424
789	240.4872	839	255.7272	889	270.9672	939	286.2072	989	301.4472
790	240.7920	840	256.0320	890	271.2720	940	286.5120	990	301.7520
791	241.0968	841	256.3368	891	271.5768	941	286.8168	991	302.0568
792	241.4016	842	256.6416	892	271.8816	942	287.1216	992	302.3616
793	241.7064	843	256.9464	893	272.1864	943	287.4264	993	302.6664
794	242.0112	844	257.2512	894	272.4912	944	287.7312	994	302.9712
795	242.3160	845	257.5560	895	272.7960	945	288.0360	995	303.2760
796	242.6208	846	257.8608	896	273.1008	946	288.3408	996	303.5808
797	242.9256	847	258.1656	897	273.4056	947	288.6456	997	303.8856
798	243.2304	848	258.4704	898	273.7104	948	288.9504	998	304.1904
799	243.5352	849	258.7752	899	274.0152	949	289.2552	999	304.4952

*Table D-8a. Conversion Meters to Feet (US Survey Foot)**

Lengths: 1 to 1,000 units, Reduction factor: 1 meter = 3.280833333 feet)

Meters	Feet	Meters	Feet	Meters	Feet	Meters	Feet	Meters	Feet
0		50	164.04167	100	328.08333	150	402.12500	200	656.16677
1	3.28083	1	167.32250	1	331.36417	1	495.40583	1	659.44750
2	6.56167	2	170.60333	2	334.64500	2	498.68667	2	662.72833
3	9.84250	3	173.88417	3	337.92583	3	501.96750	3	666.00917
4	13.12333	4	177.16500	4	341.20667	4	505.24833	4	669.29000
5	16.40417	5	180.44583	5	344.48750	5	508.52917	5	672.57083
6	19.68500	6	183.72667	6	347.76833	6	511.81000	6	675.85167
7	22.96583	7	187.00750	7	351.04917	7	515.09083	7	679.13250
8	26.24667	8	190.28833	8	354.33000	8	518.37167	8	682.41333
9	29.52750	9	193.56917	9	357.61083	9	521.65250	9	685.69417
10	32.80833	60	196.85000	110	360.89167	160	524.93333	210	688.97500
1	36.08917	1	200.13083	1	364.17250	1	528.21417	1	692.25583
2	39.37000	2	203.41167	2	367.45333	2	531.49500	2	695.53667
3	42.65083	3	206.69250	3	370.73417	3	534.77583	3	698.81750
4	45.93167	4	209.97333	4	374.01500	4	538.05667	4	702.09833
5	49.21250	5	213.25417	5	377.29583	5	541.33750	5	705.37917
6	52.49333	6	216.53500	6	380.57667	6	544.61833	6	708.66000
7	55.77417	7	219.81583	7	383.85750	7	547.89917	7	711.94083
8	59.05500	8	223.09667	8	387.13833	8	551.18000	8	715.22167
9	62.33583	9	226.37750	9	390.41917	9	554.46083	9	718.50250
20	65.61667	70	229.65833	120	393.70000	170	557.74167	220	721.78333
1	68.89750	1	232.93917	1	396.98083	1	561.02250	1	725.06417
2	72.17833	2	236.22000	2	400.26167	2	564.30333	2	728.34500
3	75.45917	3	239.50083	3	403.54250	3	567.58417	3	731.62583
4	78.74000	4	242.78167	4	406.82333	4	570.86500	4	734.90667
5	82.02083	5	246.06250	5	410.10417	5	574.14583	5	738.18750
6	85.30167	6	249.34333	6	413.38500	6	577.42667	6	741.46833
7	88.58250	7	252.62417	7	416.66583	7	580.70750	7	744.74917
8	91.86333	8	255.90500	8	419.94667	8	583.98833	8	748.03000
9	95.14417	9	259.18583	9	423.22750	9	587.26917	9	751.31083
30	98.42500	80	262.46667	130	426.50833	180	590.55000	230	754.59167
1	101.70583	1	265.74750	1	429.78917	1	593.83083	1	757.87250
2	104.98667	2	269.02833	2	433.07000	2	597.11167	2	761.15333
3	108.26750	3	272.30917	3	436.35083	3	600.39250	3	764.43417
4	111.54833	4	275.59000	4	439.63167	4	603.67333	4	767.71500
5	114.82917	5	278.87083	5	442.91250	5	606.95417	5	770.99583
6	118.11000	6	282.15167	6	446.19333	6	610.23500	6	774.27667
7	121.39083	7	285.43250	7	449.47417	7	613.51583	7	777.55750
8	124.67167	8	288.71333	8	452.75500	8	616.79667	8	780.83833
9	127.95250	9	291.99417	9	456.03583	9	620.07750	9	784.11917
40	131.23333	90	295.27500	140	459.31667	190	623.35833	240	787.40000
1	134.51417	1	298.55583	1	462.59750	1	626.63917	1	790.68083
2	137.79500	2	301.83667	2	465.87833	2	629.92000	2	793.96167
3	141.07583	3	305.11750	3	469.15917	3	633.20083	3	797.24250
4	144.35667	4	308.39833	4	472.44000	4	636.48167	4	800.52333
5	147.63750	5	311.67917	5	475.72083	5	639.76250	5	803.80417
6	150.91833	6	314.96000	6	479.00167	6	643.04333	6	807.08500
7	154.19917	7	318.24083	7	482.28250	7	646.32417	7	810.36583
8	157.48000	8	321.52167	8	485.56333	8	649.60500	8	813.64667
9	160.76083	9	324.80250	9	488.84417	9	652.88583	9	816.92750

*Reference is made to paragraph C-2, Appendix C.

Table D-8a. Conversion Meters to Feet (US Survey Foot)—Continued

Meters	Feet	Meters	Feet	Meters	Feet	Meters	Feet	Meters	Feet
250	820.20833	300	984.25000	350	1,148.29167	400	1,312.33333	450	1,476.37500
1	823.48917	1	987.53083	1	1,151.57250	1	1,315.61417	1	1,479.65583
2	826.77000	2	990.81167	2	1,154.85333	2	1,318.89500	2	1,482.93667
3	830.05083	3	994.09250	3	1,158.13417	3	1,332.17583	3	1,486.21750
4	833.33167	4	997.37333	4	1,161.41500	4	1,325.45667	4	1,489.49833
5	836.61250	5	1,000.65417	5	1,164.69583	5	1,328.73750	5	1,492.77917
6	839.89333	6	1,003.93500	6	1,167.97667	6	1,332.01833	6	1,496.06000
7	843.17417	7	1,007.21583	7	1,171.25750	7	1,335.29917	7	1,499.34083
8	846.45500	8	1,010.49667	8	1,174.53833	8	1,338.58000	8	1,502.62167
9	849.73583	9	1,013.77750	9	1,177.81917	9	1,341.86083	9	1,505.90250
260	853.01667	310	1,017.05833	360	1,181.10000	410	1,345.14167	460	1,509.18333
1	856.29750	1	1,020.33917	1	1,184.38083	1	1,348.42250	1	1,512.46417
2	859.57833	2	1,023.62000	2	1,187.66167	2	1,351.70333	2	1,515.74500
3	862.85917	3	1,026.90083	3	1,190.94250	3	1,354.98417	3	1,519.02583
4	866.14000	4	1,030.18167	4	1,194.22333	4	1,358.26500	4	1,522.30667
5	869.42083	5	1,033.46250	5	1,197.50417	5	1,361.54583	5	1,525.58750
6	872.70167	6	1,036.74333	6	1,200.78500	6	1,364.82667	6	1,528.86833
7	875.98250	7	1,040.02417	7	1,204.06583	7	1,368.10750	7	1,532.14917
8	879.26333	8	1,043.30500	8	1,207.34667	8	1,371.38833	8	1,535.43000
9	882.54417	9	1,046.58583	9	1,210.62750	9	1,374.66917	9	1,538.71083
270	885.82500	320	1,049.86667	370	1,213.90833	420	1,377.95000	470	1,541.99167
1	889.10583	1	1,053.14750	1	1,217.18917	1	1,381.23083	1	1,545.27250
2	892.38667	2	1,056.42833	2	1,220.47000	2	1,384.51167	2	1,548.55333
3	895.66750	3	1,059.70917	3	1,223.75083	3	1,387.79250	3	1,551.83417
4	898.94833	4	1,062.99000	4	1,227.03167	4	1,391.07333	4	1,555.11500
5	902.22917	5	1,066.27083	5	1,230.31250	5	1,394.35417	5	1,558.39583
6	905.51000	6	1,069.55167	6	1,233.59333	6	1,397.63500	6	1,561.67667
7	908.79083	7	1,072.83250	7	1,236.87417	7	1,400.91583	7	1,564.95750
8	912.07167	8	1,076.11333	8	1,240.15500	8	1,404.19667	8	1,568.23833
9	915.35250	9	1,079.39417	9	1,243.43583	9	1,407.47750	9	1,571.51917
280	918.63333	330	1,082.67500	380	1,246.71667	430	1,410.75833	480	1,574.80000
1	921.91417	1	1,085.95583	1	1,249.99750	1	1,414.03917	1	1,578.08083
2	925.19500	2	1,089.23667	2	1,253.27833	2	1,417.32000	2	1,581.36167
3	928.47583	3	1,092.51750	3	1,256.55917	3	1,420.60083	3	1,584.64250
4	931.75667	4	1,095.79833	4	1,259.84000	4	1,423.88167	4	1.587.92333
5	935.03750	5	1,099.07917	5	1,263.12083	5	1,427.16250	5	1,591.20417
6	938.31833	6	1,012.36000	6	1,266.40167	6	1,430.44333	6	1,594.48500
7	941.59917	7	1,105.64083	7	1,269.68250	7	1,433.72417	7	1,597.76583
8	944.88000	8	1,108.92167	8	1,272.96333	8	1,437.00500	8	1,601.04667
9	948.16083	9	1,112.20250	9	1,276.24417	9	1,440.28583	9	1,604.32750
290	951.44167	240	1,115.48333	390	1,279.52500	440	1,443.56667	490	1,607.60833
1	954.72250	1	1,118.76417	1	1,282.80583	1	1,446.84750	1	1,610.88917
2	958.00333	2	1,122.04500	2	1,286.08667	2	1,450.12833	2	1,614.17000
3	961.28417	3	1,125.32583	3	1,289.36750	3	1,453.40917	3	1,617.45083
4	964.56500	4	1,128.60667	4	1,292.64833	4	1,456.69000	4	1,620.73167
5	967.84583	5	1,131.88750	5	1,295.92917	5	1,459.97083	5	1,624.01250
6	971.12667	6	1,135.16833	6	1,299.21000	6	1,463.25167	6	1,627.29333
7	974.40750	7	1,138.44917	7	1,302.49083	7	1,466.53250	7	1,630.57417
8	977.68833	8	1,141.73000	8	1,305.77167	8	1,469.81333	8	1,633.85500
9	980.96917	9	1,145.01083	9	1,309.05250	9	1,473.09417	9	1,637.13583

Table D–8a. Conversion Meters to Feet (US Survey Foot)—Continued

Meters	Feet	Meters	Feet	Meters	Feet	Meters	Feet	Meters	Feet
500	1,640.41667	550	1,804.45833	600	1,968.50000	650	2,132.54167	700	2,296.58333
1	1,643.69750	1	1,807.73917	1	1,971.78083	1	2,135.82250	1	2,299.86417
2	1,646.97833	2	1,811.02000	2	1,975.06167	2	2,139.10333	2	2,303.14500
3	1,650.25917	3	1,814.30083	3	1,978.34250	3	2,142.38417	3	2,306.42583
4	1,653.54000	4	1,817.58167	4	1,981.62333	4	2,145.66500	4	2,309.70667
5	1,656.82083	5	1,820.86250	5	1,984.90417	5	2,148.94583	5	2,312.98750
6	1,660.10167	6	1,824.14333	6	1,988.18500	6	2,152.22667	6	2,316.26833
7	1,663.38250	7	1,827.42417	7	1,991.46583	7	2,155.50750	7	2,319.54917
8	1,666.66333	8	1,830.70500	8	1,994.74667	8	2,158.78833	8	2,322.83000
9	1,669.94417	9	1,833.98583	9	1,998.02750	9	2,162.06917	9	2,226.11083
510	1,673.22500	560	1,837.26667	610	2,001.30833	660	2,165.35000	710	2,329.39167
1	1,676.50583	1	1,840.54750	1	2,004.58917	1	2,168.63083	1	2,332.67250
2	1,679.78667	2	1,843.82833	2	2,007.87000	2	2,171.91167	2	2,335.95333
3	1,683.06750	3	1,847.10917	3	2,011.15083	3	2,175.19250	3	2,339.23417
4	1,686.34833	4	1,850.39000	4	2,014.43167	4	2,178.47333	4	2,342.51500
5	1,689.62917	5	1,853.67083	5	2,017.71250	5	2,181.75417	5	2,345.79583
6	1,692.91000	6	1,856.95167	6	2,020.99333	6	2,185.03500	6	2,349.07667
7	1,696.19083	7	1,860.23250	7	2,024.27417	7	2,188.31583	7	2,352.35750
8	1,699.47167	8	1,863.51333	8	2,027.55500	8	2,191.59667	8	2,355.63833
9	1,702.75250	9	1,866.79417	9	2,030.83583	9	2,194.87750	9	2,358.91917
520	1,706.03333	570	1,870.07500	620	2,034.11667	670	2,198.15833	720	2,362.20000
1	1,709.31417	1	1,873.35583	1	2,037.39750	1	2,201.43917	1	2,365.48083
2	1,712.59500	2	1,876.63667	2	2,040.67833	2	2,204.72000	2	2,368.76167
3	1,715.87583	3	1,879.91750	3	2,043.95917	3	2,208.00083	3	2,372.04250
4	1,719.15667	4	1,883.19833	4	2,047.24000	4	2,211.28167	4	2,375.32333
5	1,722.43750	5	1,886.47917	5	2,050.52083	5	2,214.56250	5	2,378.60417
6	1,725.71833	6	1,889.76000	6	2,053.80167	6	2,217.84333	6	2,381.88500
7	1,728.99917	7	1,893.04083	7	2,057.08250	7	2,221.12417	7	2,385.16583
8	1,732.28000	8	1,896.32167	8	2,060.36333	8	2,224.40500	8	2,388.44667
9	1,735.56083	9	1,899.60250	9	2,063.64417	9	2,227.68583	9	2,391.72750
530	1,738.84167	580	1,902.88333	630	2,066.92500	680	2,230.96667	730	2,395.00833
1	1,742.12250	1	1,906.16417	1	2,070.20583	1	2,234.24750	1	2,398.28917
2	1,745.40333	2	1,909.44500	2	2,073.48667	2	2,237.52833	2	2,401.57000
3	1,748.68417	3	1,912.72583	3	2,076.76750	3	2,240.80917	3	2,404.85083
4	1,751.96500	4	1,916.00667	4	2,080.04833	4	2,244.09000	4	2,408.13167
5	1,755.24583	5	1,919.28750	5	2,083.32917	5	2,247.37083	5	2,411.41250
6	1,758.52667	6	1,922.56833	6	2,086.61000	6	2,250.65167	6	2,414.69333
7	1,761.80750	7	1,925.84917	7	2,089.89083	7	2,253.93250	7	2,417.97417
8	1,765.08833	8	1,929.13000	8	2,093.17167	8	2,257.21333	8	2,421.25500
9	1,768.36917	9	1,932.41083	9	2,096.45250	9	2,260.49417	9	2,424.53583
540	1,771.65000	590	1,935.69167	640	2,099.73333	690	2,263.77550	740	2,427.81667
1	1,774.93083	1	1,938.97250	1	2,103.01417	1	2,267.05583	1	2,431.09750
2	1,778.21167	2	1,942.25333	2	2,106.29500	2	2,270.33667	2	2,434.37833
3	1,781.49250	3	1,945.53417	3	2,109.57583	3	2,273.61750	3	2,437.65917
4	1,784.77333	4	1,948.81500	4	2,112.85667	4	2,276.89833	4	2,440.94000
5	1,788.05417	5	1,952.09583	5	2,116.13750	5	2,280.17917	5	2,444.22083
6	1,791.33500	6	1,955.37667	6	2,119.41833	6	2,283.46000	6	2,447.50167
7	1,794.61583	7	1,958.65750	7	2,122.69917	7	2,286.74083	7	2,450.78250
8	1,797.89667	8	1,961.93833	8	2,125.98000	8	2,290.02167	8	2,454.06333
9	1,801.17750	9	1,965.21917	9	2,129.26083	9	2,293.30250	9	2,457.34417

Table D–8a. Conversion Meters to Feet (US Survey Foot)—Continued

Meters	Feet	Meters	Feet	Meters	Feet	Meters	Feet	Meters	Feet
750	2,460.62500	800	2,624.66667	850	2,788.70833	900	2,952.75000	950	3,116.79167
1	2,463.90583	1	2,627.94750	1	2,791.98917	1	2,956.03083	1	3,120.07250
2	2,467.18667	2	2,631.22833	2	2,795.27000	2	2,959.31167	2	3,123.35333
3	2,470.46750	3	2,634.50917	3	2,798.55083	3	2,962.59250	3	3,126.63417
4	2,473.74833	4	2,637.79000	4	2,801.83167	4	2,965.87333	4	3,129.91500
5	2,477.02917	5	2,641.07083	5	2,805.11250	5	2,969.15417	5	3,133.19583
6	2,480.31000	6	2,644.35167	6	2,808.39333	6	2,972.43500	6	3,136.47667
7	2,483.59083	7	2,647.63250	7	2,811.67417	7	2,975.71583	7	3,139.75750
8	2,486.87167	8	2,650.91333	8	2,814.95500	8	2,978.99667	8	3,143.03833
9	2,490.15250	9	2,654.19417	9	2,818.23583	9	2,982.27750	9	3,146.31917
760	2,493.43333	810	2,657.47500	860	2,821.51667	910	2,985.55833	960	3,149.60000
1	2,496.71417	1	2,660.75583	1	2,824.79750	1	2,988.83917	1	3,152.88083
2	2,499.99500	2	2,664.03667	2	2,828.07833	2	2,992.12000	2	3,156.16167
3	2,503.27583	3	2,667.31750	3	2,831.35917	3	2,995.40083	3	3,159.44250
4	2,506.55667	4	2,670.59833	4	2,834.64000	4	2,998.68167	4	3,162.72333
5	2,509.83750	5	2,673.87917	5	2,837.92083	5	3,001.96250	5	3,166.00417
6	2,513.11833	6	2,677.16000	6	2,841.20167	6	3,005.24333	6	3,169.28500
7	2,516.39917	7	2,680.44083	7	2,844.48250	7	3,008.52417	7	3,172.56583
8	2,519.68000	8	2,683.72167	8	2,847.76333	8	3,011.80500	8	3,175.84667
9	2,522.96083	9	2,687.00250	9	2,851.04417	9	3,015.08583	9	3,179.12750
770	2,526.24167	820	2,690.28333	870	2,854.32500	920	3,018.36667	970	3,182.40833
1	2,529.52250	1	2,693.56417	1	2,857.60583	1	3,021.64750	1	3,185.68917
2	2,532.80333	2	2,696.84500	2	2,860.88667	2	3,024.92833	2	3,188.97000
3	2,536.08417	3	2,700.12583	3	2,864.16750	3	3,028.20917	3	3,192.25083
4	2,539.36500	4	2,703.40667	4	2,867.44833	4	3,031.49000	4	3,195.53167
5	2,542.64583	5	2,706.68750	5	2,870.72917	5	3,034.77083	5	3,198.81250
6	2,545.92667	6	2,709.96833	6	2,874.01000	6	3,038.05167	6	3,202.09333
7	2,549.20750	7	2,713.24917	7	2,877.29083	7	3,041.33250	7	3,205.37417
8	2,552.48833	8	2,716.53000	8	2,880.57167	8	3,044.61333	8	3,208.65500
9	2,555.76917	9	2,719.81083	9	2,883.85250	9	3,047.89417	9	3,211.93583
780	2,559.05000	830	2,723.09167	880	2,887.13333	930	3,051.17500	980	3,215.21667
1	2,562.33083	1	2,726.37250	1	2,890.41417	1	3,054.45583	1	3,218.49750
2	2,565.61167	2	2,729.65333	2	2,893.69500	2	3,057.73667	2	3,221.77833
3	2,568.89250	3	2,732.93417	3	2,896.97583	3	3,061.01750	3	2,225.05917
4	2,572.17333	4	2,736.21500	4	2,900.25667	4	3,064.29833	4	3,228.34000
5	2,575.45417	5	2,739.49583	5	2,903.53750	5	3,067.57917	5	3,231.62083
6	2,578.73500	6	2,742.77667	6	2,906.81833	6	3,070.86000	6	3,234.90167
7	2,582.01583	7	2,746.05750	7	2,910.09917	7	3,074.14083	7	3,238.18250
8	2,585.29667	8	2,749.33833	8	2,913.38000	8	3,077.42167	8	3,241.46333
9	2,588.57750	9	2,752.61917	9	2,916.66083	9	3,080.70250	9	3,244.74417
790	2,591.85833	840	2,755.90000	890	2,919.94167	940	3,083.98333	990	3,248.02500
1	2,595.13917	1	2,759.18083	1	2,923.22250	1	3,087.26417	1	3,251.30583
2	2,598.42000	2	2,762.46167	2	2,926.50333	2	3,090.54500	2	2,254.58667
3	2,601.70083	3	2,765.74250	3	2,929.78417	3	3,093.82583	3	3,257.86750
4	2,604.98167	4	2,769.02333	4	2,933.06500	4	3,097.10667	4	3,261.14833
5	2,608.26250	5	2,772.30417	5	2,936.34583	5	3,100.38750	5	3,264.42917
6	2,611.54333	6	2,775.58500	6	2,939.62667	6	3,103.66833	6	3,267.71000
7	2,614.82417	7	2,778.86583	7	2,942.90750	7	3,106.94917	7	3,270.99083
8	2,618.10500	8	2,782.14667	8	2,946.18833	8	3,110.23000	8	3,274.27167
9	2,621.38583	9	2,785.42750	9	2,949.46917	9	3,113.51083	9	3,277.55250

*Table D–8b. Conversion Feet to Meters (US Survey Foot)**

(Lengths: 1 to 1,000 units. Reduction factor: 1 foot = 0.3048006096 meter)

Feet	Meters	Feet	Meters	Feet	Meters	Feet	Meters	Feet	Meters
0	0.0	50	15.24003	100	30.48006	150	45.72009	200	60.96012
1	0.30480	1	15.54483	1	30.78486	1	45.02489	1	61.26492
2	0.60960	2	15.84963	2	31.08966	2	46.32969	2	61.56972
3	0.91440	3	16.15443	3	31.39446	3	46.63449	3	61.87452
4	1.21920	4	16.45923	4	31.69926	4	46.93929	4	62.17932
5	1.52400	5	16.76403	5	32.00406	5	47.24409	5	62.48412
6	1.82880	6	17.06883	6	32.30886	6	47.54890	6	62.78893
7	2.13360	7	17.37363	7	32.61367	7	47.85370	7	63.09373
8	2.43840	8	17.67844	8	32.91847	8	48.15850	8	63.39853
9	2.74321	9	17.98324	9	33.22327	9	48.46330	9	63.70333
10	3.04801	60	18.28804	110	33.52807	160	48.76810	210	64.00813
1	3.35281	1	18.59284	1	33.83287	1	49.07290	1	64.31293
2	3.65761	2	18.89764	2	34.13767	2	49.37770	2	64.61773
3	3.96241	3	19.20244	3	34.44247	3	49.68250	3	64.92253
4	4.26721	4	19.50724	4	34.74727	4	49.98730	4	65.22733
5	4.57201	5	19.81204	5	35.05207	5	50.29210	5	65.53213
6	4.87681	6	20.11684	6	35.35687	6	50.59690	6	65.83693
7	5.18161	7	20.42164	7	35.66167	7	50.90170	7	66.14173
8	5.48641	8	20.72644	8	35.96647	8	51.20650	8	66.44653
9	5.79121	9	21.03124	9	36.27127	9	51.51130	9	66.75133
20	6.09601	70	21.33604	120	36.57607	170	51.81610	220	67.05613
1	6.40081	1	21.64084	1	36.88087	1	52.12090	1	67.36093
2	6.70561	2	21.94564	2	37.18567	2	52.42570	2	67.66574
3	7.01041	3	22.25044	3	37.49047	3	52.73051	3	67.97054
4	7.31521	4	22.55525	4	37.79528	4	53.03531	4	68.27534
5	7.62002	5	22.86005	5	38.10008	5	53.34011	5	68.58014
6	7.92482	6	23.16485	6	38.40488	6	53.64491	6	68.88494
7	8.22962	7	23.46965	7	38.70968	7	53.94971	7	69.18974
8	8.53442	8	23.77445	8	39.01448	8	54.25451	8	69.49454
9	8.83922	9	24.07925	9	39.31928	9	54.55931	9	69.79934
30	9.14402	80	24.38405	130	39.62408	180	54.86411	230	70.10414
1	9.44882	1	24.68885	1	39.92888	1	55.16891	1	70.40894
2	9.75362	2	24.99365	2	40.23368	2	55.47371	2	70.71374
3	10.05842	3	25.29845	3	40.53848	3	55.77851	3	71.01854
4	10.36322	4	25.60325	4	40.84328	4	56.08331	4	71.32334
5	10.66802	5	25.90805	5	41.14808	5	56.38811	5	71.62814
6	10.97282	6	26.21285	6	41.45288	6	56.69291	6	71.93294
7	11.27762	7	26.51765	7	41.75768	7	56.99771	7	72.23774
8	11.58242	8	26.82245	8	42.06248	8	57.30251	8	72.54255
9	11.88722	9	27.12725	9	42.36728	9	57.60732	9	72.84735
40	12.19202	90	27.43205	140	42.67209	190	57.91212	240	73.15215
1	12.49682	1	27.73686	1	42.97689	1	58.21692	1	73.45695
2	12.80163	2	28.04166	2	43.28169	2	58.52172	2	73.76175
3	13.10643	3	28.34646	3	43.58649	3	58.82652	3	74.06655
4	13.41123	4	28.65126	4	43.89129	4	59.13132	4	74.37135
5	13,71603	5	28.95606	5	44.19609	5	59.43612	5	74.67615
6	14.02083	6	29.26086	6	44.50089	6	59.74092	6	74.98095
7	14.32563	7	29.56566	7	44.80569	7	60.04572	7	75.28575
8	14.63043	8	29.87046	8	45.11049	8	60.35052	8	75.59055
9	14.93523	9	30.17526	9	45.41529	9	60.65532	9	75.89535

*** Reference is made to paragraph C–2, Appendix C.**

Table D-8b. Conversion Feet to Meters (US Survey Foot)—Continued

Feet	Meters	Feet	Meters	Feet	Meters	Feet	Meters	Feet	Meters
250	76.20015	300	91.44018	350	106.68021	400	121.92024	450	137.16027
1	76.50495	1	91.74498	1	106.98501	1	122.22504	1	137.46507
2	76.80975	2	92.04978	2	107.28981	2	122.52985	2	137.76988
3	77.11455	3	92.35458	3	107.59462	3	122.83465	3	138.07468
4	77.41935	4	92.65939	4	107.89942	4	123.13945	4	138.37948
5	77.72416	5	92.96419	5	108.20422	5	123.44425	5	138.68428
6	78.02896	6	93.26899	6	108.50902	6	123.74905	6	138.98908
7	78.33376	7	93.57379	7	108.81382	7	124.05385	7	139.29388
8	78.63856	8	93.87859	8	109.11862	8	124.35865	8	139.59868
9	78.94336	9	94.18339	9	109.42342	9	124.66345	9	139.90348
260	79.24816	310	94.48819	360	109.72822	410	124.96825	460	140.20828
1	79.55296	1	94.79299	1	110.03302	1	125.27305	1	140.51308
2	79.85776	2	95.09779	2	110.33782	2	125.57785	2	140.81788
3	80.16256	3	95.40259	3	110.64262	3	125.88265	3	141.12268
4	80.46736	4	95.70739	4	110.94742	4	126.18745	4	141.42748
5	80.77216	5	96.01219	5	111.25222	5	126.49225	5	141.73228
6	81.07696	6	96.31699	6	111.55702	6	126.79705	6	142.03708
7	81.38176	7	96.62179	7	111.86182	7	127.10185	7	142.34188
8	81.68656	8	96.92659	8	112.16662	8	127.40665	8	142.64669
9	81.99136	9	97.23139	9	112.47142	9	127.71146	9	142.95149
270	82.29616	320	97.53620	370	112.77623	420	128.01626	470	143.25629
1	82.60097	1	97.84100	1	113.08103	1	128.32106	1	143.56109
2	82.90577	2	98.14580	2	113.38583	2	128.62586	2	143.86589
3	83.21057	3	98.45060	3	113.69063	3	128.93066	3	144.17069
4	83.51537	4	98.75540	4	113.99543	4	129.23546	4	144.47549
5	83.82017	5	99.06020	5	114.30023	5	129.54026	5	144.78029
6	84.12497	6	99.36500	6	114.60503	6	129.84506	6	145.08509
7	84.42977	7	99.66980	7	114.90983	7	130.14986	7	145.38989
8	84.73457	8	99.97460	8	115.21463	8	130.45466	8	145.69469
9	85.03937	9	100.27940	9	115.51943	9	130.75946	9	145.99949
280	85.34417	330	100.58420	380	115.82423	430	131.06426	480	146.30429
1	85.64897	1	100.88900	1	116.12903	1	131.36906	1	146.60909
2	85.95377	2	101.19380	2	116.43383	2	131.67386	2	146.91389
3	86.25857	3	101.49860	3	116.73863	3	131.97866	3	147.21869
4	86.56337	4	101.80340	4	117.04343	4	132.28346	4	147.52350
5	86.86817	5	102.10820	5	117.34823	5	132.58827	5	147.82830
6	87.17297	6	102.41300	6	117.65304	6	132.89307	6	148.13310
7	87.47777	7	102.71781	7	117.95784	7	133.19787	7	148.43790
8	87.78258	8	103.02261	8	118.26264	8	133.50267	8	148.74270
9	88.08738	9	103.32741	9	118.56744	9	133.80747	9	149.04750
290	88.39218	340	103.63221	390	118.87224	440	134.11227	490	149.35230
1	88.69698	1	103.93701	1	119.17704	1	134.41707	1	149.65710
2	89.00178	2	104.24181	2	119.48184	2	134.72187	2	149.96190
3	89.30658	3	104.54661	3	119.78664	3	135.02667	3	150.26670
4	89.61138	4	104.85141	4	120.09144	4	135.33147	4	150.57150
5	89.91618	5	105.15621	5	120.39624	5	135.63627	5	150.87630
6	90.22098	6	105.46101	6	120.70104	6	135.94107	6	151.18110
7	90.52578	7	105.76581	7	121.00584	7	136.24587	7	151.48590
8	90.83058	8	106.07061	8	121.31064	8	136.55067	8	151.79070
9	91.13538	9	106.37541	9	121.61544	9	136.85547	9	152.09550

Table D–8b. Conversion Feet to Meters (U.S. Survey Foot)—Continued

Feet	Meters	Feet	Meters	Feet	Meters	Feet	Meters	Feet	Meters
500	152.40030	550	167.64034	600	182.88037	650	198.12040	700	213.36043
1	152.70511	1	167.94514	1	183.18517	1	198.42520	1	213.66523
2	153.00991	2	168.24994	2	183.48997	2	198.73000	2	213.97003
3	153.31471	3	168.55474	3	183.79477	3	199.03480	3	214.27483
4	153.61951	4	168.85954	4	184.09957	4	199.33960	4	214.57963
5	153.92431	5	169.16434	5	184.40437	5	199.64440	5	214.88443
6	154.22911	6	169.46914	6	184.70917	6	199.94920	6	215.18923
7	154.53391	7	169.77394	7	185.01397	7	200.25400	7	215.49403
8	154.83871	8	170.07874	8	185.31877	8	200.55880	8	215.79883
9	155.14351	9	170.38354	9	185.62357	9	200.86360	9	216.10363
510	155.44831	560	170.68834	610	185.92837	660	201.16840	710	216.40843
1	155.75311	1	170.99314	1	186.23317	1	201.47320	1	216.71323
2	156.05791	2	171.29794	2	186.53797	2	201.77800	2	217.01803
3	156.36271	3	171.60274	3	186.84277	3	202.08280	3	217.32283
4	156.66751	4	171.90754	4	187.14757	4	202.38760	4	217.62764
5	156.97231	5	172.21234	5	187.45237	5	202.69241	5	217.93244
6	157.27711	6	172.51715	6	187.75718	6	202.99721	6	218.23724
7	157.58192	7	172.81295	7	188.06198	7	203.30201	7	218.54204
8	157.88672	8	173.12675	8	188.36678	8	203.60681	8	218.84684
9	158.19152	9	173.43155	9	188.67158	9	203.91161	9	219.15164
520	158.49632	570	173.73635	620	188.97638	670	204.21641	720	219.45644
1	158.80112	1	174.04115	1	189.28118	1	204.52121	1	219.76124
2	159.10592	2	174.34595	2	189.58598	2	204.82601	2	220.06604
3	159.41072	3	174.65075	3	189.89078	3	205.13081	3	220.37084
4	159.71552	4	174.95555	4	190.19558	4	205.43561	4	220.67564
5	160.02032	5	175.26035	5	190.50038	5	205.74041	5	220.98044
6	160.32512	6	175.56515	6	190.80518	6	206.04521	6	221.28524
7	160.62992	7	175.86995	7	191.10998	7	206.35001	7	221.59004
8	160.93472	8	176.17475	8	191.41478	8	206.65481	8	221.89484
9	161.23952	9	176.47955	9	191.71958	9	206.95961	9	222.19964
530	161.54432	580	176.78435	630	192.02438	680	207.26441	730	222.50445
1	161.84912	1	177.08915	1	192.32918	1	207.56922	1	222.80925
2	162.15392	2	177.39395	2	192.63399	2	207.87402	2	223.11405
3	162.45872	3	177.69876	3	192.93879	3	208.17882	3	223.41885
4	162.76353	4	178.00356	4	193.24359	4	208.48362	4	223.72365
5	163.06833	5	178.30836	5	193.54839	5	208.78842	5	224.02845
6	163.37313	6	178.61316	6	193.85319	6	209.09322	6	224.33325
7	163.67793	7	178.91796	7	194.15799	7	209.39802	7	224.63805
8	163.98273	8	179.22276	8	194.46279	8	209.70282	8	224.94285
9	164.28753	9	179.52756	9	194.76759	9	210.00762	9	225.24765
540	164.59233	590	179.83236	640	195.07239	690	210.31242	740	225.55245
1	164.89713	1	180.13716	1	195.37719	1	210.61722	1	225.85725
2	165.20193	2	180.44196	2	195.68199	2	210.92202	2	226.16205
3	165.50673	3	180.74676	3	195.98679	3	211.22682	3	226.46685
4	165.81153	4	181.05156	4	196.29159	4	211.53162	4	226.77165
5	166.11633	5	181.35636	5	196.59639	5	211.83642	5	227.07645
6	166.42113	6	181.66116	6	196.90119	6	212.14122	6	227.38125
7	166.72593	7	181.96596	7	197.20599	7	212.44602	7	227.68606
8	167.03073	8	182.27076	8	197.51080	8	212.75083	8	227.99086
9	167.33553	9	182.57557	9	197.81560	9	213.05563	9	228.29566

Table D-8b. Conversion Feet to Meters (US Survey Foot)—Continued

Feet	Meters	Feet	Meters	Feet	Meters	Feet	Meters	Feet	Meters
750	228.60046	800	243.84049	850	259.08052	900	274.32055	950	289.56058
1	228.90526	1	244.14529	1	259.38532	1	274.62535	1	289.86538
2	229.21006	2	244.45009	2	259.69012	2	274.93015	2	290.17018
3	229.51486	3	244.75489	3	259.99492	3	275.23495	3	290.47498
4	229.81966	4	245.05969	4	260.29972	4	275.53975	4	290.77978
5	230.12446	5	245.36449	5	260.60452	5	275.84455	5	291.08458
6	230.42926	6	245.66929	6	260.90932	6	276.14935	6	291.38938
7	230.73406	7	245.97409	7	261.21412	7	276.45415	7	291.69418
8	231.03886	8	246.27889	8	261.51892	8	276.75895	8	291.99898
9	231.34366	9	246.58369	9	261.82372	9	277.06375	9	292.30378
760	231.64846	810	246.88849	860	262.12852	910	277.36855	960	292.60859
1	231.95326	1	247.19329	1	262.43332	1	277.67336	1	292.91339
2	232.25806	2	247.49809	2	262.73813	2	277.97816	2	293.21819
3	232.56287	3	247.80290	3	263.04293	3	278.28296	3	293.52299
4	232.86767	4	248.10770	4	263.34773	4	278.58776	4	293.82779
5	233.17247	5	248.41250	5	263.65253	5	278.89256	5	294.13259
6	233.47727	6	248.71730	6	263.95733	6	279.19736	6	294.43739
7	233.78207	7	249.02210	7	264.26213	7	279.50216	7	294.74219
8	234.08687	8	249.32690	8	264.56693	8	279.80696	8	295.04699
9	234.39167	9	249.63170	9	264.87173	9	280.11176	9	295.35179
770	234.69647	820	249.93650	870	265.17653	920	280.41656	970	295.65659
1	235.00127	1	250.24130	1	265.48133	1	280.72136	1	295.96139
2	235.30607	2	250.54610	2	265.78613	2	281.02616	2	296.26619
3	235.61087	3	250.85090	3	266.09093	3	281.33096	3	296.57099
4	235.91567	4	251.15570	4	266.39573	4	281.63576	4	296.87579
5	236.22047	5	251.46050	5	266.70053	5	281.94056	5	297.18059
6	236.52527	6	251.76530	6	267.00533	6	282.24536	6	297.48539
7	236.83007	7	252.07010	7	267.31013	7	282.55017	7	297.79020
8	237.13487	8	252.37490	8	267.61494	8	282.85497	8	298.09500
9	237.43967	9	252.67971	9	267.91974	9	283.15977	9	298.39980
780	237.74448	830	252.98451	880	268.22454	930	283.46457	980	298.70460
1	238.04928	1	253.28931	1	268.52934	1	283.76937	1	299.00940
2	238.35408	2	253.59411	2	268.83414	2	284.07417	2	299.31420
3	238.65888	3	253.89891	3	269.13894	3	284.37897	3	299.61900
4	238.96368	4	254.20371	4	269.44374	4	284.68377	4	299.92380
5	239.26848	5	254.50851	5	269.74854	5	286.98857	5	300.22860
6	239.57328	6	254.81331	6	270.05334	6	285.29337	6	300.53340
7	239.87808	7	255.11811	7	270.35814	7	285.59817	7	300.83820
8	240.18288	8	255.42291	8	270.66294	8	285.90297	8	301.14300
9	240.48768	9	255.72771	9	270.96774	9	286.20777	9	301.44780
790	240.79248	840	256.03251	890	271.27254	940	286.51257	990	301.75260
1	241.09728	1	256.33731	1	271.57734	1	286.81737	1	302.05740
2	241.40208	2	256.64211	2	271.88214	2	287.12217	2	302.36220
3	241.70688	3	256.94691	3	272.18694	3	287.42697	3	302.66701
4	242.01168	4	257.25171	4	272.49174	4	287.73178	4	302.97181
5	242.31648	5	257.55652	5	272.79655	5	288.03658	5	303.27661
6	242.62129	6	257.86132	6	273.10135	6	288.34138	6	303.58141
7	242.92609	7	258.16612	7	273.40615	7	288.64618	7	303.88621
8	243.23089	8	258.47092	8	273.71095	8	188.95098	8	304.19101
9	243.53569	9	258.77572	9	274.01575	9	289.25578	9	304.49581

Table D-9. Conversion Celsius to Fahrenheit and Inverse

$$°F. = 9/5° C. + 32 \quad °C = \frac{5(°F. - 32)}{9}$$

°C	°F	°C	°F	°C	°F	°C	°F	°C	°F
−40	−40.0	−30	−22.0	−20	−4.0	−10	14.0	0	32.0
−39	−38.2	−29	−20.2	−19	−2.2	−9	15.8	1	33.8
−38	−36.4	−28	−18.4	−18	−0.4	−8	17.6	2	35.6
−37	−34.6	−27	−16.6	−17	1.4	−7	19.4	3	37.4
−36	−32.8	−26	−14.8	−16	3.2	−6	21.2	4	39.2
−35	−31.0	−25	−13.0	−15	5.0	−5	23.0	5	41.0
−34	−29.2	−24	−11.2	−14	6.8	−4	24.8	6	42.8
−33	−27.4	−23	−9.4	−13	8.6	−3	26.6	7	44.6
−32	−25.6	−22	−7.6	−12	10.4	−2	28.4	8	46.4
−31	−23.8	−21	−5.8	−11	12.2	−1	30.2	9	48.2
10	50.0	20	68.0	30	86.0	40	104.0	50	122.0
11	51.8	21	69.8	31	87.8	41	105.8	51	123.8
12	53.6	22	71.6	32	89.6	42	107.6	52	125.6
13	55.4	23	73.4	33	91.4	43	109.4	53	127.4
14	57.2	24	75.2	34	93.2	44	111.2	54	129.2
15	59.0	25	77.0	35	95.0	45	113.0	55	131.0
16	60.8	26	78.8	36	96.8	46	114.8	56	132.8
17	62.6	27	80.6	37	98.6	47	116.6	57	134.6
18	64.4	28	82.4	38	100.4	48	118.4	58	136.4
19	66.2	29	84.2	39	102.2	49	120.2	59	138.2

Table D–10. Temperature Corrections, Steel Tape

Correction for 100 units
$Ct = .645x10^{-5}(T°1 - T°2)$

Temp. °F		Correction	Temp. °F		Correction	Temp. °F		Correction
−	+		−	+		−	+	
68	68	.0000	48	88	.0129	28	108	.0258
67	69	.0006	47	89	.0135	27	109	.0264
66	70	.0013	46	90	.0142	26	110	.0271
65	71	.0019	45	91	.0148	25	111	.0277
64	72	.0026	44	92	.0155	24	112	.0284
63	73	.0032	43	93	.0161	23	113	.0290
62	74	.0039	42	94	.0168	22	114	.0297
61	75	.0045	41	95	.0174	21	115	.0303
60	76	.0052	40	96	.0181	20	116	.0310
59	77	.0058	39	97	.0187	19	117	.0316
58	78	.0064	38	98	.0194	18	118	.0322
57	79	.0071	37	99	.0200	17	119	.0329
56	80	.0077	36	100	.0206	16	120	.0335
55	81	.0084	35	101	.0213	15	121	.0342
54	82	.0090	34	102	.0219	14	122	.0348
53	83	.0097	33	103	.0226	13	123	.0355
52	84	.0103	32	104	.0232	12	124	.0361
51	85	.0110	31	105	.0239	11	125	.0368
50	86	.0116	30	106	.0245	10	126	.0374
49	87	.0123	29	107	.0252	9	127	.0381

Notes.

1. For temperatures above 68° the corrections are plus, for temperatures below 68° the corrections are minus.
2. In using the table, take average temperature for the course to determine the correction for a 100-unit length then multiply the result so obtained by the number of 100 units in the course to determine the sum correction for the entire course.
3. *Example*, average temperature for course 53°F.

Taped distance for course ---------- 572.346 meters
Temperature correction = −.0097x5.7 ---------- −0.055 meters

Corrected distance ---------- 572.291 meters

Table D–11. Inclination Correction, Percent of Slope

Correction for unit 50

Versine 1 of unit 50. Correction $= \frac{\text{Distance}}{50}$ x Versine

%	0	1	2	3	4	5	6	7	8	9
0	.0000	.0000	.0001	.0002	.0004	.0007	.0009	.0013	.0016	.0021
1	.0025	.0031	.0036	.0043	.0049	.0057	.0065	.0073	.0082	.0091
2	.0100	.0111	.0121	.0133	.0145	.0157	.0170	.0183	.0197	.0211
3	.0225	.0241	.0257	.0273	.0290	.0307	.0325	.0343	.0361	.0381
4	.0401	.0421	.0442	.0464	.0485	.0507	.0530	.0553	.0577	.0602
5	.0626	.0652	.0678	.0703	.0731	.0758	.0785	.0814	.0843	.0872
6	.0902	.0931	.0963	.0994	.1025	.1058	.1091	.1125	.1158	.1193
7	.1228	.1263	.1299	.1336	.1372	.1409	.1448	.1485	.1524	.1564
8	.1603	.1642	.1686	.1726	.1768	.1811	.1855	.1897	.1916	.1986
9	.2030	.2076	.2122	.2169	.2215	.2263	.2311	.2358	.2408	.2458
10	.2509	.2557	.2609	.2661	.2714	.2765	.2818	.2873	.2927	.2980
11	.3036	.3092	.3149	.3203	.3261	.3319	.3378	.3435	.3495	.3555
12	.3616	.3674	.3736	.3798	.3861	.3925	.3986	.4050	.4115	.4180
13	.4246	.4309	.4376	.4444	.4512	.4580	.4646	.4716	.4786	.4856
14	.4927	.4999	.5067	.5190	.5213	.5286	.5361	.5435	.5507	.5582
15	.5666	.5735	.5812	.5890	.5968	.6047	.6122	.6202	.6282	.6363

Notes.

1. Inclination corrections are always minus and are always made for each full or partial tape length.
2. *Examples:*
   a. Grade 6.6%
      Tape length ---------- 50.000 meters
      Correction ---------- −0.109 meter
      Corrected length ---------- −49.891 meters
   b. Grade 5.8%
      Tape length ---------- 26.328 meters
      Correction $-.0843 \times \frac{26}{50} =$ ---------- −0.042 meter
      Corrected length ---------- 26.286 meters

APPENDIX E

SUMMARY OF COMMERCIAL EQUIPMENT AVAILABLE FOR SURVEYORS

TAPES

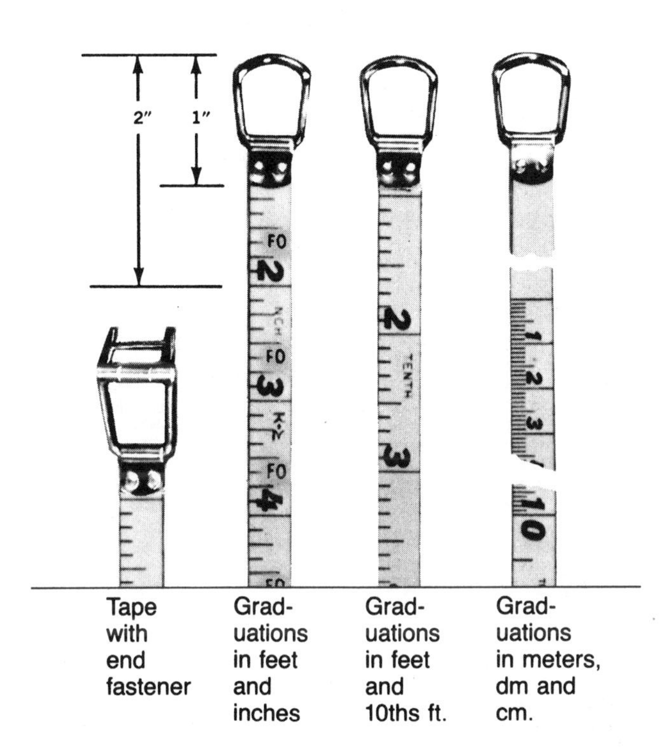

Tape with end fastener	Graduations in feet and inches	Graduations in feet and 10ths ft.	Graduations in meters, dm and cm.

GRADUATIONS

Feet, inches and 8ths.

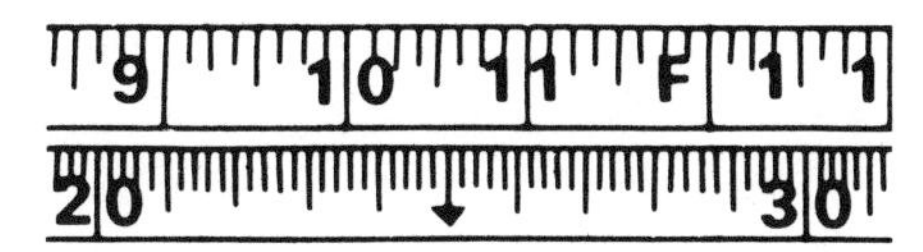

Feet, inches and 8ths;
Second side - metrics with increments in meters CM and 2MM.

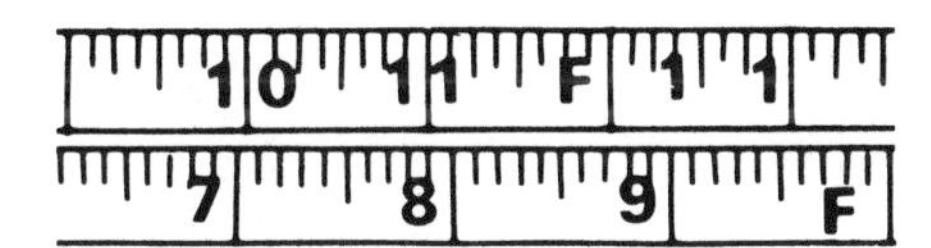

Feet, inches and 8ths.
Second side - tenths and 100ths of a foot.

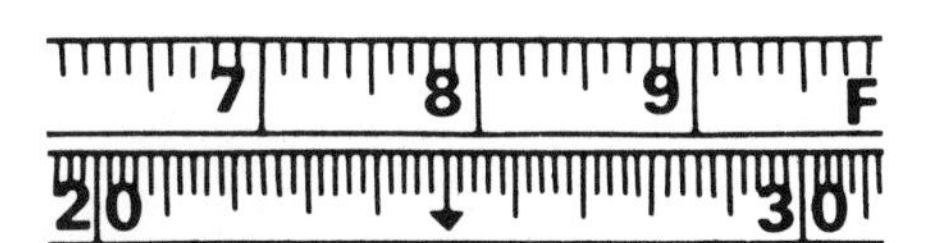

Tenths and 100ths of a foot.
Second side - metrics with increments in meters CM and 2MM.

Feet, tenths and 100ths of a foot.

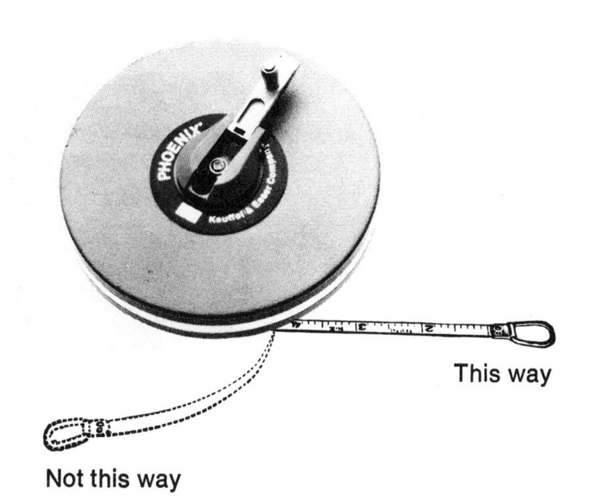

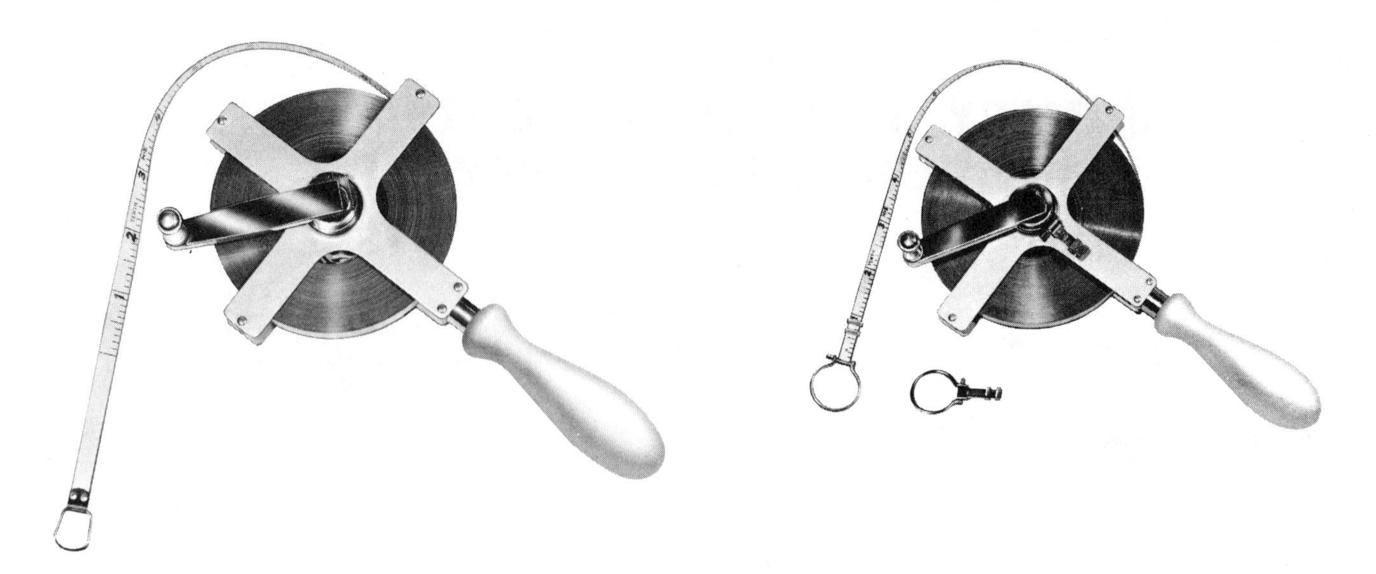

Engineers' and Surveyors' Tapes

Measuring Ruler

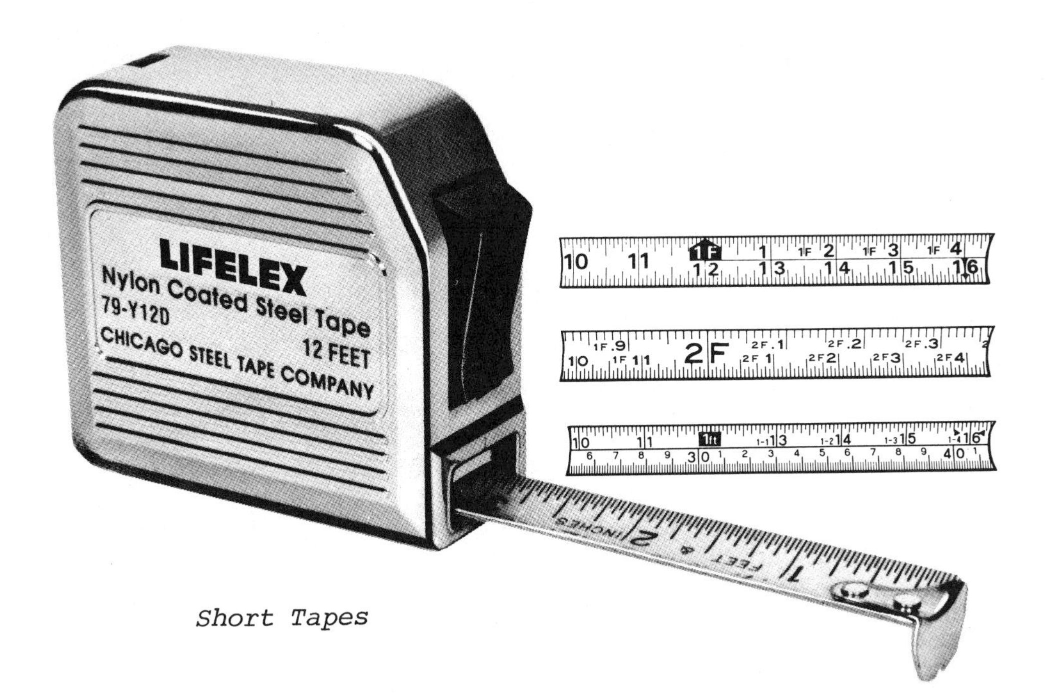

Short Tapes

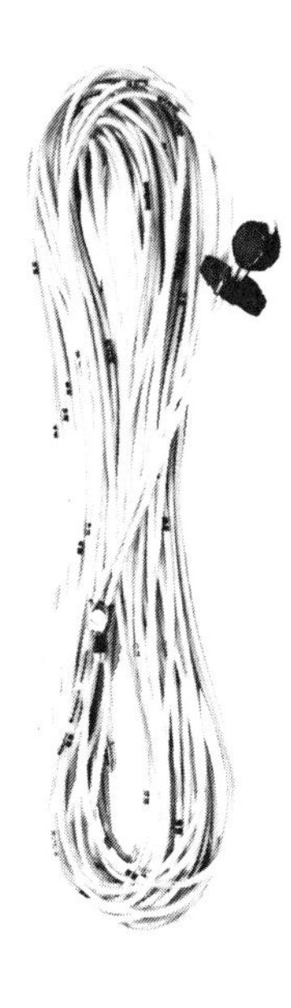

Rope Chains

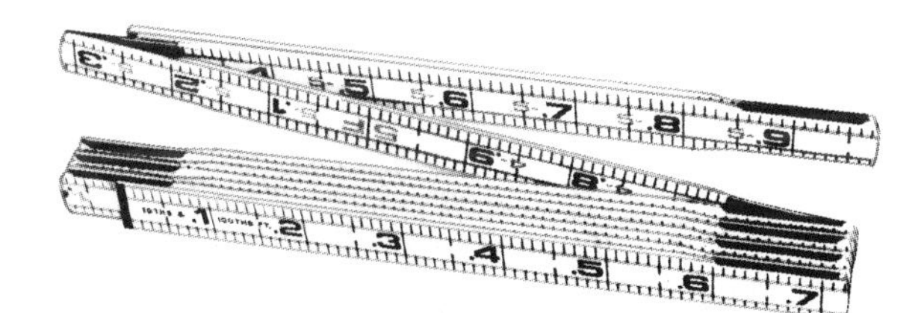

Wood Folding Rules

Pocket Tapes

Fiberglass Tapes

REELS FOR BAND CHAINS

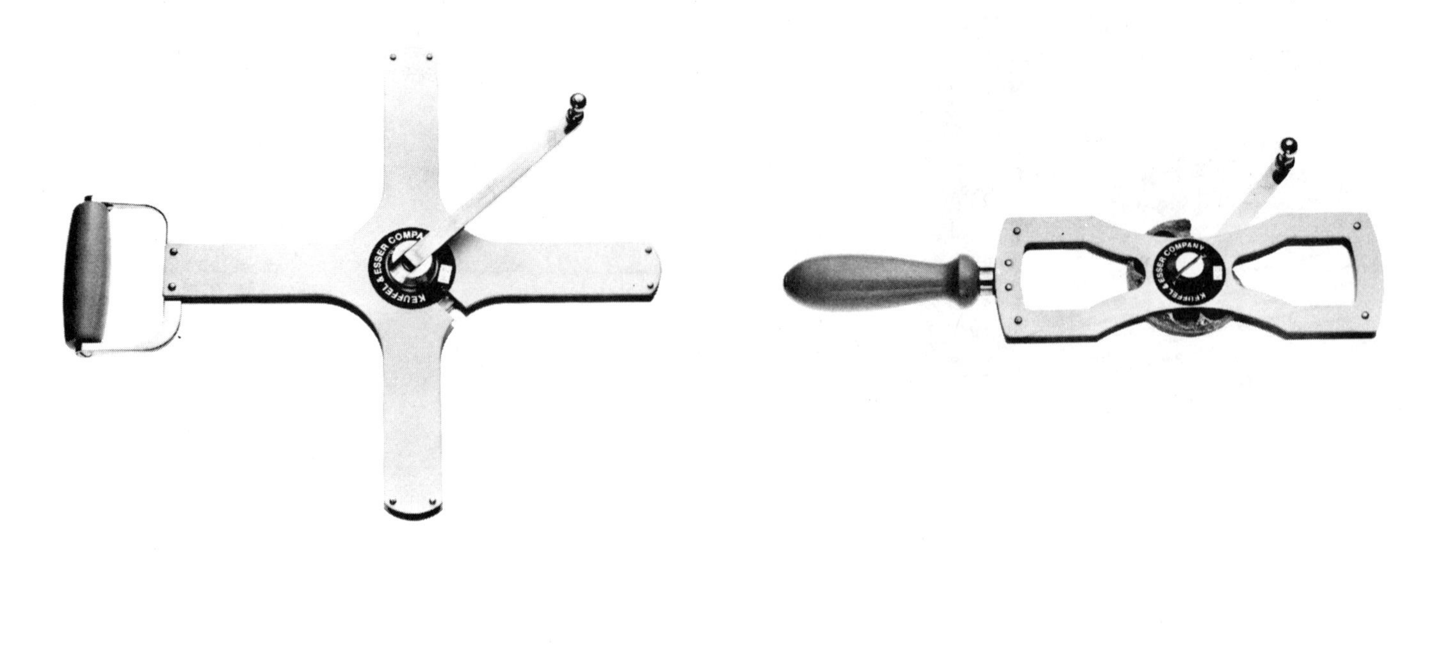

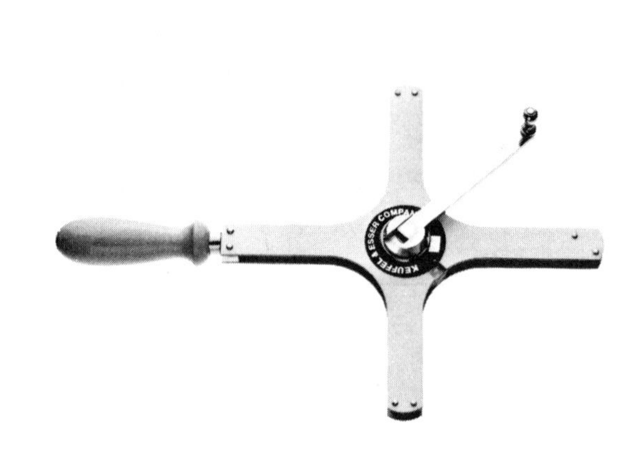

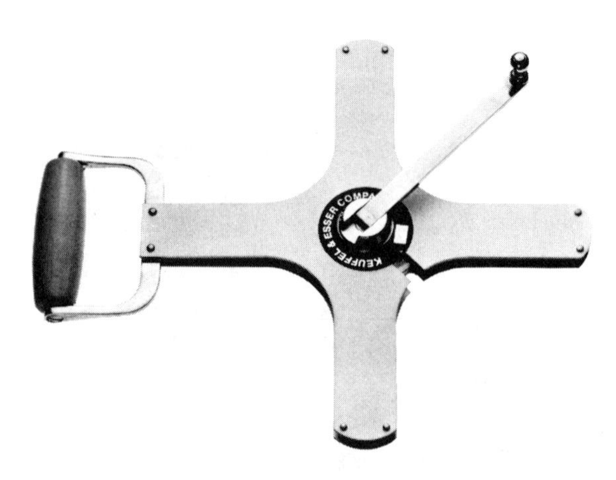

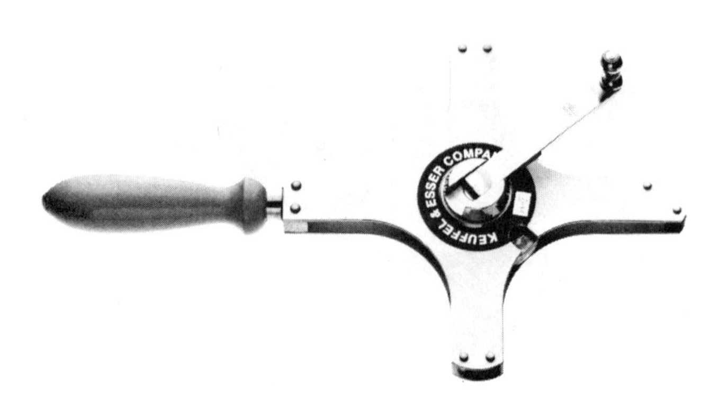

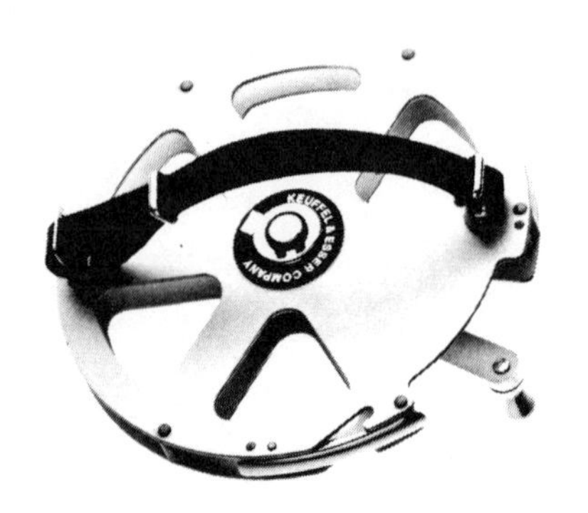

DISTANCE MEASURING WHEELS

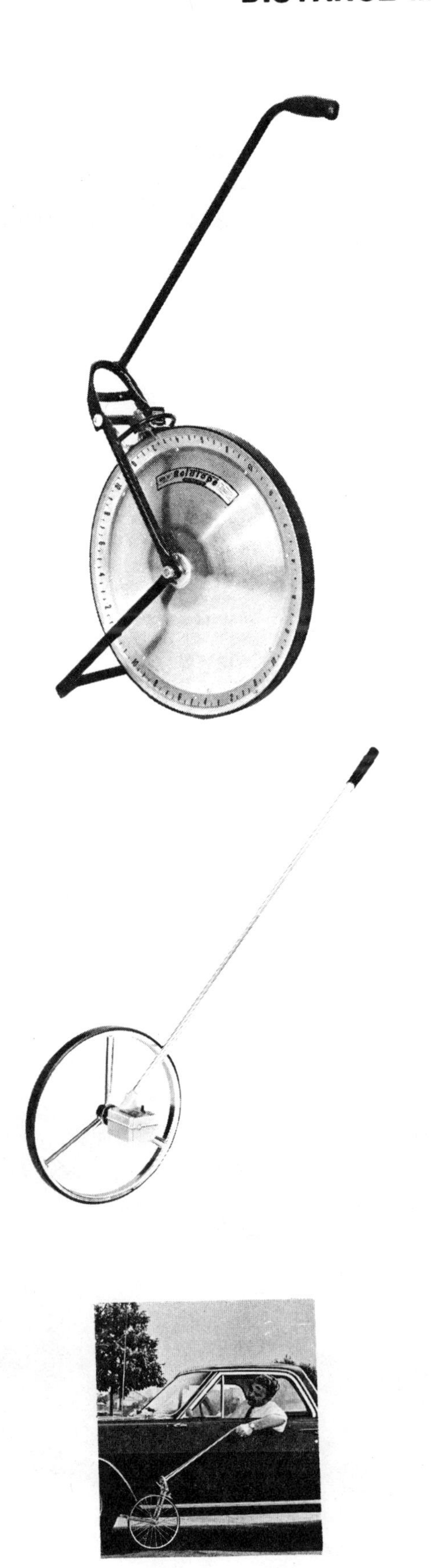

Lens-Displacing Type Range Finder

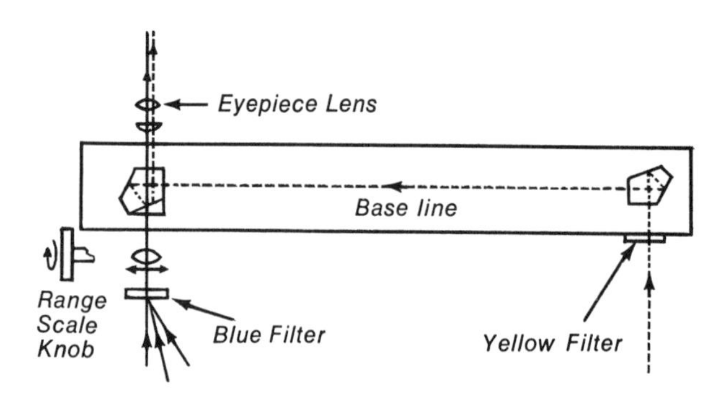

Range Finder Schematic

ELECTRONIC DISTANCE MEASURING EQUIPMENT

The modulated light beam is directed onto a retro-reflector positioned at the point to which the measurement is to be made. The reflected light beam from the retro-reflector returns to the instrument, where it is converted to an electrical signal. The phase relationship between the transmitted and reflected beams of light are compared by the units self-contained computer processing circuitry and converted into a direct readout of distance.

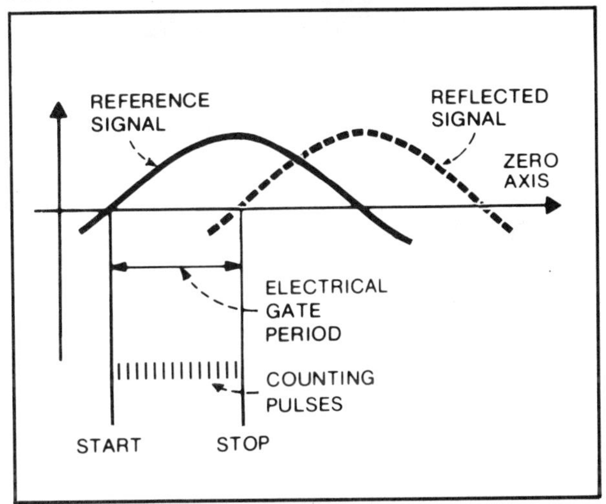

Prism cluster arrangement for a long-range measurement.

Circular Prisms

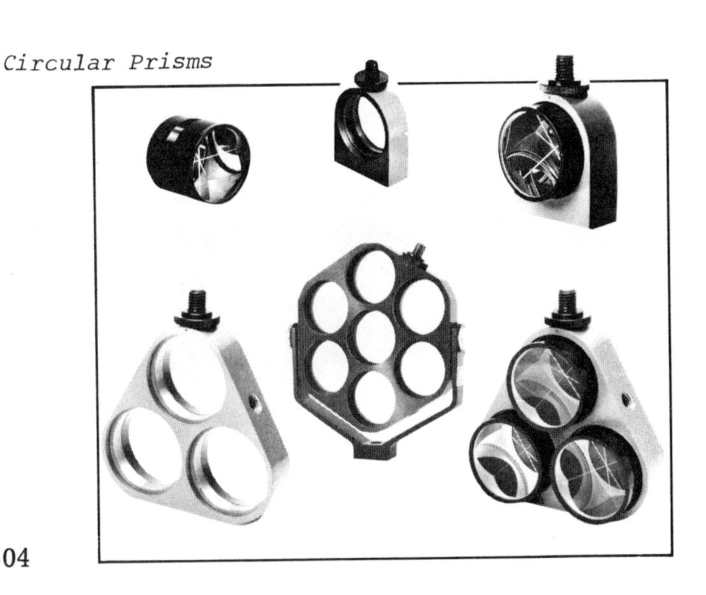

COMPASSES

INCHES & 1/32
CM & MM
TYPE 15 T
THE RANGER
Made in Finland
SUUNTO Co
LEFT
PC (%)
RIGHT
SUUNTO HELSINKI PATENT
Ushikata
MODEL TRACON

HAND LEVELS

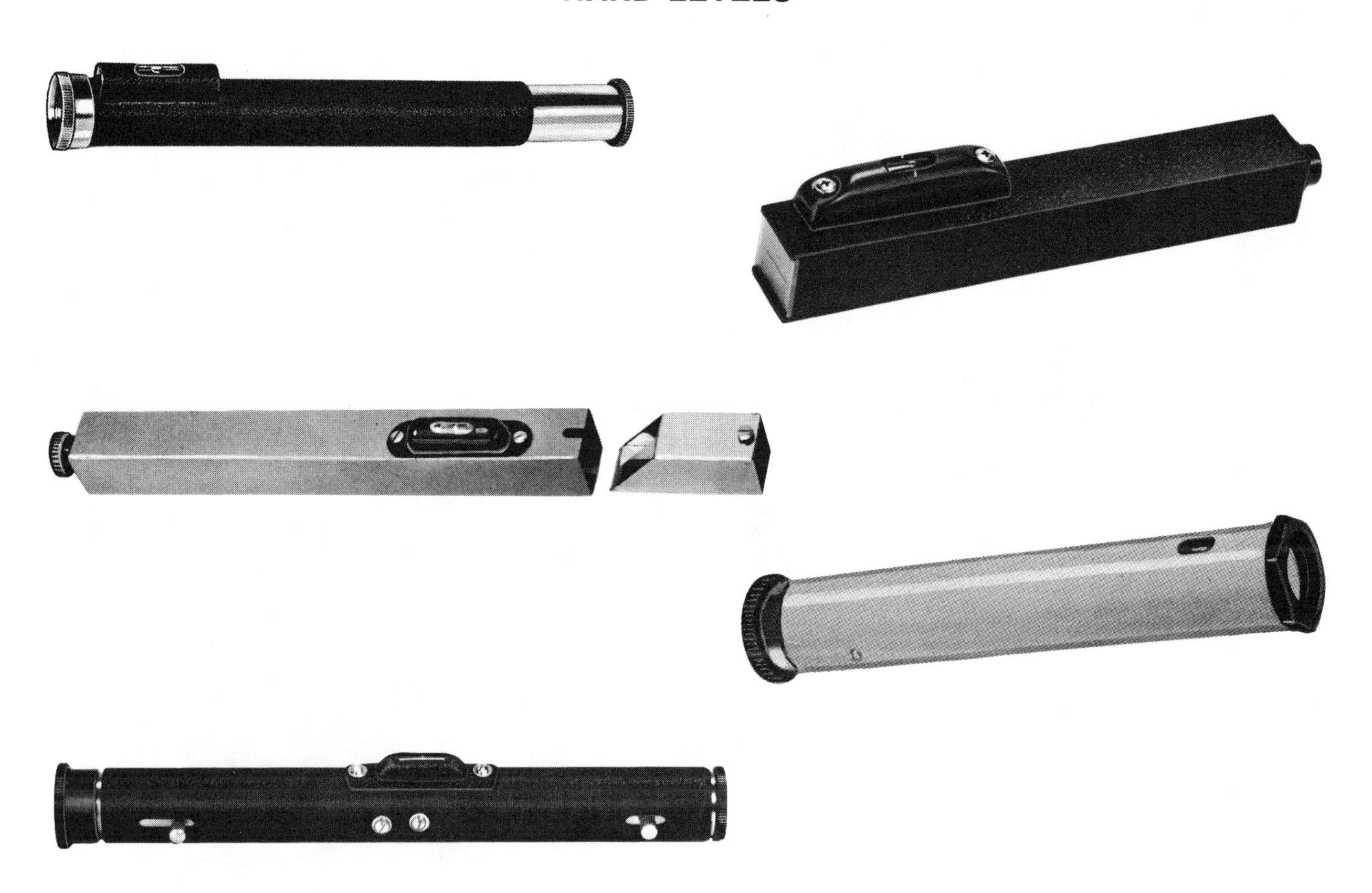

ABNEY LEVELS

OPTICAL HAND CLINOMETER

AUTOMATIC LEVELS

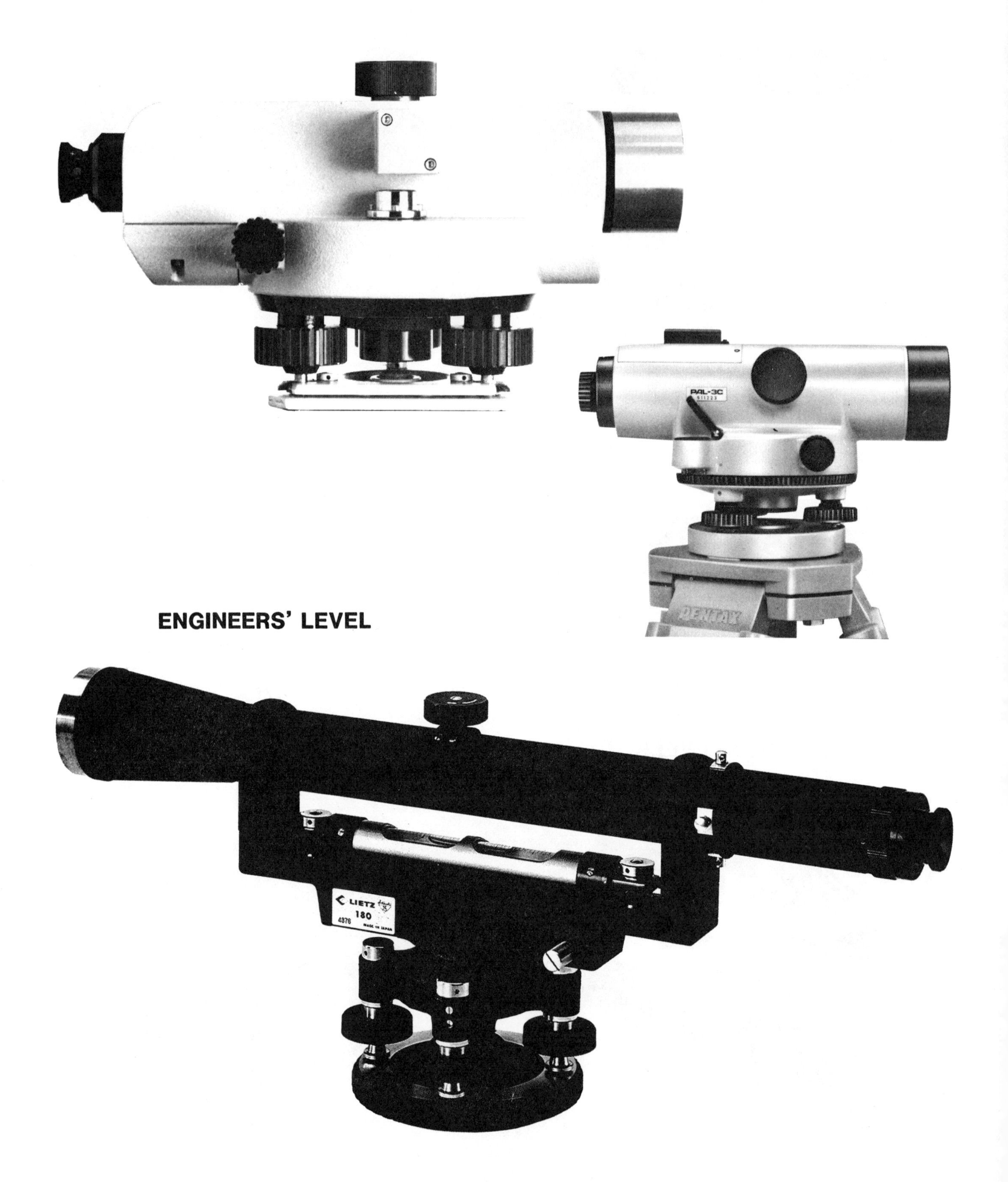

ENGINEERS' LEVEL

LEVEL-TRANSITS

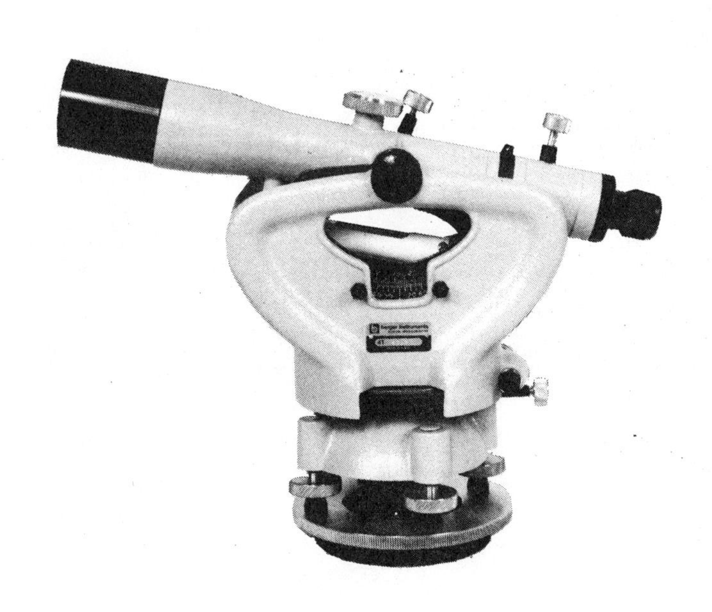

LEVELING RODS AND TARGETS

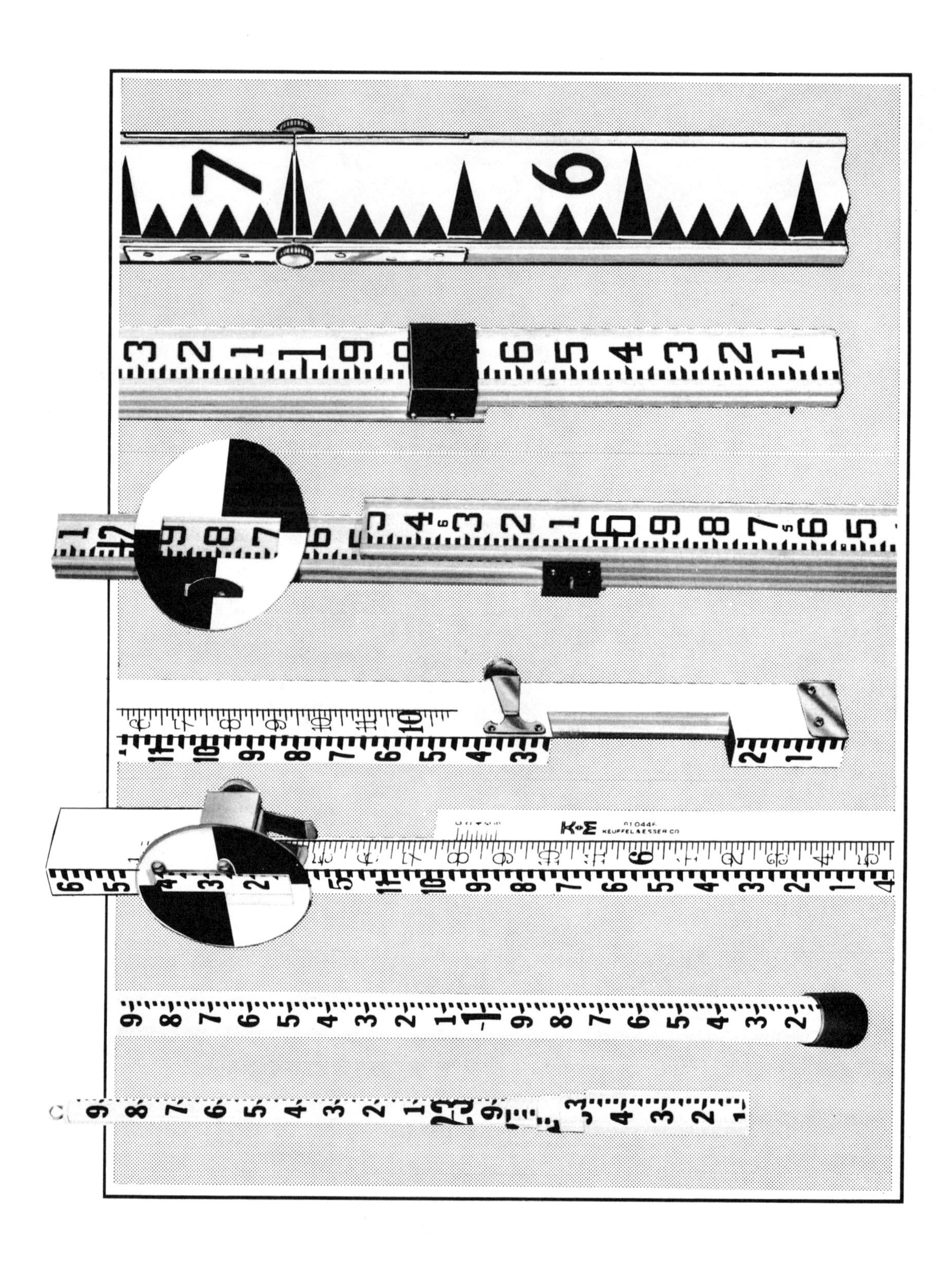

TARGET

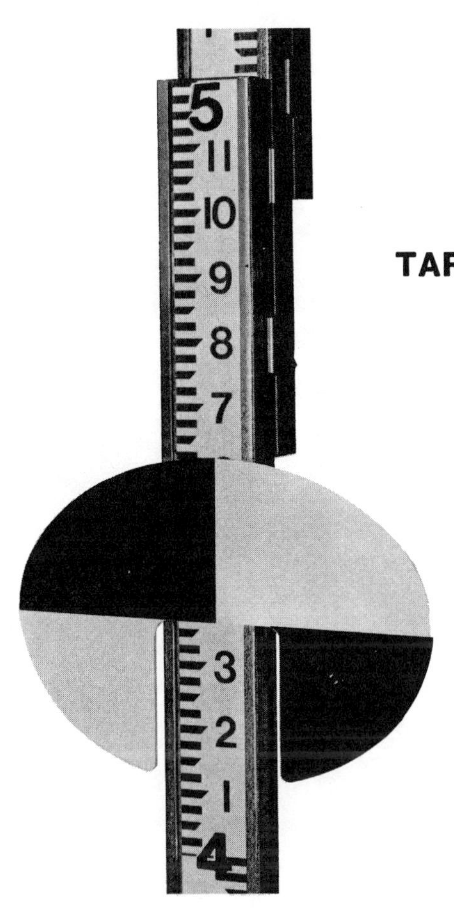

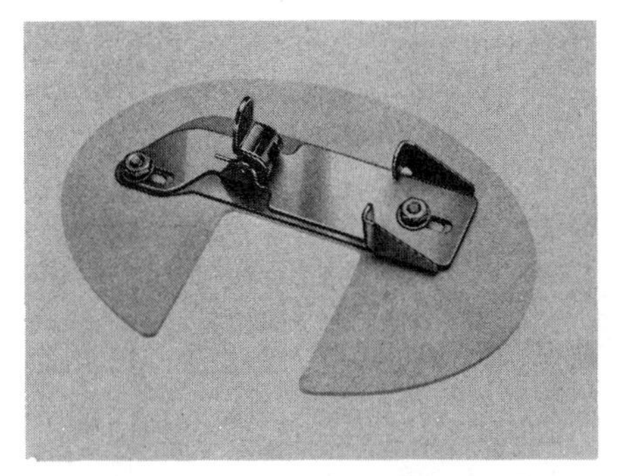

The back of the target has a nut and bolt slide for easy adjustability.

Latch-Type Clamp

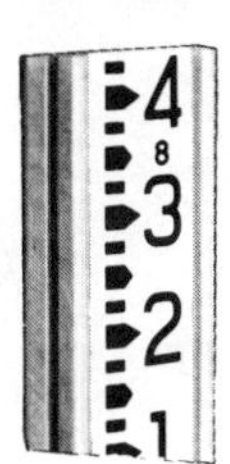

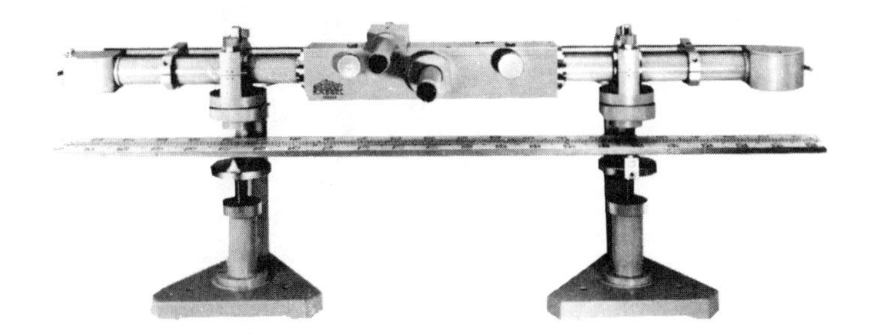

Double Image Comparator - Primarily used to check the accuracy of tapes, scales and leveling rods with line graduations.

Sectional View

RANGING POLES

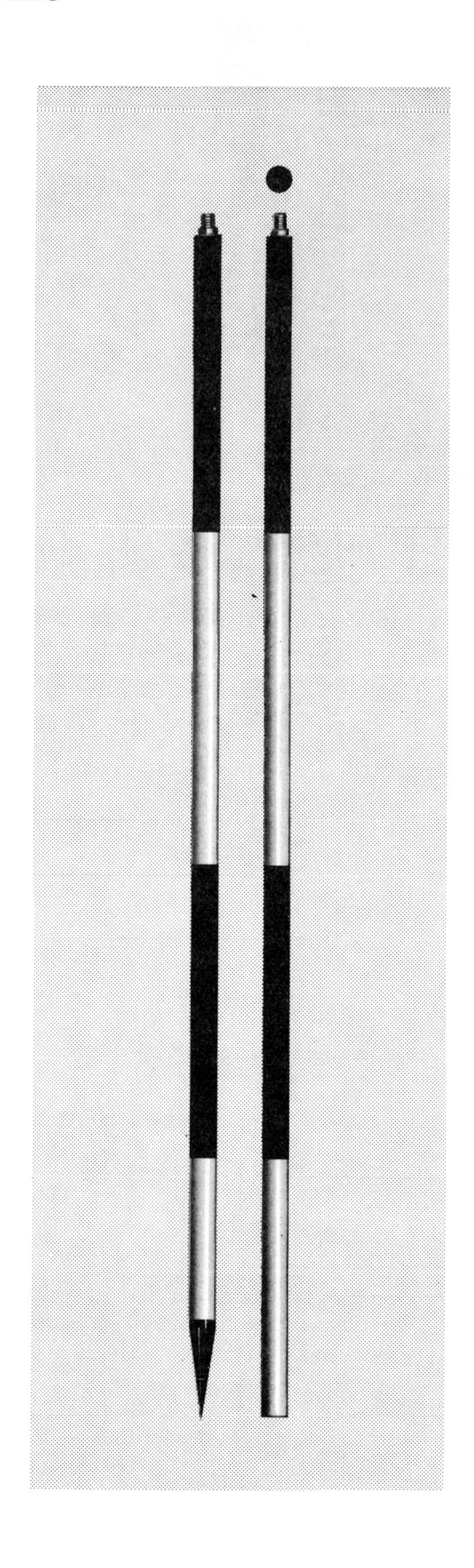

TRANSITS

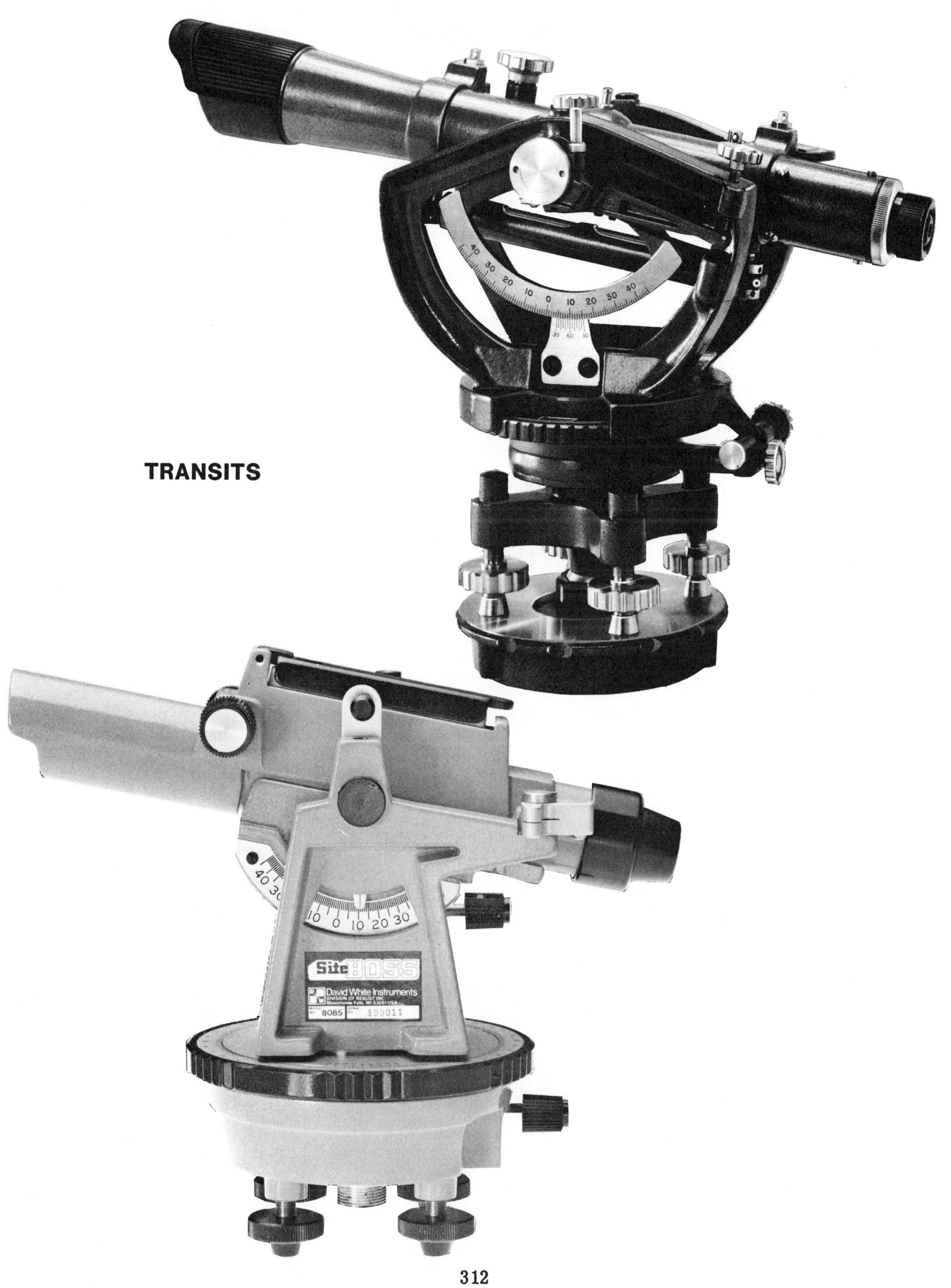
Site BOSS
David White Instruments
DIVISION OF REALIST INC
8085
300011

EXAMPLES OF GRADUATION STYLES

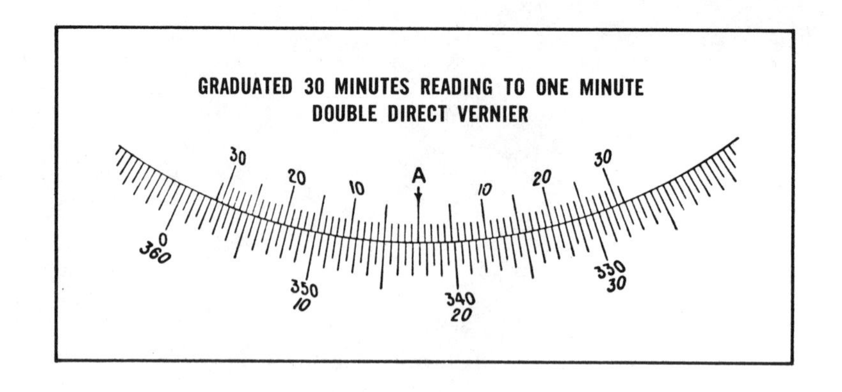

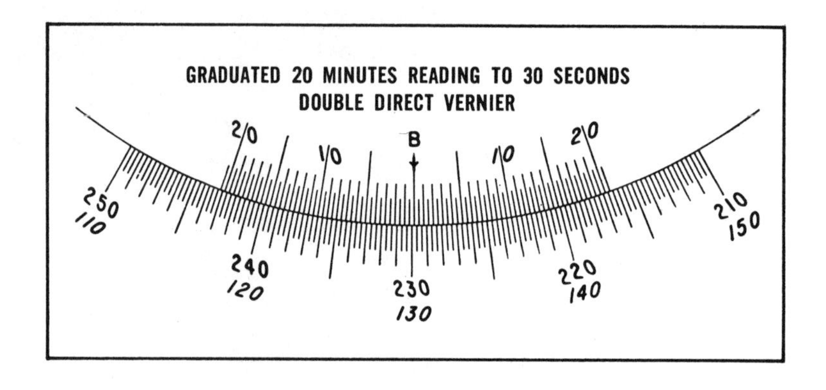

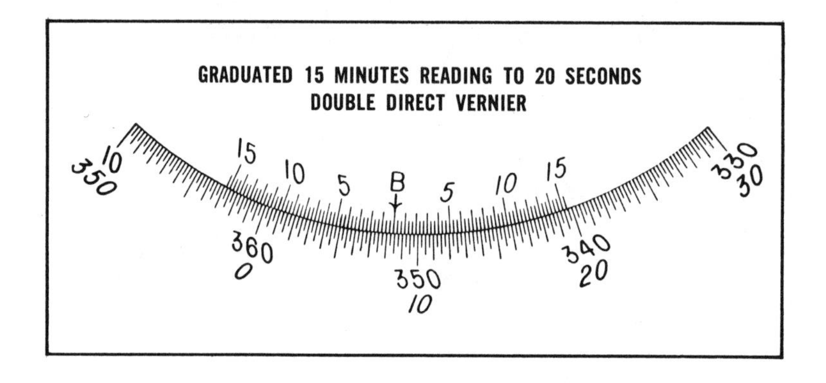

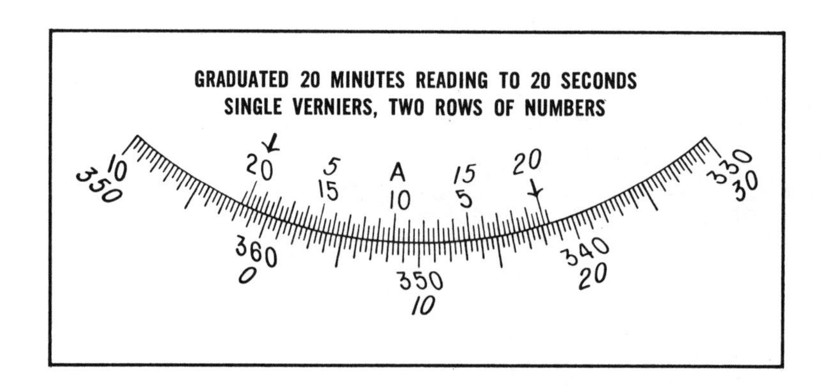

POCKET AND LASER TRANSITS

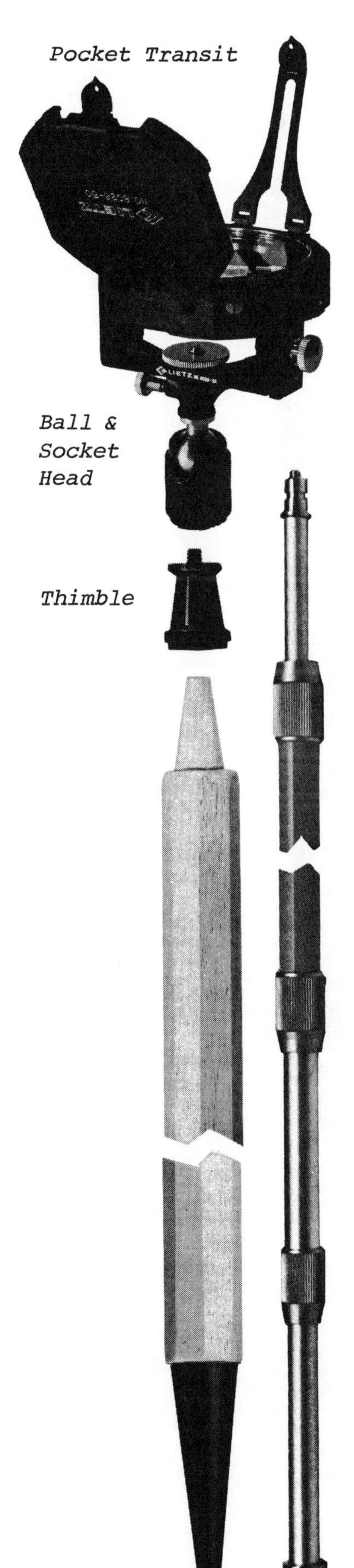

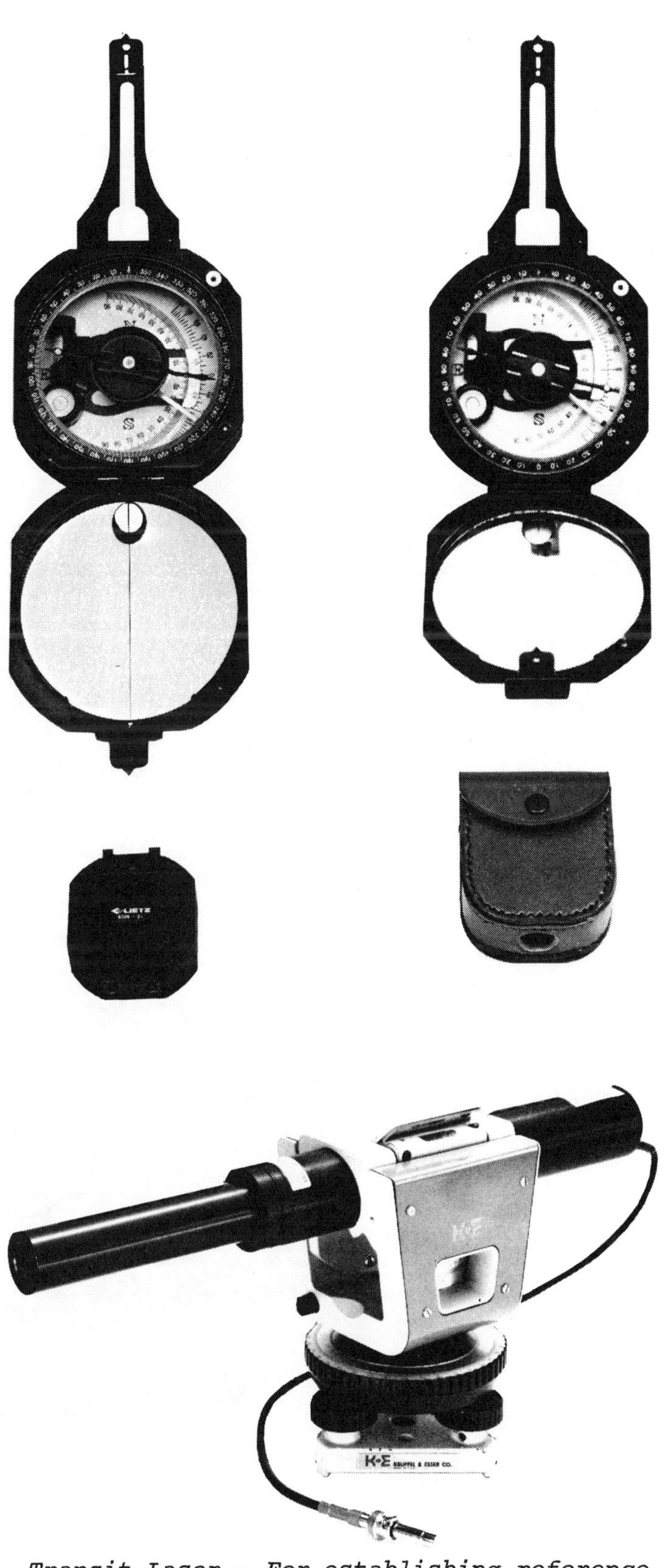

Transit Laser - For establishing reference lines of sight.

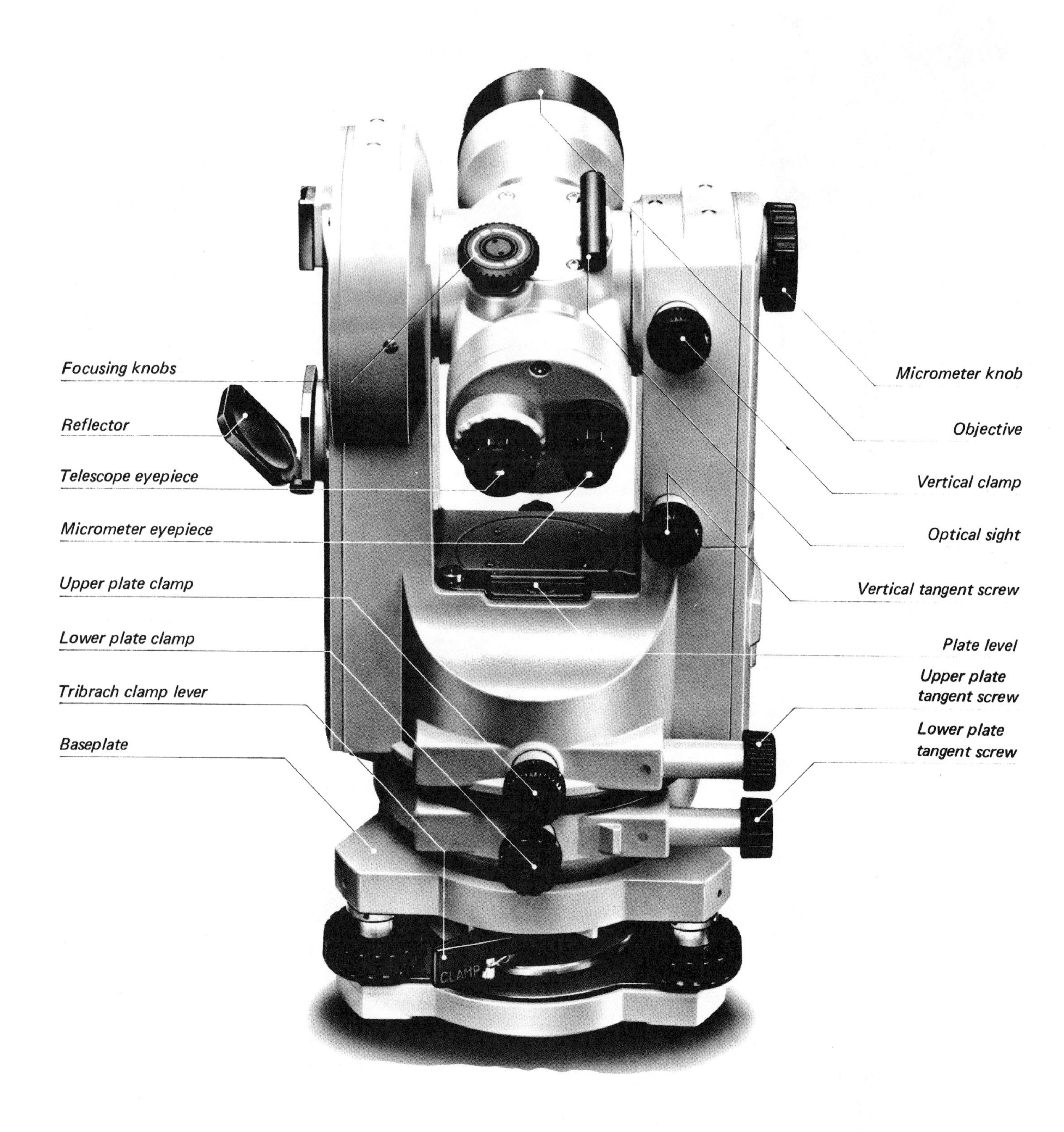
Focusing knobs
Reflector
Telescope eyepiece
Micrometer eyepiece
Upper plate clamp
Lower plate clamp
Tribrach clamp lever
Baseplate
Micrometer knob
Objective
Vertical clamp
Optical sight
Vertical tangent screw
Plate level
Upper plate tangent screw
Lower plate tangent screw
CLAMP

ENGINEER'S THEODOLITE

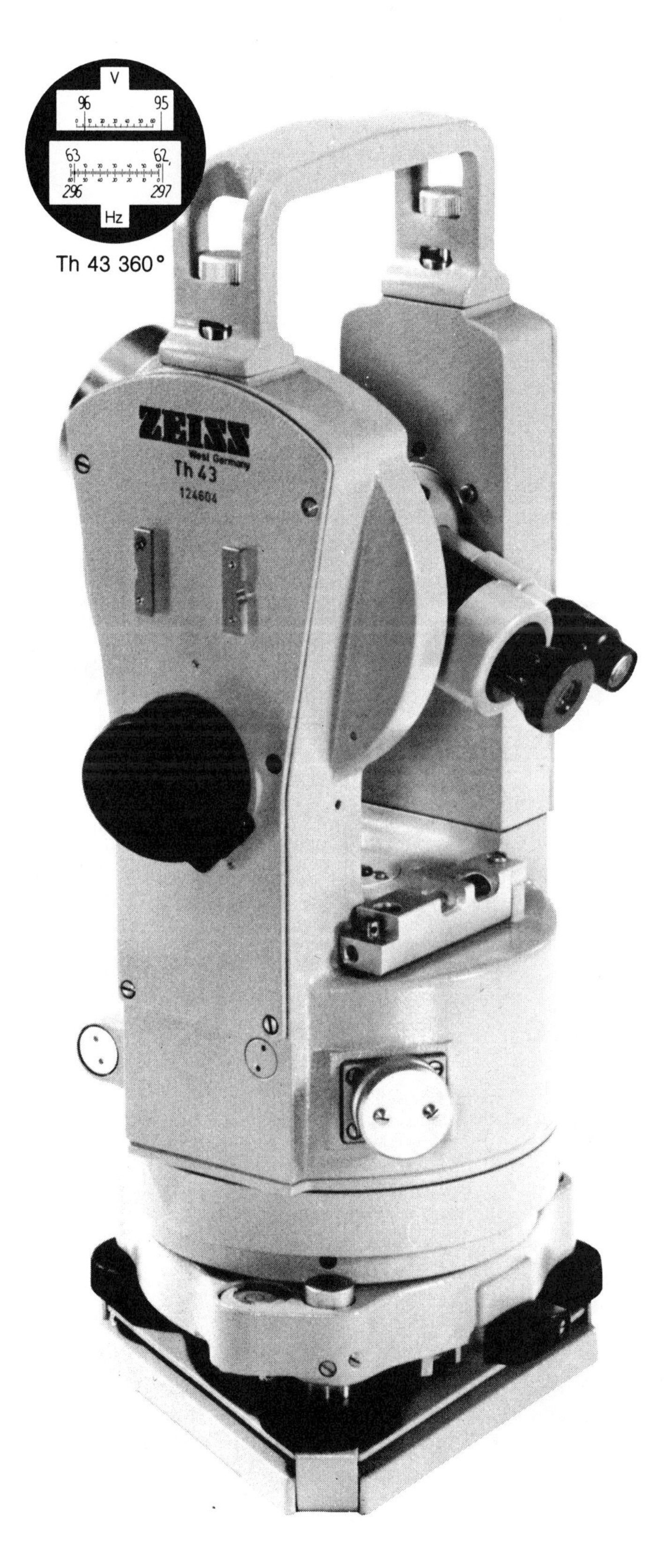

Th 43 360°

Accessories for Theodolites

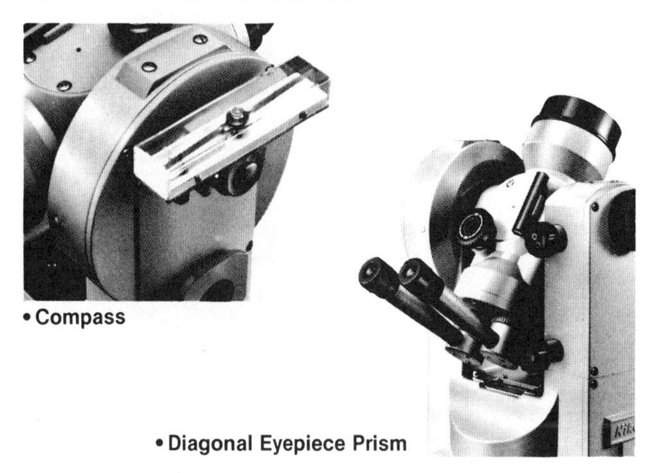

• Compass

• Diagonal Eyepiece Prism

• Optical Plummet Tribrach

• Traversing Target Set

• Plastic Carrying Case with Shoulder Strap

SURVEYOR'S ACCESSORIES

BAROMETERS/ALTIMETERS

THERMOMETERS

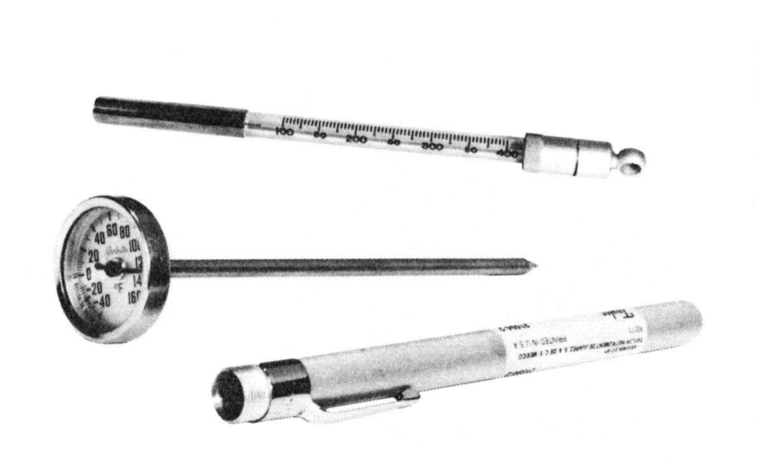

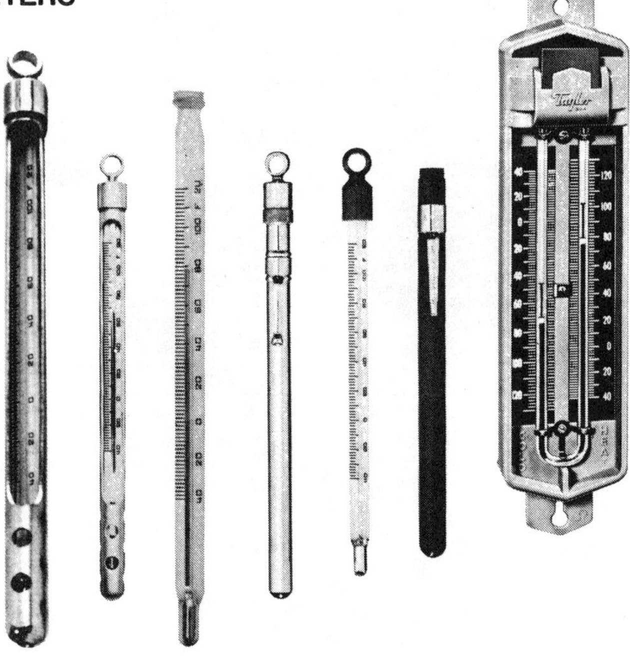

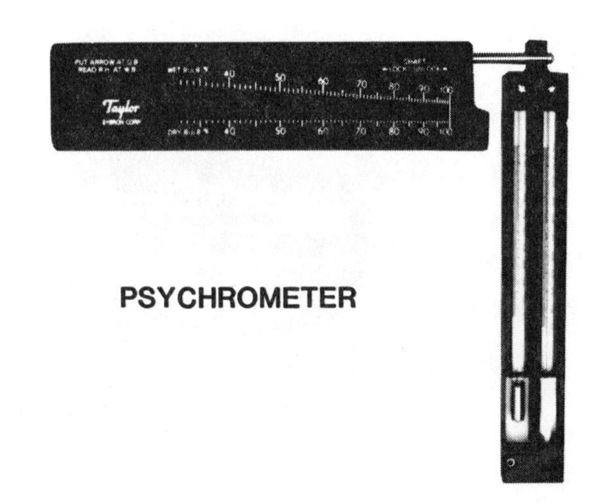

PSYCHROMETER

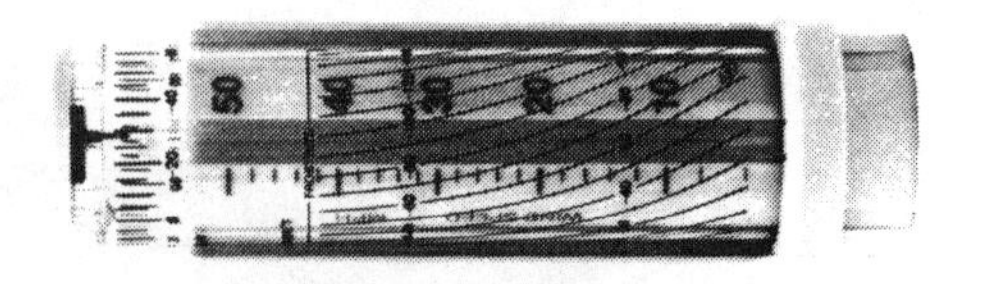

WIND CHILL & SPEED METER

TAPE ACCESSORIES

Chaining Pin Carrying Ring

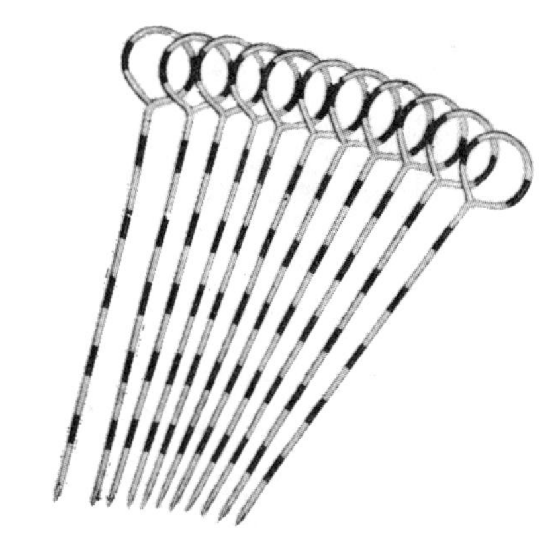

Chaining Pin Set(Surveyors Arrows)

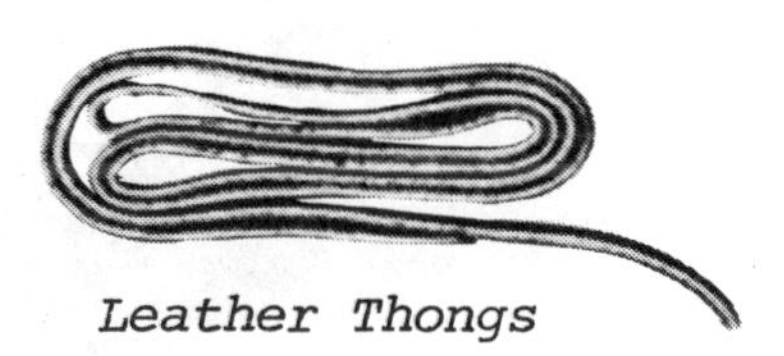

Leather Thongs

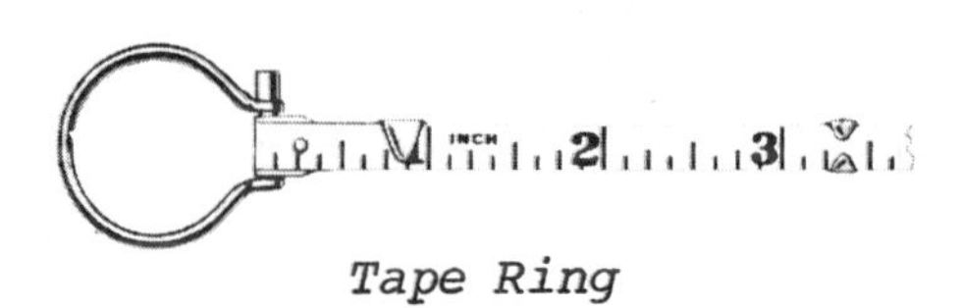

Tape Ring

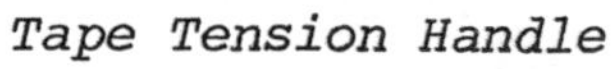

Tape Tension Handle

Tape Repair Kit

Tape Clamp Handle

MARKERS

Flat Survey Markers

Domed Survey Markers

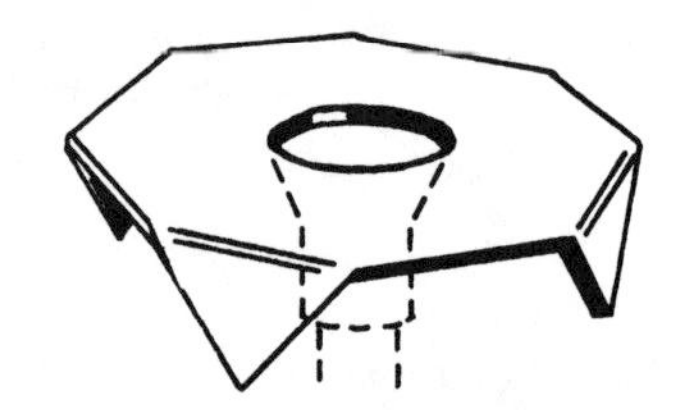

Adapter for Copperweld Survey Markers

Surveyors Flat Shiners

Imprinted Small Survey Tags

Flagging Tape

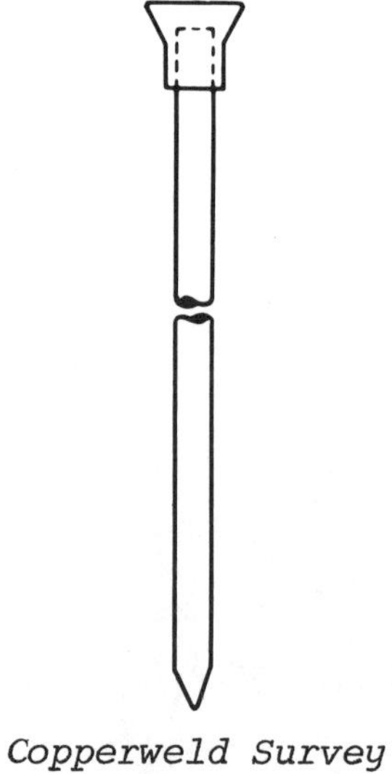

Copperweld Survey Markers

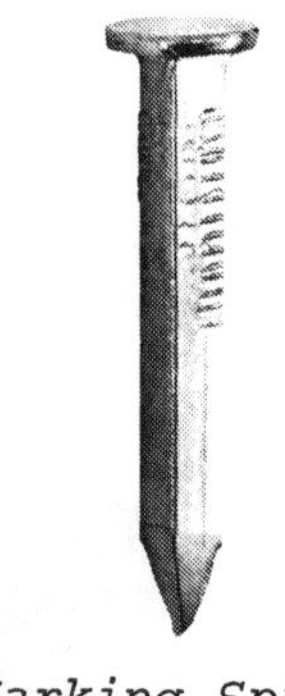

Marking Spikes

Stake Tacks

Masonry Nails

Nail Markers

SURVEYORS HAND TOOLS

Machete

Single Jack Sledge

Bush Hook

Straight Claw Hammer

Sandvik Bush Axe

Gad

Gad Pry Bar

Rock/Concrete Chisel

Pick Hammer

Concrete Drill

Hand Axe

Safety Goggles

Broad Hatchet

Timber Scriber

TRIPODS AND ACCESSORIES

Tripod Adapter

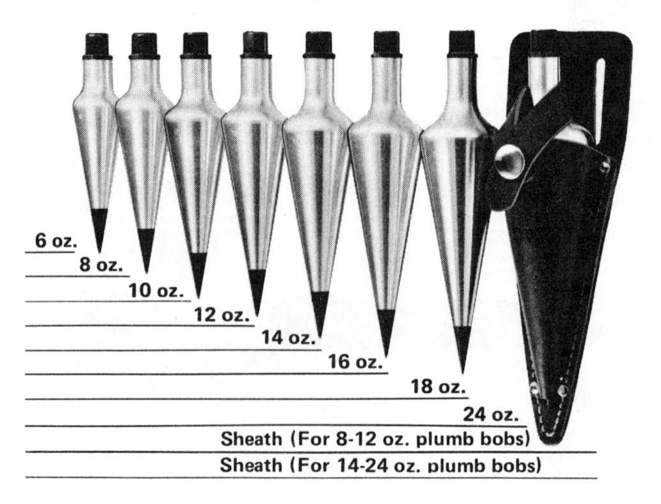

Plumb Bobs

Storage area for a spare point is within the cap.

PLANIMETERS

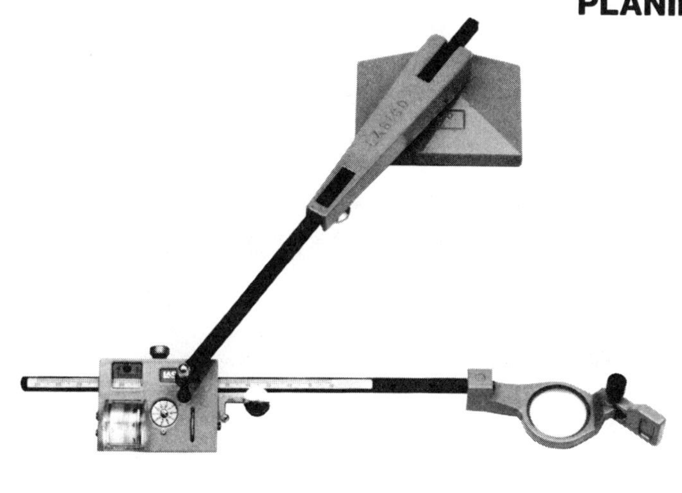

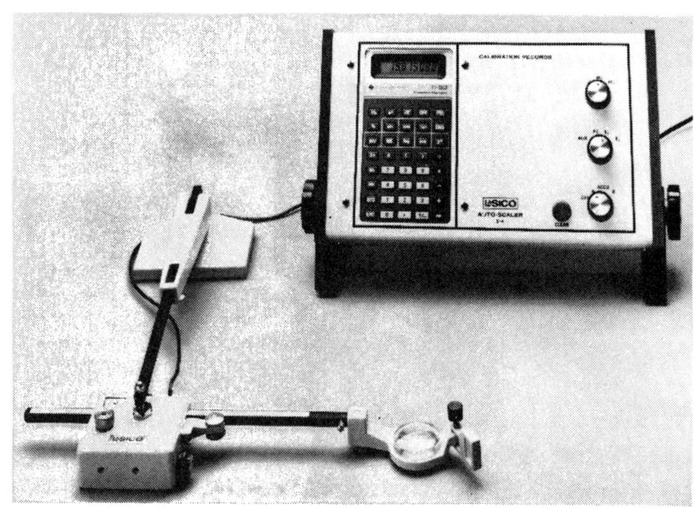

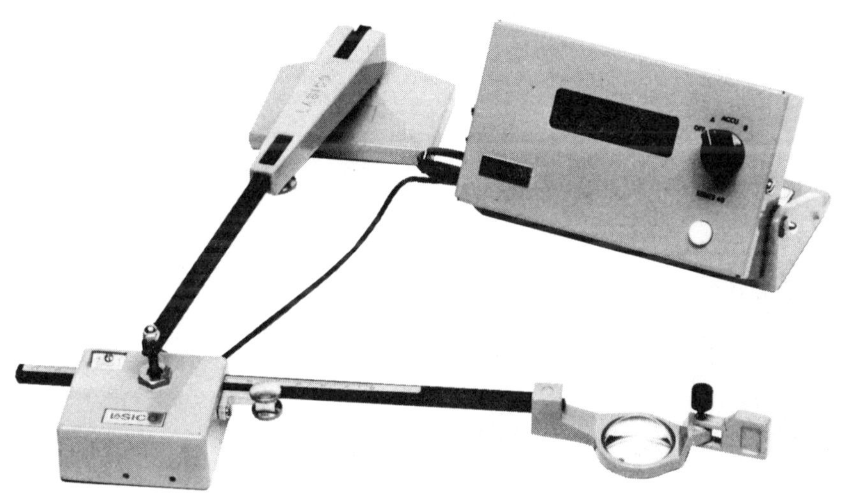

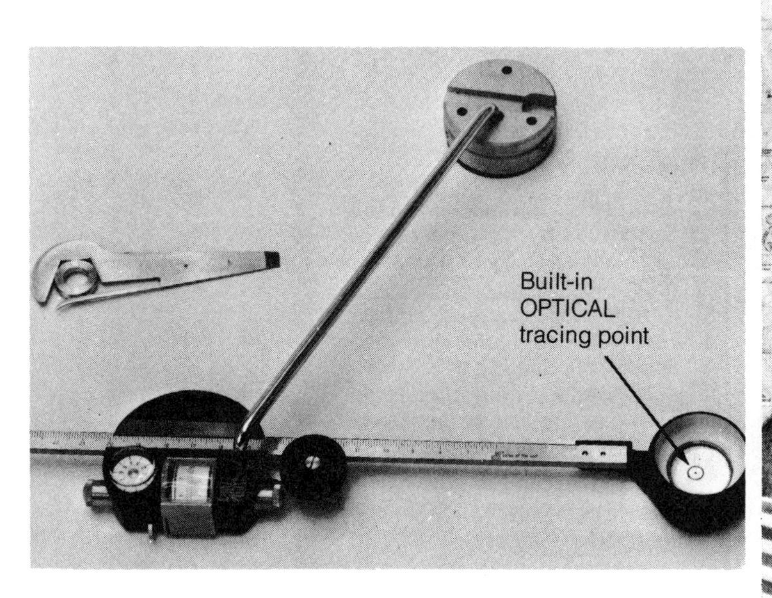

STEREOSCOPE

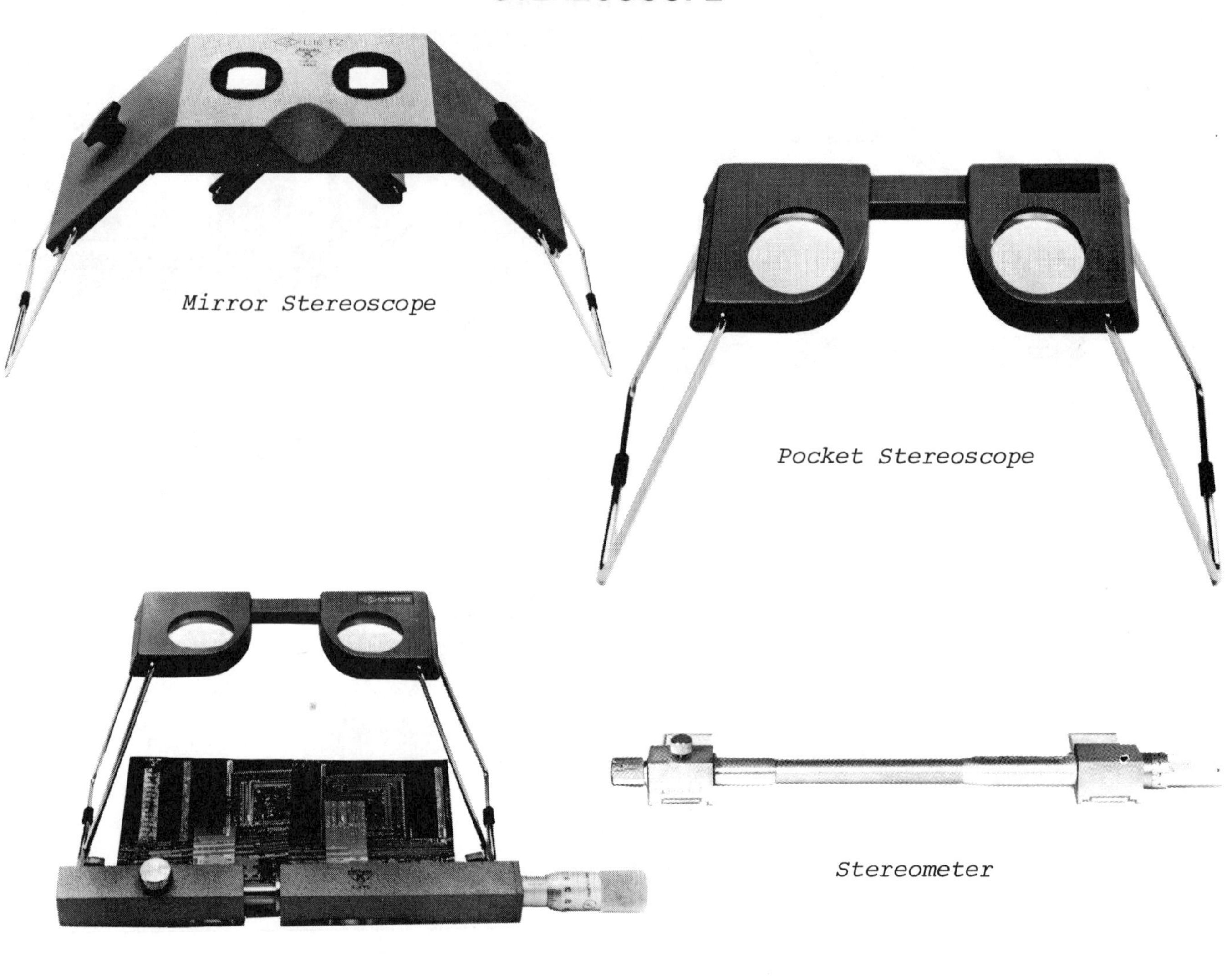

Mirror Stereoscope

Pocket Stereoscope

Stereometer

Height Finder Attachment

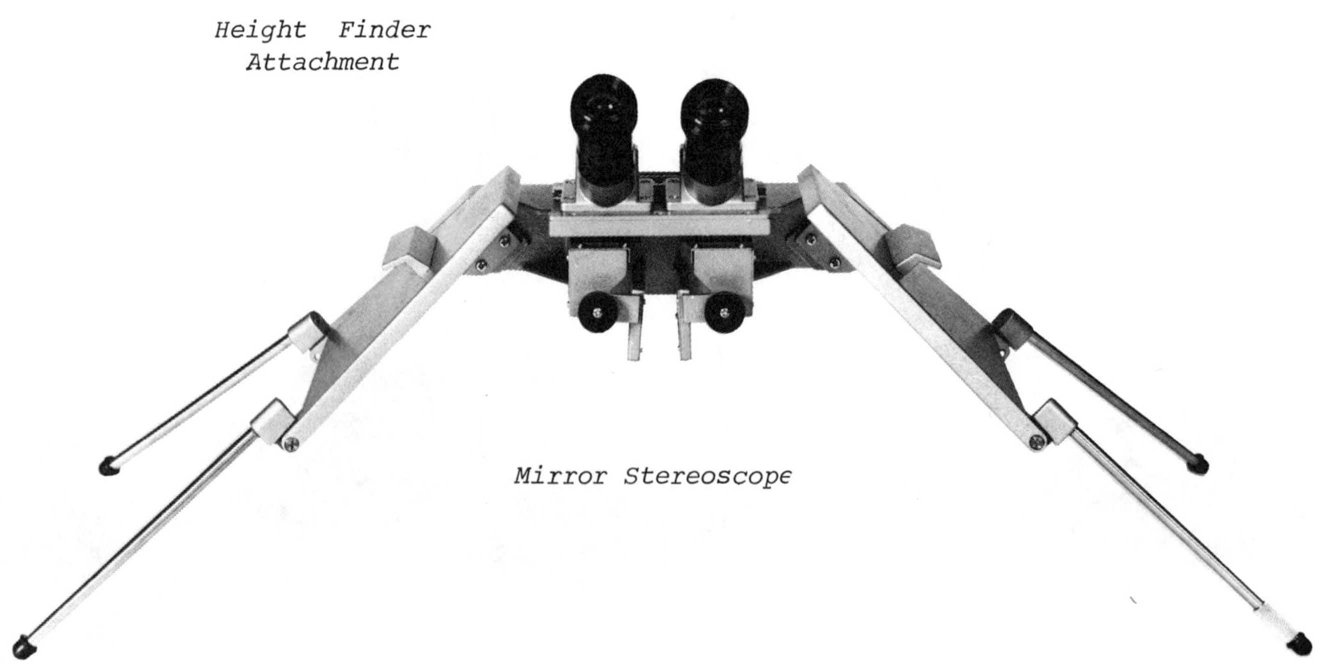

Mirror Stereoscope

Tree Calipers

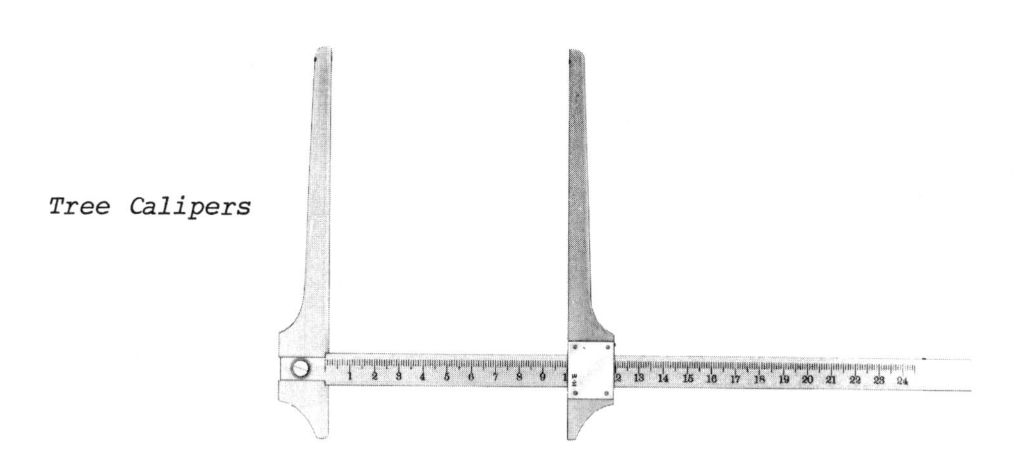

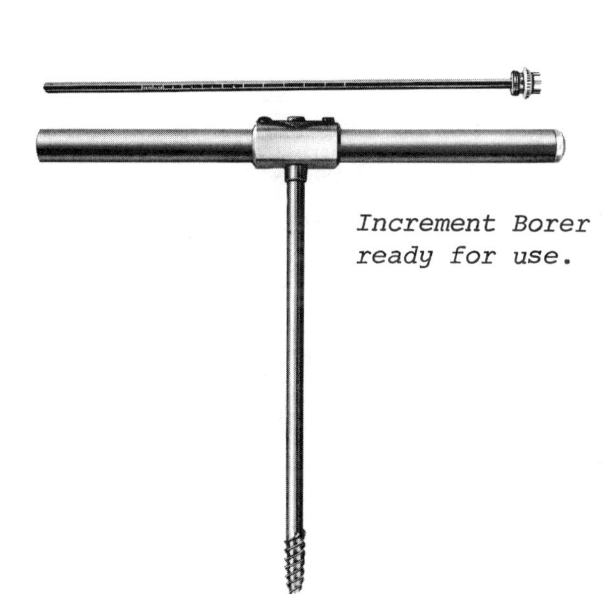

Increment Borer ready for use.

Strobe Light Prism

Magnifiers

EXAMPLES OF RETICLES

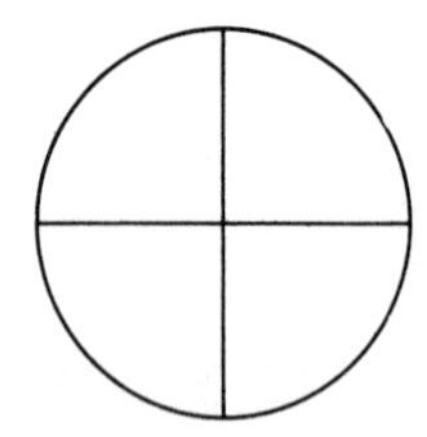

Plain cross lines.

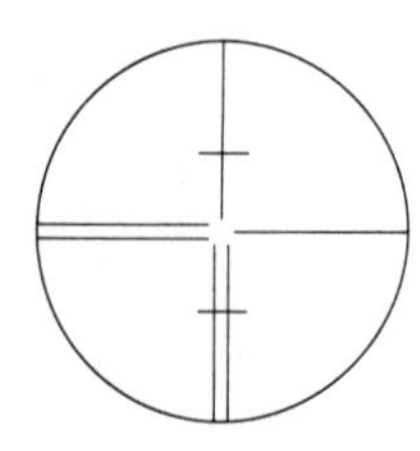

Stadia lines spaced 1:100 vertically.

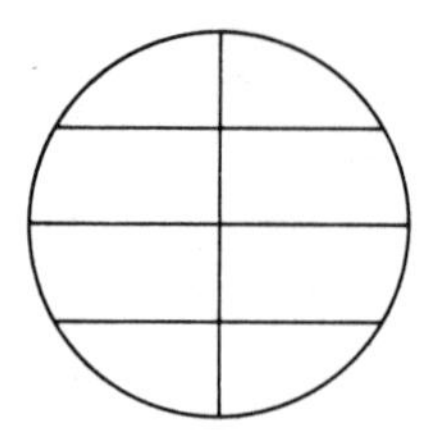

Extended stadia lines 1:100.

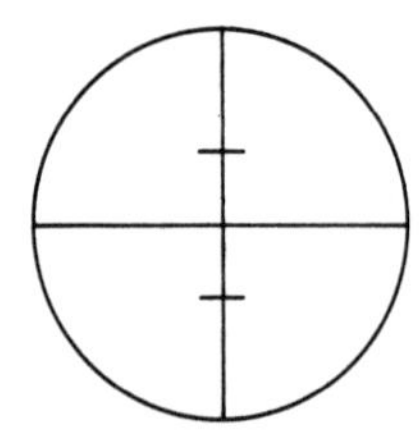

Cross lines with stadia spaced 1:100

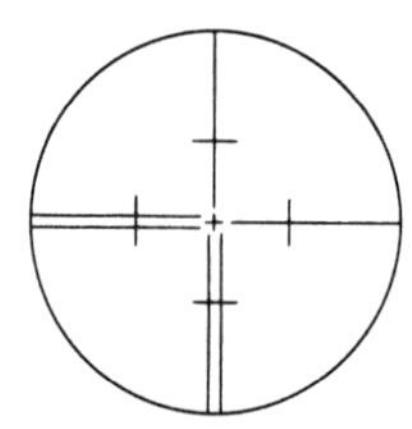

Parallel horizontal and vertical lines combined with regular cross-lines. Stadia lines are spaced 1:100 horizontally and vertically.

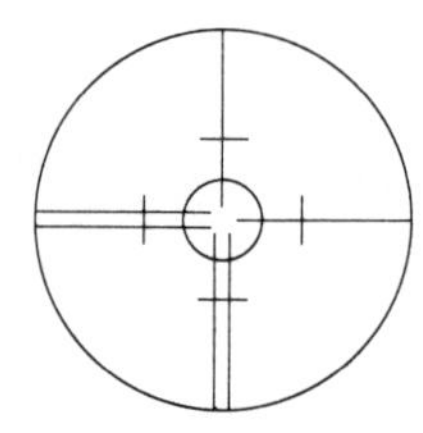

Solar reticle - with a circle 15' 45" semi-diameter for centering the suns image.

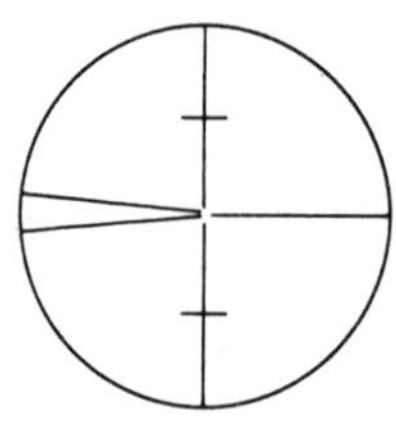

Wedge shape on the left side for use with precise leveling rods. Stadia lines 1:100.

ENGINEERS' FIELD BOOKS

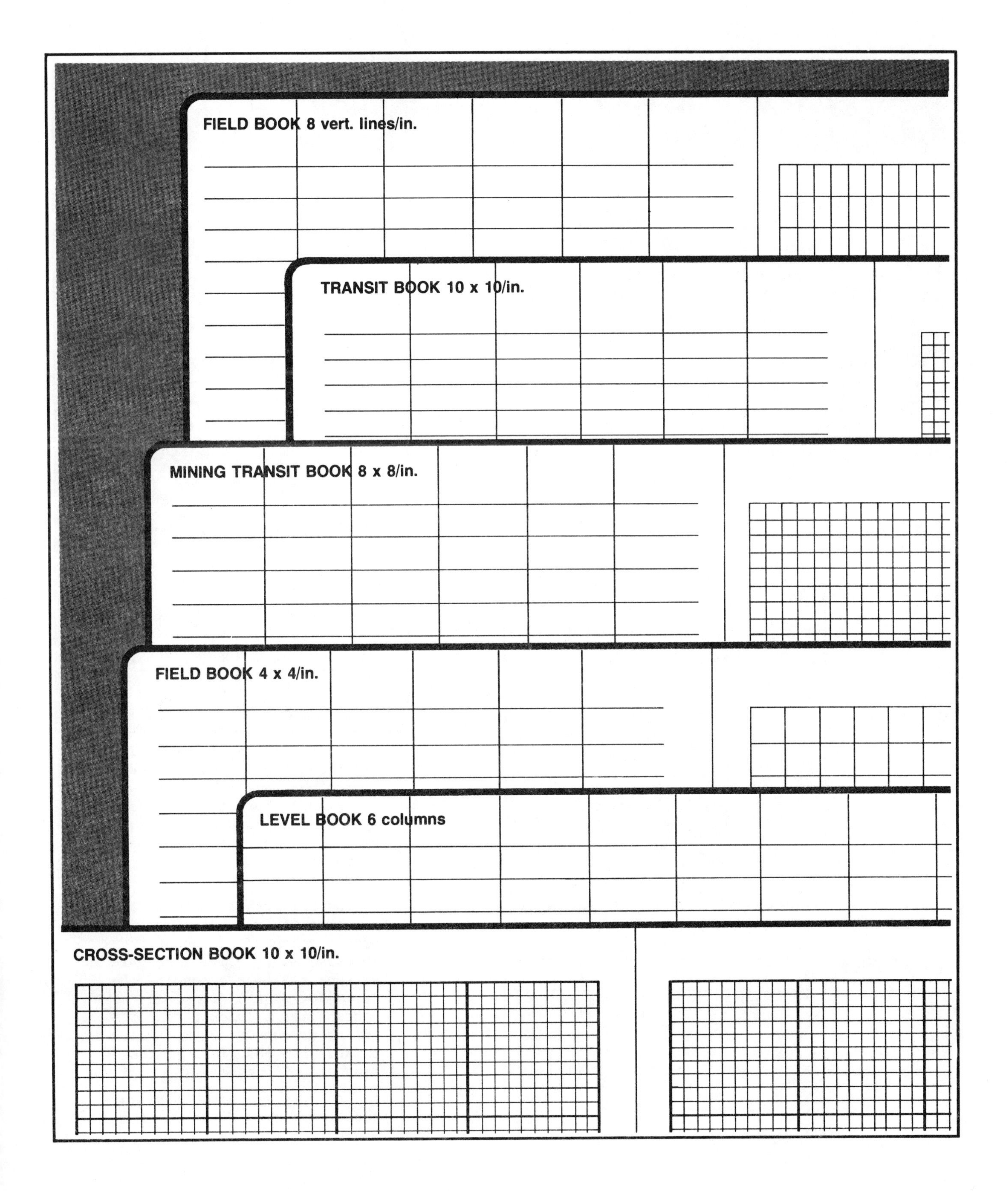

MAGNETIC LOCATORS

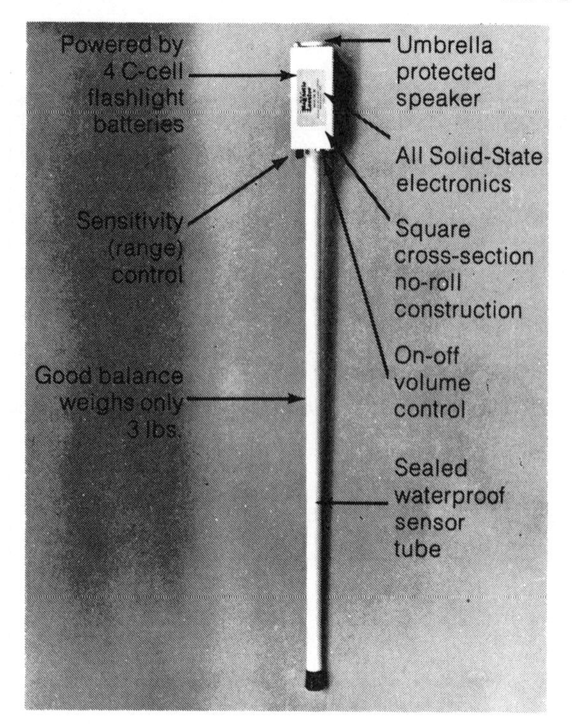

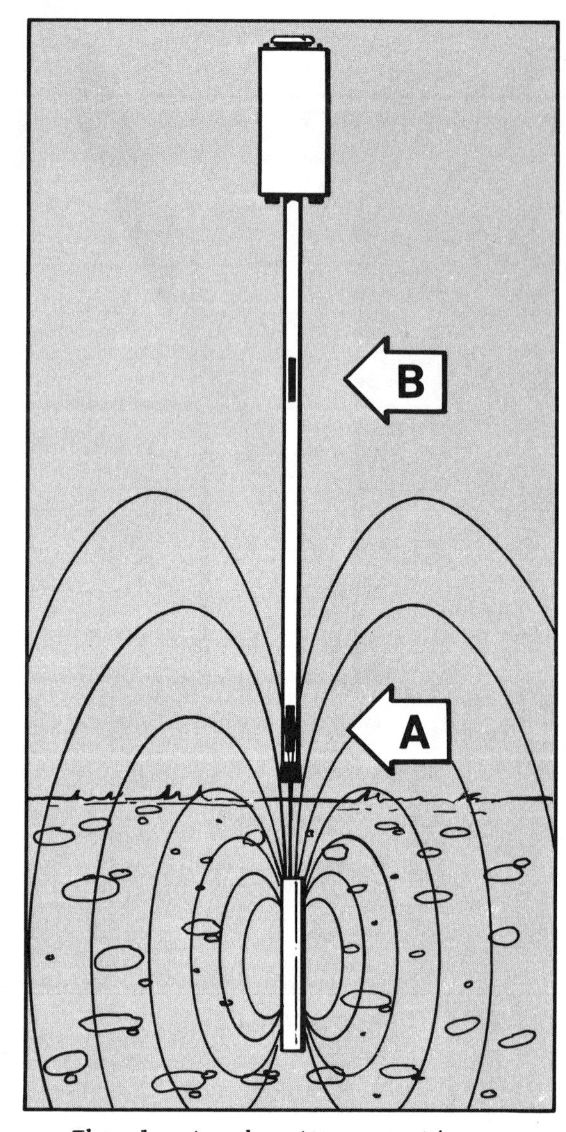

The locator has two magnetic-field sensors, A and B spaced 20 inches apart. The frequency of the signal present on the loud-speaker changes when the magnetic field at one sensor becomes stronger than at the other sensor.

In the illustration above, the magnetic field of the iron marker is stronger at sensor A (because it is closer) than it is at sensor B. As a result, the frequency of the signal on the loudspeaker is higher than its idling frequency, 65 Hz, which exists when the field is the same at both sensors.

Two sensors are used to balance out the effect of the Earth's magnetic field.

4-8 feet (4" pipe)

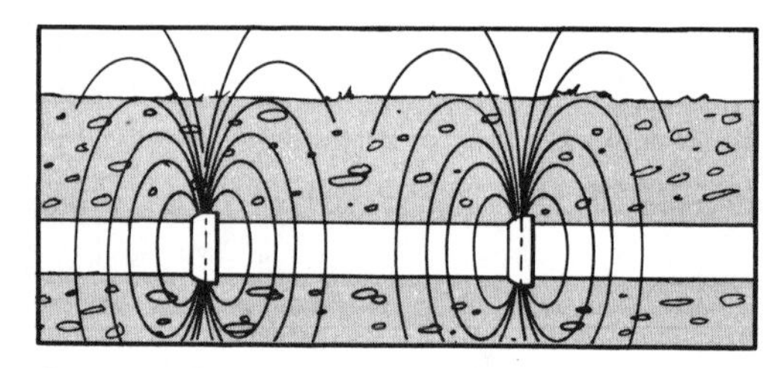

Cast Iron Water and Gas Lines

6-8 feet

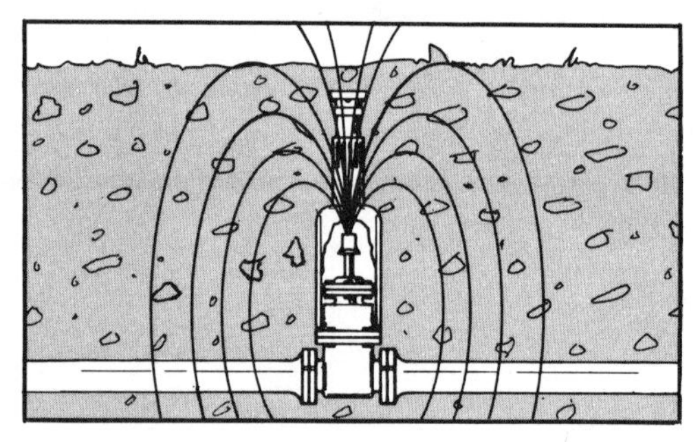

Iron Valve and Curb Boxes

10-15 feet

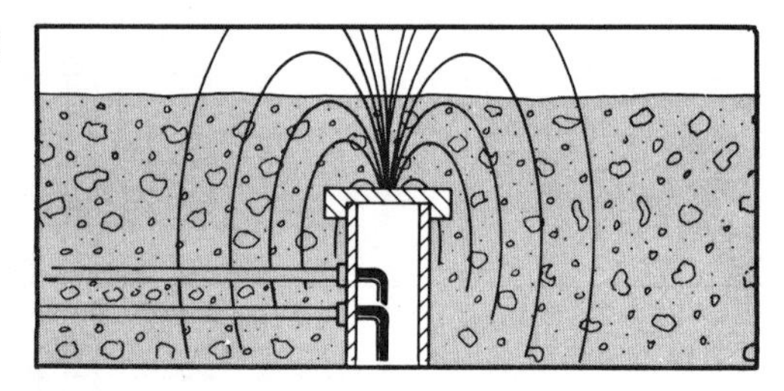

Well Casings

APPENDIX F
GLOSSARY

This glossary has been prepared to provide the surveyor with a ready reference of terms. This glossary contains those definitions that are used by the field surveyor and in the mapping field.

Aberration of light—(*astronomic*)—The apparent displacement in position of a stellar body due to the velocity of light combined with the motion of the earth itself.

Accidental error—Any small error accidentally incurred in a measurement. Unlike systematic errors, accidental errors are not governed by fixed laws. The theory of probability is based on the occurrence of these errors, which are likely to be positive as negative.

Accuracy—The degree of conformity with a standard, or the degree of perfection attained in a measurement. Accuracy relates to the quality of a result and is distinguished from precision which relates to the quality of the operation by which the result is obtained.

Actual error—The difference between the accepted value and the measured value of a physical quantity.

Adjusted position—An adjusted value for the horizontal or vertical position of a survey station, in which discrepancies due to errors in the observed data are removed. This adjustment forms a coordinated and correlated system of stations.

Altimeter—An aneroid barometer used for the measurement of approximate elevations or approximate differences of elevation.

Altitude—The vertical angle measured between the plane of the observer's true horizon and a line to the object.

Angle of depression—A negative altitude.

Angle of elevation—A positive altitude.

Arithmetical mean—The value obtained by dividing the sum of a series of values by the number of values in the series.

Astronomical latitude—The angle between the plumb line and the plane of celestial equator, Also defined as the angle between the plane of the horizon and the axis of rotation of the earth. Astronomical latitude applies only to positions on the earth and is reckoned from the astronomic equator (0°), north and south through 90°. Astronomical latitude is the latitude which results directly from observations of celestial bodies, uncorrected for deflection of the vertical.

Astronomical longitude—The angle between the plane of the celestial meridian and the plane of an initial meridian, arbitrarily chosen. Astronomical longitude is the longitude which results directly from observations on celestial bodies, uncorrected for deflection of the vertical.

Astronomical triangle—The triangle on the celestial sphere formed by arcs of great circles connecting the celestial pole, the zenith, and a celestial body. The angles of the astronomical triangles are: At the pole, the hour angle; at the celestial body, the parallactic angle; at the zenith, the azimuth angle. The sides are: Pole to zenith, the co-latitude; zenith to celestial body, the zenith distance; and celestial body to pole, the polar distance.

Azimuth (*geodetic*)—For the geodesic line from A to B, the angle between the tangent to the meridian at A and the tangent to the geodesic line at A. This angle is measured clockwise from south.

Azimuth (*grid*)—The angle in the plane of projection between a straight line and the central meridian (axis of Y) of a plane-rectangular coordinate system.

Azimuth—(Surveying) The horizontal direction of a line measured clockwise from a reference plane, usually the meridian. Often called *forward azimuth* to differentiate from *back azimuth*.

Azimuth angle—(astronomy) The angle less than 180° between the plane of the celestial meridian and the vertical plane containing the observed object, reckoned from the direction of the elevated pole. In astronomic work, the azimuth angle is the spherical angle at the zenith in the astronomical triangle which is composed of the pole, the zenith, and the star. In geodetic work, it is the horizontal angle between the celestial pole and the observed terrestrial object.

Azimuth mark—A marked point or adjacent station visible from an occupied station, the azimuth to which is determined for use in dependent surveys.

Backsight—(1) In traversing, a backsight (BS) is a sight on a previously established traverse or triangulation station, which is not the closing sight on the traverse. (2) In leveling, a backsight is a reading on a rod held on a point whose elevation has been previously determined, and which is not the closing sight of a level line.

Base line—A surveyed line established with more than usual care, to which surveys are referred for coordination and correlation.

Base line (triangulation)—The side of one of a series of connected triangles, the length of which is measured with prescribed accuracy and precision, and from which the lengths of the other triangle sides are obtained by computation.

Base net—A small net of geometric figures used to expand from a base line to a line of the main scheme of a triangulation net.

Basic control—Horizontal and vertical control of third- or higher-order accuracy, determined in the field and permanently marked or monumented, that is required to control further surveys.

Bearing—The direction of a line within a quadrant, with respect to the meridian. Bearings are measured clockwise or counterclockwise from north or south, depending on the quadrant.

Bench mark—A relatively permanent object, natural or artificial, bearing a marked point whose elevation above or below an adopted datum is known. Usually designated as a BM, such a mark is sometimes further qualified as a PBM (permanent bench mark), or as a TBM (temporary bench mark).

Cadastral survey—A survey relating to land boundaries and subdivisions, made to create units suitable for transfer or to define the limitations of title. The term cadastral survey is now used to designate the surveys of the public lands of the United States, including retracement surveys for the identification and resurveys for the restoration of property lines; the term can also be applied properly to corresponding surveys outside the public lands, although such surveys are usually termed land surveys through preference.

Celestial equator—A great circle on the celestial sphere on which any point is equidistant from the celestial poles. The plane of the earth's equator, if extended, would coincide with that of the celestial equator.

Celestial meridian—A vertical circle, passing through both celestial poles, the plane of which is perpendicular to the celestial equator.

Celestial pole—A reference point located at the point of intersection of an indefinite extension of the earth's axis of rotation and the apparent celestial sphere.

Celestial sphere—An imaginary sphere of infinite radius with the earth as a center. It rotates from east to west on a prolongation of the earth's axis.

Central meridian—1. The line of longitude at the center of a projection. Generally the basis for constructing the projection. 2 (state plane-coordinate system). The meridian used as the axis of Y for computing projection tables for a state coordinate system. The central meridian of the system usually passes close to the center of figure of the area or zone for which the tables are computed.

Chronograph—An instrument for producing a graphical record of time as shown by a clock or other device. Also capable of recording the occurrence of other events simultaneously upon the proper signal so that these events can be scaled with relation to the time base.

Chronometer—A portable timekeeper with compensated balance, capable of showing time with extreme precision and accuracy.

Circle position—A prescribed setting (reading) of the horizontal circle of a direction theodolite, to be used for the observation on the initial station of a series of stations that are to be observed.

Circuit closure—In leveling, the amount by which the algebraic sum of the measured differences of elevation around a circuit fails to equal zero.

Circumpolar star—A star in any given latitude which never goes below the horizon; hence its polar distance must be less than the given latitude. In astronomy only those stars with a polar distance of less than 10° are considered in practical problems.

Closed traverse—A traverse that starts and ends at the same point, or at stations whose positions have been determined by other surveys. See loop traverse.

Collimation—The line of sight or aiming line of the instrument when coincident with the physical alinement of the instrument; thus collimation error is the angle between the line of collimation (line of sight) of a telescope and the collimation axis of the instrument.

Control—1. The coordinated and correlated dimensional data used in geodesy and cartography to determine the positions and elevations of points on the earth's surface or on a cartographic representation of that surface. 2. A collective term for a system of marks or objects on the earth or on a map or a photograph whose positions or elevation, or both, have been or will be determined.

Control survey—A survey which provides positions (horizontal or vertical) of points to which supplementary surveys are adjusted.

Coordinates—Linear or angular quantities, or both, which designate the position of a point in relation to a given reference frame. There are two general divisions of coordinates used in surveying: polar coordinates and rectangular coordinates. These may be subdivided into three classes: plane coordinates, spherical coordinates, and space coordinates.

Culmination, transit—The instant when any point on the celestial sphere is on the meridian of an observer. When it is on that half of the meridian containing the zenith, it is called the upper transit; when it is on the other half, it is called the lower transit.

Datum—1. (general) Any numerical or geometrical quantity or set of such quantities which may serve as a reference or base for other quantities. 2. (geodetic) A reference surface consisting of five quantities: the latitude and longitude of an initial point, the azimuth of a line from this point, and the parameters of the reference ellipsoid. It forms the basis for the computation of horizontal-control surveys in which the curvature of the earth is considered. 3. (leveling) A level surface to which elevations are referred, usually, but not always, mean sea level.

Declination—1. In a system of polar or spherical coordinates, the angle at the origin between a line to a point and the equatorial plane, measured in a plane perpendicular to the equatorial plane. 2. The arc between the equator and the point measured on a great circle perpendicular to the equator. 3. (astronomy) The angular distance to a body on the celestial sphere measured north or south through 90° from the celestial equator along the hour circle of the body. Comparable to latitude on the terrestrial sphere. 4. Often used as a shortened term for magnetic declination although this use is not preferred.

Deflection of the vertical—The angular difference, at any place, between the upward direction of a plumb line (the vertical) and the perpendicular (the normal) to the reference spheroid. This difference seldom exceeds 30 seconds. Often expressed in two components, meridian and prime vertical.

Departure—In a plane survey, the difference between the eastings of the two ends of the line, which may be either plus or minus. This value is symbolized by Δ E.

Direct leveling—The determination of differences of elevation by means of a continuous series of short horizontal lines. Vertical distances from these lines to adjacent ground marks are determined by direct observations on graduated rods with a leveling instrument equipped with a spirit level.

Direct reading—The reading of the horizontal or vertical circle of a theodolite or engineer transit with the telescope direct. In field notes, a direct reading is indicated by the letter D preceding the observed value.

Direction instrument theodolite—A theodolite in which the graduated horizontal circle remains fixed during a series of observations, the telescope being pointed on a number of signals or objects in succession, and the direction of each read on the circle, usually by means of micrometer microscopes. Direction instrument theodolites are used almost exclusively in first- and second-order triangulation.

Distance angle—An angle in a triangle opposite a side used as a base in the solution of the triangle, or a side whose length is to be computed.

Ecliptic—The great circle on the celestial sphere which the sun appears to describe in its annual motion among the stars. It is inclined to the celestial equator at an angle of about 23° 27′.

Elevation—Vertical distance from a datum, usually mean sea level, to a point or object on the earth's surface. Not to be confused with altitude which refers to points or objects above the earth's surface.

Elongation—That point in the apparent movement of a circumpolar star when it has reached the extreme position east or west of the meridian.

Emulsion—A suspension of either light-sensitive silver salts, Diazos, or photopolymers, in a colloidal medium which is used for coating films, plates, and papers.

Ephemeris time—The uniform measure of time defined by the laws of dynamics and determined in principal from the orbital motions of the planets, specifically in the orbital motion of the earth as represented by Newcomb's "Tables of the Sun".

Equation: angle—A condition equation which expresses the relationship between the sum of the measured angles of a closed figure and the theoretical value of that sum, the unknowns being the corrections to the observed directions or angles, depending upon which are used in the adjustment.

Sometimes called a triangle equation, an angle equation is used to make the sum of the three observed angles of a triangle, with corrections applied, equal to 180° plus the spherical excess of the triangle.

Equation, azimuth—A condition equation which expresses the relationship between the fixed azimuths of two lines which are connected by triangulation or traverse.

When a survey (triangulation or traverse) connects two lines whose azimuths are fixed by direct observation or by previous surveys, an azimuth equation is used to make the azimuth of either line as computed through the adjusted survey from the other line agree with its azimuth as previously fixed.

Equation, condition—An equation which expresses exactly certain relationships that must exist among related quantities, which are not independent of one another, exist a priori, and are separate from relationships demanded by observation.

Example: in measuring the angles of a triangle, no condition exists until all three angles are measured. The condition equation will then express the condition that the three measured angles plus certain corrections must equal 180° plus the spherical excess of the triangle. The various condition equations set up in survey work are defined under terms which are descriptive of the conditions: as, angle equation, side equation, length equation, latitude equation, longitude equation, azimuth equation.

Equation, correlate—An equation derived from an observation or condition equation, employing undetermined multipliers, and expressing the condition that the sum of the squares of the residuals (or corrections) resulting from the application of these multipliers to the observation or condition equations shall be a minimum.

In the least-squares adjustment of triangulation, correlate equations are formed directly from the observation or condition equations, there being as many correlate equations as there are corrections to be determined, but only as many undetermined multipliers (correlates or correlatives) as there are observation or condition equations. From these correlate equations, the normal equations are formed, equal in number to the undetermined multipliers which constitute the unknowns in the normal equations. The solution of the normal equations determines values for the multipliers which, when substituted in the correlate equations, give values for the corrections which will satisfy the observation or condition equations, make the observations and their functions consistent among themselves, and at the same time make the adjusted values the most probable that can be derived from the given observations.

Equation, error—The probability equation which expresses the laws of the occurrence of accidental errors.

The error equation is the basis of the method of least squares, used in the adjustment of observations for determining the most probable value of a result from those observations.

Equation, Laplace—The equation which expresses the relationship between astronomic and geodetic azimuths in terms of astronomic and geodetic longitudes and geodetic latitude A usual form is $\alpha_A - \alpha_G = -(\lambda_A - \lambda_G) \sin \phi_G$ i which α_A and λ_A are astronomic azimuth and longitude, α_G, λ_G and ϕ_G are geodetic azimuth, longitude, and latitude respectively. The signs depend upon convention. As written above, north latitudes and west longitudes are considered postive.

The Laplace condition (expressed by the Laplace equation) arises from the fact that a deflection of the vertical in the plane of the prime vertical will give a difference between astronomic and geodetic longitude and between astronomic and geodetic azimuth; or, conversely, that the observed differences between astronomic and geodetic values of the longitude and of the azimuth may both be used to determine the deflection in the plane of the prime vertical. This deflection cannot have two values at the same place, and the imposition of the Laplace condition forces a consistency between them. Since longitudes can be carried through triangulation with very good accuracy and azimuths cannot, the practice is to compute the value of the deflection for azimuth from the deflection in the prime vertical obtained from the longitude. The value of the azimuth thus obtained is known as a Laplace azimuth.

Equation, latitude—A condition equation which expresses the relationship between the fixed latitudes of two points which are connected by triangulation or traverse.

When a survey (triangulation or traverse) connects two points whose latitudes have been fixed by direct observation or by previous surveys, a latitude equation is used to make the latitude of either point as computed through the survey from the other point agree with its latitude as previously fixed.

Equation, length—A condition equation which expresses the relationship between the fixed lengths of two lines which are connected by triangulation.

When a section of triangulation connects two lines whose lengths are fixed by direct measurement or by previous triangulation, a length equation is used to make the length of either line as computed through the adjusted triangulation from the other line agree with its length as previously fixed.

Equation: longitude—A condition equation which expresses the relationship between the fixed longitudes of two points which are connected by triangulation of traverse.

When a survey (triangulation or traverse) connects two points whose longitudes have been fixed by direct observation or by previous surveys, a longitude equation is used to make the longitude of either point as computed through the survey from the other point agree with its longitude as previously fixed.

Equation: normal—An equation derived from observation or condition equations or from correlate equations, expressing the condition that the sum of the squares of the residuals (or corrections) resulting from the substitution in the observation or condition equations of factors obtained from the normal equations either directly or through the correlate equations shall be a minumum.

In a least-squares adjustment, corrections are desired to observed values which are connected together by a series of observation or condition equations, the number of such equations being smaller than the number of observed values on which they depend. The basic equations are transformed into normal equations, either directly or through the medium of correlate equations, which contain the same number of unknowns as there are equations. Factors obtained from the solution of normal equations, either directly or through the correlate equations, are applied to the observation or condition equations to obtain the desired corrections.

Equation: observation—A condition equation which connects interrelated unknowns by means of an observed function.

Example: An angle equation: the unknowns are the corrections to the observed angles; the function is the sum of those observed angles expressed as a closing error.

Alternate definition: A condition equation connecting the function observed and the unknown quantity whose value is sought.

Equation: side—A condition which expresses the relationship between the various sides in a triangulation figure as they may be derived by computation from one another.

A side equation is used to make the computed length of a triangle side the same for all routes through the triangulation whereby it may be derived.

Equation of time—The algebraic difference in hour angle between apparent solar time and mean solar time, usually labeled + or − as it is to be applied to mean solar time to obtain apparent solar time.

Equinox—One of the two points of intersection of the ecliptic and the celestial equator, occupied by the sun when its declination is 0°.

Error—1. The difference between an observed or true value of that quantity. 2. A class of small inaccuracies due to imperfections in equipment or techniques, surrounding conditions, or human limitations; not to be confused with blunders or mistakes.

Error of closure—1. (general) The amount by which a quantity obtained by a series of related measurements differs from the true or fixed value of the same quantity. 2. (angles) The amount by which the actual sum of a series of angles fails to equal the theoretically exact value of that sum. 3. (azimuth) The amount by which two values of the azimuth of a line, derived by different surveys or along different routes, fail to be exactly equal to each other. 4. (leveling) The amount by which two values of the elevation of the same bench mark, derived by different surveys or through different survey routes or by independent observations, fail to be exactly equal to each other. 5. (loop) The error in the closure of a survey on itself. 6. (horizon) The amount by which the sum of a series of adjacent measured horizontal angles around a point fails to equal exactly 360°. Measurement of the last angle of the series is called closing the horizon; sometimes called closure of horizon; 7. (triangle) The amount by which the sum of the three observed angles of a triangle fails to equal exactly 180° plus the spherical excess of the triangle. 8. (traverse) The amount by which a value of the position of a traverse station, as obtained by computation through a traverse, fails to agree with another value of the same station as determined by a different set of observations or routes of survey.

Fixed elevation—An elevation which has been adopted, either as a result of tide observations or previous adjustment of spirit leveling, and which is held at its accepted value in any subsequent adjustment.

Foresight—1. An observation of the distance and direction to the next instrument station. 2. (transit traverse) A point set ahead to be used for reference when resetting the transit or line or when verifying the alinement. 3. (leveling) The reading on a rod that is held at a point whose elevation is to be determined.

Frequency—The number of complete cycles per second existing in any form of wave motion.

Geodesic line—A line of shortest distance between any two points on any mathematically defined surface. A geodesic line is a line of double curvature, and usually lies between the two normal section lines which the two points determine. If the two terminal points are in nearly the same latitude, the geodesic line may cross one of the normal section lines. It should be noted that, except along the equator and along the meridians, the geodesic line is not a plane curve and cannot be sighted over directly. However, for conventional triangulation the lengths and directions of geodesic lines differ inappreciably from corresponding pairs of normal section lines.

Geodesy—The science which treats the determination of the size and figure of the earth (geoid) by such direct measurements as triangulation, leveling, and gravimetric observations; which determines the external gravitational field of the earth, and, to a limited degree the internal structure.

Geodetic control—A system of horizontal and/or vertical control stations that have been established and adjusted by geodetic methods and in which the shape and size of the earth (geoid) have been considered in position computations.

Geodetic datum—A datum consisting of five quantities; the latitude and the longitude of an initial point, the azimuth of a line from this point, and two constants necessary to define the terrestrial spheroid.

It forms the basis for the computation of horizontal control surveys in which the curvature of the earth is considered.

Geodetic latitude—The angle which the normal at a point on the reference spheroid makes with the plane of the geodetic equator. Geodetic latitudes are reckoned from the equator, but in the horizontal-control survey of the United States they are computed from the latitude of station Meades Ranch as prescribed in the North American Datum of 1927.

Geodetic leveling—Spirit leveling of a high order of accuracy, usually extended over large areas, to furnish accurate vertical control as a basis for the control in the vertical dimension for all surveying and mapping operations. Spirit leveling follows the geoid and its associated level surfaces which are irregular, rather than any mathematically determined spheroid or ellipsoid and associated regular level surfaces.

Geodetic longitude—The angle between the plane of the geodetic meridian and the plane of an initial meridian, arbitrarily chosen. A geodetic longitude can be measured by the angle at the pole of rotation of the reference spheroid between the local and initial meridians, or by the arc of the geodetic equator intercepted by those meridians. In the United States, geodetic longitudes are numbered from the Meridian of Greenwich, but are computed from the meridian of station Meades Ranch as prescribed in the North American Datum of 1927. A geodetic longitude differs from the corresponding astronomical longitude by the amount of the prime vertical component of the local deflection of the vertical divided by the cosine of the latitude.

Geographic coordinates—An inclusive term generally used to designate both geodetic coordinates and astronomical coordinates.

Geographic latitude—A general term applying to both astronomic and geodetic latitudes.

Geoid—The figure of the earth considered as a sea-level surface extended continuously through the continents. The actual geoid is an equipotential surface coincident with mean sea level to which, at every point, the plumb line (direction in which gravity acts) is perpendicular. It is the geoid which is obtained from observed deflections of the vertical and is the surface of reference for astronomical observations and for geodetic leveling.

Gravimeter—A weighing device or instrument of sufficient sensitivity to register variations in the weight of a constant mass when the mass is moved from place to place on the earth and thereby is subjected to the influence of gravity at those places.

Gravitation—The acceleration produced by the mutual attraction of two masses, directed along the line joining their centers of masses, and of magnitude inversely proportional to the square of the distance between the two centers of mass.

Gravity—Viewed from a frame of reference fixed in the earth, acceleration imparted by the earth to a mass which is rotating the earth. Since the earth is rotating, the acceleration observed as gravity is the resultant of the acceleration of gravitation and the centrifugal acceleration arising from this rotation and the use of an earthbound rotating frame of reference. It is directed normal to sea level and to its geopotential surfaces.

Hachures—A method of portraying relief by short, wedge-shaped marks radiating from high elevations and following the direction of slope to the lowland.

Height of instrument—1. (spirit leveling) The height of the line of sight of a leveling instrument above the adopted datum. 2. (stadia surveying) The height of the center of the telescope (horizontal axis) of transit or telescopic alidade above the ground or station mark. 3. (trigonometrical leveling) The height of the center of the theodolite (horizontal axis) above the ground or station mark.

Heliotrope—An instrument composed of one or more plane mirrors so mounted at the station being sighted upon, that the sun's rays can be reflected to any one observing station.

Horizon—A great circle on the celestial sphere whose plane is perpendicular to the direction of the plumb line.

Horizontal control—Control which determines horizontal positions only, with respect to parallels and meridians or to other lines of reference.

Horizontal refraction—A natural error in surveying which is the result of the horizontal bending of light rays between a target and an observing instrument. Usually caused by the differences in density of the air along the path of the light rays, resulting from temperature variations.

Hour angle—Angular distance west of a celestial meridian or hour circle; the arc of the celestial equator, or the angle at the celestial pole, between the upper branch of a celestial meridian or hour circle and the hour circle of a celestial body or the vernal equinox, measured westward through 360°.

Hour circle—Any great circle on the celestial sphere whose plane is perpendicular to the plane of the celestial equator.

Intersection—A method of determining the horizontal position of a point by observations from two or more points of known position, thus measuring directions that intersect at the station being located. A station whose horizontal position is located by intersection is known as an intersection station.

Isogonic chart—A chart of which the chief feature is a system of isogonic lines, each for a different value of the magnetic declination.

Isogonic line—A line drawn on a map or chart joining points of equal magnetic variation.

Laplace azimuth—A geodetic azimuth derived from an astronomic azimuth by use of the Laplace equation.

Laplace condition—The Laplace condition, expressed by the Laplace equation, arises from the fact that a deflection of the vertical in the plane of the prime vertical will give a difference between astronomic and geodetic longitude and between astronomic and geodetic azimuth; or, conversely, that the observed differences between astronomic and geodetic values of the longitude and of the azimuth may both be used to determine the deflection in the plane of the prime vertical.

Laplace equation—The equation which expresses the relationship between astronomic and geodetic azimuths in terms of astronomic and geodetic longitudes and geodetic latitude. Thus Laplace correction = $(\lambda_A - \lambda_G)$ Sin ϕ_G.

Laplace station—A triangulation or traverse station at which a Laplace azimuth is determined. At a Laplace station both astronomic longitude and astronomic azimuth are determined.

Latitude, astronomic—The angle between the plumb line and the plane of the celestial equator. Also defined as the angle between the plane of the horizon and the axis of rotation of the earth. Astronomic latitude is the latitude which results directly from observations on celestial bodies, uncorrected for deflection of the vertical (station error) which, in the United States, may amount to as much as 25″. Astronomic latitude applies only to positions on the earth, and is reckoned from the astronomic equator (0°), north and south through 90°.

Latitude, geodetic—The angle which the normal to the spheroid at a point makes with the plane of the geodetic equator. Geodetic latitudes are reckoned from the equator, but in the horizontal control survey of the United States they are computed from the latitude of station Meades Ranch as prescribed in the North American datum of 1927. In recording a geodetic position, it is essential that the geodetic datum on which it is based be also stated. A geodetic latitude differs from the corresponding astronomic latitude by the amount of the meridional component of the local deflection of the vertical (station error), which in this country may amount to more than 25″.

Level datum—A level surface to which elevations are referred. The generally adopted level datum for leveling in the United States is mean sea level. For local surveys, an arbitrary level datum is often adopted and defined in terms of an assumed elevation for some physical mark (bench mark).

Level net—Lines of spirit leveling connected together to form a system of loops or circuits extending over an area.

Line of sight—1. The straight line between two points. This line is in the direction of a great circle, but does not follow the curvature of the earth. 2. The line extending from an instrument along which distant objects are seen, when viewed with a telescope or other sighting device.

Local hour angle—Angular distance measured on the celestial equator between the celestial meridian and the hour circle that passes through the object. The local hour angle represents physically the amount of rotation of the celestial sphere, since the object was last on the observer's celestial meridian, and is always measured westward 0° to 360° from the celestial meridian.

Longitude, astronomic—The angle between the plane of the celestial meridian and the plane of an initial meridian, arbitrarily chosen.

Longitude, geodetic—The angle between the plane of the geodetic meridian and the plane of an initial meridian, arbitrarily chosen.

Loop traverse—A closed traverse that starts and ends at the same station.

Main-scheme station—A station through which the basic survey computations are carried, also called a principal station. The main-scheme stations serve for the continued extension of the survey.

Mean Sea Level (MSL)—The mean surface water level determined by averaging heights at all stages of the tide over a 19-year period. Often used as a reference for general leveling operations.

Meridian—A north-south line from which longitudes and azimuths are reckoned; or a plane, normal to the geoid or shperoid, defining such a line.

Meridian angle—Angular distance east or west of the local celestial meridian; the arc of the celestial equator, or the angle at the celestial pole, between the upper branch of the local celestial meridian and the hour circle of a celestial body, measured eastward or westward from the local celestial meridian through 180°, and labeled E or W to indicate the direction of measurement.

Monument—Any object or collection of objects that indicate the position on the ground of a survey station. In most surveys, the term monument usually refers to a stone or concrete station marker containing a special bronze plate on which the exact station point is marked.

Nadir—The point on the terrestrial sphere directly beneath the observer and directly opposite to the zenith; the lowest point.

North American Datum of 1927—The initial point of this datum is located at Meades Ranch, Kans. Based on the Clarke spheroid of 1866, the geodetic positions of this system are derived from a readjustment of the triangulation of the entire country, in which Laplace azimuths were introduced.

Observer's meridian—A celestial meridian passing through the zenith at the point of observation and the celestial poles.

Occultation—1. (astronomy) The disappearance of a celestial body behind another body of larger apparent size. When the moon passes between the observer and a star, the star is said to be occulted. 2. (survey) Name applied to a geodetic survey technique which employs the principle of occultation where repeated observations are made on an unknown position, accurately timed with similar observations at another unknown station, and mathematically reducing this data to determine the exact geodetic position of the unknown stations.

Occupied station—A traverse or triangulation station over which a theodolite or an engineer transit is set up for the measurement of angles at this station. Also, a station at which angles have been so measured.

Offset line—A supplementary line close to and roughly parallel with a main line, to which it is referred by measured offsets. Where the line for which data are desired is in such position that it is difficult to measure over it, the required data are obtained by running an offset line in a convenient location and measuring offsets from it to salient points on the other line.

Open traverse—A survey traverse which begins from a station of known or adopted position, but does not end upon such a station.

Order of accuracy—A mathematical ratio defining the general accuracy of the measurements made in a survey. The order of accuracy of surveys are divided into four classes labeled: first order, second order, third order, and fourth or lower order.

Parallax—1. The apparent displacement of the position of a body, with respect to a reference point or system, caused by a shift in the point of observation. 2. The apparent displacement between objects on the earth's surface due to their difference in elevation.

Permanent bench mark (PBM)—A bench mark of as nearly permanent character as it is practicable to establish. Usually designated simply as a bench mark or BM. A permanent bench mark is intended to maintain its elevation with reference to an adopted datum without change over a long period of time.

Personal equation—The time interval between the sensory perception of a phenomenon and the motor reaction thereto. A personal equation may be either positive or negative, as an observer may anticipate the occurrance of an event, or wait until he actually sees it occur before making a record. This is a systematic error, treated as the constant type.

Personal error—An error caused by an individual's personal habits, his inability to perceive or measure dimensional values exactly, or by his tendency to react mentally and physically in a uniform manner under similar conditions. Contrasted with blunder; mistake.

Picture point—In surveying, a terrain feature that is easily identified on an aerial photograph and whose horizontal or vertical position or both have been determined by survey measurements. Picture points are marked on the aerial photographs by the surveyor, and are used by the photomapper.

Plumb line—1. The line of force in the geopotential field. The continuous curve to which the direction of gravity is everywhere tangential. 2. The line indicated by a plumb-bob cord.

Prime meridian—The meridian of longitude 0°, used as the origin for measurement of longitude. The meridian of Greenwich, England, is almost universally used for this purpose.

Prime vertical—The vertical circle through the east and west points of the horizon. It may be true, magnetic, compass, or grid depending upon which east or west points are involved.

Right ascension—The angular distance measured eastward on the equator from the vernal equinox to the hour circle through the celestial body, from 0 to 24 hours.

Sea Level Datum of 1929—The current standard datum for geodetic leveling in the United States, based on tidal observations over a number of years at various tide stations along the coasts.

Sexagesimal system—A system of notation by increments of 60; as the division of the circle into 360°, each degree into 60 minutes, and each minute into 60 seconds.

Sidereal day—The interval of time from a transit of the (true) vernal equinox across a given meridian to its next successive transit across the same meridian.

Sidereal time—Time based upon the rotation of the earth relative to the vernal equinox.

Solar day—1. The interval of time from the transit of either the sun or the mean sun across a given meridian to the next successive transit of the same body across the same meridian. 2. The duration of one rotation of the sun.

Solar time—1. Time based upon the rotation of the earth relative to the sun. 2. Time on the sun.

Spheroid—1. (general) Any figure differing slightly from a sphere. 2. (geodesy) A mathematical figure closely approaching the geoid in form and size and used as a surface of reference for geodetic surveys.

Target—1. Any object, point, etc., toward which something is directed. 2. An object which reflects a sufficient amount of a radiated signal to produce an echo signal on detection equipment. 3. The distinctive marking or instrumentation of a ground point to aid in its identification on a photograph. In photogrammetry, target designates a material marking so arranged and placed on the ground as to form a distinctive pattern over a geodetic or other control-point marker, on a property corner or line, or at the position of an identifying point above an underground facility or feature. A target is also the image pattern on aerial photographs of the actual mark placed on the ground prior to photography.

Tidal bench mark—A bench mark set to reference a tide staff at a tidal station and the elevation of which is determined with relation to the local tidal datum.

Tidal datum—Specific tide levels which are used as surfaces of reference for depth measurements in the sea and as a base for the determination of elevation on land. Many different datums have been used, particularly for leveling operations.

Transit—The apparent passage of a star or other celestial body across a defined line of the celestial sphere, as a meridian, prime vertical, or almucantar. The apparent passage of a star or other celestial body across a line in the recticle of a telescope, or some line of sight. The apparent passage of a smaller celestial body across the disk of a larger celestial body. The transit of a star across the meridian occurs at the moment of its culmination, and the two terms are sometimes used as having identical meanings; such usage is not correct, even where the instrument is in perfect adjustment. At the poles, a star may have no culmination but it will transit the meridians.

Vernal equinox—That point of intersection of the ecliptic and the celestial equator, occupied by the sun as it changes from south to north declination, on or about March 21. Same as first of Aries; first point of Aries; March equinox.

Vertical circle—1. A great circle of the celestial sphere, through the zenith and nadir. Vertical circles are perpendicular to the horizon. 2. A graduated disk mounted on an instrument in such a manner that the plane of its graduated surface can be placed in a vertical plane. It is primarily used for measuring vertical angles in astronomical and geodetic work

Vertical control—The measurements taken by surveying methods for the determination of elevation only with respect to an imaginary level surface, usually mean sea level.

Vertical-control datum—Any level surface (as, for example, mean sea level) taken as a surface of reference from which to reckon elevations. Although a level surface is not a plane, the vertical-control datum is frequently referred to as the datum plane.

Zenith—The point where an infinite extension of a plumb (vertical) line, at the observer's position, pierces the celestial sphere above the observer's head.

Zenith distance—The complement of the altitude, the angular distance from the zenith of the celestial body measured along a vertical circle.

INDEX

Abbreviations and symbols, field notes, 5
Abney topographic hand level, 20
Abutting, 222
Accidental errors, 249
Accuracy, 249
Adjustment of leveling, 160
Adjustments, engineer level:
- Horizontal crosshairs, 24
- Level vial, 24
- Line of sight, 25

Adjustments, one-minute theodolite:
- Compass, 67
- Horizontal collimation, 66
- Optical plumb, 67
- Plate level, 66
- Telescope level, 67
- Screws, 67
- Vertical collimation, 66
- Verticality, crosswire, 67

Adjustments, one-second theodolite:
- Horizontal collimation, 74
- Optical plumb, 73
- Plate level, 73
- Screws, 73
- Vertical collimation, 74
- Vertical crosswire, 74

Adjustments, planetable:
- Circular level, 137
- Collimation, crosswires, 137
- Self-indexing, 138
- Stadia arc, 137
- Striding level, 137
- Vertical arc, 137
- Verticality, crosswires, 137

Adjustments, transit:
- Crosswires, 58
- Plate level, 57
- Standards, 59
- Telescope level, 59
- Vertical circle vernier, 60
- Vertical wire, 58

Adjustments, traverse:
- Angles, 192
- Azimuths, 193
- Compass rule, 193
- Departures, 193
- Distances, 192
- Latitudes, 192

Adjustment, weighted observations, 251
Adverse possession, 222
Agonic line, 176
Airborne surveys:
- Hiran, 201
- Shiran, 201
- Shoran, 201

Alidade:
- Description, 132
- Open-sight, 132
- Telescopic, 132

Altimeter:
- Adjustment, 39
- Calibration, 40
- Care, 38
- Description, 37
- Graduation, 38
- Principle, 38
- Reading, 38

Angle adjustment, 181
Angle, observing:
- Affecting factors, 180
- Directing method, 180
- Party organization, 177
- Procedure, 177
- Repitition method, 179

Angles:
- Deflection, 173
- Interior and exterior, 173
- Station, 172

Angular, units of measure:
- Centesimal, 258
- Comparison, 258
- Mils, 258
- Sexigesimal, 257

Arc, stadia, 135
Area determination:
- Coordinate method, 215
- Double-meridian-distance, 216
- General, 215
- Irregular area, 216

Irregular interval, 219
Regular area, 215
Segment of circles, 219
Simpson's one-third rule, 218
Trapezoidal rule, 217
Area, unit of measure, 256
Arrows, taping, 100
Astronomic azimuth, 174
Astronomic surveys, 2, 202
Auxiliary survey network, 3
Azimuth:
Astronomic, 174
Geodetic, 174
Grid, 174
Magnetic, 174
True, 174

Back azimuth, 173
Backsight, 150
Barometric leveling:
Calibration record, 167
Eccentric points, 167
General, 165
Leapfrog method, 171
Reading scale, 166
Single base method, 171
Terrain, 166
Time of readings, 166
Two-base method, 169
Weather, 166
Base lines, 198
Basic control network, 3
Basic control surveys, 1
Bearing, 174
Bench marks, 158
Boundaries:
Laws, 222
Private land, 223
Resurveys, 221
Braced triangulation, 194
Breaking tape, 115
Broken bounderies, 236
Brunton pocket transit, 84

Calibration, tape, 99
Calls, 222
Cardinal lines, 228
Centering instrument, 16
Cestesimal system, 258
C-factor, 28, 33
Circle left, 78
Circle right, 78
Circular level adjustment, 14
Circular sector, 219
Circular segment, 219
Clearance, vertical, 246
Clinometer, 84
Closed traverse:
Connecting, 182
Loop, 182
Closing corners, 235
Coincidence bubble, 68
Coincidence method, 70
Compass:
Brunton pocket, 84
Care, 85
Engineer transit, 84
Lensatic, 83
Planetable, 84
Theodolite, 84
Compass rule method, 193
Computations, general, 6, 8
Computation, traverse:
Adjustments, 192
Area, 192
Azimuth, 190
Coordinates, grid, 190
Departures, 190
Latitudes, 190
Linear error of closure, 192
Computations, triangulation, 198
Connecting traverse, 182
Construction surveys, 1
Construction taping, 117
Contouring, 144
Control systems, detail, 211
Conventional triangulation, 194
Coordinate method, 215
Coordinates of detail, 214
Coordinate system, 220
Correction for curvature and refraction (Table 8-1), 153
Corrections, taping:
Atmospheric effects, 118
Length of tape, 117
Method of support, 118
Slope, 118
Crosshair, 10
Crosswire, 10
Crystal relationship (Table 6-1), 122
Curvature and refraction, 152
Customary system, 256

Declination, magnetic, 175

Deeds and records:
- Grantee indexes, 223
- Grantor indexes, 223
- Registry of deeds, 223
- Use, 223

Deflection angles, 173
Degree of accuracy:
- Angular, 253
- Horizontal distance, 253
- Leveling, 253
- Traverse, 253

Departures, 190, 193
Detail locating:
- Angle and distance, 211
- Control systems, 211
- Coordinate method, 214
- Distance, 213
- Intersection, 214
- Perpendicular offset, 213
- Range tie, 212
- Swing-offset, 213
- Two distance, 213

Difference in elevation, 149
Differential leveling:
- Adjustment, 160
- Bench marks, 158
- Curvature and refraction, 152
- Equipment, 153
- General, 150
- Party organization, 152
- Procedure, 153
- Reciprocal, 156
- Recording, 155
- Water transfer method, 157

Direction, 172
Direction method, 180
Direct leveling, 149
Direct vernier, 56
Discrepancy, 250
Division of lands:
- Establishing section lines, 232
- Location of islands, 232
- Numbering, 229
- Sections, 228
- Subdividing sections, 230
- Townships, 228
- Townsites, 232
- Tracts, 228

Double-Meridian-Distance method, 216
Double proportionate measurement, 234
Double zenith distance, 177
Dumpy level, 21

Easting, 192

Elevation, 149
Engineering surveys, 2
Engineer level:
- Adjustment, 24
- Care, 21
- Description, 21

Errors:
- Probable, 252
- Residual, 250
- Resultant, 250
- Sources:
  - Environmental, 249
  - Instrumental, 249
  - Personal, 249
- Types:
  - Accidental, 249
  - Comparison, 249
  - Systematic, 249

Error, sources:
- Environmental, 249
- Instrumental, 249
- Personal, 249

Establishing section lines, 232
Explement angle, 172
Extending a straight line, 241
Exterior angles, 173

Factors affecting fieldwork:
- Accuracy, 4
- Equipment, 4
- Personnel, 4
- Purpose, 4
- Speed, 5
- Systematic procedure, 4
- Weather and terrain, 4

Falling, 232
Field classification surveys, 2
Field notes:
- Abbreviations and symbols, 5
- Computing, 6
- Corrections, 5
- Descriptions, 5
- Notes, 5
- Recording, 5
- Sketches, 5
- Tabulation, 5

Field work:
- Factors affecting, 4
- Notes, 5
- Operations, 3

Foresight, 150
Forward azimuth, 173
Fractional lots, 232
Frequency, radio wave, 122

Geodetic azimuth, 174
Geodetic level, 30
Geodetic surveys, 1, 2
Geodimeter, model 2A:
  Accessories, 111
  Calibration, 112
  Care, 112
  Description, 111
  Reflector unit, 111
  Use, 111
Geodimeter, types:
  Model 2A, 111
  Model 4D, 110
  Model 6, 110
  Model 8, 110
Geodolite, 111
Grads, 258
Graduations, tape, 98
Gravity surveys, 2, 200
Grid azimuth, 174
Grid coordinates, 190
Ground swing, 123
Guide meridians, 228
Gunter's chain, 228
Gyro-azimuth surveying instrument:
  Description, 81
  Operation, 81

Hand levels:
  Abney topographic, 20
  Locator, 20
Hand signals, 6
Height of instrument, 150
Heliotrope, 88
Horizontal angles, laying off, 242
Horizontal taping:
  Breaking tape, 115
  Distance, 116
  Full tape lengths, 115
  Leveling, 115
  Plumbing, 114
Hub, station marker, 184
Hydrographic surveys, 2

Indefinite meridian, 221
Indirect leveling, 149
Initial points, 227
Initial setting of circle for repetitions
  (Table 9-1), 180
Interior angles, 173
International system, 256
Intersection, 214
Invar tape, 96
Irregular area:
  Irregular interval, 219
  Regular interval:
    Simpson's one-third rule, 218
    Trapezoidal rule, 217
Irregular interval, 219
Isogonic chart, 176
Isogonic line, 176

Lamp, signal, 89
Land parcels:
  Coordinate system, 220
  Metes and bounds, 220
  Plane coordinate system, 220
  Rectangular system, 220
Land surveys:
  Descriptions, 224
  Division of lands, 228
  Establishing section lines, 232
  General, 220
  Laws, 222
  Location of islands, 232
  Monuments, 233
  Numbering, 229
  Parcels, 220
  Plans, 224
  Plats, 236
  Private lands, 223
  Public lands, 226
  Responsibilities, 222
  Restoration, corners, 234
  Resurveys, 221
  Subdividing of sections, 230
  Townsites, 232
  United States system, 226
Laser ranger, 110
Latitudes, 190, 193
Law of sines, 198
Laws relating to bounderies:
  Abutting, 222
  Adverse possession, 222
  Calls, 222
  Errors, 222
  Function, 222
  Usage rulings, 222
Laying off a horizontal angle, 242
Lay off given distance, 241
Leapfrog method, 171
Level accessories:

Pedestals, 37
Precise rod, 34
Rod levels, 36
Rod targets, 36
Stadia rod, 36
Stakes, 36
Standard rod, 34
Turning pins, 37
Umbrella, 37
Leveling:
Alidade, 165
Barometric, 165
Description, 149
Differential, 150
Orders of accuracy, 149
Reciprocal, 156
Stadia, 164
Short base, 165
Trigonometric, 161
Types, 149
Leveling heads:
Four-screw type, 13
Three-screw type, 13
Leveling instrument:
Circular bubble, 14
Four-screw, one level, 13
Four-screw, two levels, 14
Three-screw, one level, 14
Levels:
Abney topographic, 20
Engineer, 21
Precise (Wild N3), 30
Semiprecise, 25
Levels, surface, 149
Level vial:
Alinement, 12
Calibration, 12
Description, 12
Sensitivity, 12
Lightwave instruments:
Accessories, 111
Calibration, 112
Care, 112
General, 108
Theory, 109
Types, 109
Lightwave measuring:
Accuracies, 124
Computation, 127
Introduction, 124
Meteorological effects, 126
Recording, 126
Topographic effects, 126
Use, 126
Linear conversion factors:
International (Table C-1), 257
U. S. Survey foot (Table C-2), 257
Linear error of closure, 192
Linear units of measure, 256
Location of Detail:
Angle and distance, 212
Angle methods, 214
Control systems, 211
Coordinates, 214
Distance methods, 213
General, 211
Intersection, 214
Perpendicular offsets, 213
Radiation, 211
Range ties, 212
Resection, 214
Short base, 212
Swing offsets, 213
Location of islands, 232
Loop traverse, 182
Low coefficient tapes, 96

Magnetic azimuth, 174
Magnetic declination, 175
Map plotting, strip topography, 148
Mapping surveys, 1
Meander corners, 233
Measuring vertical distance, 246
Meridians and base lines (Table 14-1), 227
Metallic tapes, 96
Meteorological equipment:
Altimeter, 37
General, 37
Psychrometer, 40
Thermometer, 42
Metes and bounds, 220
Micro-chain (MC-8), 104
Microwave instruments:
Accessories, 107
Adjustments, 106
Auxiliary equipment, 107
Calibration, 106
Care, 105
General, 102
Theory, 103
Types, 104
Microwave measuring:
Accuracies, 122
Computation, 124
Introduction, 121
Meteorological effects, 123
Recording, 124
Theory, electronic, 121

Topographic effects, 123
Use, 123
Mil system, 258
Mistakes, 249
Monument markings (Table 14-2), 233
Monuments:
Marking, 233
Meander corners, 233
Post and tree, 233
Stone, 233

Nadir, 177
Networks, survey, 3
North American datum, 3
Northings, 192
Numbering, land surveys:
Sections, 230
Townships, 229
Numbers, rounding off, 9

Obliterated corners, 221
Observation affecting factors:
Differential temperatures, 180
Environmental errors, 180
Horizontal refraction, 180
Instrumental errors, 180
Personal errors, 180
Phase, 181
Office compilation, planetable, 146
Office work:
Computation, 8
Operations, 6
One-minute theodolite:
Accessories, 63
Adjustment, 66
Care, 65
Description, 60
Horizontal motion, 60
Levels, 63
Reading:
Degree graduated, 64
Mil graduated, 64
Telescope, 63
Vertical motion, 62
One-second theodolite:
Accessories, 70
Adjustment, 73
Care, 73
Description, 68
Horizontal motion, 68
Levels, 68
Optical system, 70
Reading:
Horizontal, 70
Vertical, 72
Telescope, 68
Vertical motion, 68
One-tenth second theodolite, 79
One-wire leveling:
Procedure, 154
Recording, 155
Open sight alidade, 132
Open traverse, 182
Optical plumbing assembly, 16
Orders of accuracy, 253
Orienting planetable:
Backsighting, 139
Fundemental method, 140
Magnetic compass, 139
Resection, 140

Parabolic segment, 219
Parallax, 18
Pedestals, turning, 37
Perpendicular offsets, 213
Phase electronic, 122
Philadelphia rod, 34
Pins, turning, 37
Plan and description preparation:
Description, 224
Plan, 224
Plane coordinate system, 220
Plane surveys, 1, 3
Planetable, 132
Planetable surveys:
Compilation, 146
Contouring, 144
Description, 138
Hints, 146
Orienting, 139
Party organization, 139
Plotting detail, 144
Point location, 143
Traverse, 143
Planetable equipment:
Adjustments, 137
Care, 136
Description, 132
Introduction, 132
Stadia arc, 135
Plats, 236
Plotting detail, 144

Plumb bobs, 16
Plummet, optical, 73
Plus station, 184
Pocket tapes, 98
Point location:
    Combined method, 143
    Intersection, 143
    Radiation, 143
    Resection, 143
Point of origin, 221
Practical problems:
    Extending a straight line, 241
    Laying off a horizontal angle, 242
    Lay off given distance, 241
    Measuring vertical distance, 246
    Setting up between two points, 245
    Two lines intersecting, 246
    Typical problems (Fig. A-7), 247
Precise level (N3):
    Adjustments, 33
    Care, 31
    Circular bubble, 32
    Collimation (C), 33
    Cross error, 34
    Description, 30
    General, 30
    Stadia constant, 33
Precise rod, 34
Precision, 249
Pressure, units of measure, 259
Private land surveys:
    Deeds and records, 223
    Functions, 223
    Plans and descriptions, 224
Principal lines:
    Base lines, 228
    Meridian, 228
Probability:
    Most probable value, 250
    Theory, 250
Probable error, 252
Probable error factors (Table B-1), 252
Profile leveling, 160
Proportionate measurement:
    Double, 234
    Single, 235
Psychrometer:
    Automatic, 40
    Sling, 42
Public domain, 227
Public land surveys:
    Division of land, 228
    Establishing section lines, 232
    Islands, 232
    Monuments, 233
    Numbering, 229
    Plats, 236
    Public domain, 226
    Restoration corners, 234
    Subdividing sections, 230
    Townsites, 232
    United States system, 226

Radiation, 143, 211
Radio waves, 121
Range, 229
Range lines, 228
Range master, 110
Range poles, 85
Range ties, 212
Ranging-in, 245
Reciprocal leveling, 156
Rectangular grid, 174
Rectangular system, 220
Reflector unit, 111
Refraction, 152
Regular area determination:
    Coordinate method, 215
    DMD method, 216
    Rectangle, 215
    Triangle, 215
Regular interval, 217
Repair, tape, 100
Repetition method, 179
Representative fraction, 138
Resection, 143, 214
Residual error, 250
Responsibilities, land surveys, 222
Restoration of lost corners:
    Double proportionate measurement, 234
    Single proportionate measurement, 235
Resultant error, 250
Resurveys of old bounderies:
    Distance failure, 221
    Indefinite meridian, 221
    Obliterated corners, 221
    Point of origin, 221
Retrograde vernier, 56
Ripa, 222
Rod:
    Levels, 36
    Precise, 34
    Stadia, 36
    Standard, 34
    Targets, 36
Rounding off number, 9

Sattelite surveys:
BC-4 camera, 201
SECOR, 201
Scissor clamp, 101
Sections, 228
Segment of circles, 219
Semiprecise level:
Adjustments, 30
Care, 26
C-factor, 28
Description, 25
Instrument test, 26
Stadia constant, 28
Setting up between two points, 245
Sexagesimal system, 257
Short base expansion ratio (Table 6-2), 131
Short base method, 130
Sideshots, 188
Signal lights, 89
Significant figures:
Decimals, 9
Measured quantity, 9
Surveying, 9
Zeros, 9
Simpson's one-third rule, 218
Single base method, 171
Single proportionate measurement, 235
Slope taping, 114
Sodano azimuth surveys, 201
Special surveys:
Airborne, 201
Astronomic, 202
Gravity, 200
Satellite, 201
Sodano azimuth, 201
Trilateration, 200
Spring balance, 101
Stadia:
Accuracy, 129
Measurement, 129
Theory, 127
Stadia arc:
Horizontal scale, 136
Vertical angle scale, 136
Vertical scale, 135
Stadia constant determination, 28, 33
Stadia rod, 36
Standardization tapes, 99
Standard parallel, 228
Station adjustment, 181
Station angles, 172
Station marks, traverse:
Numbering, 184
Permanent, 184
Reference, 185
Temporary, 184
Steel tape, 96
Stool, taping, 100
Straight line extending, 241
Strength of figure, 196
Striding level, 68
Strip topography and profiles:
Method, 147
Plotting, 148
Procedures, 147
Subdividing sections:
Fractional lots, 232
Quarter-sections, 232
Section, 230
Subtense method, 130
Supplementary network, 3
Surveying, 1
Surveying classifications:
Functional:
Astronomic, 2
Basic control, 1
Construction, 1
Engineering, 2
Field classification, 2
Gravity, 2
Hydrographic, 2
Land, 2
Satellite, 2
Special, 2
Topographic, 1
Technical:
Geodetic, 1
Plane, 1
Surveying instrument, electronic (Model 99), 104
Survey networks:
Auxiliary, 3
Basic, 3
Supplementary, 3
Survey targets, 85
Swing ground, 123
Swing offsets, 213
Symbols and abbreviations, 5
Systematic errors, 249

Tables:
Beaufort scale of wind factors (Table 3-1), 39
Conversion, Celsius to Fahrenheit and inverse (Table D-9), 295
Conversion, feet to meters (SI) (Table D-7), 283
Conversion, feet to meters (U. S. Survey foot) (Table D-8b), 291

Conversion, meters to feet (SI) (Table D-7a), 279
Conversion, meters to feet (U. S. Survey foot) (Table D-8a), 287
Conversion of degrees to mils and inverse (Table D-1), 261
Conversion for curvature and refraction (leveling) (Table 8-1), 153
Crystal relationship (Table 6-1), 122
DE, horizontal distance and gradients (Table D-3), 267
DE, slope distance and gradients (Table D-4), 268
Horizontal distance, slope distance and gradients (Table D-5), 269
Inclination correction, percent of slope (Table D-11), 297
Inclination corrections, 50 meter tape (Table D-2), 263
Initial settings of circle for repetitions (Table 9-1), 180
Linear conversion factors (SI) (Table C-1), 257
Linear conversion factors (U. S. Survey foot) (Table C-2), 257
Meridians and base lines of rectangular surveys (Table 14-1), 227
Monument markings (Table 14-2), 233
Probable-error factors (Table B-1), 252
Short base expansion ratio (Table 6-2), 131
Stadia reduction (Table D-6), 270
Temperature correction, steel tape (Table D-10), 296
Tapes:
Accessories, 100
Description:
Low coefficient, 96
Metallic, 96
Pocket, 98
Steel, 96
Graduation, 98
Repair, 100
Standardization, 99
Taping:
Corrections, 117
Horizontal, 114
Introduction, 113
Mistakes and errors, 119
Party, 113
Precautions, 121
Recording, 119
Slope, 114
Use, 116
Taping arrows, 100
Taping stools, 100
Target, rod, 36
Target, set, 86
Telescopes:
Care, 18
Eyepiece, 17
Objective lens, 18
Reticle, 17
Telescopic alidade, 132
Tellurometer (MRA 301), 105
Temperature, units of measure, 258
Tension scale, 101
Theodolite:
One-minute, 60
One-second, 68
One-tenth-second, 79
Two-tenth-second, 74
Thermometer, 42
Three-wire leveling:
Procedure, 154
Tier, 229
Time, units of measure:
Angles, 258
Difference, 258
Notation, 258
Topographic surveys, 1
Township lines, 228
Townships, 228
Townsites, 232
Transit, engineer:
Adjustments, 57
Care, 56
Circles and verniers, 51
Description, 48
General, 48
Trapezoidal rule, 217
Traverse:
Accuracy, 183
Adjustments, 192
Azimuth extending, 189
Classification, 182
Computation, 190
General, 182
Horizontal angles, 189
Linear measurements, 188
Notekeeping, 189
Party organization, 186
Plus station, 184
Reconnaissance, 183
Route selection, 184
Starting and closing points, 183
Station marks, 184
Stations, 188
Ties to existing controls, 187
Vertical angles, 189
Traverse taping, 116

Triangle of error, 142
Triangulation:
- Accuracy, 195
- Base line, 198
- Braced, 194
- Computation, 198
- Conventional, 194
- General, 194
- Introduction, 194
- Observation, 198
- Party organization, 196
- Reconnaissance, 195
- Recording, 198
- Station markers, 196
- Strength of figure, 196
- Targets, 197

Triangulation taping, 116
Tribrach assembly, 64
Trigonometric leveling:
- Alidade method, 165
- Description, 161
- Electronic method, 165
- General, 161
- Limiting factors, 164
- Party organization, 164
- Short base method, 165
- Stadia method, 164

Trilateration, 200
Tripod:
- Care, 11
- Setting up, 11
- Types, 10

True azimuth, 174
Turning pins, 37
Turning plate, 37
Turning points, 150
Two-base method:
- Procedure, 169
- Recording, 170

Two lines intersecting, 246
Two peg test, 25
Two-tenth-second theodolite:
- Adjustment, 79
- Calibration, 79
- Care, 79
- Description, 74
- General, 74
- Reading:
  - Horizontal, 77
  - Vertical, 78

Umbrella, 37
United States system:
- Cardinal lines, 228
- Initial points, 227
- Principal line, 227
- Responsibility, 226
- System of surveying, 227

Units of measurement:
- Angular, 257
- Area, 256
- General:
  - Customary system, 256
  - International system, 256
- Introduction, 256
- Linear, 256
- Pressure, 259
- Temperature, 258
- Time, 258
- Volume, 257
- Weight, 258

Universal equipment:
- Leveling head, 13
- Level vial, 11
- Optical plumbing, 16
- Plumb bobs, 16
- Telescopes, 17
- Tripod, 10

Usage rulings, 222

Verniers, reading:
- Circle, 56
- Direct, 56
- Retrograde, 56

Vertical angle, 176
Voice signals, 6
Volume, unit of measure, 257

Water level transfer method, 157
Weighted observation:
- Assignment of weights, 251
- Theory, 251
- Use of weights, 251

Weight, units of measure, 258
Wiggling-in, 245

Zenith, 177
Zenith distance, 177